D1472201

Second Edition

Math Lit

A Pathway to College Mathematics

Kathleen Almy
Rock Valley College

Heather Foes
Rock Valley College

PEARSON

Boston   Columbus   Indianapolis   New York   San Francisco
Amsterdam   Cape Town   Dubai   London   Madrid   Milan   Munich   Paris   Montréal   Toronto
Delhi   Mexico City   São Paulo   Sydney   Hong Kong   Seoul   Singapore   Taipei   Tokyo

Editorial Director: Chris Hoag
Editor in Chief: Michael Hirsch
Senior Acquisitions Editor: Rachel Ross
Editorial Assistant: Megan Tripp
Project Manager: Sherry Berg
Project Management Team Lead: Karen Wernholm
Media Producer: Marielle Guiney
Senior Content Developer, TestGen: John Flanagan
TestGen Content Manager: Marty Wright
MathXL Executive Content Manager: Rebecca Williams
Math XL, Associate Content Manager: Eric Gregg
Product Marketing Manager: Alicia Frankel
Field Marketing Managers: Jennifer Crum and Lauren Schur
Marketing Assistant: Alexandra Habashi
Senior Author Support/Technology Specialist: Joe Vetere
Manager, Rights and Permissions: Gina Cheselka
Manufacturing Buyer: Carol Melville, RR Donnelley
Program Design Lead: Barbara Atkinson
Text Design, Production Coordination, Composition, and Illustrations: Cenveo® Publisher Services
Cover Design: Studio Montage
Cover Image: New York City in Manhattan on High Line Park, Sean Pavone Photo/Fotolia

Credits appear following the Acknowledgments on page xxiv, and constitute a continuation of
the copyright page.

PEARSON, ALWAYS LEARNING, and MYMATHLAB are exclusive trademarks owned by
Pearson Education, Inc. or its affiliates in the U.S. and/or other countries.

Unless otherwise indicated herein, any third-party trademarks that may appear in this work are
the property of their respective owners and any references to third-party trademarks, logos or
other trade dress are for demonstrative or descriptive purposes only. Such references are not
intended to imply any sponsorship, endorsement, authorization, or promotion of Pearson's
products by the owners of such marks, or any relationship between the owner and Pearson
Education, Inc. or its affiliates, authors, licensees or distributors.

Library of Congress Cataloging-in-Publication Data
Names: Almy, Kathleen. | Foes, Heather.
Title: Math lit : a pathway to college mathematics / Kathleen Almy, Rock Valley
College: Heather Foes, Rock Valley College.
Description: 2nd edition. | Boston : Pearson, [2017] | Includes index.
Identifiers: LCCN 2016014009 | ISBN 9780134304083
Subjects: LCSH: Algebra—Textbooks. | Problem solving—Textbooks.
Classification: LCC QA152.3 .A497 2017 | DDC 512.9—dc23
LC record available at http://lccn.loc.gov/2016014009

1 16

www.pearsonhighered.com

Annotated Instructor's Edition
ISBN 10: 0-13-443895-7
ISBN 13: 978-0-13-443895-5

ISBN 10: 0-13-443311-4
ISBN 13: 978-0-13-443311-0

To develop a complete mind:

Study the science of art;

Study the art of science.

Learn how to see.

Realize that everything connects to everything else.

—Leonardo da Vinci

Contents

Letter to the Student   viii
Letter to the Instructor   ix
Changes to the Second Edition   xvi
Instructor Guide   xxv

CYCLE ONE Where do we start?

Cycle 1 establishes the foundation for the course by introducing important vocabulary, skills, and concepts for future cycles. It develops essential numeric and algebraic skills while incorporating the use of technology. The sections are organized around the cycle question, "Where do we start?", which encourages conversations about how to begin a difficult problem or start a new course.

Self-Assessment: Preview 2

1.1 The BP Oil Spill: Focus Problem 3

**1.2 Getting Started:
Reviewing Prealgebra** 5
- Use Venn diagrams
- Write equivalent fractions
- Add, subtract, multiply, and divide fractions
- Solve applied problems involving fractions

1.3 Hello, My Name Is: Graphing Points 17
- Plot ordered pairs
- Determine the coordinates of a point

**1.4 A Tale of Two Numbers:
Ratios and Proportions** 25
- Interpret ratios
- Scale ratios to produce equivalent ratios
- Determine if quantities are proportional

Getting Ready for Section 1.5 33

1.5 Chances Are: Probability Basics 34
- Find relative frequencies
- Find and interpret experimental probabilities

**1.6 It's All Relative:
Understanding Integers** 40
- Interpret signed number situations
- Find the opposite and absolute value of a number

1.7 Sign and Size: Integer Operations 47
- Add, subtract, multiply, and divide signed numbers
- Solve applied problems that involve signed-number addition, subtraction, multiplication, and division

1.8 An Ounce of Prevention: Means 56
- Find and interpret the mean of a set of numbers
- Use means in applied problems

Mid-cycle Recap 64

**1.9 Picture This:
Making and Interpreting Graphs** 65
- Create and interpret pie graphs
- Create and interpret bar graphs
- Create and interpret line graphs

1.10 Two by Two: Scatterplots 79
- Create and interpret scatterplots
- Sketch a curve to best fit a scatterplot

Getting Ready for Section 1.11 92

1.11 Multiply vs. Divide: Converting Units 93
- Convert units by multiplying or dividing

1.12 Up and Down: Percent Change 99
- Apply a percent change
- Find percent change
- Interpret percent change

1.13 The X Factor: Algebraic Terminology 110
- Differentiate between variables and constants
- Differentiate between expressions and equations
- Differentiate between factors and terms

**1.14 General Number:
Recognizing Patterns** 117
- Make conjectures and generalize patterns
- Identify and use arithmetic and geometric sequences

**1.15 Social Network:
Linear and Exponential Change** 124
- Model change with linear and exponential functions

**1.16 Infinity and Beyond:
Perimeter and Area** 137
- Calculate perimeter and area

Cycle 1 Study Sheet 146
Self-Assessment: Review 147
Cycle 1 Wrap-Up 148

CYCLE TWO How does that work?

Cycle 2 contains many topics typically seen in a beginning algebra course. It develops a deeper understanding of operations that is then used to simplify expressions and solve equations. The cycle question, "How does that work?", focuses the sections on understanding how and why procedures work.

Self-Assessment: Preview   160

2.1 Predicting a Child's Height:
Focus Problem   161

Getting Ready for Section 2.2   163

2.2 Rule of Thumb: Weighted Means   165
- Find and interpret weighted means
- Find the median and mode of a data set

2.3 Measure Up: Basic Exponent Rules   176
- Apply basic exponent rules
- Use geometric formulas

2.4 Count Up: Adding Polynomials   184
- Identify and add like terms

2.5 A Winning Formula:
Applying Order of Operations   190
- Use the order of operations to simplify expressions
- Evaluate formulas and expressions

2.6 Does Order Matter?:
Rewriting Expressions   197
- Apply the commutative and associative properties

2.7 Fair Share: Distributive Property   202
- Apply the distributive property to expressions
- Use the distributive property in applied contexts

2.8 Seat Yourself:
Equivalent Expressions   210
- Write an expression to represent a scenario
- Determine if two expressions are equivalent by using the commutative, associative, and distributive properties

2.9 Parts of Speech:
Using Operations Correctly   217
- Distinguish between an operator and an object
- Determine the object on which an operator is acting
- Recognize when the distributive property can be applied

Mid-cycle Recap   223

2.10 A Fine Balance: Verifying Solutions   224
- Verify a solution to an equation

2.11 Separate but Equal:
Solving Simple Equations   231
- Write and solve one-step equations

2.12 A State of Equality:
More Equation Solving   241
- Solve two-step and multi-step linear equations

2.13 Quarter Wing Night:
Writing and Solving Equations   251
- Write an equation to model a situation
- Solve a problem numerically and algebraically

2.14 Outwit and Outlast:
Using Proportions   257
- Write and solve proportions

2.15 Three of a Kind:
Pythagorean Theorem   264
- Use the Pythagorean theorem to find the length of a side in a right triangle
- Solve problems using the Pythagorean theorem

Getting Ready for Section 2.16   271

2.16 What Are the Odds?:
Theoretical Probability   273
- Calculate theoretical probabilities
- Compare theoretical and experimental probabilities

2.17 Size Up: Volume and Surface Area   280
- Calculate volume and surface area

Cycle 2 Study Sheet   286
Self-Assessment: Review   287
Cycle 2 Wrap-Up   288

CYCLE THREE When is it worth it?

Cycle 3 focuses on writing and graphing linear functions as well as factoring quadratic expressions and solving quadratic equations. It also extends earlier content with equation solving and exponential functions. The cycle question, "When is it worth it?", facilitates a discussion about the use of numeric, graphic, and algebraic techniques.

Self-Assessment: Preview   298

3.1 **Deciding to Run: Focus Problem**   299

3.2 **What's Trending: Correlation**   301
- Determine if data has a positive or negative linear correlation
- Graph the equation of the trendline
- Use the equation of the trendline to make predictions

3.3 **Constant Change: Slope**   312
- Find the slope of a line from points, tables, and graphs
- Interpret the slope as a rate of change

3.4 **Shortest Path: Distance Formula**   324
- Use the distance formula to find the distance between two points

3.5 **More or Less: Linear Relationships**   331
- Make comparisons using equations, tables, and graphs

3.6 **Get in Line: Slope-Intercept Form**   341
- Find and interpret the slope and y-intercept from a linear equation
- Graph a line using a table and using the slope and y-intercept

3.7 **Chain, Chain, Chain: Writing Linear Equations**   352
- Write the equation of a line using a point and the slope
- Write the equation of a line using two points
- Create a linear model in an applied problem

Getting Ready for Section 3.8   361

3.8 **Going Viral: Exponential Functions**   363
- Write the equation of an exponential function using a starting value and rate of change
- Model with exponential functions
- Graph exponential functions

Mid-cycle Recap   372

3.9 **Untangling the Knot: Solving Nonlinear Equations** 373
- Solve nonlinear equations

Getting Ready for Section 3.10   379

3.10 **Hot and Cold: Rewriting Formulas**   381
- Solve an equation for a specified variable

3.11 **A Common Goal: Greatest Common Factors**   388
- Factor an expression using the greatest common factor

3.12 **Thinking Outside the Box: Factoring Quadratic Expressions** 396
- Factor quadratic expressions

3.13 **A Formula for Success: The Quadratic Formula**   408
- Use the quadratic formula to solve equations

3.14 **Systematic Thinking: Graphing and Substitution**   414
- Solve a 2×2 linear system of equations by graphing
- Solve a 2×2 linear system of equations by substitution

3.15 **Opposites Attract: Elimination**   427
- Solve a 2×2 linear system of equations by elimination

3.16 **The Turning Point: Quadratic Functions**   435
- Identify a quadratic pattern in data
- Find the vertex of a parabola

Cycle 3 Study Sheet   442
Self-Assessment: Review   443
Cycle 3 Wrap-Up   444

CYCLE FOUR What else can we do?

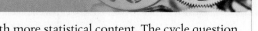

Cycle 4 includes some topics often found in intermediate algebra along with more statistical content. The cycle question, "What else can we do?", refers to additional objectives such as negative exponents, variation, and functions.

Self-Assessment: Preview   456

**4.1 Measuring Temperature Variability:
Focus Problem**   457

**4.2 A Matter of Change:
Dimensional Analysis**   458
- Convert units using dimensional analysis

Getting Ready for Section 4.3   465

**4.3 Little Giants:
Scientific and Engineering Notation**   467
- Convert numbers between scientific and standard notation
- Convert numbers into and out of engineering notation

**4.4 A Model Approach:
Negative Exponents**   476
- Use exponent rules to simplify expressions that have negative exponents

**4.5 Variation on a Theme:
Standard Deviation**   484
- Find the standard deviation of a data set
- Interpret the standard deviation of a data set

**4.6 An Order of Magnitude:
Understanding Logarithmic Scales**   495
- Interpret logarithmic scales

Mid-cycle Recap   504

**4.7 Straight to the Point:
Direct Variation**   505
- Identify direct variation from a graph, table, or equation
- Write models for direct variation problems
- Solve direct variation problems

Getting Ready for Section 4.8   515

4.8 Gas Up and Go: Inverse Variation   517
- Identify inverse variation from a table
- Write models for inverse variation problems
- Solve inverse variation problems

**4.9 Ghost in the Machine:
Function Notation**   527
- Use function notation
- Find a function input or output given the other

**4.10 What's Your Function?:
Vertical Line Test, Domain, and Range**   533
- Apply the vertical line test
- Find the domain and range from a graph

**4.11 An Important Point:
Vertex Form of a Quadratic Function**   541
- Identify the vertex of a quadratic function in vertex form
- Graph a quadratic function in vertex form
- Write the vertex form of a quadratic function given the vertex and a point

**4.12 A Survey of Trig:
Trigonometric Functions**   556
- Write the six trigonometric ratios for an acute angle
- Use trigonometric functions to find the measures of sides and angles of a right triangle

Cycle 4 Study Sheet   565
Self-Assessment: Review   566
Cycle 4 Wrap-Up   567

Excel Appendix (online in MyMathLab only)
Answers (online in MyMathLab only)
Index   574
Applications Index   582

Letter to the Student

Welcome to a new pathway to college mathematics! This book is not simply a repetition of the math you had in high school. Instead, it has been written to prepare you for success in your next mathematics course, such as statistics or quantitative literacy. While *Math Lit: A Pathway to College Mathematics, Second Edition*, does include the algebra you will need to be successful in a follow-up course, it should be immediately apparent that this is not an algebra book. Instead, the focus is on solving realistic problems, gaining number sense, and building mathematical literacy. Algebra is a tool that we use when needed, but it is not the focus of the book. When we study algebraic ideas, our goal will be to understand how and when to use those ideas.

Some topics in the book might be familiar to you, but the approach to the topics and the order in which they are presented will likely not be what you have seen in the past. We have spiraled topics throughout the book, introducing a topic and then revisiting it several times, going a little more in depth each time. This approach will help you achieve a better understanding of the material, which in turn will help you to remember it longer and make use of it when needed. Most topics are explored in context before all the procedures and details are given. The goal of this book is not simply for you to gain a set of skills. Additionally, we want to work on your understanding of concepts so that they can be used to solve other problems. It's not just "Can I perform this skill?", but instead "Do I understand what I'm doing?" and "Can I use this skill?"

Our approach will require that you are an active participant in class activities and your own learning. Skill-based homework can be done in MyMathLab®, which will offer you instant feedback, multiple tries on problems, and learning aids to be used when and if you need them. Conceptual homework can be done in the worktext after you have completed the MyMathLab® problems, thus allowing you to work with the subtle details of the content and improve your understanding beyond the procedural level.

When using this book, you will likely work harder than you have before, but you may also gain more than you have in previous math classes. During your journey, you will learn more than just mathematics. You will learn how to be successful in college and beyond, and what you specifically need to do to achieve your goals.

We hope you will bring an open mind to this course and a willingness to explore how things work. If you are willing to work, your work will often be rewarded with the feeling of accomplishment that comes from truly figuring something out for yourself!

We hope you enjoy *Math Lit* and find the experience rewarding. Let's get started.

It is not knowledge, but the act of learning, not possession but the act of getting there, which grants the greatest enjoyment.

—Carl Friedrich Gauss

Kathleen Almy

Heather Foes

Letter to the Instructor

A New Pathway that Looks Forward to College Math

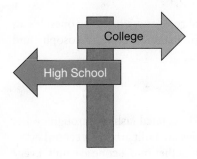

This text supports pathways courses such as Mathematical Literacy for College Students (MLCS), which typically is a one-semester course for non-math and non-science majors. MLCS integrates numeracy, proportional reasoning, algebraic reasoning, and functions. The focus of this course is on developing mathematical maturity through problem solving, critical thinking, writing, and communicating mathematics. Additionally, college success components are integrated with the mathematical topics. During this course, students will develop conceptual and procedural tools that support the use of key mathematical concepts in a variety of contexts, including statistics and geometry. Upon completion of the course, students will be prepared for a statistics course, a general education mathematics course, a quantitative literacy course, or further algebra. The prerequisite to MLCS is basic math or prealgebra. The course can be added to the traditional developmental algebra sequence as an alternative pathway, or it can be used to replace beginning algebra.

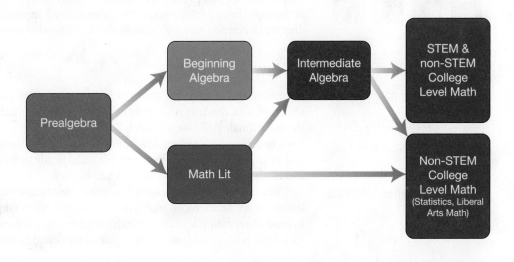

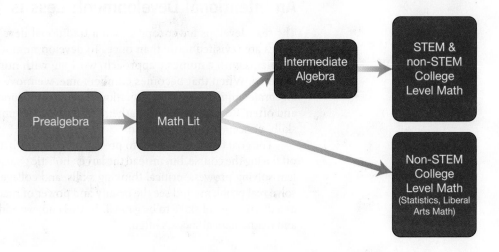

The title of this book, *Math Lit*, is an abbreviation of the course's name, "Mathematical Literacy for College Students." It is also a play on words, giving a nod to the approach taken in the text: the connection between mathematics and the written word. The majority of problems in the book are presented in a verbal format and are not overly formulated with mathematical commands like "solve" or "simplify." Additionally, each cycle contains two articles to support and develop mathematical content.

A New Paradigm

In every facet of the book, we threw out the traditional rules and wrote new ones. It's an unconventional approach, but one that works. Let's explore the structure, philosophy, and features of the book.

The Use of Cycles

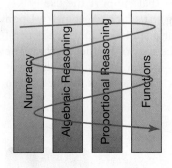

Instead of linear chapters, content is developed in an integrated fashion through cycles, which increase in depth as the course progresses. Cycles are built around a central question that underlies the activities and ties the sections together into a cohesive unit. Every cycle includes components from each of these four strands: numeracy, proportional reasoning, algebraic reasoning, and functions. Geometry, statistics, student success, and mathematical success are recurring themes in each cycle. A topic is usually not developed to completion the first time it is presented. Often we revisit a skill or concept but go deeper than we had previously in order to apply skills previously gained as well as to look at the topic from a fresh perspective. This layered approach helps the developmental student make connections and transfer the new knowledge into long-term memory.

A New Approach

Most developmental math students have had years of mathematics and, often, algebra. Our goal with this text is to approach developmental math in a way that is different from the traditional, skill-based model. To do so, we have used a different section structure, topic ordering, and content development than a traditional text uses. While this approach will be unfamiliar at first, students will quickly adjust to seeing mathematics differently, and many times they end up liking what they see.

The content, techniques, and contexts chosen are meant to be relevant for an adult learner who needs more mathematical skills before taking a college-level math class. Instead of repeating topics seen in high school in the same way and order, we approach each topic with two questions: how does it work and how can I use it?

The difficulty lies not so much in developing new ideas as in escaping from old ones.

—John Maynard Keynes

An Intentional Development: Less Is More

The text develops fewer topics than a traditional developmental math text, but many topics are revisited more than once. To develop number sense, we begin solving many problems with a numeric approach, working with numbers as long as it makes sense to do so. When that becomes cumbersome, we move on to other techniques such as algebraic approaches using equations or graphs. Proportional reasoning is used early and often. Regardless of the mathematical topic, we emphasize three core components: skills, concepts, and applications.

The goal of the text is not to present students with a list of skills that can be checked off during the course, but instead the larger, holistic goals of mathematical maturity, problem solving prowess, critical thinking skills, and college readiness. We want students to solve real problems and see the beauty and power of mathematics instead of seeing math as a discrete list of skills to be gained. To this end, we address the ideas of student success and mathematical success often.

These tenets are embodied throughout the text:

- Depth over breadth
- Quality over quantity
- Why before how

Processes are valued, but answers and methods need not be "one size fits all." This theme will appear often in sections and assessments. Metacognition is emphasized throughout the text so that students understand how they think and learn. To do this and solidify mathematical understanding, we tackle each problem from multiple perspectives. This way, students learn more about the skills and processes and also find the techniques that make the most sense to them.

The Role of Algebra

The developmental student needs not only enough new and different concepts and/or contexts to remain engaged and interested but also some familiar concepts to ease their anxieties. With this in mind, all algebra topics are approached in novel ways. Algebra itself is not the goal, but is rather another method that can be used to solve problems. This philosophy is very inviting to students. When they are not required to use algebra, students will often choose to use it because of its organizational and time-saving strengths.

If covered in its totality, the text will expose students to nearly all beginning algebra concepts. Students will also see many intermediate algebra ideas like quadratic and exponential functions, logarithmic scales, and function concepts like domain and range. With this approach, students are prepared to transition to a traditional intermediate algebra course if needed.

Factoring is included with a focus of using it for rewriting expressions with the greatest common factor or understanding the process of factoring trinomials. Additional factoring practice is available in MyMathLab to support students heading to intermediate algebra or non-STEM college-level courses that require factoring.

The Role of Technology

The approach to technology in this text is the same approach used in the workplace: use the tool that makes the most sense for the job at hand. Sometimes a scientific or graphing calculator is needed. Other times Microsoft Excel® is used for a spreadsheet. When mental calculations are possible and faster, we encourage students to use them. And sometimes the best tool is just a pencil. *Tech Tips* are provided throughout the text to guide students with the use of technology.

A spreadsheet icon is included next to problems that are well suited to be solved with Excel. However, the use of spreadsheets is optional and can be easily omitted.

Instructions that are provided for graphing calculators reference TI calculators. Spreadsheet instructions support Microsoft Excel®. You can adjust the instructions if you are using a different brand of calculator or software.

Thinking Outside the Box, Writing Inside It

The text uses a conversational style that is designed to be readable and inviting. It is presented in worktext format to fulfill a primary goal of the course: student engagement. We have found through years of classroom testing that students enjoy using a worktext that seamlessly integrates theory with their notes. This format allows students to write only what is needed and thus to pay more attention to doing math than taking notes. Since key terms, examples, and procedures are called out in boxes and in a summary at the end of each cycle, students have a useful reference when studying. Instead of a book that is not opened or used, the worktext becomes a dynamic document that is added to in each class period.

Having knowledge but lacking the power to express it clearly is no better than never having any ideas at all.

—Pericles

A Flexible Approach

The text supports activity-based courses as well as courses that include all direct instruction or some direct instruction with some group work. The included Instructor Guide provides more information on ways to use the text with particular formats. MyMathLab® is used for homework and supports classroom activities, but it does not replace instruction. It does, however, provide the additional practice with skills that some students need but may not get in the classroom due to time constraints.

Productive Struggle

All truths are easy to understand once they are discovered; the point is to discover them.
—Galileo Galilei

We need to allow students to think, struggle, and question. It's at those moments that they make connections and develop understanding. Countries with successful mathematics education regularly encourage this process instead of reducing each topic to a set of procedures. Allowing students to struggle in a productive way will also increase their persistence and perseverance, two components necessary for success not only in this course, but also in college as a whole.

We are asking students to accept a new way of learning and a new interaction between them and the instructor. Likewise, we ask the instructor to learn a new dynamic with students, which at times is unfamiliar. Similar to what we ask of students, we ask you to accept this temporary discomfort, as the rewards will be plentiful and evident in short order.

How Will Students Transition to a Traditional Environment?

Students usually transition very well from problem-solving environments like the one encouraged by this book to more traditional classrooms. After an experience like an MLCS course, students will have strategies for facing challenges, evaluating their work, and solving nonroutine problems. A traditional environment, with a linear progression of topics, can be easier, although less engaging, than what they experience in MLCS. Further, that traditional environment is what they have likely had in most of their previous classes. Thus this transition is a return to the familiar, not the unfamiliar.

Built by Faculty for Faculty

This textbook presents a new way of teaching and learning, so both the instructor and the student might have an initial period of adjustment. Acknowledging this, we designed the text to assist the instructor and the student with its nontraditional structure. Each section has a predictable rhythm, as does each cycle. Additionally, all sections are annotated with answers as well as notes to the instructor regarding pedagogy, common issues, and strategies to consider. We have also created an Instructor Guide to assist you with teaching your course.

Organization of the Book and Sections

This book is divided into cycles instead of chapters since there is a rhythm to each cycle of introducing a focus problem, building up skills and understanding, connecting concepts, and then wrapping up with a test and a solution to the focus problem. Instead of a linear, skill-based approach, we move through a variety of problems while simultaneously completing a cycle. Composed of open-ended problems, worked-out examples, interactive problems, and assessments, the cycles provide a rich and interesting arena for developing mathematical muscle.

Each cycle is divided into halves. Each half contains a set of cohesive sections to develop mathematical skills and concepts. A *Mid-Cycle Recap* is provided at the halfway point of the cycle as a way for students to check on their progress. This approach is used to provide predictability and structure. Additionally, the *Cycle Wrap-Up* helps students to reflect on what they have learned to date and to practice questions that are at test-level difficulty.

Gears are a recurring graphic used to convey the visual progression through the various components creating each cycle. Together, they turn to move forward mathematical problem solving and student success. This is accomplished by carefully designed tasks, explorations, and instruction that are assembled and paced in an effective way for the developmental learner.

Section Format

Because the goal of the course is for students to develop mathematical maturity more than a specific skill set, the goal of the sections is the creation of an experience. Over time, content is spiraled to elicit connections and deepen understanding across contexts. By the end of the semester, students should feel that they have partaken in an adventure whose journey was just as valuable as the destination.

Many sections economize on precious class time to accomplish more than just mathematical goals. Often, a skill will be developed while exploring additional concepts or problem-solving strategies. Developmental students usually need time to work on and make sense of ideas. Depth, not breadth, is the goal.

The sections within the cycles are divided into the following sections:

Explore — An interesting problem opens the section and sets the stage for the new material.

Discover — A new theory is presented, with examples and practice problems.

Connect — A problem connects the content of the section to past or future sections.

Reflect — An opportunity is given to look back at what has been learned.

Additionally, some sections have a preview assignment prior to the section known as *Getting Ready for the Section*. Each *Getting Ready* assignment includes an article from the popular press with accompanying questions related to the mathematics in the article.

Focus Problems

The value of a problem is not so much coming up with the answer as in the ideas and attempted ideas it forces on the would-be solver.
—I.N. Herstein

For each cycle, there is a focus problem that students can solve in groups. The focus problems are current, real-world issues that are not boiled down and formulated for students. They are challenging, but ultimately engaging and accessible. A section at the start of each cycle introduces students to the focus problem and how it relates to the content of the cycle. Periodic sticky notes are provided to help students make progress solving the focus problem, writing up a solution, and making sense of the solution prior to the cycle test. Focus problems allow students to apply knowledge gained during the cycle to a larger, more involved problem that does not have just one correct solution. Students learn to work with a group, understand new contexts, and write a solution in a coherent way.

Additional focus problems are available in MyMathLab. Writing templates for students and grading rubrics are included along with a detailed sample solution. More information about using the focus problems is provided in the Instructor Guide.

Section Features

Within the sections, the following special features recur:

How It Works

Boxes summarize procedures and provide worked-out examples of them.

To evaluate a formula:

1. Replace the variables in the formula with the appropriate numerical values.
    a. When replacing a variable with a negative number, use parentheses around the number.
    b. State the units with the numerical values if appropriate.
2. Use the order of operations rules to simplify the expression.
3. State the result, with units if appropriate.

EXAMPLE: Use the formula $F = \frac{9}{5}C + 32°$ to convert 100° Celsius to Fahrenheit.

Begin by replacing C with 100° and then simplify.

$$F = \frac{9}{5}(100°) + 32°$$
$$F = 180° + 32°$$
$$F = 212°$$

So 100° Celsius is the same as 212° Fahrenheit.

Look It Up

Short definition boxes present key terms, along with an example.

Ratio and Rate

A **ratio** is a comparison of two numbers and can be written in three different ways. For example, if the student-to-faculty ratio at a college is 40 to 1, the ratio can also be written as 40:1 or $\frac{40}{1}$.

When the quantities being compared in the ratio have different units, the ratio is called a **rate**.

FOR EXAMPLE, the rate 30 mpg (miles per gallon) is a comparison of the two quantities 30 miles and 1 gallon. It is perhaps easier to see this as a rate when it is written in fraction form as $\frac{30 \text{ miles}}{1 \text{ gallon}}$. Since the units (miles and gallons) are different, we refer to this ratio as a rate.

Tech TIP

Notes contain helpful hints for using calculators or other technologies.

Tech TIP

You can find exponents and roots on a scientific calculator. Depending on your calculator, you might have a generic exponent button y^x or a button like $\wedge$ to indicate an exponent. You should also have a root button like $\sqrt[x]{y}$. Experiment with your calculator now so you know how to evaluate an expression with exponents.

Remember?
Reminders are included for skills that have been presented already but might need to be reviewed.

Remember?
To increase a number by 20%, multiply by 1.20.

NEW
Worked-out Examples
Detailed examples illustrate theory in action and demonstrate problem solving techniques.

EXAMPLE 3

Kudzu vines are invasive and can exhibit incredible growth in favorable conditions. Suppose a vine's length is given as $L = 525 + 9t$, where L is in inches and t is the number of days since July 1. Identify and interpret the slope and y-intercept of this linear function.

SOLUTION

The slope is the coefficient of the independent variable, t.

$$m = 9 = \frac{\text{change in } L}{\text{change in } t} = \frac{9 \text{ inches}}{1 \text{ day}}$$

The slope tells us that the vine is growing at a rate of 9 inches per day. The y-intercept is the ordered pair (0, 525), which tells us the vine was 525 inches long on July 1.

Sticky Note
Brief notes are written to students with study tips and other information to address both student and mathematical success.

It is common to write a polynomial with the terms in decreasing order of degree. The previous result had the term x^2 with degree 2 first, the term $5x$ with degree 1 next, and finally the constant term 6 with degree 0.

NEW
Spreadsheet icons
These icons identify problems well suited to use with Microsoft Excel.

4. a. At the same university as mentioned earlier, the biology department has another course with four sections per semester, Anatomy and Physiology. The class sizes are 60, 63, 68, and 81. Find the standard deviation for the class sizes.

Test Prep

The end-of-cycle *Wrap-Up* contains a five-step plan to help students study for the cycle test. Included in the *Wrap-Up* are a cycle study sheet; a skill review in MyMathLab®; a vocabulary review; and practice application problems. The last part of the *Wrap-Up* is devoted to helping students overcome test anxiety by having them place themselves in a realistic test situation.

NEW
Study Sheet
This page functions like a note card that a student might make before a test to highlight key ideas.

Changes to the Second Edition

We listened to our users and made a second edition with the goals of simplicity, functionality, and versatility. Nearly all content from the first edition is included in the second edition, but it is often updated, condensed, or simplified to improve its function and ease of use. Many additional contexts, topics, and problems have been included to address user requests.

Simplified structure

- The cycle and section structures have been simplified.
  - Cycles are divided into halves with a *Mid-Cycle Recap* to assess student understanding.
- Sections have a more consistent format and length.
  - Some short sections have been expanded or combined with compatible content.
  - Lengthy sections have been divided into two sections.
- All key sections are in the first three cycles for use with 4-credit-hour courses.
- Cross references between sections have been reduced to allow more flexibility.
- Section subtitles and objectives (as Instructor Notes) have been added at the beginning of each section to clarify section goals.
- Guiding questions are included in each *Discover* to help students focus on important ideas.
- Answers have been removed from the back of the book to a tab in MyMathLab that can be revealed to or hidden from students.

Improved functionality and versatility

- Alternate focus problems are available in MyMathLab for instructors who want more variety. Additionally, there is only one section per cycle devoted to the focus problem. In the place of previous sections are periodic sticky notes to remind students what to do next to solve their focus problem.
- Text can be taught using groups, direct instruction, or a combination of both.
  - Group/whole class/individual icons and estimated times for each part of a section have been removed to allow for greater flexibility.
- Simplified Instructor Guide at the beginning of the text contains information on teaching in various formats (groups, online, etc.).
- Instructor Notes are streamlined and reduced.
- Text can be more easily used in face-to-face, hybrid, or online formats with flexible student resources including:
  - Even more worked-out examples and additional exposition throughout
  - Additional videos in MyMathLab
- Worktext pages have improved usability.
  - Printable homework pages in MyMathLab make homework collection easier without loss of content from the text
  - Larger spaces and blanks for students to write in answers
  - Increased graph size

Improved and new features

- *Tech Tips* include information about TI graphing calculators and Microsoft Excel.
- *Looking forward, looking back* problems are included in every homework assignment to preview important ideas and practice skills and concepts that students often find challenging.

- A Cycle *Study Sheet* replaces the Cycle Profile at the beginning of the *Cycle Wrap-Up*. The *Study Sheet* mimics a notecard that students would make to study for a test with key skills, concepts, and a few examples for illustration.
- The *Self-Assessment: Review* includes section numbers for easy reference when students study for a test.

Improved content development

- Student success ideas are addressed in sticky notes to be present but not obtrusive.
- Mathematical success is addressed in problems and content development, tips in sticky notes, and homework problems where students find errors in a student's work.
- Emphasis on units has been increased.
- Use of patterns and functions has been increased.

Updated and additional content

- Many traditional algebra topics have been added.
  - Factoring out the GCF has been made into its own section (Section 3.11).
    - Factoring by grouping has been included with factoring out the GCF.
  - Factoring coverage has been increased with an additional section (Section 3.12).
    - Topics addressed include trinomial factoring, difference of two squares, and sum and difference of two cubes.
  - Quadratic coverage has been increased to include quadratic patterns (Section 3.16), zeros of quadratic functions (Section 3.13), solving quadratics by factoring (Section 3.12), the quadratic formula (Section 3.13), and the vertex form of quadratic functions (Section 4.11).
  - Function coverage has been increased to include function notation (Section 4.9), zeros (Section 3.13), domain, range, and the vertical line test (Section 4.10).
  - Exponential function coverage has been increased to include more about writing and graphing exponential functions (Section 3.8).
  - Additional algebraic concepts have been added.
    - Midpoint formula (Section 1.8)
    - Arithmetic and geometric sequences (Section 1.14)
    - Rational exponents of the form $1/n$ (Section 2.3)
    - FOIL (Section 2.7)
    - Clearing fractions and decimals in linear equations (Section 2.13)
    - Point-slope form of a linear equation (Section 3.10)
    - Zero-product property (Section 3.12)
    - Zeros of functions (Section 3.13)
    - Solving systems of equations with elimination (Section 3.15)
    - Logarithmic scales (Section 4.6)
  - Geometry content has been increased with the addition of new sections (Section 2.17 – Volume and Surface Area, Section 4.12 – Trigonometric Functions) as well as additional problems that use geometric ideas as the context.
- A greater emphasis on statistical literacy is included.
  - Statistical topics appear in each cycle, including measures of center and spread, correlation, residuals, z-scores, probability, and graphs.
    - Residuals have been moved to be with correlation and regression (Section 3.2).
    - Standard deviation and z-scores have been combined into one section (Section 4.5).

◦ Statistics content has been increased with the addition of another probability section and with the use of statistics ideas as the context of a problem.

▪ Two-way tables in probability problems

▪ Interpreting probabilities out of 10, 100, or 1,000

▪ Quantitative and qualitative variables

◦ Preparedness for a statistics course is developed through reading, interpreting notation, the use of technology, emphasis on numeracy, interpretation of slope and *y*-intercept of lines, equation solving, rewriting formulas, and the emphasis on relationships between variables.

• Emphasis on and practice with fractions throughout have been increased.

• New *Getting Ready for the Section* articles and problems are included.

• More uses of Excel have been incorporated with the inclusion of an icon near any problem or procedure well suited for use with Excel. Additional Excel functions have been added to the Excel appendix available in MyMathLab.

• Important content has been moved earlier in the text.

◦ Integers, means, and experimental probability are covered in Cycle 1 instead of Cycle 2.

◦ Dimensional analysis is previewed when unit conversions are addressed the first time (Section 1.11).

◦ Equation solving, including the algebraic solving of proportions, is covered in Cycle 2 instead of Cycles 3 and 4.

◦ Theoretical probability and volume and surface area are covered in Cycle 2.

◦ Writing equations of lines and exponential functions is covered in Cycle 3 instead of Cycle 4.

◦ Factoring and systems of equations are covered in Cycle 3 instead of Cycle 4.

• The rigor of topics in the fourth cycle has been increased for schools wanting more difficult topics and/or a 5- or 6-credit course.

◦ Topics in Cycle 4 can be used in their current location as a separate cycle or inserted earlier in the book to delve deeper into a particular topic.

▪ Section 4.2 (Dimensional Analysis) can be completed after Cycle 2.

▪ Sections 4.3 (Scientific Notation) and 4.4 (Negative Exponents) can be used after Cycle 2 and Section 4.2.

▪ Section 4.5 (Standard Deviation) can be used after Section 3.2.

▪ Section 4.6 (Logarithmic Scales) can be used after Section 4.4.

▪ Sections 4.7 (Direct Variation) and 4.8 (Inverse Variation) can be used after Section 3.8.

▪ Sections 4.9 (Function Notation) and 4.10 (Domain and Range) can be used after Cycle 3.

▪ Section 4.11 (Vertex Form of a Quadratic Function) can be used after Section 3.16.

▪ Section 4.12 (Right-Triangle Trigonometry) can be used after Cycle 2.

Improved MyMathLab course

• An expanded video program supports online and hybrid formats

• More assignable problems are available including more "Skills" exercises both from the worktext, and many of the "Concepts and Applications" exercises from the worktext.

• Downloadable section homework sets are available in MyMathLab for students to print and turn in without loss of content from the text itself

• Premade tests can be assigned

MyMathLab® and Conceptual Homework: The Best of Both Worlds

This book and the accompanying MyMathLab course contain two different types of homework problems: exercises that address skills and problems that address concepts and applications. The two types of homework problems work together to help students develop the skills necessary to solve applied problems. A common approach to homework involves two steps:

1. Complete exercises in MyMathLab to master skills
2. Solve problems in the worktext homework, many of which are now also available to be assigned in MyMathLab, to check skill mastery and apply skills

Exercises in MyMathLab Only

MyMathLab provides carefully chosen exercises to allow students to practice all the skills necessary for solving applied problems. Students get immediate feedback, allowing them to master the skills of the section.

We recommend that students begin each homework assignment by completing these exercises in MyMathLab.

These problems are available only in MyMathLab and not the worktext. You will find these exercises in the MyMathLab assignment manager, designated with "MML Only" (e.g. MML Only 1.2.1). Additionally, sample homework assignments have been created for each section and are available in MyMathLab for instructor convenience.

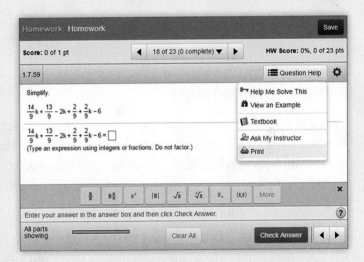

Skills Exercises in the Worktext

Two *Skills* exercises are available at the start of each homework assignment in the worktext. These exercises provide the opportunity for students to work a few exercises like ones in MyMathLab without learning aids. They also allow students to have a record of their work on paper to review for a test.

We recommend that students complete the *Skills* exercises after they complete the MyMathLab exercises. Algorithmic versions of these problems are available for instructors to assign via MyMathLab. They are designated with a "Text Skills" (e.g. Text Skills 1.2.1).

Skills MyMathLab

First complete the MyMathLab homework online. Then work the two exercises to check your understanding.

1. Are $10xy^2$ and $10x^2y$ like terms? Why or why not?

   No; they do not have the same variables raised to exactly the same exponents.

2. Simplify each expression.

   a. $-11x^2 + 11xy + 22xy - 12x^2 + 2$

   $-23x^2 + 33xy + 2$

   b. $(-11x^2)(11xy)(22xy)(-12x^2)(2)$

   $63,888x^6y^2$

Concepts and Applications Problems in the Worktext

For each section in the worktext, the *Skills* problems are followed by *Concepts and Applications* problems that focus more on applications and help students connect and apply ideas. Most conceptual homework assignments include only 3–10 problems, which increases the importance of each and every problem in the assignment.

Algorithmic versions of many of the *Concepts and Applications* problems are available for instructors to assign via MyMathLab as well. You will find these problems in the MyMathLab assignment manager following the above-mentioned "Text Skills" exercises and designated with a "Text C&Apps" (e.g. Text C&Apps 1.2.3).

Concepts and Applications

Complete the problems to practice applying the skills and concepts learned in the section.

3. A student was asked to simplify two expressions by combining like terms. Explain the mistake in each problem and then find the correct answer.

   a. $-3xy + 5x + 6y - 8xy$
      $= -3xy + 11xy - 8xy$
      $= 8xy - 8xy$
      $= 0$

   The student added $5x$ and $6y$ to get $11xy$, but x and y are not like terms. They should not be combined. The correct answer is $-11xy + 5x + 6y$.

   b. $15x^2 + 10x - 21x - 18x^2$
      $= 15x^2 - 11x - 18x^2$

   The expression still has like terms ($15x^2$ and $-18x^2$) that should be combined. The correct answer is $-3x^2 - 11x$.

4. Give an example of an expression that contains like terms and another that does not.
   Answers will vary.

Sample Assignments

There is a sample homework assignment for each section. These sample assignments initially contain the MyMathLab Only exercises. If you would like to have students complete their homework assignment entirely in MyMathLab, these sample homework assignments can easily be modified by adding any of the following:

- Additional MyMathLab Only exercises ("MML Only")
- *Skills* exercises from the worktext ("Text Skills")
- *Concepts and Applications* problems from the worktext ("Text C&Apps")

Creating an assignment in this manner emulates the approach used when combining online and text homework, but has the added convenience of all grading being completed in MyMathLab.

With this new edition, the MyMathLab course now contains a larger pool of exercises, which supports a variety of course formats.

Sample Quizzes and Tests

There is a sample quiz to accompany the *Mid-Cycle Recap* in each cycle. There is also a sample test for each cycle to support instructors who want to give tests online.

Resources for Success

MyMathLab is available to accompany Pearson's market-leading text offerings. This text's flavor and approach are tightly integrated throughout the accompanying MyMathLab course, giving students a consistent tone, voice, and teaching method that make learning the material as seamless as possible.

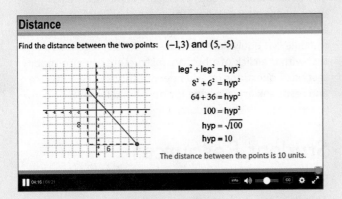

Videos for Students

Videos in MyMathLab give students even more resources for outside the classroom, and give instructors added flexibility and resources for various class formats. These videos can be used to supplement learning as students practice, or they can be assigned to prepare students for classroom activities.

- Videos walk students through examples, giving them an extra opportunity to reinforce and practice skills and concepts.
- Videos cover some "Look It Up" and "How It Works" features, allowing students to follow along with the content from the text.

Instructor Resource Videos

These videos, produced by the authors, present helpful suggestions based on their own experiences teaching Math Literacy for College Students. Including topics such as a typical class, focus problems, and group work, these videos provide the guidance an instructor might find useful in teaching this course, or with this text, for the first time.

Using groups to your advantage

- **Group quizzes for problems**
- **Time during a test or quiz with group**
- **Focus problem grades can be shared**

Learning Catalytics

Integrated into MyMathLab, Learning Catalytics uses students' devices, such as tablets, smart phones, or computers, for an engagement, assessment, and classroom response system. Pearson-created questions for developmental math topics are available to allow you to take advantage of this exciting technology immediately. Additionally, "Explore" questions related to the worktext are also available to give instructors a starting point for using this feature with Learning Catalytics. Search the question library for MLAF and the section number, for example MLAF312 for Section 3.12.

Instructors can

- Pose a variety of open-ended questions to help develop critical-thinking skills.
- Use real-time data to adjust instructional strategy and improve engagement.
- Manage student interactions by automatically grouping students for discussion, teamwork, and peer-to-peer learning.

Resources for Success

With MyMathLab, students and instructors get a robust course-delivery system, the full Almy/Foes eText, and many assignable exercises and media assets. Additionally, MyMathLab houses these extra instructor and student resources, making the entire set of resources available in one easy-to-access online location.

Instructor Resources

Annotated Instructor's Edition

This version of the text includes answers to all exercises presented in the book, as well as helpful teaching tips and the Instructor Guide.

Instructor's Solutions Manual

This online manual contains fully worked-out solutions to all text homework exercises. Available in MyMathLab and to download from www.pearsonhighered.com.

Instructor's Resource Manual

To provide additional support for instructors, this manual includes the following resources:

- Two quizzes for each cycle half
- One test for each half cycle
- Two tests for each whole cycle
- Two final exams

Available in MyMathLab and to download from www.pearsonhighered.com.

PowerPoint Lecture Slides

Available in MyMathLab and to download from www.pearsonhighered.com. These fully editable lecture slides include key concepts and examples for use in a lecture setting.

TestGen®

TestGen (www.pearsoned.com/testgen) enables instructors to build, edit, print, and administer tests using a computerized bank of questions. TestGen is algorithmically based, allowing instructors to create multiple but equivalent versions of the same question or test with the click of a button. Instructors can also modify test bank questions or add new questions. The software and test bank are available for download from www.pearsonhighered.com.

Student Resources

Student Success Module in MyMathLab

This new interactive module is available in the left-hand navigation of MyMathLab and includes videos, activities, and post-tests for these three student-success areas:

- **Math-Reading Connections,** including topics such as "Using Word Clues" and "Looking for Patterns."
- **Study Skills,** including topics such as "Time Management" and "Preparing for and Taking Exams."
- **College Success,** including topics such as "College Transition" and "Online Learning."

Instructors can assign these videos and/or activities as media assignments, along with prebuilt post-tests to make sure students learn and understand how to improve their skills in these areas. Instructors can integrate these assignments with their traditional MyMathLab homework to incorporate student success topics into their course, as they deem appropriate.

Acknowledgments

We are very grateful for all the wonderful and thoughtful feedback we have received as we developed *Math Lit*. Many thanks are due to the following people for their participation in focus groups, reviewer projects, and class testing. Asterisks are placed beside this edition's reviewers.

Debra Adiletta,
Monroe Community College

Mark Alexander,
Kapi'olani Community College

Chandra Noel Allen,
Guilford Technical Community College

Alvina J. Atkinson, Ph.D.,
Georgia Gwinnett College

Susan Barbitta,
Guilford Technical Community College

Susan L. Beane,
University of Houston–Downtown

Gale Brewer,
Amarillo College

Rogelio N. Briones,
Los Medanos College

Sam Bundy,
Kaskaskia College

*Maureen Burt,
Mohave Community College

Andrea Cahan,
Everett Community College

Robert Cappetta,
College of DuPage

Edie Carter,
Amarillo College

Jimmy Chang,
St. Petersburg College

Roberta Christie,
Shawnee Community College

Dianna Cichocki,
Erie Community College–South Campus

John Climent,
Cecil College

*Greg Cripe,
Spokane Falls Community College

Addie Davis, Ph.D.,
Olive-Harvey College

*Tamie Dickson,
Reading Area Community College

Deborah Lynn Doucette,
Erie Community College–North Campus

*Nicole Duvernay,
Spokane Community College

Dennis C. Ebersole,
Northampton Community College

Sarah Endel,
Rochester Community and Technical College

Ellen England,
Frederick Community College

Melissa Gaddini,
Robert Morris University

*Eric Gilbertsen,
Montana State University Billings

Kelly Groginski,
Onondaga Community College

Craig Hardesty,
Hillsborough Community College

*Julie Hartzler,
Des Moines Area Community College

Loye Henrikson,
State Fair Community College

Tonja Hester,
Amarillo College

Jennifer Hill,
College of Lake County

Andrea Hoagland,
Lansing Community College

*Erica Hotsinpiller,
College of DuPage

Jennifer Howard,
Western Kentucky University

Hans Humbarger,
Scott Community College

Sonya Johnson,
Central Piedmont Community College

Kimberly Dawn Jones,
Dakota State University

Maryann E. Justinger, Ed.D.,
Erie Community College–South Campus

Brian Karasek,
South Mountain Community College

John Keating,
Massasoit Community College

Elizabeth Kiedaisch,
College of DuPage

Clay King,
McLennan Community College

Mickey Koerner,
Des Moines Area Community College

Monica Kurth,
Scott Community College

Julianne M. Labbiento,
Lehigh Carbon Community College

Myrna La Rosa,
Triton College

Perri Larson,
Palomar College

Rowan Lindley,
Westchester Community College

LaRonda Lowery,
Robeson Community College

*Laura Lynch,
College of Coastal Georgia

Shanna Manny,
Northern Arizona University

Ilva Mariani,
Cerritos College

Aimee W. Martin,
Amarillo College

Connie McLean,
Black Hawk College

Rinav C. Mehta,
Central Piedmont Community College

Valerie Melvin,
Cape Fear Community College

*Patricia Miceli,
Wilbur Wright College

Curtis Mitchell,
Kirkwood Community College

Ruth Mortha,
Malcolm X College

Cindy Moss,
Skyline College

Catherine Moushon,
Elgin Community College

Glynis B. Mullins,
Pitt Community College

Julius Nadas,
Wilbur Wright College

Denise Nunley,
Scottsdale Community College

Louise Olshan,
County College of Morris

Kate Patterson,
Central Wyoming College

*Andrea Payne,
Montana State University Billings

Kathryn Pearson,
Hudson Valley Community College

*Mark Pelfrey,
Southwestern Michigan College

Dawn Peterson,
Illinois Central College

*Dan Petrak,
Des Moines Area Community College

Davidson Pierre,
State College of Florida

Matthew S. Pitassi,
Rio Hondo College

*Thomas Pulver,
Waubonsee Community College

*Rebekah Reger,
*City College at Montana State University
Billings*

*Pat Rhodes,
Treasure Valley Community College

Christopher Riola,
Moraine Valley Community College

Sylvester Roebuck,
Olive Harvey College

*Doug Roth,
Pikes Peak Community College

Jack Rotman,
Lansing Community College

Andrew H. S. Russell,
*Queensborough Community College,
C.U.N.Y*

*Jo Lynn Sedgwick,
Waubonsee Community College

Brenda Shepard,
Lake Michigan College

*Robin Shipkosky,
Southwestern Michigan College

Pavel Sikorskii,
Michigan State University

Craig Slocum,
Moraine Valley Community College

Pam Smith,
Jefferson Community College

Kelly Steinmetz,
Dakota State University

*Chairsty Stewart,
Montana State University Billings

Deborah Strance,
Allan Hancock College

Sara Taylor,
Dutchess Community College

Janet E. Teeguarden,
Ivy Tech Community College

Ivan Temesvari,
Oakton Community College

*Kelly Thannum,
Illinois Central College

*Cynthia Torgison,
Palomar College

*Diane Veneziale,
Burlington County College

Meredith Watts,
Massachusetts Bay Community College

*Greg Wheaton,
Elgin Community College

Sheela L. Whelan,
Westchester Community College

Erin Wilding-Martin,
Parkland College

Alma Wlazlinski,
McLennan Community College

Additionally, we would like to thank Paul McCombs and Ryan Cromar for ensuring the accuracy of our text; John Orr and Marilyn Dwyer at Cenveo, who oversaw the production.

Our thanks would not be complete without acknowledging the efforts of everyone at Pearson Education: Michael Hirsch, our editor in chief; Rachel Ross, our editor; Megan Tripp, editorial assistant, Alicia Frankel, our marketing manager; Alexandra Habashi, marketing assistant; Sherry Berg, our project manager; Marielle Guiney, our media producer; Rebecca Williams and Eric Gregg, our MathXL content managers.

We would like to thank our families for their support while we wrote this book. We would also like to thank the faculty who used the first edition and provided valuable feedback for this edition.

Credits

Photo Credits

Instructor Guide

This guide provides information to help you teach *Math Lit* successfully and in a manner that honors your strengths and preferences. Topics addressed include:

- Book and section structure
- Customizing your course
- Teaching with *Math Lit*
- Course formats
- Assessment
- Focus problems
- Group work
- Training for *Math Lit*

Book and Section Structure

Math Lit uses cycles that are nonlinear units of integrated content built around a central question. A cycle is arranged as follows:

- A **Self-Assessment** previews the topics of the cycle and provides an opportunity for students to gauge their understanding.
- The first section introduces a **focus problem** and cycle question.
- Sections of content are separated into halves by a **Mid-Cycle Recap**, which provides a chance for students to check their understanding at the halfway point of the cycle.
- A **Self-Assessment** reviews the topics of the cycle and provides an opportunity for students to determine the progress they have made.
- A **Cycle Wrap-Up** provides five action steps to help students study effectively for a test.

Each section has a problem-solving focus, often using contextual problems. Sections have the following flow:

- *Explore* an interesting problem.
- *Discover* theory and skills while practicing new skills.
- *Connect* new concepts to other contexts or new algebraic ideas.
- *Reflect* on the goals of the section.

Homework in MyMathLab and/or the worktext closes each section so that students can master skills and apply them to conceptual problems. See page xix in the Preface for more information about the MyMathLab course for *Math Lit*.

Customizing Your Course

Math Lit contains nearly all topics typically found in a beginning algebra textbook along with many intermediate algebra topics. Additionally, some statistical and geometric topics are included. Cycles 1–3 contain the core content that is suitable for a 4-credit-hour course. Since the sections in Cycle 1 are essential to many sections that appear later in the book, we advise that you cover the majority, if not all, of Cycle 1. Each section of the text takes approximately one hour to complete. More information on time management is provided in the *Course Formats* section of this guide.

If you are teaching a 5- or 6-credit-hour course or want greater depth and rigor, consider using sections from Cycle 4. This cycle provides extensions on many previously developed topics as well as several new topics. Topics in Cycle 4 can be used in their current location as a separate cycle or inserted earlier in the book to delve more deeply into a particular topic. Here are some ways to customize your course with the content found in Cycle 4:

- Section 4.2 (Dimensional Analysis) can be completed after Cycle 2.
- Sections 4.3 (Scientific Notation) and 4.4 (Negative Exponents) can be used after Cycle 2 and Section 4.2.
- Section 4.5 (Standard Deviation) can be used after Section 3.2.
- Section 4.6 (Logarithmic Scales) can be used after Section 4.4.
- Sections 4.7 (Direct Variation) and 4.8 (Inverse Variation) can be used after Section 3.8.
- Sections 4.9 (Function Notation) and 4.10 (Domain and Range) can be used after Cycle 3.
- Section 4.11 (Vertex Form of a Quadratic Function) can be used after Section 3.16.
- Section 4.12 (Right-Triangle Trigonometry) can be used after Cycle 2.

If you would like to use some of the sections in Cycle 4 to customize a 4-credit-hour course, consider omitting sections in the second half of Cycle 3 that are not needed for your follow-up courses. Use the desired sections from Cycle 4 in their place.

Prerequisite knowledge

This textbook assumes that students have successfully completed a prealgebra course or have placed into beginning algebra. Specifically, students should know how to work with whole numbers, fractions, decimals, and percents and also have basic knowledge of geometry. Because fractions and percents are often troublesome to students, they are reviewed early and used often throughout the text.

Teaching with *Math Lit*

Math Lit uses an active approach to learning, whether students work in groups, as a whole class, or alone. The hallmark of that approach is solving problems, not just exercises, that have depth and rigor while integrating multiple concepts. Many, but not all, problems are contextual.

A variety of contexts, including science, finance, and sports, are used in the text. Students do not need to have prior understanding of a context to successfully solve problems using it. They just need to be willing to read and apply their skills to a new situation.

Sometimes students get frustrated when they work on problems that are challenging and unfamiliar. While this can be uncomfortable, the experience is good for students. It helps them see that real problems aren't formulaic and teaches them patience and persistence.

Before teaching a section

To teach a section effectively, it is essential that you understand its goals. At the beginning of each section is an Instructor Note with the section objectives. (For a complete list of section objectives for a cycle, consult the Table of Contents.) This allows you to understand the intent of the section without spoiling the punchline of the section for students. For students to become better problem solvers, they need to solve problems without knowing the kind of problem it is (e.g., like terms). To help students learn the content and understand the goals of the section without providing all the objectives, guiding questions are stated at the beginning of each *Discover* portion of a section. These questions are similar, but not identical, to the section objectives and are written in language for students instead of instructors. In the *Reflect*, students will see the same section objectives that were stated for the instructor at the beginning of the section.

Emphasis on group work

If you are teaching your course in a face-to-face format, you can use the text in a variety of ways depending on how much or little group work you want to use. You can use a

> The only way to learn mathematics is to do mathematics.
> —Paul Halmos

consistent amount of group work for the entire course, cycle, or section, or you can flex throughout the course.

All group work:	Students work together throughout each class period, regularly reporting results.
Some group work:	Some portions of the section are completed as a class, while others are completed in groups.
No group work:	All of the section is completed as a whole class through lecture or direct instruction with interaction.

Using the four components to a section

Each section contains four components: *Explore, Discover, Connect,* and *Reflect.* Each component can be used in various course formats.

Explore: The *Explore* portion is an interesting problem or set of problems that is typically in paragraph form and is not scaffolded. It introduces students to a situation and asks them to use skills and concepts they already have. This portion sets the stage for the content to be developed. The *Explore* problems are well suited to group work.

> The best teachers are those who show you where to look, but don't tell you what to see.
> —Alexandra K. Trenfor

Amount of group work used in class	Ways to use the *Explore* portion of the section	Approximate amount of time needed
None	Give students time to work on the *Explore* problems alone before discussing the solution.	5–10 minutes
Some or all	Give student groups time to work together to solve the problem. Solutions can be reported by each group. Discuss the problems and students' solutions.	10–15 minutes

Learning Catalytics (available in MyMathLab) can be used to increase student interaction and accountability. Students can work individually or in groups to solve problems related to the *Explore* problems in each section and submit their answers through Learning Catalytics.

Discover: Skills, vocabulary, and theory are developed in the *Discover* portion of the section. Worked-out examples and problems for students to work are used throughout. Boxes with summaries of procedures and definitions are included for reference. The *Discover* portion can be used in a variety of ways depending on your class format.

Amount of group work used in class	Ways to use the *Discover* portion of the section	Approximate amount of time needed
None	Use the *Discover* portion as a guide for the examples to discuss and problems to complete with the class.	20–25 minutes
Some	If you would like to use direct instruction but still want some class interaction, consider having students work in groups for the numbered problems in the *Discover* portion and work as a class to address all other content.	25–35 minutes
All	Have students read through the explanations and examples and solve the numbered problems together. You might want to have students report on their results regularly. See the *Group Work* section of this guide for more tips.	25–35 minutes

Connect: The *Connect* portion of the section provides students with another new problem to apply what they learned in the *Discover* portion. Some *Connect* portions use an applied context, while others extend the content from the *Discover* to another mathematical concept. The *Connect* problems are well suited to group work.

Amount of group work used in class	Ways to use the *Connect* portion of the section	Approximate amount of time needed
None	Give students time to work on the *Connect* problems alone before discussing the solution.	5–10 minutes
Some or all	Give student groups time to work together to solve the problem. Solutions can be reported by each group. Discuss the problems and students' solutions.	10–15 minutes

Reflect: The *Reflect* portion wraps up the section. Students reflect on the point of the section and what they should have learned during it. Regardless of the amount of group work used in class, discuss the *Reflect* summary and section objectives. This should take approximately 5 minutes.

In summary, a typical section would look like the following depending on the amount of group work used:

All group work

- Student groups work on *Explore* problems before they are discussed (10–15 minutes)
- Student groups read explanations and work numbered problems for *Discover* section (25–35 minutes)
- Student groups work on *Connect* problems before they are discussed (10–15 minutes)
- Discuss the *Reflect* summary with the class (5 minutes)

Some group work

- Student groups work on *Explore* problems before they are discussed (10–15 minutes)
- Student groups work numbered problems; all other *Discover* content is discussed as a class (25–35 minutes)
- Student groups work on *Connect* problems before they are discussed (10–15 minutes)
- Discuss the *Reflect* summary with the class (5 minutes)

No group work

- Students work alone on *Explore* problems before they are discussed (5–10 minutes)
- Discuss *Discover* examples and summaries and complete problems as a class (20–25 minutes)
- Students work alone on *Connect* problems before they are discussed (5–10 minutes)
- Discuss the *Reflect* summary with the class (5 minutes)

Transitioning between section components: Regardless of how you structure the class to do each portion of a section, it is important to help students transition between the portions of the section. We suggest summarizing the findings of the *Explore* problems and then explaining how they lead into the *Discover* portion. After students complete the *Connect* problems, summarize the findings and how they relate to the goals of the section, as stated in the *Reflect* portion. Doing so ties the components of the section together into a cohesive whole.

Using the section features

Each section contains one or more features that summarize important concepts and definitions, provide examples of problems worked out, or state the steps for a procedure. If you are using all group work, students can read through these features within their groups. Otherwise, here are some ways these features can be used with some or no group work class formats.

Feature	Ways to use the feature
Look It Up	Discuss definitions and the included example.
How It Works	Suggest as a reference for students when they study or discuss the steps for a procedure and the included example. Work the example with students or do the *Need more practice?* problems if they are provided. Students can complete the *Need more practice?* problems alone, in groups, or as a class.
Worked-out examples	Discuss the example and techniques used to solve the problem or work the example with students. Ask students questions about the problem and its solution to go deeper into the concepts and make connections.
	Worked-out examples provide support to online students, those who miss class, and/or those who want an example to reference or an explanation for classes doing all group work.

Time management

Most sections can be taught in one hour, but they can take more or less time depending on how much group work and direct instruction you use. The first section of each cycle, which introduces a focus problem, can be covered in 20–30 minutes. The *Cycle Wrap-Up* can be done in as little as 25 minutes, if it is only overviewed, or as long as an hour if students begin working on it in class.

To gain more class time for assessments, discussion of homework problems, additional content, or problem solving, consider these options:

- Use class time on the focus problem only to explore it at the beginning of the cycle and debrief at the end of the cycle.
- Assign the *Explore* to be done before class.
- Do the *Discover* as a whole class and put limits on group work time.
- Omit the *Connect* or assign it as part of the homework.
- Limit test review using the *Cycle Wrap-Up* to 20–25 minutes.

Course Formats

In addition to face-to-face classes, *Math Lit* can also be used with online and hybrid course formats.

Hybrid format

The *Explore* and *Connect* problems could be solved during class time, particularly in groups. They could serve as excellent opening and closing problems for students in class. Students could watch the videos that accompany the *Discover* portion of the section and complete the problems in it outside of class.

Online format

The *Explore* and *Connect* problems could be solved individually or in online groups of students. A discussion board could be used for discussion and/or solutions. The *Discover* portions could be worked through individually by the online student with video support.

Assessment

Math Lit uses a variety of assessments, each with the same emphasis on skill exercises and problems on concepts and applications. In addition to homework, other kinds of assessments are available. Each can be done partly in MyMathLab (for skills) and partly in the worktext (for concepts and applications) or entirely in MyMathLab. For more information on the homework and assessment setup in the worktext and MyMathLab, see page xix in the Preface.

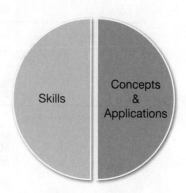

Mid-Cycle Recaps are one-page assessments on the first half of each cycle. An optional MyMathLab quiz is available and pre-built for each Mid-Cycle Recap. Additional quizzes are available in the Instructor Resource Manual located in the Instructor Resource section of MyMathLab.

Tests can be given on paper or in MyMathLab. Sample online tests are available in MyMathLab. Sample paper tests are available in the Instructor Resource Manual located in the Instructor Resource section of MyMathLab.

Focus problems, which are open-ended problems that include multiple topics from a cycle, can be used as group assessments or individual assessments of a student's problem-solving and writing abilities.

Assessment tips

- Assign the *Self-Assessment: Preview* that opens each cycle before the first day of a new cycle. It will help students to preview the cycle's topics.
- Collect homework from the worktext to encourage accountability.
- Give a test after a half cycle if a full cycle is too large.
- Allow students to complete a quiz in groups.
- Allow students discuss a test with a group for 5 minutes with pencils down.
- Include a test question on the focus problem to hold each student accountable for understanding the focus problem.

Allowing students to work together in a group on a quiz or part of a test encourages productivity and interaction and does not have to lead to grade inflation. To reduce the possibility of grade inflation, limit the number of points students can earn with their group.

Focus Problems

Cycle projects, called focus problems, are available to extend a student's problem-solving abilities and connect the cycle's content. They are best solved over time throughout the cycle. The focus problems don't always have exact solutions, and a good estimate is sometimes the best we can achieve. Each focus problem has an open-ended component for which there is more than one correct answer. Many focus problems are well suited to being solved with Excel.

Learning without thought is labor lost; thought without learning is perilous.

—Confucius

The goals of the focus problems are:

- **Establish** the cycle question
- **Develop** patient problem solvers
- **Improve** written communication skills
- **Connect** cycle skills and concepts

This guide includes an explanation of ways to use these problems, if you choose to, and the available resources for you and students. Additional focus problems are available for each cycle in MyMathLab.

Each student can complete a focus problem and submit it for an individual grade. However, the focus problems work well in groups. Because they are involved and have multiple tasks, the work can be split among group members. Roles can be assigned as a way of dividing the work. Possible roles include researcher, rough draft writer, problem solver, and accuracy checker. Working with other students can help students get through the problems more successfully.

Using the focus problems

The focus problems can be solved individually or in groups outside of class and submitted as an assignment. This approach does not take any class time. Or class time can be used to help students get started, make progress, and debrief the solution.

Progress can be made on the problems over four class sessions: initial work session, two check-in sessions, and a debriefing session. Periodic sticky notes are provided throughout the cycle to remind students to work on the focus problem and what they should be doing to solve it. These notes can be found near the *Reflect* box at the halfway point of each cycle half and in the last section of each cycle.

Focus Problem Session	Tasks	Approximate amount of time needed
1	Students read the focus problem and begin making a list of tasks they need to accomplish. Tasks can be divided among group members. Roles can be chosen and contact information exchanged. Group meeting times outside of class can be established.	20–30 minutes
2	Groups check in to share progress and determine what needs to be done next and who will do it.	20–30 minutes
3	Groups work on writing the solution to the problem. Students should check each other's mathematical work and proofread the document for grammatical and spelling errors. Consider passing out a grading rubric and writing template, available in MyMathLab, to help students as they begin to form their solution.	20–30 minutes
4	A model solution, either a sample solution or one from a group, is shared and discussed. Graded papers are returned and evaluated for strengths and weaknesses.	20–30 minutes

Focus problem tips

- For more structure, consider asking students to turn in something at the two intermediate check-in points.
- If students are struggling to find time to meet outside of class, encourage them to meet asynchronously using Google docs.

- When students are working on the focus problems, resist the urge to give definite answers. Answer questions with questions.
- Encourage students to use the writing templates if they are struggling to write a solution.
- To gain more class time, only do focus problem sessions #1 and #4 in class.
- Include a test question on the focus problem, worth 5–10% of the test score, to keep students accountable.
- Consider having students grade their focus problem solutions using a rubric when the solution is debriefed.

The whole is greater than the sum of its parts.

—Aristotle

Group Work

Working in groups can help students solve more involved problems and deepen their understanding because students talk about mathematics and hear other students' ideas and approaches. The collective effect is often greater than what could be achieved by working alone. Active learning with groups usually takes longer, but the rewards make up for the time used. When students are talking about mathematics and explaining their thought processes, their learning deepens. Noise will come with that, but it is not necessarily a negative if students are productive.

Even with all the possible benefits, group work can sometimes pose challenges for both students and instructors. This portion of the Instructor Guide will provide the tools you need to be successful if you choose to teach with groups. Included are tips we have learned from teaching *Math Lit* for several years as well as research-based strategies and resources for implementing group work effectively.

In addition to this guide, the first half of Cycle 1 contains timely tips about groups in the Instructor Notes to help you as you begin teaching the course. During the first cycle, you can determine which techniques you prefer and where you like to use them.

What makes group work effective?

For students to work well together in groups, they need to have positive interdependence, promotive interaction, individual accountability, appropriate use of social skills, and group processing (Johnson, Johnson, and Smith 2014, 93–95). This means that students feel that they need each other to succeed. They encourage each other and speak in supportive ways. They are responsible to each other. When they use appropriate social skills as a group, communication and conflict are dealt with effectively and assessed regularly so that interaction improves over time. Suggestions for creating these conditions are included in this guide.

There are things we can do as instructors to create groups that function well. But the content also needs to support active learning. Having creative, challenging tasks that give students something to talk about will encourage them to work together (Frey and Fisher 2010). The sections of *Math Lit* have been designed with this in mind. For example, *Explore* and *Connect* problems are challenging and less scaffolded in order to support cooperative learning. The focus problems are involved problems that need to be solved over time. Using groups for these features is optimal because students are less resistant to group work when they can see a need for it.

How can you make group work a positive experience?

You play a large part in creating and guiding groups so that they are successful and productive. To this end, we will discuss the following tasks for an instructor:

- Forming groups
- Starting group work
- Creating interaction
- Managing groups
- Encouraging accountability

Forming groups

The size and construction of groups should be chosen based on the tasks the groups are to complete. Groups of 2 to 5 students are effective in encouraging interaction (Frey and Fisher 2010). In this book, the tasks are designed so that groups of 3 to 4 students work well. Prior to the first class meeting, create groups of 3 to 4 students. We suggest using these initial groups throughout the entire first cycle and then changing groups at the beginning of each new cycle. Heterogeneous groups are best to prevent concentrating resources in a single group (Frey and Fisher 2010). When forming groups, incorporate an even mix of genders if possible. At the beginning of the course, you may not know your students and their abilities well enough to ensure a mix of abilities in the groups. That can be done better in Cycle 2. At that point, consider personalities and ability levels in addition to genders. While it is ideal to have stronger students work with weaker students, it is just as important to group students who have compatible personalities.

We like to create groups at the beginning of a cycle and then change them only when a new cycle starts. However, a cycle is not the only length of time students can stay in one group. Other options include changing groups every week or every half cycle. The key is to change groups regularly, since too much familiarity can lead to off-task behavior. You might need to change groups that are not working well.

Starting group work

It is critical that students understand from the beginning that the course will operate with a set of expectations that might be different from what they have had in previous courses. If you plan to use groups for a significant portion of the course, it is important to explain to students on the first day that groups will be used to help them learn the content. It is not uncommon for students to be reluctant, but that reluctance can lessen as students see how the content works and how groups help them be successful.

On the **first day** before the first activity, provide students with the names of the members of each group and the place where each group should sit. You could post a list of the members in each group or call out names. Consider numbering the groups and having each group number sit in the same location throughout the term. For example, you could have group 1 always sit in the front right area of the room. When Cycle 2 begins and you assign students to their groups, they will automatically know where to sit based on the number of the group they are given.

Since the first day sets the tone for a course, consider using the syllabus to convey how the course will operate. To encourage interaction, give each group one copy of the syllabus and 10 questions to answer. A syllabus quiz can be used for individual accountability.

To help groups function well, it is important to **establish clear expectations** and teach them from the beginning of the course. Even strong students do not always know how to work effectively in groups. Therefore, it is necessary to provide students with ground rules for effective group work and to teach students how to function in groups. These are some ground rules we have used to support the five elements of successful group work that Johnson et al. (2014, 93–95) recommend. We suggest giving these rules to students the first time groups are formed and then reminding them of these rules each time the groups change. If you keep the same groups for one cycle, the groups will become cohesive, and the group ground rules will have to be addressed only once in the cycle.

- Listen as well as talk.
- Compromise and show respect.
- Share group tasks in class.
- Be accountable for out-of-class work.

Additionally, we ask students to do the following each time a new group is formed:

- Get contact information for each group member (first and last names, phone number, e-mail address).
- Designate a "group manager" to do such things as mediate conflict, keep the group on task, pull in members who are not participating, and so on.

The group manager is not the group leader; rather, his/her role is an additional one that keeps the group on course. A good person to be the group manager is not someone who likes to lead. It is better to have someone who has a strong enough personality to say the difficult things but who can also let others take the reins. A skillful manager will mediate and head off conflict; such managers do not tend to create conflict. This is a position in which more mature adult students often shine. Consider choosing a group manager for the first cycle after groups have worked together for a week. Another option is to allow groups to choose their own manager or allow students to volunteer for the role.

In addition to the role of manager, you may want to give students **roles** while they work together to encourage interaction and accountability. Some options include recorder (writes the group's work), reporter (explains the group's results), and content expert (offers explanations and reasoning). If a group has only three members, the recorder and reporter can be the same student.

Everyone in the classroom has a role in making groups function well. It is helpful to **teach students what they can do** to be a helpful part of their group. We have found that when students have the following five traits, they usually enjoy group work more and other students enjoy working with them. The experience is more positive and productive for everyone.

Prepared	Completes assignments prior to class
Present	Attends class regularly
Positive	Speaks with supportive, noncritical words and works through conflict
Participating	Works during class and interacts with the group
Productive	Stays on task while working

The beginning of the course is a good time to address the issue of a member **not pulling his or her weight** before it becomes a problem. Discuss with the class ways to deal with this issue and allow them to choose a consequence they feel is appropriate. For example, a group could decide not to put a member's name on a group paper, in effect ensuring that person earns a zero. The group members could refuse to share information with a member who had not been coming to class. The class needs to come to a consensus on consequences that are fair.

Creating interaction

There are numerous ways to **structure group work** that will encourage interaction beyond "turn to your partner and work on the problem." Here are some methods we have found useful. Consider incorporating some of these in each section in which you use group work. They work particularly well with *Explore* and *Connect* problems. For more techniques, please see the References list at the end of this part of the Instructor Guide.

Pencils down

- With no pencils or pens, students work individually on the problem for a few minutes. Then they come together with the group and talk without writing. Once the instructor says so, groups can write a solution.

One, then all

- Each group member spends 5 minutes working on and writing a solution. Once the instructor says so, the group members come together to compare write-ups. Each student should share something or ask a question if he/she has nothing to share. From the individual solutions, one solution is written that should be better than the individual responses.

Divide and conquer

- Divide a large task among the groups, with each group doing something different from the others. The results are collated.

Roundtable

- A paper with the problem being solved is passed around the group. Each student adds something to the paper before passing it along. Option: Do not allow students to talk while they work.

Brainstorming

- Instead of solving the problem, each group generates a list of ideas of ways to tackle the problem. Ideas are collected from the groups. The class evaluates the ideas and determines which ones to use. The problem can then be solved as a class or within groups.

Questions

- Instead of solving the problem, the group generates a list of questions they think will help them solve the problem. Questions are collected from the groups. Any student who has an answer to a question can offer it. The problem can then be solved as a class or within groups.

Before discussing the solution to a problem as a class, you might have groups report their results. Options include:

- Verbally (if answer is brief)
- On the board (if answer can be written easily)
- Using a document camera or projector (if transferring the answer would be too time-consuming)
- Using Learning Catalytics

Managing groups

Just as students have a role in the success of group work, so does the instructor. Being an engaged presence helps maintain progress and productivity (Frey and Fisher 2010). Here are some **general guidelines for instructors** we have found to be useful in a *Math Lit* class:

1. Answer questions with questions.
2. Don't write for students.
3. Encourage students to work with their group instead of relying on you.

Following these guidelines conveys to students that you will help them but not solve problems for them.

Before students work on a problem in a group, explain the task and how they will be completing it as a group, such as using one of the ideas in the previous part of this guide on creating interaction. Tell them what you want to see while they're working, how long they have to work, and what they should have produced by the end of that time.

While students are working, walk around looking for groups that are getting off task. Sometimes a group may need direction or a reminder to focus on the task at hand. If productivity is low for most groups, it may be because they are unable to get started on the problem. Sometimes a hint or some instruction is necessary to ensure that groups continue making progress. Students need to be adequately challenged but not overwhelmed. Knowing when to intervene and when to allow students to struggle can be difficult, but you will quickly gain confidence with this skill as you teach the course.

An important idea for students, instructors, and even tutors to keep in mind is that helping does not mean giving answers. Ask questions and give hints instead. Also, encourage students to use the "3 before me" rule. That is, they must talk to at least 3 students from their group or another group before asking the instructor a question.

Even with the best group setup, guidelines, and direction, it is possible to have **issues within groups** that hinder them from working well together. As a first measure, facilitate communication between the group members but encourage them to work out their own problems. If a student comes to you privately with concerns about his/her group, ask the student what *he/she* can do to help the group function better. Each student has a role, just like the other group members, in the group dynamic. Although a group can be disbanded, we've found that the best solution is often to keep the group intact until the next cycle and then separate students who do not work well together into different groups.

For more ideas on dealing with dysfunctional groups, consult "Effective Strategies for Cooperative Learning" by Felder and Brent (2001, 70–72) in the References.

Encouraging accountability

There are a variety of ways to hold each student accountable for work done within a group. Here are some options:

- Call on groups in class to encourage group members to work together. Also, call on different group members each time to boost discussion. If you are using roles, require students to vary who is the reporter.

- Attendance or participation points can be given to students based on how much and how well they work within their group.

- Instead of using roles, have each student record the work of the group and then randomly collect one paper from each group to use to share results.

- If focus problems are used, give a group grade in which everyone in the group receives the same score, and also consider including a test question that covers something from the focus problem. This will show you and the student whether he/she comprehended the problem and its solution.

Group grades can be used to hold students accountable. Some options include sharing grades, averaging grades, and having students choose how to distribute points based on individual performance. For more information on these options as well their advantages and disadvantages, consult Carnegie Mellon's "Grading Methods for Group Work" mentioned in the References.

References

In addition to the sources cited, the following articles offer more information on group work:

Carnegie Mellon. "Grading Methods for Group Work." Accessed March 18, 2016. www.cmu.edu

Felder, Richard M., and Rebecca Brent. 2001. "Effective Strategies for Cooperative Learning." *The Journal of Cooperation and Collaboration in College Teaching* 10 (2): 69–75. www4.ncsu.edu

Foundation Coalition. "Positive Interdependence, Individual Accountability, Promotive Interaction: Three Pillars of Cooperative Learning." Accessed March 18, 2016. www.uwstout.edu

Frey, Nancy, and Douglas Fisher. 2010. "Making Group Work Productive." *Educational Leadership* 68 (1). www.ascd.org

Johnson, David W., Roger T. Johnson, and Karl A. Smith. 2014. "Cooperative Learning: Improving University Instruction by Basing Practice on Validated Theory." *Journal on Excellence in University Teaching* 25 (3 & 4): 85–118. celt.miamioh.edu

Smith, Karl A., Sherri D. Sheppard, David W. Johnson, and Roger T. Johnson. 2005. "Pedagogies of Engagement: Classroom-Based Practices." *Journal of Engineering Education* 94 (1): 87–101. ProQuest Education Journals. www.engr.wisc.edu

Training for *Math Lit*

Instructors do not necessarily need specific training to teach *Math Lit*. Instructors can read this guide and look through a few sections in the text to get a feel for the flow of content and philosophy that we have used. If instructors can attend a class period of another instructor, that can increase their familiarity with the course and ease their concerns. Additionally, Instructor Notes are provided throughout the text to guide the teaching of content. Videos that offer instructor teaching tips are available in the Instructor Resource section of MyMathLab.

Math Lit

A Pathway to College Mathematics

WHERE DO WE START?

mean
graph
fraction
area
factor
ratio
function
exponential
linear
integer
equation
variable
probability
proportion
rate
SCALE
perimeter
expression
arithmetic
conversion
percent
model
geometric
term

SELF-ASSESSMENT: PREVIEW

Below is a list of objectives for this cycle. For each objective, use the boxes provided to indicate your current level of expertise.

INSTRUCTOR NOTE: Consider discussing this page when you begin the cycle. This same list is presented at the end of the cycle so that students can note their progress and identify areas that they still need to work on before the test.

SKILL or CONCEPT

Low → High

1. Simplify, add, subtract, multiply, and divide fractions.

2. Solve applied problems involving fractions.

3. Plot ordered pairs.

4. Determine the coordinates of a point.

5. Interpret ratios.

6. Scale ratios to produce equivalent ratios.

7. Determine if quantities are proportional.

8. Find and interpret experimental probabilities.

9. Interpret signed numbers in situations.

10. Add, subtract, multiply, and divide signed numbers.

11. Find and interpret the mean of a set of numbers.

12. Create and interpret pie, bar, and line graphs.

13. Create and interpret scatterplots.

14. Sketch a curve to best fit a scatterplot.

15. Convert units by multiplying or dividing.

16. Apply, find, and interpret percent change.

17. Differentiate between variables/constants, expressions/equations, and factors/terms.

18. Make conjectures and generalize patterns.

19. Model change with linear and exponential functions.

20. Calculate perimeter and area.

The problem is not that there are problems. The problem is expecting otherwise and thinking that having problems is a problem.

—Theodore Isaac Rubin

1.1 The BP Oil Spill
Focus Problem

INSTRUCTOR NOTE: Additional focus problems are available in MyMathLab. Each focus problem is accompanied by a writing template to help students form their solution. A grading rubric is also available. For more information on ways to use the focus problems, see the Instructor Guide at the beginning of the worktext.

For 87 days in the spring and summer of 2010, British Petroleum's Deepwater Horizon spilled millions of gallons of crude oil into the Gulf of Mexico, resulting in one of the largest oil spills in the history of the United States and one of the largest unintentional oil spills in history.

In addition to the cost of cleaning up the spill and compensation to those affected, BP was also responsible for penalties under the Clean Water Act, which calls for fines of $1,100 per barrel, for simple negligence, up to $3,000 per barrel, for gross negligence. The federal prosecutors argued that the fines should go as high as $4,300 under the Environmental Protection Agency rules.

An important part of determining the appropriate fine was estimating the size of the spill. Experts offered differing opinions on the flow rate of the oil from the undersea well. Initially, an expert from BP insisted that the oil was flowing at a rate of only 5,000 barrels per day. It turned out that this was a gross underestimate of the flow rate, and the BP expert was eventually prosecuted for misleading the investigation. An expert in fluid dynamics studied the flow rate on video and computed a rate of approximately 70,000 barrels per day (+/−20%).

During the third phase of the trial, a judge was left to weigh the differing opinions and decide on a penalty for BP. A final settlement amount of $18.7 billion was eventually agreed upon. The trial took two years and was one of the most complicated civil cases in the history of the U.S. courts.

Use the fluid dynamics expert's range on the number of barrels per day, along with the allowed penalty range under the Clean Water Act and EPA guidelines, to calculate a minimum

and maximum penalty for BP. What percent of this maximum was the final settlement? Is the final amount closer to the minimum, the maximum, or the midrange (mean of the two)? Do you think the penalty imposed by the judge in the third phase of the trial was fair? Explain your response and include some information you learned from researching the topic in news reports.

What was the value of the oil BP lost in the spill at the time of the spill? What would the oil be worth now? Convert the number of barrels of oil spilled to gallons. Convert this amount to something practical to make sense of the size of the spill.

Some estimates of the size of the oil spill relied on analyzing the flow rate from the pipe as seen on video footage. If the circular opening of the pipe has a diameter of 21 inches and the oil was estimated to be flowing at 30 inches/second, how many cubic feet of oil were leaving the pipe each second? Each day? How does this estimate compare to the other estimates mentioned for gallons per day? Discuss some possible reasons for the discrepancy.

Compare the size of this oil spill to others in history to put it into perspective.

When we're trying to assign blame in an environmental disaster of unprecedented scale and there are billions of dollars on the line . . .

WHERE DO WE START?

This focus problem is an example of a real-world problem that does not have a quick, easy solution. In complicated situations, it is not always possible to get an exact answer, and often a reasonable estimate is the best that can be done. You should not expect to be able to solve this problem today, but, instead, you will return to the problem several times throughout the cycle to apply knowledge and skills as you acquire them. As you work on the problem, it will help to consider George Polya's four steps of problem solving:

1. Understand the problem: read the problem carefully, identify the task, research terminology

2. Make a plan: choose a strategy, list steps, get organized

3. Carry out the plan: implement the strategy, evaluate your progress, start over if necessary

4. Look back: assess the reasonableness of the answer, check the result with another method

You will see periodic sticky notes throughout the cycle to encourage you through these four steps and toward a solution to the focus problem.

The question "Where do we start?" will be the focus of the first cycle of the book. Often, it will be tied to a mathematical concept or skill. But the question is also larger in scope, asking you to think about how to approach not just a course but also a challenging, real-world problem.

INSTRUCTOR NOTE: If you are using groups and would like more detailed information on using them effectively, please consult the Instructor Guide at the front of the worktext. It contains techniques and resources for group work.

1.2 Getting Started: Reviewing Prealgebra

Explore

1. Find all the numbers that satisfy the following descriptions:

   - The number has seven digits.

   - The number is between 1,000 and 10,000.

   - The digit in the tenths place is 150% of the digit in the thousandths place.

   - The digit in the hundredths place is $\frac{6}{5}$ of the sum of the digit in the tenths place and the digit in the thousandths place.

   - The sum of all the digits in the number is 17.

   - None of the digits are zero.

   2211.362, 2121.362, 2112.362, 1221.362, 1212.362, 1122.362

2. Add one additional description so that there is only one possible number. What is the only number that satisfies all the descriptions?   Answers will vary.

Discover

The problem in the *Explore* challenges you to recall a variety of mathematical vocabulary words and skills. It is also a nonroutine problem that you have likely not seen before. Problems like this will be used regularly in this book to help you improve your problem-solving abilities and deepen your understanding of the content you are learning.

As you work through this section, think about these questions:

1 Which prerequisite skills do you still need to practice?

2 Why is a common denominator needed when you add or subtract fractions?

3 Do you know how to use fractions to solve applied problems?

To get a sense of where you are in relation to the skills needed for this course, next we will consider some sample problems. We can classify them into categories using a Venn diagram.

INSTRUCTOR NOTE: With group work, it is important to teach students early on that they need to ask their group members instead of you for help. Answer their questions with questions instead of providing answers. Resist the urge to do any writing on their pages.

Venn Diagram

A **Venn diagram** is a graph that helps compare and contrast sets. Each set is represented by a circle, and circles that have elements in common overlap. The circles are usually enclosed in a box, which represents a larger set and provides a space for elements that do not belong within any circle.

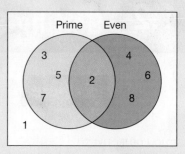

EXAMPLE: Use a Venn diagram with circles representing prime numbers and even numbers to sort the whole numbers 1 through 8.

SOLUTION:

The numbers 2, 3, 5, and 7 are prime. The numbers 2, 4, 6, and 8 are even.

Since 2 is both prime and even, it is located in the overlap of the circles.

Since 1 is neither prime nor even, it is located outside the circles.

Since 5 is prime but not even, it is located in the prime circle but outside the even circle.

A Venn diagram is used in #3 to categorize skills that you have likely learned in a prior course. It will allow you to indicate the level at which you can perform each skill shown in the table on the following page. Each circle of the diagram addresses one of the three facets of learning math discussed in this book: skills, concepts, and applications.

3. We will use the diagram that follows the table to sort the skills listed in the table. First, make sure that it is clear what each region in the diagram represents.

  a. What does Region 1 represent?   Region 1 represents skills that you can perform but cannot explain or apply to a new situation.

  b. What does Region 8 represent?   Region 8 represents skills that you cannot perform, cannot explain, and cannot apply.

  c. For each skill listed in the table, write the letter of the skill in the appropriate area on the Venn diagram. Be honest as you complete the diagram. Do not actually work the problems in the table.

As you start a course, it is important to get organized. Think about your weekly schedule. Plan for class, work, study time, meals, sleep, and free time. Bring any needed materials to each class. If you have MyMathLab homework, plan where and when you can work at a computer.

Letter	Description of Skill (performed without a calculator)	Example
A	Add/subtract/multiply/divide whole numbers	$56 + 89$; $122 - 37$; $77 \cdot 91$; $114 \div 18$
B	Add/subtract fractions (like denominators)	$\dfrac{3}{7} - \dfrac{1}{7}$
C	Add/subtract fractions (unlike denominators)	$\dfrac{1}{7} + \dfrac{3}{5}$
D	Simplify fractions	Reduce to simplest terms: $\dfrac{24}{39}$
E	Multiply/divide fractions	$\dfrac{2}{9} \cdot \dfrac{4}{5}$; $\dfrac{2}{9} \div \dfrac{1}{5}$
F	Add/subtract decimals	$0.32 + 1.047$; $2 - 0.53$
G	Multiply/divide decimals	$2.3 \cdot 8.92$; $11.4 \div 1.8$
H	Add/subtract integers	$-3 + 12$; $4 - (-8)$
I	Multiply/divide integers	$5(-6)$; $24 \div (-2)$
J	Find a percent of a number	Find 20% of 85.
K	Convert fractions to percents	Write $\dfrac{25}{52}$ as a percent.
L	Use the order of operations to simplify expressions	Simplify $3^3 - 2(5 - 9)$.
M	Solve linear equations	Solve $3x - 2 = 10$.
N	Plot ordered pairs as points	Graph $(2, -3)$.
O	Combine like terms	Simplify $-7x + 8y + 3x - 12y$.

MyMathLab

To assist you with skills that are prerequisites but may be rusty, use the Study Plan in MyMathLab.

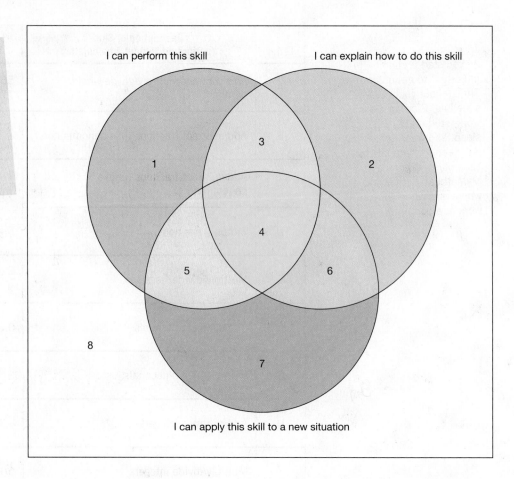

This book is written with the assumption that skills A–G are prerequisites and will not be taught directly. Skills H–O will be taught with the assumption that you have seen them before but may not have mastered them or understood them to the point of being able to apply them.

It is possible to know how to perform a skill but not know it well enough to be able to apply it to a new situation. It is also possible to be able to explain how to do a skill but not be able to perform it. In this course, you will be asked to apply any skills you learn to new situations. You will also be expected to understand a skill well enough to be able to explain it.

Since some of your prerequisite skills may be rusty, we will review some basics.

Recall that a fraction is often used to represent a part of a whole. So $\frac{1}{3}$ means one part out of three equal-sized parts that make up a whole.

If we multiply the numerator and denominator of a fraction by the same nonzero number, we create an equivalent fraction.

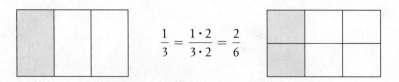

$$\frac{1}{3} = \frac{1 \cdot 2}{3 \cdot 2} = \frac{2}{6}$$

1 of 3 is equivalent to 2 of 6

Likewise, if we divide the numerator and denominator of a fraction by the same non-zero number, we create a simpler equivalent fraction. This process is known as simplifying a fraction, and it should be done at the end of calculations. The numerator and denominator become smaller, but the relative amount represented by the fraction remains the same.

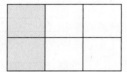 $$\frac{2}{6} = \frac{2 \div 2}{6 \div 2} = \frac{1}{3}$$

2 of 6 is equivalent to 1 of 3

Remember?

Percent means "out of one hundred."

4. In a class of 20 students, 17 of them passed a prerequisite skills quiz. What percent of the students passed the quiz? Instead of using a calculator or division to solve this problem, create an equivalent fraction. This process is also known as **scaling.**

$$\frac{17}{20} = \frac{17 \cdot 5}{20 \cdot 5} = \frac{85}{100} = 85\%$$

85% of the students passed the quiz.

Creating equivalent fractions is necessary when you need to add fractions that have different denominators. Adding and subtracting fractions can be tricky if we only think about the notation. Instead, try to focus on what the notation means.

EXAMPLE

Suppose at a college, $\frac{1}{3}$ of the freshmen start in beginning algebra and $\frac{1}{5}$ of the freshmen start in intermediate algebra. What fraction of the freshman class start in one of these two developmental math classes? To the nearest whole percent, what percent of the freshman class does this represent?

SOLUTION

INSTRUCTOR NOTE: If students are still unclear on why a common denominator is needed, a comparison to coins may help. If we want to add 15 quarters and 2 nickels and end up with 17, we have to use a trait they have a common (coins). When we count anything, the quantities must be alike to total them.

To find the answer to the first question, we need to add the fraction of freshmen in beginning algebra to the fraction of freshmen in intermediate algebra. However, $\frac{1}{3}$ and $\frac{1}{5}$ do not add to $\frac{2}{8}$. We can draw fraction pictures to verify this, or we can change the fractions into decimals.

$$\frac{1}{3} + \frac{1}{5} = 0.\overline{3} + 0.2 = 0.5\overline{3} \qquad \frac{2}{8} = 0.25$$

Since thirds and fifths are different-sized pieces of a whole, we can't add these fractions as they are. We can, however, change each into an equivalent fraction with the same denominator—15 in this case.

To find a new denominator to use, choose a common multiple of the denominators of the fractions you are adding. This will be the new common denominator. Scale each fraction to have that denominator and then add them.

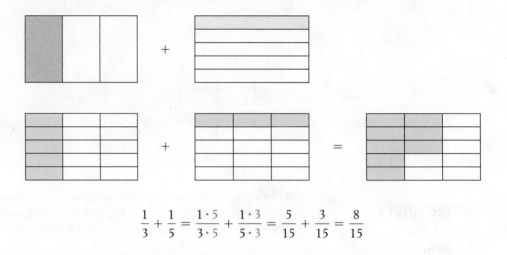

$$\frac{1}{3} + \frac{1}{5} = \frac{1 \cdot 5}{3 \cdot 5} + \frac{1 \cdot 3}{5 \cdot 3} = \frac{5}{15} + \frac{3}{15} = \frac{8}{15}$$

Remember?

When rounding, look at the digit one place to the right of the place you are rounding. If the digit to the right is 0–4, the digit in the place you are rounding will remain the same. If the digit to the right of it is 5–9, the digit you are rounding will increase by one.

The second question in the problem asks what percent the fractional result represents. Percent means out of 100. If the denominator was a factor of 100, we could change $\frac{8}{15}$ into a fraction out of 100 by scaling. Since this is not the case, we can use the fact that the fraction bar means division.

$$\frac{8}{15} = 8 \div 15 = 0.5\overline{3} = 53.\overline{3}\% \approx 53\%$$

About 53% of students at this college start in beginning or intermediate algebra.

INSTRUCTOR NOTE: If students do not remember how to multiply fractions, a review of the procedure is provided in the *How It Works* that follows. Consider drawing an area model for students after they have the solution to illustrate another method. See homework problem #11 for more information.

5. At a particular college, half of all freshmen start in developmental math. Of those who start at this level, $\frac{4}{5}$ start in beginning algebra. What is the chance that a freshman at this college starts in beginning algebra? Give your answer as a percent without using a calculator.

$$\frac{4}{5} \cdot \frac{1}{2} = \frac{4}{10} = \frac{40}{100} = 40\%$$

HOW IT WORKS

Need more practice?
Perform the operation.

1. $\dfrac{9}{17} - \dfrac{1}{17}$ 2. $3\dfrac{1}{8} + 2\dfrac{2}{3}$

3. $\dfrac{5}{3} \div \dfrac{15}{11}$

Answers:

1. $\dfrac{8}{17}$ 2. $5\dfrac{19}{24}$

3. $\dfrac{11}{9}$ or $1\dfrac{2}{9}$

Tech TIP

Learn how your particular calculator performs fraction operations.

Connect

Operations with fractions:

Always check to see if a fraction can be simplified before you write the final answer.

1. An equivalent fraction can be written with a different denominator by multiplying or dividing the numerator and denominator of a fraction by the same nonzero number.

Examples: $\dfrac{2}{3} = \dfrac{2 \cdot 5}{3 \cdot 5} = \dfrac{10}{15}$ $\dfrac{24}{36} = \dfrac{24 \div 12}{36 \div 12} = \dfrac{2}{3}$

2. To add fractions that have the same denominator, add the numerators and keep the same denominator.

Example: $\dfrac{3}{8} + \dfrac{2}{8} = \dfrac{3+2}{8} = \dfrac{5}{8}$

3. To add or subtract fractions that have different denominators, first get a common denominator. Then add or subtract the fractions by adding or subtracting the numerators and keeping the same denominator.

Example: $\dfrac{5}{8} + \dfrac{1}{12} = \dfrac{5 \cdot 3}{8 \cdot 3} + \dfrac{1 \cdot 2}{12 \cdot 2} = \dfrac{15}{24} + \dfrac{2}{24} = \dfrac{17}{24}$

4. To multiply fractions, multiply the numerators and multiply the denominators. Simplify either before or after doing the multiplication. A common denominator is not necessary.

Example: $\dfrac{2}{5} \cdot \dfrac{7}{4} = \dfrac{2 \cdot 7}{5 \cdot 4} = \dfrac{14}{20} = \dfrac{7}{10}$

5. To divide two fractions, multiply the first fraction by the reciprocal of the second fraction.

Example: $\dfrac{2}{5} \div \dfrac{7}{4} = \dfrac{2}{5} \cdot \dfrac{4}{7} = \dfrac{8}{35}$

6. To change a fraction to a decimal, first write it with a denominator that is a power of ten, or use your calculator to divide the numerator by the denominator.

Example: $\dfrac{2}{25} = \dfrac{2 \cdot 4}{25 \cdot 4} = \dfrac{8}{100} = 0.08$

$\dfrac{6}{7} = 6 \div 7 \approx 0.857$

To change a decimal to a percent, move the decimal point two places to the right.

Example: $0.857 = 85.7\%$

6. Suppose a student stands with her back against one of the classroom walls. She walks halfway to the opposite wall and then stops. She then walks half the remaining distance to the wall and stops. What fraction of the length of the room has she walked after doing this process four times? Will she ever technically reach the opposite wall?

$\dfrac{1}{2} + \dfrac{1}{4} + \dfrac{1}{8} + \dfrac{1}{16} = \dfrac{15}{16}$; no

INSTRUCTOR NOTE: The *Reflect* provides students with the objectives of the section and gives you an opportunity to clarify those objectives before students begin working on homework.

WRAP-UP

What's the point?

It is helpful to know what will and will not be taught in the course. However, there will also be new aspects to skills you may have seen in a prior course. It is necessary not only to be able to perform a skill, but also to understand it and apply it. This is particularly true with fractions. You will encounter fractions repeatedly in this course and in all your future math courses (as well as in life). It is crucial that you can work with them accurately and efficiently.

What did you learn?

How to use Venn diagrams
How to write equivalent fractions
How to add, subtract, multiply, and divide fractions
How to solve applied problems involving fractions

In this section, you *explored* a new problem, *discovered* new skills and vocabulary, *connected* the new ideas, and *reflected* on the goals of the section. This approach will be used in each section to learn the content.

1.2 Homework

Homework consists of skill exercises and problems that require a deeper understanding and using the skills learned. Both exercises and problems are important and will take time to complete. The goals of homework are understanding and application, not mimicking and memorization.

Skills MyMathLab

First complete the MyMathLab homework online. Then work the two exercises to check your understanding.

1. Perform each fraction operation and write the answer in simplest form.

   a. $\dfrac{2}{15} + \dfrac{1}{9}$ $\dfrac{11}{45}$

   b. $\dfrac{4}{21} - \dfrac{1}{12}$ $\dfrac{3}{28}$

2. Perform each fraction operation and write the answer in simplest form.

   a. $\dfrac{2}{15} \cdot \dfrac{3}{8}$ $\dfrac{1}{20}$

   b. $\dfrac{2}{15} \div \dfrac{1}{9}$ $\dfrac{6}{5}$

Concepts and Applications

Complete the problems to practice applying the skills and concepts learned in the section.

3. Consider the following Venn diagram, which compares high school and college math classes.

High School and College: Similarities and Differences

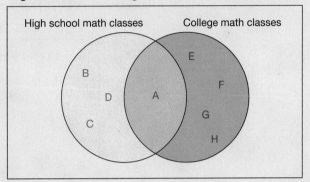

a. Fill in the Venn diagram by writing the letter of each of the following statements in the appropriate section. Since there is more than one correct way of completing the diagram, give a reason for the placement of each statement.

A. Homework must be done regularly in order for you to be successful.

B. If you miss class, it is the teacher's responsibility to make sure you complete the work eventually.

C. If you miss an exam for any reason, you can take it later.

D. If you receive a D, you can still go on to the next class.

E. Grades are based on mastery of content more than on effort.

F. You are responsible for knowing school policies, important dates, and the registration process.

G. You are expected to study 2 hours outside of class for every 1 hour in class.

H. Attendance is expected, but may not be required.

b. What does it mean if a statement is in the space where the two circles overlap?
It applies to both high school and college math classes.

c. What does it mean if a statement is outside both circles?
It does not apply to high school or college math classes.

4. In a freshman class of 2,000 students at a particular college, 1,250 are taking either a developmental math or reading class during their first semester. Of these 1,250 students, 726 are taking both a developmental math class and a developmental reading class, and 437 are taking only a developmental math class.

a. Draw a Venn diagram to represent this situation. Use two circles, one for developmental math and one for developmental reading, and enclose them in a rectangle that represents the freshman class. Write numbers in each section to indicate how many students are in that particular category.

Freshman Class

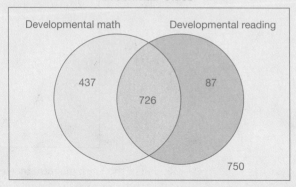

b. Use your Venn diagram to determine how many students are taking a developmental reading class but not a developmental math class.   87 students

c. Use your Venn diagram to determine how many students are taking neither developmental class.   750 students

5. A serving of a small frozen casserole is $\frac{1}{3}$ of the pan. If you eat $\frac{1}{4}$ of the pan, what fraction of a serving have you eaten?

$\frac{3}{4}$ of a serving

6. Think about the operations of adding/subtracting fractions and multiplying/dividing fractions. List some features that these operations have in common and some differences between them, and then place your statements in the correct space on the following Venn diagram.

It may be easier to make a numbered list of statements and then place the numbers in the diagram.   Answers will vary, but may include the need to get a common denominator as a difference and a need to simplify the answer as a common trait.

Fraction Operations

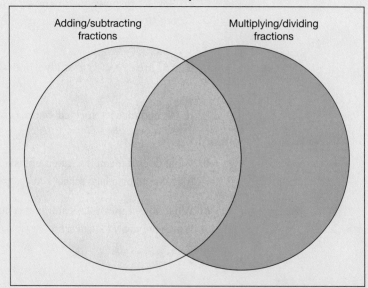

7. A school has two pathways through the math program. $\frac{2}{3}$ of students follow the first path; the rest take the second path. During one semester, 75% of the students who follow the first path pass the program, while 90% of the students who follow the second path pass it. Follow the steps to find what fraction of students pass math this semester.

   a. Find the fraction of students who take the first path and pass. $\frac{2}{3}(0.75) = \frac{1}{2}$

   b. Find the fraction of students who take the second path and pass. $\frac{1}{3}(0.90) = \frac{3}{10}$

   c. Add the two fractions. $\frac{1}{2} + \frac{3}{10} = \frac{4}{5}$

8. A cookie recipe calls for $\frac{2}{3}$ of a cup of flour and makes 2 dozen cookies.

   a. If you plan to make 8 dozen cookies, how many cups of flour do you need?

   $4 \cdot \frac{2}{3} = \frac{8}{3} = 2\frac{2}{3}$ cups of flour

   b. How many cookies can you make if you have 10 cups of flour and want to use it all?

   $10 \div \frac{2}{3} = 15$ batches of 2 dozen cookies or 30 dozen cookies

9. a. What operation is implied between the whole-number and fraction parts of the mixed number $4\frac{2}{3}$?   Addition

   b. What is the procedure for converting a mixed number to an improper fraction?
   Multiply the whole number by the denominator, add that result to the numerator, and write the total over the denominator.

   c. Illustrate the procedure by changing $4\frac{2}{3}$ to an improper fraction.

   $4\frac{2}{3} = \frac{4 \cdot 3 + 2}{3} = \frac{14}{3}$

   d. Enter the mixed number $4\frac{2}{3}$ into your calculator and get the decimal equivalent. Enter the improper fraction you found into your calculator and get the decimal equivalent. Are they the same? $4.\overline{6}$; yes

   e. Draw a picture to illustrate the mixed number $4\frac{2}{3}$.

   f. Use your picture to show how this mixed number can be written as an improper fraction and then write the improper fraction.
   $\frac{14}{3}$

10. Is it easier to add or multiply fractions? Explain your answer and give an example of each.
It is often considered easier to multiply, since you do not have to get a common denominator. Examples will vary.

11. Multiplication of fractions can be illustrated with an area model.
For example, consider $\frac{2}{3} \cdot \frac{4}{5}$.

First draw a picture to illustrate $\frac{4}{5}$ by dividing a whole into fifths and shading four of them.

Next, take $\frac{2}{3}$ of the shaded portion by dividing it into thirds and shading two of the thirds.

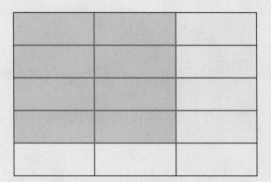

Finally, find the answer to $\frac{2}{3} \cdot \frac{4}{5}$ by counting the number of sections that are shaded twice and dividing by the total number of sections. In other words, $\frac{2}{3} \cdot \frac{4}{5} = \frac{8}{15}$.

Find the answers to the following fraction multiplication problems by drawing an area model for each.

a. $\frac{3}{5} \cdot \frac{1}{2}$ $\frac{3}{10}$ **b.** $\frac{1}{4} \cdot \frac{5}{7}$ $\frac{5}{28}$

12. Looking Forward, Looking Back If your walking pace is 12 minutes per mile, how many miles can you walk in an hour?   5 miles

1.3 Hello, My Name Is: Graphing Points

SECTION OBJECTIVES:

- Plot ordered pairs
- Determine the coordinates of a point

INSTRUCTOR NOTE: Students will need to complete the *Explore* individually whether they are working in a group or as a whole class.

An example of one student's results is shown. You might want to make a single coordinate graph for the entire class and add each person's point.

In this course, you might work in groups. Even if you do not, you will likely work in groups or teams in other courses or at your workplace. Knowing yourself and how you work can make working with others more positive and productive. There are two key components to how people work in groups: how they approach problems and how they communicate. The following activity is designed to help you see what kind of worker you are.

1. This line represents the range of problem-solving approaches, from intuitive to logical. Intuitive approaches use feelings, whereas logical approaches use reasoning and logic. Mark an X to indicate where you fall on this line.

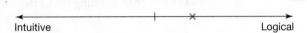

2. This line represents the range of communication styles, from outgoing to reserved. Mark an X to indicate where you fall on this line.

3. We can rearrange the two lines and have them intersect to form a two-dimensional picture. Redraw the X's you drew above on the lines shown here.

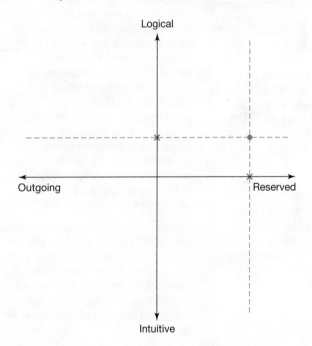

If you are working in a group, it is ideal if all members contribute and communicate effectively. If a group has too many leaders or followers, think about ways to help the group function better. That could mean asking a group member to give everyone a chance to participate, or it could mean asking someone to participate more.

4. Draw a horizontal dashed line through your X on the line for problem-solving style. Draw a vertical dashed line through your X on the line for communication style. The point where these dashed lines cross is your position on this graph. Draw a dot at this intersection. It is possible for this point to be on one of the original lines.

Discover

The graph in the *Explore* is an example of a graph on the Cartesian coordinate system. Graphs using the Cartesian coordinate system are employed often in mathematics and many other disciplines as a way to display paired data and look for relationships. In the context of how people work in groups, this graph provides a comparison of communication and work styles. Next we will learn more about graphing on the Cartesian coordinate system.

As you work through this section, think about these questions:

1. Can you explain the Cartesian coordinate system using correct terminology?

2. What do all the points on the *x*-axis have in common? All the points on the *y*-axis?

Let's begin by defining the Cartesian coordinate system and its parts.

The name "Cartesian" comes from René Descartes, who is credited with inventing this type of graphing in the 17th century.

INSTRUCTOR NOTE: Consider having students state the quadrant or axis in which their X appears in the *Explore* problems.

Cartesian Coordinate System

The **Cartesian coordinate system** is composed of two number lines called **axes** that intersect at a right angle. The horizontal number line is usually called the **x-axis**, and the vertical number line is usually called the **y-axis**. The values on the number lines increase from left to right and from bottom to top. Their point of intersection is called the **origin**. The axes divide the plane into four regions called **quadrants**, which are labeled with Roman numerals, starting in the upper right quadrant and moving counterclockwise.

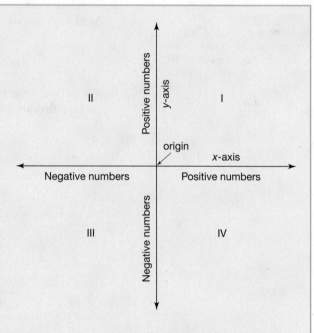

Every point on the Cartesian coordinate system has a position described by an **ordered pair**, which functions like an address. If we are using *x*- and *y*-axes, the ordered pair has the form (*x*, *y*), where *x* represents the **x-coordinate** and *y* represents the **y-coordinate**. The *x*-coordinate describes movement from the origin along the horizontal axis, and the *y*-coordinate describes movement from that position in the vertical direction. Positive numbers indicate movement right for *x* or up for *y*, and negative numbers indicate movement left for *x* or down for *y*. The origin's position is described by the ordered pair (0, 0).

FOR EXAMPLE, (2, 3) is an ordered pair that indicates a point in the first quadrant.

When there is a context for the ordered pairs, it's good form to label the axes with letters that have some meaning for the situation, such as *H* for height. Otherwise, the generic labels *x* and *y* can be used. The letters used to label the axes are an example of variables.

> **Variable**
>
> A **variable** is a letter used to represent an unknown quantity that can take on various values.
>
> **FOR EXAMPLE,** in the ordered pair (x, y), x and y are letters that represent numbers that can change.

EXAMPLE 1

Plot the point $(-3, 4)$.

SOLUTION

To plot the point $(-3, 4)$, we start at the origin, move 3 left, and then move 4 up from there. At that location, we draw a point. Notice that the point is located in the second quadrant.

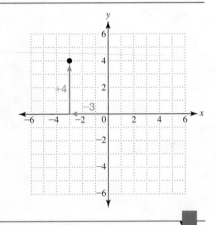

We can plot a point given its ordered pair. We can also determine the ordered pair and its location given its point on a graph.

EXAMPLE 2

For each point on the graph, state its ordered pair and the quadrant or axis where it is located.

SOLUTION

To read the coordinates of a point on a graph, we first determine the movement needed to get from the origin to the point. It is easier to think of how you would move horizontally, and then how you would move vertically, since the ordered pair is given in that order.

To get to point A from the origin, we move left 5 units and up 2 units. Point A has the ordered pair $(-5, 2)$. It is located in the second quadrant.

To get to point B from the origin, we do not move left or right. Instead, we move down 4 units from the origin. Point B has the ordered pair $(0, -4)$. It is located on the y-axis.

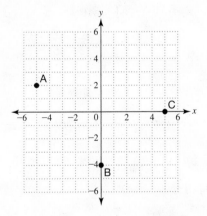

To get to point C from the origin, we move 5 units right. We do not move up or down from there. Point C has the ordered pair $(5, 0)$. It is located on the x-axis.

HOW IT WORKS

To plot an ordered pair:

1. Start at the origin, the intersection of the two axes.
2. Move horizontally using the first number in the pair. Positive numbers indicate right; negative numbers indicate left.
3. From the ending position in Step 2, move vertically using the second number in the pair. Positive numbers indicate up; negative numbers indicate down.
4. Draw a point at the location.

EXAMPLE: To plot the point for the ordered pair $(5, -2)$, start at the origin. Move right 5 and down 2. Draw a point.

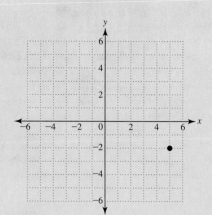

To determine the coordinates of an ordered pair:

1. Start at the origin, the intersection of the two axes.
2. Determine a path to get from the origin to the point. The number of horizontal units becomes the *x*-coordinate, and the number of vertical units becomes the *y*-coordinate.
3. Be careful to include negative signs in the coordinates if it is necessary to move left or down to get to the point from the origin.

EXAMPLE: To move from the origin to the point shown on the graph, move left one unit and up four units. The ordered pair of the point is $(-1, 4)$.

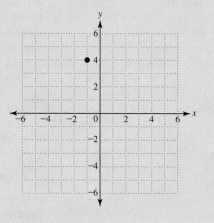

5. Plot each of the following points on the graph provided. Write the letter next to the point on the graph. For each point, state which quadrant it is in or the axis it is on.

Letter	Point	Axis/Quadrant
A	$(3, 4)$	I
B	$(5, 1.7)$	I
C	$(2, -4)$	IV
D	$\left(-\dfrac{5}{2}, -4\right)$	III
E	$(0, 1)$	*y*-axis
F	$(-3, 0)$	*x*-axis

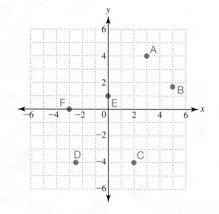

6. For each point on the graph provided, state its ordered pair and the quadrant or axis.

Letter	Point	Axis/Quadrant
A	(2, 3)	I
B	(−4, 1)	II
C	(1, −5)	IV
D	(−2, 0)	x-axis

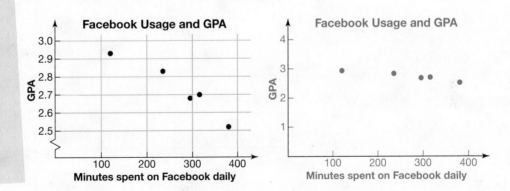

Connect

INSTRUCTOR NOTE: Giving students ground rules for effective group work can help them be productive. Even strong students do not always know how to work with others in a positive way. See the Instructor Guide for specific tips on group work guidelines.

If the minimum value on an axis is something other than zero, there should be a jagged start to the axis to show the break in the scale. You will see many graphs in the press where this is not done, however. The important point is to **always pay attention to the axes labels**, since they affect the look and impression of graphs.

7. The following graph from a study on Facebook use is called a **scatterplot**. It shows how the amount of time spent on Facebook is related to college GPA. It gives the impression that GPA falls fairly significantly as the number of minutes spent on Facebook increases. This impression is influenced in part by the scale on the y-axis, which includes the minimum and maximum values and the increment (or difference between tick marks). The scatterplot shown has a minimum GPA of 2.5, a maximum of 3.0, and an increment of 0.1. The jagged start on the y-axis indicates that the graph starts at a value other than zero.

Redraw the scatterplot with a scale of 0–4 and an increment of 1 on the y-axis. Using this choice of scale does not require a jagged start on the y-axis. Include axis labels and a title. Comment on the impression the new graph gives about the relationship between GPA and time spent on Facebook.

With the new scale, the effect of Facebook time on GPA seems much less significant.

Notice that since all the information appears in one quadrant, only that quadrant is drawn to focus the graph on what is important. Also, the researchers who conducted the study used the horizontal axis for minutes on Facebook daily and the vertical axis for GPA. They displayed the data in this way because they suspected that the time spent on Facebook affects GPA and not the reverse. You will learn how to construct scatterplots in Section 1.10.

WRAP-UP

What's the point?

In order to make a class group function well, it helps to understand how you and your peers think and communicate. We visualized this with a graph that used the Cartesian coordinate system.

What did you learn?

How to plot ordered pairs
How to determine the coordinates of a point

1.3 Homework

Skills MyMathLab

First complete the MyMathLab homework online. Then work the two exercises to check your understanding.

1. Plot $(2, -1)$. State its location (quadrant or axis).   Quadrant IV

2. State the ordered pair of the point on the grid. $(-3, 5)$

Where is it located? State the axis or quadrant.   Quadrant II

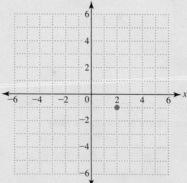

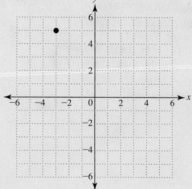

Concepts and Applications

Complete the problems to practice applying the skills and concepts learned in the section.

3. Use the *Explore* from this section to answer these questions. In each case, be descriptive in your response.

  a. What does it mean to be on the positive vertical axis?   Neither outgoing nor reserved, a logical thinker

b. Negative horizontal axis?   Neither intuitive nor logical; outgoing; uses thoughts and feelings equally to solve problems and is social in communication style

c. At the origin?   Very neutral personality, in the middle for communication and problem-solving styles; exhibits some of each characteristic

d. Quadrant III?   Outgoing and uses an intuitive approach to solve problems; creative and social.

4. Simplified graphs of constellations can be made on a Cartesian coordinate system. On the following grid, label the horizontal axis with capital letters: A, B, C, D, and so on. Label the vertical axis with whole numbers: 0, 1, 2, 3, and so on. The origin should be (A, 0). Form a constellation by drawing a dot at each pair of coordinates to represent a star and then connecting the dots with line segments to form the recognizable constellation. You might need to Google an image of the constellation if you are not familiar with the shape.

   a. Ursa Major: (M, 37); (Q, 34); (R, 34); (U, 33); (W, 35); (Z, 32); (X, 30)

   b. Ursa Minor: (R, 17); (O, 18); (N, 20); (M, 22); (K, 22); (L, 25); (N, 25)

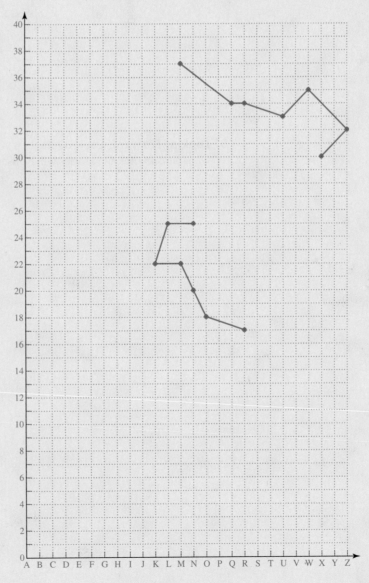

5. Consider the following data for a car traveling at 60 miles per hour. Plot the data as ordered pairs with the time as the *x*-variable and the distance as the *y*-variable.

Time in hours (*x*)	Distance in miles (*y*)
5	300
8	480
10	600
12	720
18	1,080

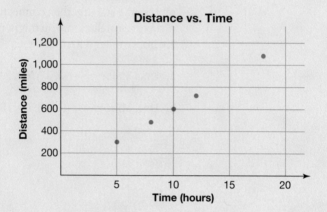

6. Looking Forward, Looking Back If you plan to consume 2,000 calories per day, what percent of your calories will be used on a 650-calorie lunch?   32.5%

1.4 A Tale of Two Numbers: Ratios and Proportions

SECTION OBJECTIVES:

- Interpret ratios
- Scale ratios to produce equivalent ratios
- Determine if quantities are proportional

Remember?

To convert a fraction to a percent, divide the numerator by the denominator and multiply the result by 100, which is the same as moving the decimal point two places to the right.

INSTRUCTOR NOTE: If you are using groups and notice that students are not interacting, consult the Instructor Guide at the beginning of the worktext for ways to encourage more collaboration within groups.

Consider the following Nutrition Facts for Junior Mints candies. Use the information on the label to answer the questions, including units in your work and final answers. Keep in mind that nutrition labels tend to list approximate values, and percents are usually rounded to whole numbers.

1. What percent of the calories in Junior Mints is from fat?

30 fat calories/170 total calories ≈ 17.6%

2. This package of Junior Mints accounts for 5% of a person's daily intake of fat for a 2,000-calorie diet. How was that number calculated?

Use 2,000-calorie guidelines.

3 fat grams from JM/65 total fat grams ≈ 4.6%, which rounds to 5%.

3. What percent of a person's daily allowance of carbohydrates does she consume if she eats this whole package of Junior Mints? Find the answer on the label and show how to calculate it, assuming a 2,000 calorie diet.

On the label, it says 12%.

Also, 35 g of carbs in JM/300 total g of carbs ≈ 11.7%, which rounds to 12%.

Nutrition Facts		
Serving Size (40g)		
Amount per Serving		
Calories 170 Calories from Fat 30		
		% Daily Value*
Total Fat 3g		5%
Saturated Fat 2.5g		13%
Trans Fat 0g		
Cholesterol 0mg		0%
Sodium 30mg		1%
Total Carbohydrate 35g		12%
Dietary Fiber 1g		4%
Sugars 32g		
Protein 1g		2%
Vitamin A 0%	•	Vitamin C 0%
Calcium 0%	•	Iron 8%

*Percent Daily Values are based on a 2,000 calorie diet. Your daily values may be higher or lower depending on your calorie needs.

		Calories	2,000	2,500
Total Fat	Less than		65g	80g
Saturated Fat	Less than		20g	25g
Cholesterol	Less than		300mg	300mg
Sodium	Less than		2,400mg	2,400mg
Total Carbohydrate			300g	375g
Dietary Fiber			25g	30g
Protein			50g	65g

Calories per gram:
Fat 9 • Carbohydrate 4 • Protein 4

Junior Mints trademark used with permission.

4. Find the number of calories per piece of Junior Mint candy if there are 16 pieces in a serving.

170 cal/16 JM = 10.625 cal/JM

5. Assume a person has a 2,500-calorie diet. What percent of his daily allowance of carbohydrates does he consume if he eats this whole package of Junior Mints? Round your answer to the nearest whole percent.

35 g of carbs in JM/375 total g of carbs ≈ 9.3%, which rounds to 9%

6. How many calories are there per gram of Junior Mints?

170 cal/40 g of JM = 4.25 cal/g of JM

7. The label says there are 30 calories from fat. How was that number calculated? How accurate is it?

$3 \text{ g of fat} \cdot \frac{9 \text{ calories}}{1 \text{ g of fat}} = 27$ calories. The result has been rounded to 30, which is a little high.

Discover

To fully understand the details on a nutrition label, you must be able to calculate ratios and make sense of them. As you progress through this section on ratios and proportions, consider the following:

1 Why is it important to include units when you write a ratio?

2 Can you tell if two ratios are equal?

3 What does it mean to say that a group is disproportionately represented?

Let's begin by defining ratio and rate.

Ratio and Rate

A **ratio** is a comparison of two numbers and can be written in three different ways. For example, if the student-to-faculty ratio at a college is 40 to 1, the ratio can also be written as 40:1 or $\frac{40}{1}$.

When the quantities being compared in the ratio have different units, the ratio is called a **rate**.

FOR EXAMPLE, the rate 30 mpg (miles per gallon) is a comparison of the two quantities 30 miles and 1 gallon. It is perhaps easier to see this as a rate when it is written in fraction form as $\frac{30 \text{ miles}}{1 \text{ gallon}}$. Since the units (miles and gallons) are different, we refer to this ratio as a rate.

Including units in a ratio can help you decide which quantity should go in the numerator and which in the denominator.

8. Suppose you have set the cruise control on a long car trip to 70 mph. Below are several interpretations of the rate 70 mph. Circle all that are correct.

a. $\dfrac{70 \text{ miles}}{1 \text{ hour}}$

b. 70 miles every hour

c. 17.5 miles in 15 minutes

d. 70 miles

e. 315 miles after 4.5 hours

f. 1 mile in 70 hours

a, b, c, e are correct.

It is common for recipes and mixtures to use the term "part" instead of a specific unit like "cup" or "tablespoon." A part is any unit of volume, such as teaspoon, cup, quart, or gallon. The person who creates the mixture can choose the unit based on the amount being mixed. If we are mixing paint for a wall, a part could be a quart. If we are mixing paint for a canvas, a part could be a 1-second squirt from a paint tube.

9. A juice drink recipe calls for 3 parts orange juice and 2 parts pineapple juice. Below are several statements. Circle all that are correct.

a. For every cup of pineapple juice, you need 1.5 cups of orange juice.

b. For every 2 quarts of pineapple juice, you need 6 quarts of orange juice.

c. When the juices are mixed together, the mixture is $\frac{3}{5}$ orange juice.

d. The ratio of pineapple juice to orange juice is 2:3.

e. If you have 15 ounces of orange juice, you need to add 10 ounces of pineapple juice.

f. If you use 4 cups of pineapple juice, you can make 6 cups of the drink.

Statements a, c, d and e are correct.

There are many instances in which we need to determine if ratios are equivalent. This can be done by simplifying the ratios or scaling one to see if it equals the other.

10. Most nutrition labels include suggested guidelines based on daily calorie totals.

*Percent Daily Values are based on a 2,000 calorie diet. Your daily values may be higher or lower depending on your calorie needs.

		Calories	2,000	2,500
Total fat	Less than	65g	80g
Saturated Fat	Less than	20g	25g
Cholesterol	Less than	300mg	300mg
Sodium	Less than	2,400mg	2,400mg
Total Carbohydrate		300g	375g
Dietary Fiber		25g	30g
Protein		50g	65g

Calories per gram:
Fat 9 • Carbohydrate 4 • Protein 4

a. Create a rate that compares suggested saturated fat grams to total calories for a 2,000-calorie diet. Simplify the fraction.

20 g of sat fat/2,000 diet cal = 1 g of sat fat/100 diet cal

b. Create a rate that compares suggested saturated fat grams to total calories for a 2,500-calorie diet. Simplify the fraction.

25 g of sat fat/2,500 diet cal = 1 g of sat fat/100 diet cal

c. Compare your answers from parts a and b. What do you notice?

The simplified fractions are the same.

d. Interpret the new simplified rate in words.

It is suggested that a diet should have 1 gram of saturated fat for every 100 calories, regardless of the calorie total.

INSTRUCTOR NOTE: Help students see two different relationships in the proportion in the *Look It Up*. They should see that each denominator in the proportion is five times the corresponding numerator. They should also see that the second fraction can be found from the first by multiplying the numerator and denominator each by 3.

Since the comparison of suggested saturated fat grams to total calories is the same for each diet (2,000 or 2,500 calories), we say that the number of saturated fat grams and the number of total calories are proportional. That is, they form the same ratio in each calorie diet.

Proportion

When two ratios are equal, they form a **proportion**. When two ratios are equal, the quantities are **in proportion** or **proportional**.

FOR EXAMPLE, the rates of 1 man for every 5 women and 3 men for every 15 women are equal to each other.

$$\frac{1 \text{ man}}{5 \text{ women}} = \frac{3 \text{ men}}{15 \text{ women}}$$

Notice that a specific ratio is maintained. The number of women is always five times the number of men. Another perspective is that when the number of men is tripled, so is the number of women.

Sometimes one ratio can be turned into another by simply scaling it. To **scale** a ratio, multiply or divide each quantity by the same nonzero number. If you can scale one ratio and turn it into another, then those two ratios are equivalent. Scaling a ratio creates an equivalent fraction.

EXAMPLE

Suppose you have been hired for a job that pays $12 per hour. Is it true that you make 20 cents per minute? Is this equivalent to making $480 in a week with five 8-hour shifts?

SOLUTION

INSTRUCTOR NOTE: Be sure to emphasize the scaling process since it is used throughout the book.

The answer can be determined by writing the 20 cents per minute as a fraction and then scaling it to turn it into a ratio of dollars per hour. We can start by multiplying the numerator and denominator each by 5 and then rewriting 100 cents as $1. Then the numerator and denominator can each be multiplied by 12 to make the denominator 1 hour.

$$\overset{\cdot\,5}{\overset{\frown}{\frac{20 \text{ cents}}{1 \text{ minute}}}} = \underset{\cdot\,5}{\underset{\smile}{\frac{100 \text{ cents}}{5 \text{ minutes}}}} = \overset{\cdot\,12}{\overset{\frown}{\frac{\$1}{5 \text{ minutes}}}} = \underset{\cdot\,12}{\underset{\smile}{\frac{\$12}{60 \text{ minutes}}}} = \frac{\$12}{1 \text{ hour}}$$

In a similar way, we can write $480 per week as a ratio and then change the one week to 40 hours. Dividing the numerator and denominator each by 40 produces the ratio of $12 per hour.

$$\frac{\$480}{1 \text{ week}} = \overset{\div\,40}{\overset{\frown}{\frac{\$480}{40 \text{ hours}}}} \underset{\div\,40}{\underset{\smile}{=}} \frac{\$12}{1 \text{ hour}}$$

The scaling process shows that making $12 per hour is equivalent to making 20 cents per minute or $480 for a week of five 8-hour shifts.

You can also check if two ratios are equivalent by converting each to a decimal. If the decimals are equal, then the ratios are equivalent and the two quantities are in proportion. If the ratios are not equivalent, we say that the quantities are disproportionate.

Notice that $12/1 hour = $12/hour and $0.20/(1/60 hour) = $12/hour. Five 8-hour shifts would total $5 \cdot 8 = 40$ hours. Then $480/40 hours = $12/hour.

So each of the ratios is equivalent to $12/hour. The money and time are in proportion to each other.

11. Nutrition guidelines suggest that we consume 25 grams of fiber for a 2,000-calorie diet and 30 grams of fiber for a 2,500-calorie diet.

    a. Is the suggested amount of fiber proportional to the number of calories? Is it almost proportional?

       25 g/2,000 calories = 0.0125 g/cal and 30 g/2,500 calories = 0.012 g/cal
       Almost proportional

b. Approximately how many grams of fiber should be recommended for a 3,000-calorie diet?   Answers will vary.

A possible answer: 25 g/2,000 cal=37.5 g/3,000 cal; about 37.5 g

c. Consider the recommendations for total fat, cholesterol, sodium, fiber, and protein shown in #10. Which of these are proportional or almost proportional to the number of calories?

Total fat, fiber, and protein are almost proportional to calories.

d. Knowing if relationships are proportional or constant allows us to make recommendations for a 3,000-calorie diet, which you will do in the following table. If you think the suggested amount of a nutrient is almost in proportion to the number of calories in a diet, use the idea of how much the number of grams or milligrams of the nutrient increases when the number of diet calories increases by 500.

Nutrient	Amount for 3,000-Calorie Diet
Total fat	95 g
Cholesterol	300 mg
Sodium	2,400 mg
Fiber	35 g
Protein	80 g

12. According to the U.S. Department of Justice, from 2003 to 2009, there were 2,931 deaths related to police arrests. Of these deaths, 41.7% were white and 31.7% were black. If the U.S. population of approximately 322 million is approximately 63% white and 13% black, are black people disproportionately likely to be killed by police? Are more black people than white people killed by police? Fully explain your reasoning and include the number of white people and black people in the population and in the death count.

Data from blogs.channel4.com

Number of white people: 202,860,000; number of white deaths: 1,222

Number of black people: 4,186,000; number of black deaths: 929

More white people than black people are killed by police. Yes, black people are disproportionately killed by police, since they make up a higher percent of the deaths than they represent in the population.

If you are struggling with a math problem, it doesn't necessarily mean something is wrong. You will have many challenging problems throughout this course that you will often be expected to work through without someone teaching you first. Learning how to work through those challenges helps you develop persistence, which will pay off in math, college courses, and your job.

Connect

INSTRUCTOR NOTE: Consider providing a reminder to students if they forget how to find the percent of a number. Percents will be addressed in greater detail in Section 1.12.

Reflect

WRAP-UP

What's the point?

Ratios and proportionality appear frequently in the news, and it is important that you can interpret and use them correctly.

What did you learn?

How to interpret ratios
How to scale ratios to produce equivalent ratios
How to determine if quantities are proportional

INSTRUCTOR NOTE: Remind students to read the *Getting Ready* article for the next section. Consider assigning the included questions as part of homework.

1.4   Homework

Skills MyMathLab

First complete the MyMathLab homework online. Then work the two exercises to check your understanding.

1. **a.** Maddy lost 18 pounds in 8 weeks. Find her rate of pounds lost per week.

   2.25 pounds per week

   b. Compare in hours: 27 hours to 3 days. Write the ratio as a fraction in lowest terms.

   $\dfrac{3}{8}$

2. **a.** Complete the following ratios by scaling up or down as necessary:

   $$\frac{2}{10} = \frac{\square}{5} = \frac{\square}{20} = \frac{\square}{100}$$

   $$\frac{2}{10} = \frac{1}{5} = \frac{4}{20} = \frac{20}{100}$$

   b. $\dfrac{2}{10}$ is what percent?   20%

Concepts and Applications

Complete the problems to practice applying the skills and concepts learned in the section.

3. Suppose the sales tax in your town is 8%. Write at least three interpretations of this ratio.

   Possible answers:

   a. You will pay 8 cents in sales tax for every dollar purchase.

   b. You will pay $8 in sales tax for every $100 purchase.

   c. You will pay $80 in sales tax for a $1,000 purchase.

4. A fifth-grader comments to his friend that the ratio of boys to girls in his class is 5 to 7. The friend replies, "Your class only has 12 students?" Explain the flaw in the second student's reasoning.

   The second student is assuming that the ratio also identifies the total number of students, but it does not. There are many different combinations of boys and girls that could give a ratio of 5 to 7. There could be 10 boys and 14 girls, for example.

5. If the ratio of boys to girls in a student's class is 5 to 7, what fraction of the class is boys? What fraction is girls?

   If the ratio of boys to girls is 5 to 7, then the ratio of boys to students is 5 to 12. So $\frac{5}{12}$ of the students are boys. The other $\frac{7}{12}$ of the students are girls.

6. Suppose the ratio of desktop to laptop computers in a school is 3 to 4. Give three different scenarios that would result in the school having this ratio of computers. For each scenario, write the ratio of desktop to laptop computers and then simplify it.

   Possible answers:

   3 desktops and 4 laptops, $\left(\frac{3}{4}\right)$

   6 desktops and 8 laptops, $\left(\frac{6}{8} = \frac{3}{4}\right)$

   9 desktops and 12 laptops, $\left(\frac{9}{12} = \frac{3}{4}\right)$

7. If a school has 70 computers in the ratio of 3 desktops for every 4 laptops, how many are desktops? Explain your answer.

   If the ratio of desktops to laptops is 3 to 4, then the ratio of desktops to total computers is 3 to 7.

   The ratio $\frac{3}{7}$ is equivalent to $\frac{30}{70}$.

   If there are 70 computers total, then 30 of them are desktops.

8. A baby panda born at a national zoo weighed 8 ounces at birth and reached 75 pounds at one year of age. What was the panda's growth rate per month? Per week?

   6.21 pounds/month

   1.43 pounds/week

9. Confirm that the quantities are proportional in the following rates: 75 miles in 5 hours, and 97.5 miles in 6.5 hours.

   75 miles/5 hours = 15 mph; 97.5 miles/6.5 hours = 15 mph. Since the rates are equal, distance and time are proportional.

10. Use the Velveeta Nutrition Facts provided to answer the questions.

    Nutrition Facts

    Serving Size 1 ounce (28g)

Amount per Serving		
Calories 85	Calories from Fat 55	
		% Daily Value*
Total Fat 6g		9%
Saturated Fat 4g		20%
Trans Fat		
Cholesterol 22mg		7%
Sodium 420mg		17%
Total Carbohydrate 3g		1%
Dietary Fiber 0g		0%
Sugars 2g		
Protein 5g		

Vitamin A	0%	• Vitamin C	0%
Calcium	13%	• Iron	0%

    *Percent Daily Values are based on a 2,000 calorie diet. Your daily values may be higher or lower depending on your calorie needs.

    VELVEETA is a registered trademark of Kraft Foods and is used with permission.

    a. What percent of the total fat is saturated fat? Round to the nearest percent.

       4/6 ≈ 67%

    b. One serving of Velveeta contains 13% of the daily recommended amount of calcium. If it is recommended that you get 1,000 mg of calcium daily, how many mg are in one serving of Velveeta?

       13% of 1,000 mg = 130 mg

    c. How many ounces of Velveeta would you need to consume to meet your total calcium needs for the day? Round to the nearest tenth if necessary.

       $\frac{100\%}{13\%}$ ≈ 7.7 servings. Each serving is 1 ounce, so 7.7 ounces.

11. When reading a nutrition label on a package of cheese ravioli, you see that there are 240 calories in 2 pieces. Write this comparison as a rate. Use scaling to find the number of calories per piece and the number of calories in a plate of 5 pieces.

$$\frac{240 \text{ calories}}{2 \text{ pieces}} = \frac{120 \text{ calories}}{1 \text{ piece}} = \frac{600 \text{ calories}}{5 \text{ pieces}}; 600 \text{ calories in a 5-piece plate.}$$

12. The racial composition in a Texas town is shown in the following table:

Race	Percent of Population
White	45
Hispanic	27
Black	18
Asian	7
Other	3

a. A local university has 2,815 Hispanic students out of a total student population of 20,250. Do the Hispanic students have proportional representation at the university?

No; the Hispanic students make up only 13.9% of the student population but 27% of the town population.

b. How many more Hispanic students would the university need in order for the Hispanic students to have proportional representation?

27% of 20,250 ≈ 5,468
5,468 − 2,815 = 2,653 students

13. One national poll showed that 451 of 1,100 registered voters planned to watch the vice-presidential debate. Another national poll conducted by a different organization showed that 920 of 2,000 registered voters planned to watch the vice-presidential debate.

a. Is the number of voters who planned to watch the debate proportional to the number of voters in the poll for these two different polls?

No; one poll showed that 41% planned to watch the debate, while the other poll showed that 46% planned to watch.

b. If you answered no to part a, how many voters would have needed to say yes in the second poll for the quantities to be proportional?

41% of 2,000 = 820 voters

14. Looking Forward, Looking Back Simplify.

a. $\dfrac{8}{11} + \dfrac{3}{5}$ $\dfrac{73}{55}$

b. $\dfrac{8}{11} - \dfrac{3}{5}$ $\dfrac{7}{55}$

c. $\dfrac{8}{11} \cdot \dfrac{3}{5}$ $\dfrac{24}{55}$

d. $\dfrac{8}{11} \div \dfrac{3}{5}$ $\dfrac{40}{33}$

Getting Ready for Section 1.5

Read the article and answer the questions that follow.

Federal Panel Finalizes Mammogram Advice That Stirred Controversy

The mammography debate heated up once again in April 2015, when the U.S. Preventive Services Task Force issued a draft of its latest breast cancer screening recommendations.

Now, after the public had a chance to comment, the influential task force has finalized the advice, reiterating that women ages 50-74 ought to receive a screening mammogram every two years. The USPSTF says that women between 40 and 49 don't get as much benefit from screening as do older women, so they should make an individual decision on when to start based on how they view the benefits and harms. (Women with a family history of breast cancer may benefit more from starting screening before age 50.)

"Our recommendations support the entire range of decisions available to women in their 40s," Kirsten Bibbins-Domingo, a physician and vice chair of the USPSTF, told Shots. Some women may choose to begin at 40 or soon after, deciding they want to lower their cancer risk as much as possible and can handle the chance of false positive results or possible overdiagnosis, when cancer is discovered that never would have been harmful to health.

Other women, she says, may opt to wait until later in their 40s or until they turn 50.

The task force's supporting materials include statistical models estimating the lifetime consequences of screening women from ages 50-74 and from 40-74. For each 1,000 women screened, the model finds that starting screening at 40 means an estimated one additional breast cancer death averted (deaths drop from eight to seven), with 576 additional false positive tests (1,529 vs. 953), 58 extra benign biopsies (204 vs. 146) and two additional overdiagnosed cases of breast cancer (21 vs. 19).

The task force also says there's not enough evidence to say whether or not women 75 and older benefit from routine screening for breast cancer. The recommendations were published Monday in the *Annals of Internal Medicine*.

Since the draft was made public last spring, the American Cancer Society changed its advice for breast cancer screening, saying that average-risk women don't need to begin annual mammograms until age 45 and can start screening every other year beginning at age 55. Other medical groups still recommend annual screening starting at 40.

While mammography guidelines differ, "it's important for women and physicians to understand how much convergence there is," says Bibbins-Domingo. The groups agree that mammography has value as a screening tool, and that the value of screening generally rises with age.

The Affordable Care Act guarantees private insurance coverage of preventive services without out-of-pocket costs for consumers if the evidence supporting the test has an A or B grade from the task force.

But Congress requires full coverage of mammography in women in their 40s, despite the C grade, which indicates there is "at least moderate certainty that the net benefit is small." In an editorial, the task force says that "coverage decisions are the domain of payers, regulators, and legislators" and that the group "cannot exaggerate our interpretation of the science to ensure coverage for a service."

Just to be clear, this ongoing debate is over screening mammography, which means looking for signs of breast cancer in healthy women who have no symptoms of the disease. No matter your age, or whether or not you've started regular screening, if you have symptoms, you need to see a doctor.

Questions

1. Screening women in their forties instead of starting at age 50 increases the number of false positive results, benign biopsies, and over diagnoses of breast cancer. According to the numbers given in the article, which of these three increased by the greatest percent?

   The number of false positives

2. Explain the difference between a false-positive result and a false-negative result.

   A false result is one that is opposite the woman's actual condition. A false-positive result occurs when a woman gets a positive result but doesn't have cancer. A false-negative result occurs when a woman gets a negative result but does have cancer.

1.5 Chances Are: Probability Basics

Explore

1. Much media attention has been focused on the use of mammography for detecting breast cancer. The following table shows, for 10,000 women in their forties, the number of women with positive vs. negative mammograms and the number who were diagnosed with cancer vs. those who were not diagnosed for 10 years.

	Positive Mammogram	Negative Mammogram	Total
Cancer	302	76	378
No Cancer	6,130	3,492	9,622
Total	6,432	3,568	10,000

a. What percent of the 10,000 women had a positive mammogram?

$$\frac{6,432}{10,000} = 0.6432 \approx 64\%$$

b. What percent of the 10,000 women had a cancer diagnosis?

$$\frac{378}{10,000} = 0.0378 \approx 3.8\%$$

c. What percent of the women who had cancer had a positive mammogram?

$$\frac{302}{378} \approx 0.7989 \approx 80\%$$

d. What percent of the women who had a positive mammogram had cancer?

$$\frac{302}{6,432} = 0.04695 \approx 4.7\%$$

e. Explain why the answers to parts c and d are not the same. What do these two percents actually say about mammograms?

Both percents include the number of cancer diagnoses with a positive mammogram. Part c considers this number out of the total of cancer diagnoses, while part d considers it out of the number of positive mammograms. So the bases for the percents are different.

Mammograms have a high rate of detecting cancer when it's there, since the answer to part c is 80%. However, only about 5% of the positive mammograms are actually cancer. So the other 95% are not cancer, which is a very high rate of false positives.

Discover

In this section, we will focus on calculating what percent of the time an event happens and then interpreting those percents as probabilities. You should try to focus on the following questions:

1 Do you remember how to find a percent?

2 Why is a relative frequency also considered a probability?

3 How is experimental probability different from theoretical probability?

The percents that you calculated in the *Explore* are considered relative frequencies because they tell us what percent of the time something happened.

LOOK IT UP

Relative Frequency

Frequency is the number of times an event occurs. **Relative frequency** is the ratio formed when the frequency is divided by the total number.

FOR EXAMPLE, if 15 women in a group of 75 were diagnosed with cancer, then the relative frequency of getting cancer in that group is $\frac{15}{75} = 0.2 = 20\%$. A relative frequency can be expressed as a fraction, decimal, or percent.

EXAMPLE

Each year the General Social Survey asks respondents if they are working full time, working part time, going to school, keeping house, and so on. In 1972, 1,615 people responded; 750 of them were working full time and 121 part time. In 2014, 2,538 people responded; 1,230 were working full time and 273 part time. Find the relative frequency of full-time workers and the relative frequency of part-time workers for each year. How do they compare?

Data from gssdataexplorer.norc.org

SOLUTION

In 1972, the relative frequency of full-time workers was $\frac{750}{1,615} \approx 0.4644 \approx 46\%$.

In 1972, the relative frequency of part-time workers was $\frac{121}{1,615} \approx 0.0749 \approx 7\%$.

In 2014, the relative frequency of full-time workers was $\frac{1,230}{2,538} \approx 0.4846 \approx 48\%$.

In 2014, the relative frequency of part-time workers was $\frac{273}{2,538} \approx 0.1076 \approx 11\%$.

There was a slightly higher percent of full-time workers in 2014 compared to 1972. The percent of part-time workers also increased over these years and by more than the increase in full-time workers (4 percentage points).

2. A national survey asked how many people were living in the respondent's household. The following table shows the results. Find the relative frequency of households with more than 8 people and interpret it.

Data from gssdataexplorer.norc.org

Household Size	1	2	3	4	5	6	7	8	9	10	11
Number of Respondents	695	997	385	261	132	47	15	1	3	1	1

$\frac{5}{2,538} \approx 0.00197 \approx 0.002$; approximately 2 out of every 1,000 households had more than 8 people.

Relative frequencies can be quite small, and it's important to be able to correctly interpret the decimals that result in a useful way. Remembering place value is the key to interpreting decimals accurately. Try rounding the decimal, and then use place value to write it as a fraction. For example, 0.00938 is approximately 0.01 or $\frac{1}{100}$.

Experimental Probability

Probability tells us the likelihood or chance of something happening. Relative frequencies that tell us what percent of the time something did happen can also be considered **experimental probabilities**. These percents can be used to say what percent of the time we expect something to happen in the future based on actual data.

FOR EXAMPLE, a recent survey showed that 40% of Americans approve of the job the President is doing. In other words, we would expect 40 out of every 100 or 4 out of every 10 or 2 out of every 5 Americans to approve of the job the President is doing. You could interpret this as an experimental probability and say that, if you randomly pick an American, you would expect to get someone who favors the President about 40% of the time.

3. Recently a poll was conducted to try to understand consumer coffee habits. Complete the following table and then do the problems that follow it.

Age	Drink at least 1 cup per day	Do not drink at least 1 cup per day	Total
18 to 44	357	387	744
45 to 64	342	177	519
65 +	196	70	266
Total	895	634	1,529

a. Find the experimental probability that someone in each age group drinks at least one cup of coffee per day. Which age group has the highest probability?

18 to 44: $\frac{357}{744} \approx 0.4798 \approx 48\%$; 45 to 54: $\frac{342}{519} \approx 0.6590 \approx 66\%$;

65+: $\frac{196}{266} \approx 0.7368 \approx 74\%$

The 65 and over age group has the highest probability of drinking at least one cup of coffee per day.

b. Find the experimental probability that someone over the age of 65 does not drink at least one cup of coffee per day. What's the fastest way to find this?

$\frac{70}{266} \approx 0.2632 \approx 26\%$. It would be faster to notice from part a that if 74% of them do drink at least one cup, then $100\% - 74\% = 26\%$ of them do not.

c. Find the overall probability that someone 18 years old or older drinks at least one cup of coffee per day.

$\frac{895}{1,529} \approx 0.5853 \approx 58\%$

INSTRUCTOR NOTE: Consider showing students the coin-flipping applet available in MyMathLab. You can use it instead of doing the physical experiment or in addition to the experiment to better make the point.

4. a. Explore the experimental probability of getting heads by flipping a fair coin and recording the number of heads in the following table. First, flip the coin 10 times and record the number of heads and the experimental probability of getting heads as a fraction and a decimal. Then, flip the coin another 10 times and record the same information for the combined 20 flips. Finally, combine your results with the results of four other students to get a total of 100 flips and record that information in the final column.

	10 flips	20 flips	100 flips
Number of Heads			
Probability of Getting Heads			

b. How do the three experimental probabilities compare?

Students will likely get different results for all three probabilities. Usually, the probability will get closer to 50% as the number of flips increases.

c. What percent of the time *should* you get heads if the coin is fair? Explain why you might not have gotten that percent for each of your experimental probabilities.

If the coin is fair, we expect to get heads 50% of the time. This is the theoretical probability and not what necessarily happens in an experiment.

You might have noticed with the coin-flipping exercise that experimental probabilities are not always the same as what should happen in theory. **Theoretical probabilities**, which tell us what should happen based on theory and not experimental results, will be discussed more in Cycle 2.

Connect

5. The following table shows a person's probability of experiencing certain events in the United States in 2014.

Contracting Ebola in America	1 in 13.3 million
Dying in a plane crash	1 in 11 million
Dying from a lightning strike	1 in 9.6 million
Dying from a bee sting	1 in 5.2 million
Killed by a shark	1 in 3.7 million
Killed in a car accident	1 in 9100

Data from www.npr.org

In some math classes, the teacher does not explain all new concepts. This may require an adjustment if you are used to lectures, but give it a chance. Many things you learned in your life, you learned by doing instead of someone telling you how to do them. Experience, including making mistakes, can be a powerful teacher.

a. Determine approximately how many people experienced each event, assuming the population of the United States at the time was approximately 320 million.

Contracting Ebola in America:	24
Dying in a plane crash:	29
Dying from a lightning strike:	33
Dying from a bee sting:	62
Killed by a shark:	86
Killed in a car accident:	35,165

b. How many times greater was the probability of dying from a bee sting compared to contracting Ebola?   Approximately 2.6 times greater

Reflect

WRAP-UP

What's the point?

If you know what percent of the time something happens, you can use that relative frequency to make predictions about the likelihood of future events.

What did you learn?

How to find relative frequencies
How to find and interpret experimental probabilities

Focus Problem Check-in:
Use what you have learned so far in this cycle to work toward a solution of the focus problem. If you are working in a group, work with your group members to create a list of any remaining tasks that need to be completed. Determine who will do each task, what can be done now, and what has to wait until you have learned more.

1.5 Homework

Skills MyMathLab

First complete the MyMathLab homework online. Then work the two exercises to check your understanding.

1. According to a recent poll, 4,060 out of 14,500 people in the United States were classified as obese based on their body mass index. What's the relative frequency of obesity according to this poll?

$$\frac{4,060}{14,500} = 0.28 = 28\%$$

2. Of the 1,200 respondents to a survey, 456 claimed to be worried about money. What is the likelihood that a person selected at random will not be worried about money?

$$\frac{744}{1,200} = 0.62 = 62\%$$

Concepts and Applications

Complete the problems to practice applying the skills and concepts learned in the section.

3. If there were 4 million births in a particular country and 2,247 sets of naturally occurring triplets, what is the experimental probability of having triplets naturally? Express your answer as a decimal and then provide an interpretation.

$$\frac{2,247}{4,000,000} = 0.00056175 \approx 0.0006; \text{ approximately 6 out of every 10,000 births are}$$
naturally occurring triplets.

4. The following table shows the number of survey respondents who completed each level of education. Change each frequency to a relative frequency. Complete the Total column.

Data from gssdataexplorer.norc.org

	Less Than High School	High School	Junior College	Bachelor's Degree	Graduate Degree	Total
Frequency	330	1,269	186	472	281	2,538
Relative Frequency	13.0%	50%	7.3%	18.6%	11.1%	100%

5. A home pregnancy tests yields the following results. Complete the table and answer the questions that follow.

	Pregnant	Not Pregnant	Total
Positive Test Result	16	4	20
Negative Test Result	6	74	80
Total	22	78	100

a. What is the false-positive rate for this test?

$$\frac{4}{78} \approx 0.05 = 5\%$$

b. What is the false-negative rate for this test?

$$\frac{6}{22} \approx 0.27 = 27\%$$

c. The accuracy rate of a test is the percent of cases that are correctly identified. What is the accuracy rate for this test?

$$\frac{90}{100} = 0.90 = 90\%$$

6. Repeat the process used to collect data in class with rolling a fair die. Roll a standard six-sided die and record the number of 3's from 10, 20, and 40 rolls. Then calculate the experimental probability of rolling a 3 in each of those cases. Record your results in the table. How do the experimental probabilities change with the number of rolls? Do they get closer to what you'd expect?

	10 rolls	20 rolls	40 rolls
Number of 3's			
Probability of rolling a 3			

Answers will vary.

7. **Looking Forward, Looking Back** You work 40 hours per week and spend 5 hours per week commuting to and from work. Your friend works 56 hours per week and spends 7 hours per week commuting to and from work. Do you each have the same ratio of commute time to work hours per week? Are these two quantities in proportion?

Yes; both are in the ratio 1:8.

1.6 It's All Relative: Understanding Integers

Explore

SECTION OBJECTIVES:
• Interpret signed number situations
• Find the opposite and absolute value of a number

INSTRUCTOR NOTE: If you are using groups and notice that some students do not carry their weight, consider calling on different members of the group during the whole class discussion. For more information on ways to encourage group accountability, see the Instructor Guide at the beginning of the worktext.

1. Atoms contain protons, neutrons, and electrons. Each proton has a positive charge, each electron has a negative charge, and neutrons have no charge. An atom, by definition, is neutral and has the same number of protons and electrons. An atom can gain or lose electrons and become a charged particle called an ion. A positive ion has more protons than electrons, while a negative ion has more electrons than protons. Use this information about atoms and ions to complete the Charge from Protons and Charge from Electrons columns in the following table.

	Symbol	Number of Protons	Number of Electrons	Charge from Protons	Charge from Electrons
Sodium atom	Na	11	11	11	−11
Sodium ion	Na$^+$	11	10	11	−10
Chlorine atom	Cl	17	17	17	−17
Chloride ion	Cl$^-$	17	18	17	−18
Aluminum atom	Al	13	13	13	−13
Aluminum ion	Al^{3+}	13	10	13	−10

2. The abbreviation for an ion includes either a positive or a negative exponent that represents the net charge. If the net charge of an ion is +1 or −1, only a positive or negative sign is shown, since the 1 is implied. Because the net charge of an atom is 0, no exponent is written. Explain how to determine the net charge, which is shown on the atom or ion's symbol in the table above.

The net charge of an element is the sum of the charge from protons and the charge from electrons.

INSTRUCTOR NOTE: Since the table in #3 does not have as many columns as the previous table, students will have to keep track of positive and negative values.

3. An atom can become a charged ion only by gaining or losing electrons, not protons. An atom and an ion of the same element always have the same number of protons. Use this information to complete the following chart.

	Symbol	Number of Protons	Number of Electrons	Net Charge
Hydrogen atom	H	1	1	0
Hydrogen ion	H$^+$	1	0	+1
Oxygen atom	O	8	8	0
Oxygen ion	O^{2-}	8	10	−2
Calcium atom	Ca	20	20	0
Calcium ion	Ca^{2+}	20	18	+2

Discover

Negative numbers may seem abstract since they don't represent physical counts. However, there are naturally occurring situations in which we have less than zero, such as debt, and negative numbers allow us to describe such situations. All numbers were created to deal with issues that occur in life, whether we are dealing with debt or counting a set of objects.

As you work through this section, think about these questions:

1 Do you understand the difference between the sign and size of a number?

2 Do you know when two negatives make a positive?

4. Describe the kind of numbers you would use to answer each question.

a. How many cars are in the parking lot?   Whole numbers: 0, 1, 2, 3, . . .

b. What is the depth of a submarine (relative to sea level) if it is descending at a rate of 200 feet per hour?   Negative numbers

c. Beth purchases a candy bar to share with her children and husband. How much will each person get?   Fractions or decimals

d. Jake uses a spreadsheet to keep track of his finances. He lists all purchases and bills for the month as well as any income. How much money does he have left after paying his bills?   Decimals

Historically, sets of numbers have been invented to help solve problems or satisfy needs. Then notation was developed to record them and their operations.

LOOK IT UP

Negative Numbers and Absolute Value

A **negative number** is a number less than zero. We indicate a negative number by writing a negative sign before it. Positive numbers may have a plus sign in front of them, but usually they do not have a sign before them. Zero is neither positive nor negative, but every real number other than zero has a sign. Think of each number as consisting of two parts: its sign and its size.

Sign   Size

The "size of a number" is the informal way of saying how far the number is from zero on a number line. In mathematics, we refer to the size of a number as its **absolute value**. To indicate absolute value, we draw vertical bars before and after the number.

FOR EXAMPLE, the absolute value of -5 is 5, since -5 is five units from zero. The absolute value of -5 is written as $|-5| = 5$.

Two numbers that have the same size but opposite signs are called **opposites**. For example, 17 and -17 are opposites. Notice that 17 and -17 are both 17 units from zero on a number line but are on opposite sides of zero.

-17 can be read as negative 17 or the opposite of 17, so the symbol for opposite is the negative sign.

The set of whole numbers, together with their opposites, form the **integers**. When we draw a number line, we often label it with integers.

5. a. What is the opposite of -8?   8

b. Find $|-11|$.   11

c. What is the sign of 24? The size?   Sign is positive; size or absolute value is 24

d. Read and simplify: $-(-6)$ The opposite of -6; 6

e. Read and simplify: $-|-6|$ The opposite of the absolute value of -6; -6

When a situation requires a negative number, there are many ways to express it. On a budget sheet, you might see negative quantities written in red or parentheses and positive numbers in black. In other cases, the negative may be implied but not shown.

6. The following examples do not use negative numbers, but each has a positive or negative meaning. Determine whether the meaning is positive or negative. If it is negative, write it using a negative number.

a. 100 feet below sea level   Negative, -100

b. \$500 in the black   Positive

7. The following situations contain negative numbers used in real-life situations. Interpret what each one means.

a. $-30°F$ 30 degrees Fahrenheit below zero

b. A poll has a $+/-3\%$ margin of error   For poll results, a range of likely values is found by taking the percent in the poll and adding and subtracting the margin of error.

8. Write a sentence to explain Lily Lake's enrollment.

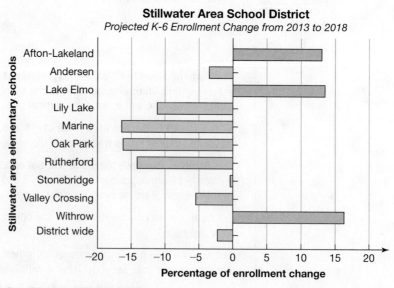

Stillwater Area School District
Projected K-6 Enrollment Change from 2013 to 2018

Data from www.stillwater.k12.mn.us

Between 2013 and 2018, Lily Lake's enrollment is expected to drop approximately 12%.

Connect

The periodic table of elements provides helpful information about each element. The atomic number, which is the number of protons, is always included along with the atomic mass. The atomic mass, rounded to the nearest whole number, is equal to the number of protons plus the number of neutrons.

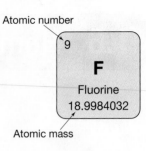

Atomic number

F
Fluorine
18.9984032

Atomic mass

Using this information and the periodic table entry for fluorine, we can determine that fluorine has a total of 19 protons and neutrons. Since the atomic number is 9, there are 9 protons. There must then be $19 - 9 = 10$ neutrons. Since fluorine is a neutral atom, it also has 9 electrons.

9. For each element shown in the periodic table entries that follow, find the number of protons, neutrons, and electrons in one atom.

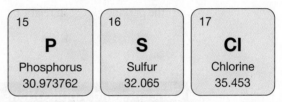

15	16	17
P	**S**	**Cl**
Phosphorus	Sulfur	Chlorine
30.973762	32.065	35.453

Phosphorus: Mass ≈ 31, P = 15, N = 16, E = 15
Sulfur: Mass ≈ 32, P = 16, N = 16, E = 16
Chlorine: Mass ≈ 35, P = 17, N = 18, E = 17

Reflect

WRAP-UP

What's the point?

Negative numbers are commonly used and provide a way of describing relative position. Often they describe a number less than zero, such as temperature. However, they can also describe being below a reference point (such as sea level) just as positive numbers describe being above it.

What did you learn?

- How to interpret signed number situations
- How to find the opposite and absolute value of a number

One component to success in a math class is ATTITUDE, which includes your beliefs about yourself and your abilities as well as how you approach situations. Here are some ways to demonstrate a good attitude in a math class:

1. Attend and participate in every class.
2. Complete all assignments on time.
3. Ask questions when you are confused.
4. Be willing to learn new material.
5. Do not give up when you are frustrated.
6. Work with your classmates and instructor.

1.6 Homework

Skills MyMathLab

First complete the MyMathLab homework online. Then work the two exercises to check your understanding.

1. Simplify: $-|-15|$ -15

2. The parking garage of a hotel is two floors below the lobby. Indicate the location of the parking garage in relation to the lobby using a signed number. -2

Concepts and Applications

Complete the problems to practice applying the skills and concepts learned in the section.

3. The following descriptions do not use negative numbers, but each has a positive or negative meaning. Determine whether the meaning is positive or negative. If it is negative, write it using a negative number.

   a. $10,000 in debt
      Negative, $-$10,000

   b. 400 BC
      Negative, -400

4. The following situations contain negative numbers used in real-life situations. Interpret what each one means.

   a. -20 rushing yards in a football game
      Loss of 20 yards

   b. Stock market performance for the day: -120.62 points
      Loss of 120.62 points for the day

5. Use the graphic to determine which model of iPhone generated the greatest percent change in stock value within 30 days of Apple's announcement of the phone.

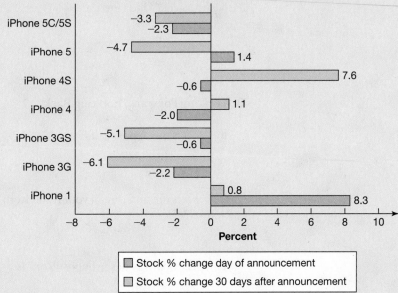

Apple Stock Reaction to iPhone Announcements

Data from www.forbes.com

iPhone 4S

6. Graph the following numbers on the number line provided. Make a dot at the appropriate location and write the number above it.

$$4.5, \ -2, \ \frac{2}{3}, \ -6.7, \ \frac{-16}{5}$$

7. As we move to the right on a number line, the numbers increase. If two numbers are graphed on a number line, the larger one is on the right. Graph each pair of numbers and then circle the larger one. In the box, between the numbers, write > or <.

a. $6\frac{1}{4}$ ☐ $6\frac{2}{3}$ (box: <)

b. -3 ☐ -4 (box: >)

c. -5 ☐ -5.2 (box: >)

d. -5 ☐ -4.8 (box: <)

8. One student says −5 is bigger than −4 and uses money as the analogy: "If I owe $5, I have a bigger debt than if I owe $4." What is wrong with this argument?

The student's analogy with money is actually correct. However, that does not mean that −5 is bigger than −4. The bigger the debt, the more the person is "in the hole" and the less money he or she has. Therefore, −5 is less than −4.

9. Looking Forward, Looking Back

a. What is the opposite of 7?   −7

b. What is the opposite of your result from part a?   7

c. So the opposite of the opposite of 7 is __7__ . That is, −(−7) = __7__ .

d. Generalize this: For any number a, −(−a) = __a__ .

1.7 Sign and Size: Integer Operations

Explore

SECTION OBJECTIVES:

- Add, subtract, multiply, and divide signed numbers
- Solve applied problems that involve signed-number addition, subtraction, multiplication, and division

1. In golf, the average number of strokes a good player should need in order to complete a course is called par. Par for a whole course is calculated by adding up the par scores for each hole. Scores in golf are often expressed as some number either greater than or less than par. Assume a particular golfer has golfed 2 under par on each of 3 holes, 3 over par on each of 3 holes, and 1 under par on each of 3 holes. If par for the 9-hole course is 32, how many strokes did it take the golfer to complete the course?   32

Discover

INSTRUCTOR NOTE: Consider that the content of this section might be review for your students.

Signed numbers are commonly used when discussing debt, cold temperatures, or poor stock performance. Before we consider computations with signed numbers, we will start with situations and interpret them. This will allow us to find a result and learn the notation for writing out computations.

As you work through this section, think about these questions:

1 When calculating with signed numbers, do you know if a result should be positive or negative?

2 Do you know *why* a result is positive or negative?

3 Do you know when parentheses are necessary and when they are not?

INSTRUCTOR NOTE: In #2, parentheses are used to avoid writing $+ - 42.68$. The issue of when parentheses matter will continue to be addressed in section problems and homework.

2. In the following table, read each problem statement and then complete the other columns. Use the example provided in the first row of the table as a guide.

Problem Statement	Picture	Calculation	Result & Interpretation
It is 34 degrees below zero, but the temperature will increase 12 degrees throughout the day. What will the temperature be at the end of the day?	Mark -34 on the line. Draw an arrow to the right 12 spaces, landing at -22. ⊢————12————→ -34 .... -22	$-34 + 12$	-22 It will be 22 degrees below zero at the end of the day.
If a civilization began in 100 BC and lasted 450 years, when did it end?	Mark -100 on the line. Draw an arrow to the right and indicate that its length is 450. We land at 350. ⊢————450————→ -100 350	$-100 + 450$ or $450 - 100$	350 AD The civilization ended in 350 AD.
The stock market has already lost 12 points for the day but, in the next hour, drops 42.68 points more. What is the total loss so far?	Mark -12 on the line. Draw an arrow to the left and indicate its length is 42.68. We land at -54.68. ←————42.68————⊣ -54.68 .... -12	$-12 - 42.68$ or $-12 + (-42.68)$	-54.68 The total loss is 54.68 points.

Notice in #2 that we are adding signed numbers in each scenario. When we add two negative numbers, the result is even more negative. The size of the result is the sum of the sizes of the numbers in the calculation. When we add numbers with different signs, the sign of the result is determined by the sign of the number in the calculation that has the largest size. To get the size of the result, we subtract the sizes.

HOW IT WORKS

To add two numbers with the same sign:

Add the sizes; keep the sign.

EXAMPLE: Add −15 and −40.

To find the sum, add the sizes (15 + 40) and keep the sign (negative). The result is −55.

$$-15 + (-40) = -(15 + 40)$$
$$= -55$$

To add two numbers with different signs:

Subtract the sizes (larger—smaller); keep the sign of the number that has the larger size.

EXAMPLE: Add −22 and 30.

To find the sum, subtract the sizes (30 − 22) and keep the sign of the number whose size is larger. Since 30 is larger than 22, keep its sign, which is positive. The final result is 8.

$$-22 + 30 = +(30 - 22)$$
$$= +8$$

So far, we have learned two rules about adding signed numbers.

• When adding two numbers with different signs, **subtract** the sizes.

• When adding two numbers with the same sign, **add** the sizes.

Adding two negative numbers is just like adding two positive numbers, but the numbers are on a different side of the number line.

> When adding signed numbers, keep this in mind:
> 1. Different signs, think difference
> 2. Same signs, think sum

3. For each problem, compute the result without a calculator. If you are unsure how to proceed, put the problem into a familiar context, such as money or temperatures, or use a number line.

a. −58 + (−18)   −76

b. 58 + 18   76

c. −58 + 18   −40

d. 58 + (−18)   40

If we were to use a number line to calculate #3a, we would move left 18 spaces from −58 to land on −76. We would move the same way on the number line from −58 to find −58 − 18, arriving at −76 again. Since −58 − 18 and −58 + (−18) yield the same answer, they are equivalent expressions. Specifically, subtracting 18 yielded the same answer as adding its opposite of −18.

In general, subtracting a number is the same as adding its opposite. That is, $a - b = a + (-b)$. Consider the context of money to make sense of this. When you gain debt (add a negative amount), you lose money.

We can also use this idea with calculations like $5 - (-1)$. Since subtraction is the same as adding the opposite, we can rewrite the calculation as $5 + (+1) = 6$. A number line can be used to confirm this result. On a number line, the difference of two numbers is the distance between them. So $5 - (-1)$ asks us to find the distance from -1 to 5. To do so, we find the distance from -1 to 0 and from 0 to 5 and add them.

$$5 - (-1) = 5 + (+1) = 6$$

Generalizing these findings, we have the following rule:

HOW IT WORKS

To change subtraction into addition:

Instead of subtracting, add the opposite. That is, rewrite $a - b$ as $a + (-b)$, or rewrite $a - (-b)$ as $a + b$. Then add.

EXAMPLES: $-5 - 10$ can be written as $-5 + (-10) = -15$

$5 - (-10)$ can be written as $5 + (+10) = 15$

Remember that the negative sign can be thought of interchangeably as negative or opposite or subtraction.

You might consider writing a subtraction problem as addition if you find that easier to do.

4. For each problem, compute the result without a calculator. If you are unsure how to proceed, put the problem in a familiar context, such as money or temperatures, or use a number line. Also, convert subtraction problems to addition when helpful.

   a. $-120 - 7$ -127

   b. $54 + (-11)$ 43

   c. $-72 + 12$ -60

   d. $-72 - (-12)$ -60

   e. $-\dfrac{2}{3} - \left(-\dfrac{1}{4}\right)$ $-\dfrac{5}{12}$

Signed-number addition and subtraction can be challenging. Take your time, use a familiar context if it helps, and visualize the problem on a number line. Together, these techniques can give you more confidence in your work and results.

Now that we understand how to add and subtract signed numbers, we will look at other operations. Although multiplying and dividing signed numbers can occur naturally in real-life situations, we will most often see these calculations embedded in other problems or when we use formulas. We will look at some familiar situations in order to develop rules for multiplying and dividing signed numbers.

We can begin by looking at a negative number multiplied by a positive number. The parentheses in $3(-4)$ are one way to clarify that the operation is multiplication and not subtraction. A dot could also be used to indicate multiplication. Since multiplication is repeated addition, the calculation $3(-4)$ can be changed to an addition problem and simplified.

$$3(-4) = (-4) + (-4) + (-4) = -12$$

Generalizing, we can see how to multiply a positive number and a negative number without writing the calculation using addition. Instead, we can multiply the sizes. The product will be negative.

5. Use this idea to perform the first four calculations.

$$3(-7) = -21$$
$$2(-7) = -14$$
$$1(-7) = -7$$
$$0(-7) = 0$$

Note the pattern in the results and then use this pattern to perform the remaining three calculations.

$$(-1)(-7) = 7$$
$$(-2)(-7) = 14$$
$$(-3)(-7) = 21$$

Notice in the last three calculations that a negative number multiplied by a negative number results in a positive number. If we multiply the sizes of the numbers, we will have the size of the result.

Observe that $-1(5) = -5$, which is the opposite of 5. This will be true for any number, not just 5. That is, multiplying a number by -1 produces its opposite.

6. The sign rules for multiplying numbers can be used to divide them as well. One way to perform calculations such as $24 \div 3$ is to think, "What times 3 is 24?" Rewrite each division problem as a multiplication problem and then find the result.

$$36 \div 18 \qquad 18 \cdot ? = 36 \qquad \rightarrow ? = 2$$
$$-25 \div -5 \qquad -5 \cdot ? = -25 \rightarrow ? = 5$$
$$12 \div -6 \qquad -6 \cdot ? = 12 \qquad \rightarrow ? = -2$$
$$-48 \div 8 \qquad 8 \cdot ? = -48 \rightarrow ? = -6$$

Notice that to get the result, we divide the sizes of the numbers. But we also have to determine the sign, as we did with multiplication.

7. Generalize the results from #5 and #6 and complete these tables to summarize the sign rules.

$\times$	Positive	Negative
Positive	+	−
Negative	−	+

$\div$	Positive	Negative
Positive	+	−
Negative	−	+

It's tempting to say to yourself that two negatives make a positive. However, this is only true with multiplication and division, not addition.

At first glance, it might seem as though there are a lot of details to remember for signed-number operations. Look through all eight rules to form a simpler summary. If two numbers with the same sign are multiplied or divided, then the result is positive. If two numbers with different signs are multiplied or divided, then the result is negative.

HOW IT WORKS

To multiply or divide two numbers with the same sign:

Multiply or divide the sizes. The result is positive.

EXAMPLES: $(3)(6) = 18$ $35 \div 7 = 5$

$(-3)(-6) = 18$ $-35 \div (-7) = 5$

To multiply or divide two numbers with different signs:

Multiply or divide the sizes. The result is negative.

EXAMPLES: $(3)(-6) = -18$ $-35 \div 7 = -5$

$(-3)(6) = -18$ $35 \div (-7) = -5$

8. Use the rules to complete these computations without a calculator.

a. $-\dfrac{3}{8} - \dfrac{4}{17}$ $-\dfrac{83}{136}$

b. $\left(-\dfrac{3}{8}\right)\left(-\dfrac{4}{17}\right)$ $\dfrac{3}{34}$

c. $\dfrac{15}{2} \div \left(-\dfrac{30}{11}\right)$ $-\dfrac{11}{4}$

Connect

This section has focused on performing operations with integers, but it's also important to understand how the set of integers relates to other types of numbers.

LOOK IT UP

Real Numbers

The set of **real numbers** includes all the numbers that can be represented on a number line. For example, 3, 0, $\frac{1}{2}$, -5.7, and π are real numbers. The set of real numbers contains each of the following sets of numbers:

The set of **counting numbers**, $\{1, 2, 3, \ldots\}$, is used to count objects.

The set of **whole numbers**, $\{0, 1, 2, 3, \ldots\}$, consists of the counting numbers and zero.

The set of **integers**, $\{\ldots, -3, -2, -1, 0, 1, 2, 3, \ldots\}$, consists of the whole numbers and their opposites.

Rational numbers are numbers that can be written as a quotient of two integers. Examples include $\frac{2}{3}$, -3, 1.2, $4\frac{1}{2}$, 0.

Irrational numbers are numbers that cannot be written as a quotient of two integers. Examples include $\sqrt{5}$ and π.

9. a. Find all numbers that give a result of 25 when squared. That is, find all numbers that can be inserted inside both sets of parentheses to make this equation true: ( ) ( ) = ( )². 5, –5

Tech TIP

Most scientific calculators will give an error for the square root of a negative number unless the calculator has a complex number mode.

b. If you are asked to find $\sqrt{b}$, this means to find the *non-negative* number that can be squared to get *b*. Find $\sqrt{25}$.   5

c. If you were asked to find a number whose square is -25, how would you respond?
There is no real number whose square is –25.

d. Use your calculator or guess and check to find the value of $\sqrt{-25}$.
This is not a real number. Some calculators might return the imaginary number 5*i*.

Since square roots of negative numbers are not real numbers, a new set of numbers, called **imaginary numbers**, is needed to express them. Imaginary numbers have the form *bi*, where *b* is a real number and *i* is the square root of -1. Imaginary numbers are part of the **complex numbers**, which have the form $a + bi$, where *a* and *b* are real numbers. This book focuses primarily on real numbers.

Reflect

WRAP-UP

What's the point?

When performing a calculation involving signed numbers, we can use familiar contexts and pictures to make sense of the problem. To perform the computation, we have to find both the size of the result and its sign. If you understand the reasoning for the procedures, you will have fewer rules to memorize.

What did you learn?

How to add, subtract, multiply, and divide signed numbers
How to solve applied problems that involve signed-number addition, subtraction, multiplication, and division

Another component to success in a math class is ABILITY, which includes your skills and knowledge. Here are some actions of students with strong math ability:

1. Use a calculator when necessary but not for every calculation.
2. Estimate results before performing calculations.
3. Check answers for reasonableness.
4. Write meaningful answers with units.
5. Read critically and completely.

1.7 Homework

Skills MyMathLab

First complete the MyMathLab homework online. Then work the two exercises to check your understanding.

1. Add or subtract.

   a. $5.2 - (-1.1)$ 6.3

   b. $-13 + (-31)$ -44

2. Multiply or divide.

   a. $(0)(-15)$ 0

   b. $\dfrac{-15}{3}$ -5

   c. $(-4)(-2)(-7)$ -56

Concepts and Applications

Complete the problems to practice applying the skills and concepts learned in the section.

3. For each situation, write a computation and find the result. Interpret your answer.

   a. Beth already owes Joe $10 but asks to borrow $25 more. How much will she owe him in total?

   $-10 - 25$ or $-10 + (-25)$.

   Result: -35

   She will owe him $35.

   b. Lora has $11,500 in debt but receives $20,000 through an inheritance. How much will she have after the debt is paid?

   $-11,500 + 20,000$ or $20,000 - 11,500$

   Result: 8,500

   She will have $8,500.

4. A football team gained 6 yards on the first down, lost 15 yards on the second down, and gained 12 yards on the third down. How many yards does it need to gain on the fourth down to have a 10-yard gain (known as getting a first down) from its starting position?

   $6 - 15 + 12 = 3$. The team needs to get 7 more yards on the fourth down to have a 10-yard gain (or get a first down).

5. a. If it is 22 degrees Fahrenheit in the morning but rises to 47 degrees by noon, what is the change in temperature? Write a calculation, find the result, and interpret it. Your result should reflect the numerical change and direction—in this case, rising.

47 − 22 = 25; 25-degree rise in temperature

b. Generalize how you wrote the calculation. That is, in what order is the subtraction done? You will need this idea to complete the next two problems.

Ending temperature − starting temperature

c. If it is 5 degrees Fahrenheit below zero when you wake up in the morning, but the temperature is 26 degrees above zero by the time you leave work, what is the change in temperature? Write a calculation, find the result, and interpret it.

26 − (−5) = 26 + 5 = 31; 31-degree rise in temperature

d. If it is 26 degrees Fahrenheit now but will drop to 5 degrees below zero overnight, what is the change in temperature? Write a calculation, find the result, and interpret it.

−5 − 26 = −31; 31-degree drop in temperature

6. a. Give an example in which a mathematical operation of two negative numbers results in a positive number, and an example in which a mathematical operation of two negative numbers results in a negative number.

Example: $(−4) \times (−4) = 16$

Example: $(−4) + (−4) = −8$

b. Can we always conclude that performing a mathematical operation on two negative numbers results in a positive number?

No

7. The role of parentheses is important in mathematics. Sometimes they make a big difference, whereas other times they can be removed without changing the problem. Work through each of the following problems. Then rewrite each without parentheses, compute, and determine whether the parentheses made a difference.

Problem with parentheses	Answer	Rewritten without parentheses	New Result	Do parentheses matter?
a. 2(−4)	−8	2 − 4	−2	Yes
b. (−2) + (−4)	−6	−2 + −4	−6	No
c. 2 − (4)	−2	2 − 4	−2	No
d. (−2)(−4)	8	−2 − 4	−6	Yes

8. When trying to find the solution to the problem 6 ÷ 3, we can think "3 × ? = 6?" Let's apply this idea to the following problems. Rewrite each division problem as a multiplication problem and then find the solution. Verify your answer by using your calculator. For the answer that results in an error on the calculator, write "undefined."

a. 0 ÷ 6   6 × ? = 0   ? = 0

b. 6 ÷ 0   0 × ? = 6   undefined

c. Try a few more similar problems in which zero is divided by a number or a number is divided by zero. What conclusions can be drawn?

We cannot divide by zero. Zero divided by any number except zero is zero.

9. For each problem, state *positive*, *negative*, or *it depends*. If you state *it depends*, give two examples to support your answer.

   a. Positive + Positive   Positive

   b. Positive + Negative   It depends   Possible examples: 3 + (−7) = −4; 7 + (−3) = 4

   c. Negative + Negative   Negative

   d. Positive − Negative   Positive

   e. Positive × Positive   Positive

   f. Positive × Negative   Negative

   g. Negative − Positive   Negative

10. Suppose your friend reads the following instructions to you from a book of math tricks:

    Pick a number.

    Subtract 1.

    Multiply by 3.

    Add 6.

    Divide by 3.

    Subtract the original number.

    She then correctly guesses that your final number is 1. You repeat this and arrive at the same result.

    Try this once with a negative number and once with a fraction. Do you get 1 each time as an answer? Write out your calculation steps.

    Yes. The result is always 1, even if you start with a negative number or a fraction.

11. **Looking Forward, Looking Back** Find the mean of this set of numbers: −8, −30, 15, −$\frac{4}{5}$, −1.2.

    −5

1.8   An Ounce of Prevention: Means

Explore

SECTION OBJECTIVES:

- Find and interpret the mean of a set of numbers
- Use means in applied problems

1. A student is taking a class that has five exams, each worth 100 points. After four exams, her average in the class is 80%. She tells her friend, "If I get a 100 on the last exam, my grade will be an A. 80 and 100 average to be a 90, which is an A." Her friend replies, "No, it's not even possible for you to get an A in the class." Who is right and why?

The friend is correct. Currently, the student has 80% of 400 points, or 320 points. To earn an A in the class, she needs 90% of 500 points, or 450 points. She would need to earn 130 points on the last exam, which is not possible.

2. Suppose her friend scores 90 out of 100 points on a test in a different class. What is his average in the course after the test if he had 60% of 100 possible points prior to the test? What if he had 60% of 900 possible points prior to the test?

75%, 63%

Discover

INSTRUCTOR NOTE: Most students know how to calculate a mean, but struggle to interpret its meaning. This section will equally emphasize both the calculation and interpretation of the mean.

The expression "an ounce of prevention is worth a pound of cure" smartly describes how it is often easier and simpler to prevent a problem than it is to deal with the consequences later. The situations in the *Explore* problems illustrate the impact of scores on a grade, early vs. later in a course. Most grades are found by calculating an average or mean, and understanding how averages are calculated can help you use them to your advantage. In this section, we will learn more about the mean as a measure of the center of a set of numbers.

As you work through this section, think about these questions:

1 Can you determine when you are being asked to find a mean even if it is not stated as such?

2 Do you understand how to interpret a mean as a leveling-off value or a balancing point?

Let's begin by defining the mean.

Mean

The **mean** or average of a set of numbers is one measure of the set's center or middle. It is the sum of the values divided by the number of values.

FOR EXAMPLE, if Peyton Manning threw a football 85 times for 769 yards total, then he averaged $\frac{769}{85} = 9.05$ yards per throw. The average tells us that Manning could have achieved this same total by throwing a perfectly consistent 9.05 yards per throw. That is not necessarily what happened because some throws were likely shorter or longer than 9.05 yards. The average is a redistribution of the set of numbers equally.

EXAMPLE 1

The graph shows enrollment in Chicago community colleges in 2014. What is the average enrollment?

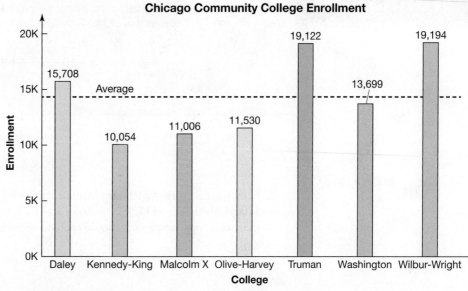

Chicago Community College Enrollment

Data from www.iccb.org

SOLUTION

To find the mean or average of a set of numbers, we add all the numbers and then divide the total by the number of values. The enrollment for each community college in Chicago is listed above its bar. Totaling these values, we get 100,313 students. Dividing by the number of colleges, 7, we get an average of approximately 14,330 students per college.

Notice the dashed line at the average in the graph. If the areas in the bars that are above the dashed line were cut up and rearranged, they would fill the empty space in the bars below the line to bring the shorter bars up to the line. The average represents the value at which the bars would level off and be the same height.

HOW IT WORKS

Need more practice?
Mindy loses 4 pounds the first week of a diet, 0.8 pound the second week, 1.2 pounds the third week, and 1.5 pounds the fourth week and gains 1 pound the fifth week. On average, how much did she gain or lose each week?

Answer: She lost on average 1.3 pounds per week.

To find the mean of a set of values:

1. Add the values.
2. Divide the total by the number of values.

EXAMPLE: John buys dinner at the same restaurant three times in one month with bills of $24, $38, and $40. What is his average dinner bill?

His average dinner bill is
$$\frac{\$24 + \$38 + \$40}{3} = \frac{\$102}{3} = \$34/\text{meal}.$$
This means that if he had spent the same total of $102 on three meals but paid the same amount each time, he would have spent $34 on each dinner.

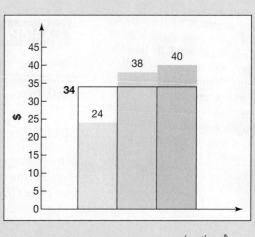

(*continued*)

Notice in the previous graph that the area in the bars for 38 and 40 that is above the line at 34 would fill in the space below the line but above the bar for 24.

The mean can also be thought of as the balancing point, or fulcrum, of a teeter-totter. The one number, 24, below the mean, balances the two numbers above the mean, 38 and 40, because it is so much smaller than the mean.

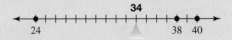

3. Eight servers work at a restaurant and earn the following tips on a particular night: $100, $70.50, $72, $145, $50.44, $66.12, $88.19, $112. If their practice is to evenly share their tips, how much should each server receive?

$88.03

4. Two months into a biology course, Erin has taken three quizzes with these scores (out of 10 points each): 8, 8, 5. What is her average quiz score? Suppose, on the fourth quiz, she scores 3 points out of 10. Compute her new average. How does this low score affect her grade? Why is that?

7; 6; the score of 3 lowers her average. Since she scored below her old average of 7 on the fourth quiz, it brought down her average.

One of the takeaway points from this section is that some things are difficult to undo. Whether it is debt or low grades, it can be easy to get to an unwanted place but hard to get out of it. Take the time now to earn any available points in the course and address any issues that may be causing lower scores. It is much easier to fix problems now than at the end of the course.

We have seen how to find and interpret the mean of a set of numbers. We can also complete a set of numbers so that it has a particular mean.

EXAMPLE 2

Continuing with the scenario in #4, Erin is unhappy with her current average and decides that she wants to raise her grade to a C (7 out of 10). If the fifth quiz is also worth 10 points, what will she need to score on it to bring her average back to a C?

SOLUTION

Before using only numbers to solve the problem, let's use a picture approach. We can visualize the current four quiz scores using squares, stacked vertically, to represent the quiz points (Figure 1).

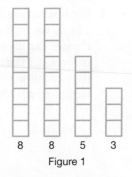

8   8   5   3

Figure 1

Erin's goal is an average of 7 on five quizzes, meaning that we need five stacks each 7 squares tall. We can rearrange the current squares so that the two tallest stacks have only 7 squares each. We can take one square off the first stack and one off the second stack, and add them both to the third stack. But we need four more squares to add onto the fourth stack, and we still need 7 for the last stack (Figure 2). So we need 11 more squares, or quiz points, than we had when we began (Figure 3).

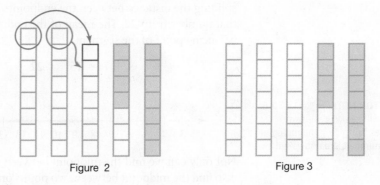

Figure 2   Figure 3

We can check our answer by averaging the five scores: $\frac{8 + 8 + 5 + 3 + 11}{5} = 7$

Scoring 11 points may be possible if the instructor offers a bonus point on the quiz, but it is unlikely, since Erin never scored that high on the earlier quizzes in the course.

We can also solve this problem by using numbers instead of pictures. Currently Erin has $8 + 8 + 5 + 3 = 24$ points. She would like her average on five quiz scores to be 7 points. An average is found by totaling the five scores and dividing the result by 5.

$$Average = \frac{\text{Total of 5 quiz scores}}{5} = 7$$

$$\frac{?}{5} = 7$$

$$? = 5 \cdot 7 = 35$$

The number divided by 5 that equals 7 is 35. So the total of the five quiz scores must be 35. Since she has 24 points currently, she will need $35 - 24 = 11$ additional points.

On a balance, the two scores below the mean will be balanced by the three scores above the mean. The two low scores are together 6 points from the mean of 7, since the 3 is four points below and the 5 is two points below the mean. The three high scores are also together 6 points from the mean, one point above for each of the 8's and four points for the 11.

5. Ben is searching for reviews on a book he is interested in reading. He sees that there are 12 reviews, each ranging from 0 to 5 stars. The average rating of the book is 3.75 stars. He sees that there are four reviews at 5 stars each, four reviews at 4 stars each, two reviews at 3 stars each, and one review at 2 stars. He sees the first 11 reviews on one page of the website, but the last review is on the next page. Without clicking to the next page, what is the rating of the last review?

1

Connect

The midpoint of a line segment is the point that is halfway between the endpoints of the line segment. For example, on a number line, the number that is halfway between 8 and 22 is 15. The midpoint is the same distance, 7 units, from each endpoint and can be found by splitting the distance between the endpoints and adding that number to 8 or subtracting that number from 22. The mean of the endpoints also gives the midpoint, since it is the balancing point of the segment.

$$8 + 7 = 15 \qquad 22 - 7 = 15 \qquad \frac{8 + 22}{2} = 15$$

INSTRUCTOR NOTE: In Section 2.8, students will see why the two approaches for finding the midpoint between numbers are equivalent.

Not only can we find the midpoint between two numbers on a number line, but we can also find the midpoint between two points on a coordinate system.

Midpoint

The **midpoint** of the line segment between two points (a, b) and (c, d) is found by averaging the x-coordinates and averaging the y-coordinates of the two points. That is, the midpoint is $\left(\frac{a + c}{2}, \frac{b + d}{2}\right)$.

EXAMPLE: Find the midpoint of the line segment between $(-3, 9)$ and $(2, -5)$.

SOLUTION: Averaging the x-coordinates and y-coordinates, we find the midpoint is $\left(\frac{-3 + 2}{2}, \frac{9 + (-5)}{2}\right) = \left(-\frac{1}{2}, \frac{4}{2}\right) = \left(-\frac{1}{2}, 2\right)$. Notice that the midpoint of the line segment does appear to be in the middle of the line segment.

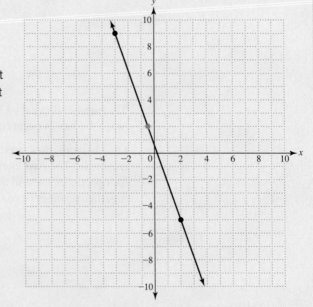

After finding a midpoint, consider drawing a quick sketch to verify that it is halfway between the two points.

6. Find the midpoint of the line segment that connects $\left(-\frac{3}{4}, -5\right)$ and $\left(\frac{1}{2}, 4\right)$. Plot both points and their midpoint on the graph provided.

$\left(-\frac{1}{8}, -\frac{1}{2}\right)$

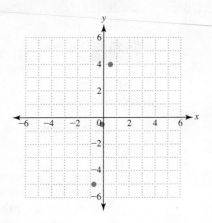

Reflect

WRAP-UP

What's the point?

The mean is one measure of center. Although we need to be able to calculate the mean for a set of numbers, it is also just as important to understand the meaning of it. The mean can be thought of as a leveling-off value or a balancing point.

What did you learn?

- Find and interpret the mean of a set of numbers
- Use means in applied problems

An important component of success in a math class is AGILITY, or how fast or easily you can do something. Being agile in mathematics means that you remember skills and are able to apply them to different situations. If you are agile, you can make connections between topics and explain concepts to someone else.

1.8 Homework

Skills MyMathLab

First complete the MyMathLab homework online. Then work the two exercises to check your understanding.

1. Find the mean of these numbers: $-2, 7, 8, 4, -1, -10, 1$ 1

2. A student has a class with four tests, each worth 100 points, and has earned these scores on the first three tests: 75, 74, 71. What is her average now? What score must she earn on the fourth test to get a B (80) in the class?

Her current average is 73.3. She must earn 100 points on the last test to have a B average.

Concepts and Applications

Complete the problems to practice applying the skills and concepts learned in the section.

3. Suppose you have three quizzes and want an average of 7 points. List at least five ways this average can be achieved. Each score can be no higher than 10 points.

Any three scores, between 0 and 10 inclusive, that add to 21 will work.

4. Suppose you have lunch with four friends, and the total check (with tax and tip) is $155. How much should you each pay if you split the check evenly?

You each pay $31.

5. Consider the following two sets of incomes. Each income is in thousands of dollars per year.

Group 1: 32, 36, 38, 39, 42, 43, 44, 47, 49, 50

Group 2: 32, 36, 38, 39, 42, 43, 44, 47, 49, 150

a. Find the average or mean salary for each group.

Group 1: $42,000/year;   Group 2: $52,000/year

b. How does an extreme value affect the mean?

The mean is pulled in the direction of the extreme value.

c. Do you think the mean of the second group is a good indication of the typical salary in that group? Explain.

No. When there is an extreme value, the mean is not always a good representative of the values. Only one salary in Group 2 is above the mean of $52,000, while the other nine are below it.

6. a. Create a set of numbers with five different values that have a mean of 50.

Answers will vary.

b. Add 10 to each of your five numbers. List the new values.

Answers will vary.

c. Find the new mean.

60

d. What do you notice?

When 10 is added to each number, the mean also increases by 10.

d. If a teacher wants to increase a class test average of 62 to 70, what could he or she do to each student's test score to achieve that?

Add 8 points to each student's test score.

7. a. A student scores 50 and 100 on two tests in a class. What is his average?

75

b. Another student scores 75 and 75 on two tests in a class. What is her average?

75

c. The mean gives us information about the center of a set of numbers. What does it not describe?

It does not tell whether the numbers are close to the mean or spread out.

8. Find the point at which these numbers would balance: 7, 13, 6, 14, 5, 15

10

9. A line segment has an endpoint at $(-3, 9)$, and its midpoint is at $(3, 3)$. What is the ordered pair for the other endpoint of the line segment?

$(9, -3)$

10. Looking Forward, Looking Back Simplify.

a. $-\dfrac{9}{7} + \dfrac{8}{3}$ $\dfrac{29}{21}$

b. $-\dfrac{9}{7} - \dfrac{8}{3}$ $-\dfrac{83}{21}$

c. $-\dfrac{9}{7} \cdot \dfrac{8}{3}$ $-\dfrac{24}{7}$

d. $-\dfrac{9}{7} \div \dfrac{8}{3}$ $-\dfrac{27}{56}$

There are specific tasks you can do to improve your chances of success in a math class.

1. Attend and participate in every class. Persevere even through challenging content.

2. Work 2 hours outside of class for every 1 hour in class.

3. Read your class notes each day and highlight important parts.

4. Do all assigned homework.

5. Seek help from your instructor if needed.

Mid-Cycle Recap

Complete the problems to practice applying the skills and concepts learned in the first half of the cycle.

Skills

1. Convert a salary of $50,000/year to dollars per hour, assuming that the worker works fifty 40-hour weeks. $25/hour

Concepts and Applications

2. Jack is driving on a highway through Wyoming, from Casper to Cheyenne, with the cruise control set at 75 mph.

   a. 75 mph is a rate. What two quantities are being compared?

   75 miles and 1 hour

   b. Jack uses an online map service to determine how long the 179-mile trip will take. It says the trip should take 2 hours, 37 minutes. What average speed is assumed by the online map program?

   68.4 mph

3. Suppose you owe $50 to each of four friends, $25 to each of three friends, and $100 to your parents. If two friends each owe you $75, what is your net debt?

   $225 in debt

4. **a.** Will the average of a list of negative numbers always be negative? Explain.

   Yes; the sum of negative numbers will always be negative. The average will be the negative sum divided by a positive number and will always be negative.

   b. Create two different sets of six numbers so that each set has an average of zero.

   Answers will vary.

   Possible answers:

   0, 0, 0, 0, 0, 0

   −3, −2, −1, 1, 2, 3

1.9 Picture This: Making and Interpreting Graphs

Explore

Use the summary graphs from a personal fitness tracker to answer the questions that follow. Round each percent to the nearest whole percent.

SECTION OBJECTIVES:
- Create and interpret pie graphs
- Create and interpret bar graphs
- Create and interpret line graphs

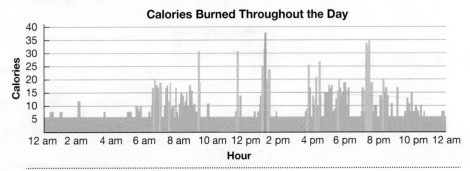

Calories Burned Throughout the Day

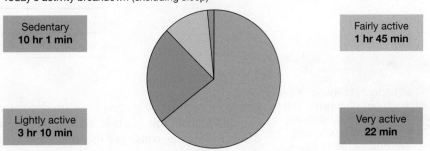

Today's activity breakdown (excluding sleep)

Sedentary
10 hr 1 min

Fairly active
1 hr 45 min

Lightly active
3 hr 10 min

Very active
22 min

1. During what percent of this person's waking hours is he very active?   2%

2. What percent of the day does he sleep?   36%

3. During which periods of the day does most of his light activity occur?
   From 6:30 to 9 a.m., from 4 to 6 p.m., around 8 p.m., and around 10 p.m.

4. What information does each graph give that the other does not?
   The bar graph shows when he exhibits the four activity levels but not the total minutes at each level. The pie graph tells how much total time he spends at each activity level but not how that time is distributed throughout his day.

Discover

Graphs provide a way to make sense of the information that surround us daily. They can summarize important details and illuminate trends. In this section, you will learn how to choose and create an appropriate graph to display information.

As you work through this section, think about these questions:

1. Can you read and understand information presented in a pie, bar, or line graph?

2. When should a bar graph be used instead of a pie graph?

3. If we have percents that add to 100%, can we always make a pie graph to display them?

Graphs are created to display data, which is information that we find from a source or collect through an experiment. Data can be **quantitative**, which means the values describe quantities that can be counted or measured. For example, heights are quantitative data, since they are measured. Data that provide labels or categories are considered **qualitative** data. This type of data is information that is a quality or attribute. For example, car names and colors are qualitative data.

When you are trying to display qualitative data and make a comparison between a part and a whole, a great way to do that is with a **pie graph**.

EXAMPLE 1

Create a pie graph to display the following data about a group of college students who use four popular cell phone carriers.

Cell Phone Carrier	Number of Subscribers
AT&T	8
Sprint	3
T-Mobile	4
Verizon	19

SOLUTION

Before we can make a pie graph, we need to determine if it should be used to display the data. The data in this example, cell phone carriers, are qualitative. Also, the college students make up the whole and the students who use exactly one of each of the carriers are the parts. Therefore a pie graph is appropriate.

The size of each sector of the pie is based on the percent the sector's category represents. To find the percent for each category, we first find the total number of subscribers, 34. Then we divide the number of subscribers in each category by the total and change the decimals to percents. Although the percents should sum to 100, they may not due to rounding. Next, we determine the size of each sector of the pie by finding the measure of its central angle. To do so, we take the sector's percent as a decimal and multiply it by 360°. When we calculate the central angle, we use the unrounded decimal.

> We often have to round percents, but you should still use the unrounded version for any further calculations, such as finding the measure of the central angle. Since the percents in the table have been rounded, they might not sum to exactly 100%. Similarly, the central angles might not sum to exactly 360°.

Cell Phone Carrier	Number of Subscribers	Percent	Central Angle
AT&T	8	$\frac{8}{34} \approx 0.24 = 24\%$	85°
Sprint	3	$\frac{3}{34} \approx 0.09 = 9\%$	32°
T-Mobile	4	$\frac{4}{34} \approx 0.12 = 12\%$	42°
Verizon	19	$\frac{19}{34} \approx 0.56 = 56\%$	201°

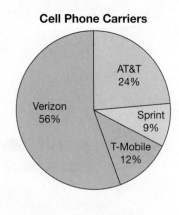

Cell Phone Carriers

Now we draw the sectors in a circle using the degrees in the central angles as a guide. Since 85° is almost 90°, the angle for the AT&T sector is almost a right angle. The central angle for Verizon, which measures 201°, is larger than a straight angle, so the sector is greater than half the circle. Each sector is labeled with a name and percent, and the pie graph is given a title. We can see from the graph that most students in this group use Verizon as their cell phone carrier.

HOW IT WORKS

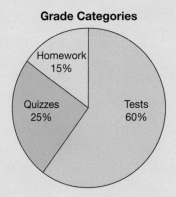

To create a pie graph:

1. Determine the percent for each category. The percents should add to 100, at least approximately.
2. Find the measure of each sector's central angle by taking the sector's percent and multiplying it by 360°.
3. Draw each sector using the central angle as a guide.
4. Label each sector with a name and percent. Give the graph a title.

Note: A pie graph should be used for qualitative data only when the categories do not overlap and when they make up the whole. A pie graph should never be used for quantitative data.

EXAMPLE: An instructor determines the final grades in her class by counting tests as 60% of the grade, quizzes as 25% of the grade, and homework as 15% of the grade. Represent the grading structure with a pie graph.

Multiplying the percent, as a decimal, for each category by 360° gives the central angles of 216° for tests, 90° for quizzes, and 54° for homework. Using these angles, the accompanying pie graph can be created. The sectors are labeled with the category names and percents, and the graph has a title.

Grade Categories

Homework 15% · Quizzes 25% · Tests 60%

Tech TIP

Each graph in this section can be made quickly on a spreadsheet program like Excel with many options for the display. Consult the Excel appendix for more information.

Don't wait until the last minute when you are using technology. Expect that issues may arise. Giving yourself enough time to complete an assignment that requires technology will reduce your frustrations.

5. Ian attends a company-sponsored workshop that is designed to increase employee efficiency. His task is to create a pie graph showing how he uses his 8 hours at work on an average day to help him analyze how he can better use his time. Represent the information in the table with a pie graph.

Ian's work hours are distributed as follows:

Categories	Hours
Meetings	1.5
Computer	2.5
Calls	1
Breaks	0.5
Lunch	1
Off-task	1.5

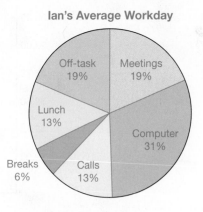

Ian's Average Workday

Off-task 19% · Meetings 19% · Lunch 13% · Computer 31% · Breaks 6% · Calls 13%

Another way to display data is with a **bar graph**. The bars on a bar graph can be horizontal or vertical, touching or spaced apart. When the bars are ordered from tallest to shortest, the bar graph is known as a **Pareto chart**. Bar graphs are a useful tool for making part-to-part comparisons, usually for qualitative data.

EXAMPLE 2

INSTRUCTOR NOTE: Consider asking students to choose another starting value and increment for the *y*-axis. What effect will those choices have on the appearance of the Pareto chart?

Create a Pareto chart to display the top 10 average gas prices, by state, in the United States.

State	Average Gas Price
Alaska	$3.424
California	$3.601
Hawaii	$3.191
Illinois	$3.163
Indiana	$2.974
Michigan	$3.006
Nevada	$3.213
Oregon	$3.003
Washington	$3.083
Wisconsin	$2.967

Data from www.gasbuddy.com

SOLUTION

We begin by reordering the gas prices from highest to lowest. The states will be listed on the horizontal axis, since the bar graph is going to be vertical. Because the gas prices range from just under $3.00 to almost $4.00, the vertical axis can start at $2.00 and have increments of $0.25. The differences in prices would be less apparent if the vertical axis started at zero.

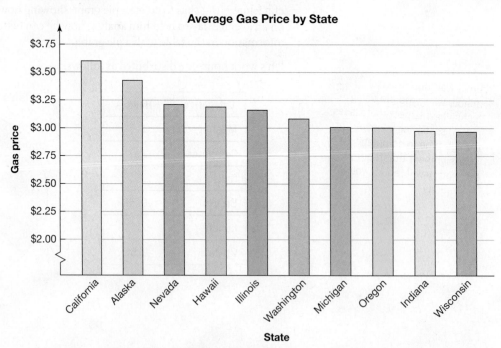

The Pareto chart makes it easy to see that the highest average gas price nationwide is in California. However, the actual average price is difficult to read exactly. To help the reader, the values of the heights of the bars can be added above the bars. It is always good form to include a title and axis labels, as shown on this graph.

HOW IT WORKS

To create a vertical bar graph:

1. Determine the categories that will be graphed and the values to be graphed for each category.
2. Draw and label the axes.
3. Use the values to draw and label tick marks on the vertical axis using a consistent increment.
4. Draw vertical bars, either touching or spaced evenly, on the horizontal axis. Use the tick marks on the vertical axis and the category values to determine the height of each bar. Write the name of the category under each bar.
5. Give the graph a title.

Note: You may want to draw tick marks on the horizontal axis before you draw the vertical bars. It is a helpful practice but one that is not always followed.

EXAMPLE: Emily receives ratings on a scale from 1 to 5 from users of an app she created. Her ratings are: 4, 5, 3, 3, 5, 4, 5, 5, 1, 2, 5, 5, 4, 4, 5, 4, 4, 3, 5, 5. Create a vertical bar graph to display how many of each rating she received.

First, find the frequency of each rating and record it in a table. Next, draw vertical bars on the horizontal axis at the ratings 1, 2, 3, 4, and 5. Since the frequencies range from 1 to 9, draw tick marks on the vertical axis at every unit starting at 0 and ending at 10. The graph shows that Emily's most common rating is 5.

Information presented as a pie graph can also be presented as a bar graph, but the reverse is not necessarily true. Even if information can be shown with either graph, pick the graph that displays the information in the clearest way.

Rating	Frequency
1	1
2	1
3	3
4	6
5	9

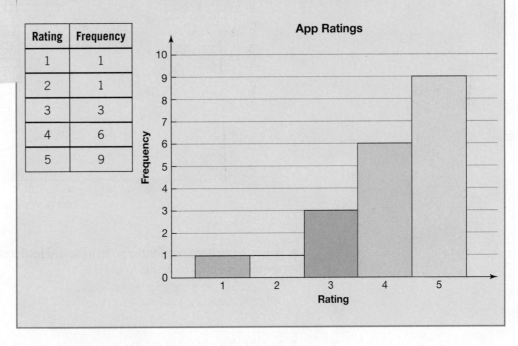

The procedure used to make a horizontal bar graph is similar to the steps presented in *How It Works*, but the axes are switched. Category labels appear on the vertical axis, and numbers for the length of the bars appear on the horizontal axis. An example of a horizontal bar graph appears in the homework.

Another way to display quantitative data is with a **line graph**, which is a series of connected, ordered pair data points. Since quantitative data has an implied order, we can connect the dots. Line graphs are often used to display data that changes over time.

6. The data that accompanies over-the-counter and prescription drugs is full of important, often mathematical, information. In the following graph, the vertical axis represents the concentration of acetaminophen in the body, measured in micrograms per milliliter of blood (µg/mL). The horizontal axis represents the number of hours that have passed since the medication was ingested. The graph is constructed in this way because the amount of acetaminophen in the bloodstream depends on how much time has passed since the medication was ingested. Since all the data values are positive, only the first quadrant is shown. The graph shows two different types of medication. Use the line graph provided to answer the following questions.

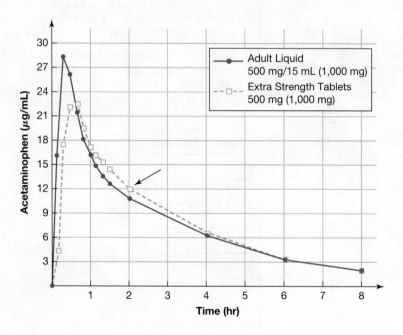

a. Write the ordered pair for the point indicated by the arrow and interpret it. Do not use abbreviations for units.

(2, 12); the concentration of acetaminophen in the body 2 hours after taking the tablets is 12 micrograms per milliliter.

b. Use the shape of the graph to describe what happens to the concentration of medication in the body over 8 hours and why a second dose is needed after 4–6 hours.

Within an hour, the concentration reaches its highest level and then decreases, getting closer and closer to zero. After 4–6 hours, there is still some of the residual drug in the body, but the levels are considerably lower than the peak levels. Notice, though, that slightly more of the drug is still in the body at this point when using the tablets instead of the liquid.

c. When do the two formulations result in the same concentration of acetaminophen in the bloodstream?

At 0 minutes, 45 minutes, and 6–8 hours

d. Based on the graph, which would you take for pain relief: liquid or tablets? Explain your choice.

Possible answers: Liquid: It gets into your body faster so you feel relief faster.

Tablets: They maintain a higher level of concentration in the blood than the liquid after the first hour.

INSTRUCTOR NOTE: Students should be careful with the units used. These units (mg/mL) are the amount of acetaminophen in the liquid, whereas the previous units (μg/mL) described the amount of acetaminophen in the blood.

e. The liquid dose is listed as 500 mg/15 mL (1,000 mg). That means there is 500 mg of acetaminophen for every 15 mL of liquid. If 15 mL is equivalent to 1 tablespoon, how many tablespoons do we need to take to ingest 1,000 mg of liquid medication?

Scaling 500 mg/15 mL to a ratio involving 1,000 mg, we need to ingest 30 mL to take 1,000 mg of the drug. Scaling again, we get

$$\frac{15 \text{ mL}}{1 \text{ tablespoon}} = \frac{30 \text{ mL}}{2 \text{ tablespoon}}.$$ So the answer is 2 tablespoons.

HOW IT WORKS

To make a line graph:

1. Determine the variables that will be graphed and the values to be graphed for each variable.
2. Draw and label the axes.
3. Use the values to draw and label tick marks using a consistent increment for each axis.
4. Plot the points and connect them.
5. Give the graph a title.

EXAMPLE: A flu outbreak occurs at an elementary school. Create a line graph to show how the number of sick students changes over time.

Because the number of students is changing based on the number of days, put the number of days since the outbreak on the horizontal axis and the number of students on the vertical axis. Tick marks are placed on the horizontal axis from 0 to 30 with increments of 5 days. Similarly, tick marks are placed on the vertical axis from 0 to 140 with increments of 20 students. Notice that the number of students who are sick increases until sometime between the 9th and 12th day, when it begins declining.

Number of Days Since Outbreak	Number of Sick Students
1	1
2	4
3	20
7	100
9	125
12	120
14	109
21	40
28	5

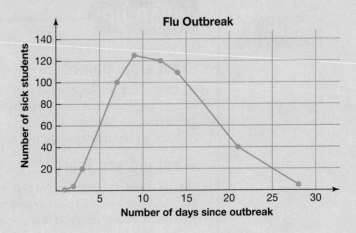

Connect

In this section, you learned how to create three graphs. While it is important to know *how* to create a graph, it is also just as important to know *when* to use a particular graph.

7. Using the statistics for the top 10 PGA Tour golfers in 2015 given in the table, create a graph to display some of the information provided. Explain why you chose the graph you did and how the graph helps make sense of the information.

RK	Player	Age	Events	Rounds	Cuts Made	Top 10	Wins	Cup Points	Earnings
1	Jordan Spieth	22	26	92	22	16	6	6,392	$12,030,465.00
2	Jason Day	27	21	76	19	12	5	6,970	$9,403,330.00
3	Bubba Watson	36	20	72	18	10	2	4,009	$6,876,797.00
4	Rickie Fowler	26	22	76	18	8	2	4,196	$5,773,430.00
5	Dustin Johnson	31	21	73	18	11	1	2,854	$5,509,466.50
6	Justin Rose	35	21	70	16	9	1	3,046	$5,462,677.00
7	Rory McIlroy	26	12	42	11	7	2	2,407	$4,863,312.00
8	Zach Johnson	39	26	91	21	11	1	2,794	$4,801,487.00
9	Henrik Stenson	39	17	64	17	9	0	4,499	$4,755,070.00
10	Jimmy Walker	37	25	90	22	6	2	2,370	$4,521,350.00

Data from espn.go.com

Possible answer: A Pareto chart can be created to highlight the players who had the most top 10 finishes. From the graph, we can see that Jordan Spieth had the most top 10 finishes by quite a bit. More than half of the other players had between 9 and 12 top 10 finishes.

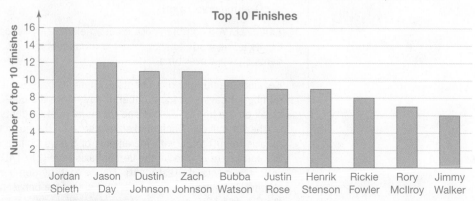

Reflect

WRAP-UP

What's the point?
Pie graphs, bar graphs, and line graphs are three common ways to display data. Interpreting them is an important skill. You need to be able to make sense of graphs that, for example, accompany your prescriptions, explain your fitness regimen, or summarize a political poll.

What did you learn?
How to create and interpret pie graphs
How to create and interpret bar graphs
How to create and interpret line graphs

Another component to success in a math class is ACHIEVEMENT, which is what you accomplish through work or talent. While natural ability helps, hard work and a strong work ethic are essential to achieve success, in or out of a math class. Being responsible also contributes to what you can achieve.

1.9 Homework

Skills MyMathLab

First complete the MyMathLab homework online. Then work the two exercises to check your understanding.

1. Pie graphs are most useful for _____part-to-whole_____ comparisons, while bar graphs are most useful for _____part-to-part_____ comparisons.

2. The bar graph shows the percent of adults in the United States who say that they have a great deal of trust in the mass media.

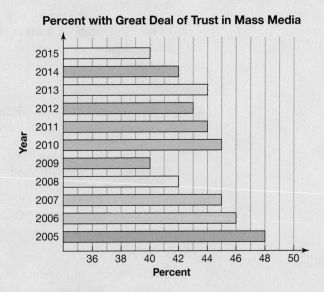

Percent with Great Deal of Trust in Mass Media

a. In which year(s) did the percent increase from the previous year? 2010, 2013

b. If there were 950 people in the sample, how many of them expressed a great deal of trust in the mass media in 2005? 456 people

Concepts and Applications

Complete the problems to practice applying the skills and concepts learned in the section.

3. Will spending too much time on Facebook have a negative impact on your performance in college? Many studies have been done to answer this and other related questions. It appears that *how* you use Facebook is as important as *how much* you use it. Consider the following pie graphs, which show the percent of students in one study who use various features on Facebook.

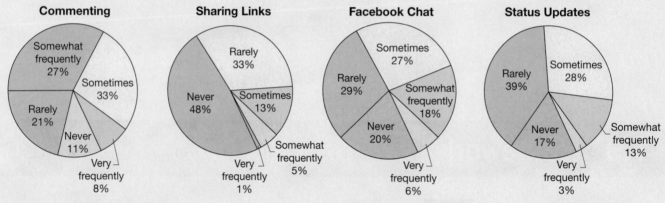

Source: theultralinx.com who got the graphic from onlineeducation.net who got the data from this study: reyjunco.com

a. If the study consisted of 1,839 college students, how many of them claimed that they comment on Facebook somewhat frequently?

0.27(1,839) = 496.53 or 497 students

b. If the study consisted of 1,839 college students, how many of them claimed that they make status updates *at most* rarely?

Include the "Never" and "Rarely" categories for a total of 56%. 0.56(1,839) = 1,029.84 or 1,030 students

c. Of the listed Facebook features, which seems to be the least popular with college students? Explain your answer choice.

Sharing Links has the smallest percent of students who claim to use it very frequently and the largest percent of students who claim to use it never.

4. Consider the following graph, which shows the educational attainment of residents of Rockford compared to the Illinois average in a recent year.

a. For which educational levels does Rockford exceed the state average?   Less than 9th grade, 9th to 12th grade (no diploma), and high school graduate (includes equivalency)

b. What is the most common educational level in the town?

High school graduate (includes equivalency)

c. Do the percents for Rockford total to 100%? Are there categories of educational achievement that have not been included on the graph?

Percents seem to total to 100%. All levels seem to be listed.

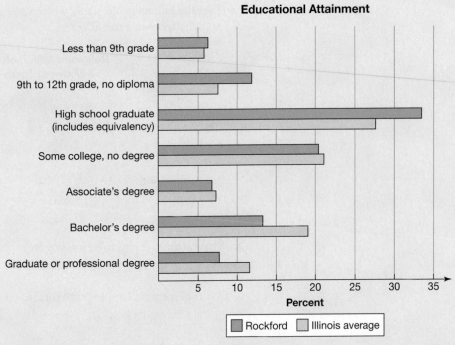

Educational Attainment

Source: census.gov

d. Represent the percent of Rockford residents with various educational levels using a pie graph.

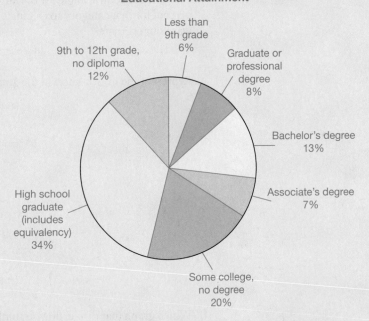

Educational Attainment

e. Which graph (bar or pie) shows the order of the various educational levels?

Bar graph

5. Consider the following pie graph, which shows the religious affiliation for residents of a Midwest town in a recent year.

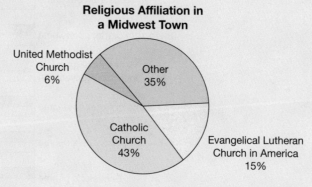

Religious Affiliation in a Midwest Town

United Methodist Church 6%

Other 35%

Catholic Church 43%

Evangelical Lutheran Church in America 15%

a. What category seems to be missing from this graph?

Residents who claim no religious affiliation

b. Do the percents in the pie graph total to 100%? If not, give a possible reason.

No; the percents total to 99%, likely due to rounding.

c. The pie graph actually represents only the 50% of residents who claim a religious affiliation. If the population of the town at the time the graph was made was 156,300 people, then how many people claimed some religious affiliation? How many claimed to be Catholic?

78,150 residents; 33,605 residents

d. Estimate the central angle for the largest sector of the pie graph. Then use the percent for that category to calculate the central angle more accurately. How close was your estimate?

Estimates around 150°; sector angle is actually 154.8°.

e. Represent the information in the graph with a bar graph.

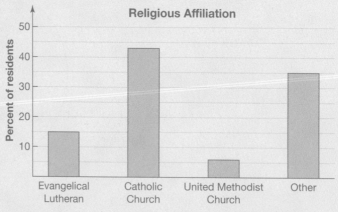

Religious Affiliation

Percent of residents

Evangelical Lutheran   Catholic Church   United Methodist Church   Other

f. Which graph (bar or pie) do you think better represents this data? Justify your choice.   Answers will vary.

6. A student teacher has the grade for each of her 22 students as a percent. Explain why a pie graph should not be used to represent this data.

Pie graphs are not used for quantitative data.

7. The graph that follows represents the mortgage rates on a 30-year fixed-rate mortgage for several weeks of a recent year.

a. Write at least one complete sentence to describe the general shape of the graph and what it tells you about mortgage rates.

Mortgage rates generally decreased during April and May, but started to rise at the beginning of June.

b. Estimate when mortgage rates reached their highest value in the time period shown.

Approximately 4/7

c. Estimate when mortgage rates reached their lowest value in the time period shown.

Approximately 6/8

d. Give the ordered pair for the point indicated by the arrow. Write a sentence to interpret the point in the context of mortgage rates.

(5/1, 4.91); on 5/1/15, the interest rate on a 30-year fixed-rate mortgage was 4.91%.

e. If the graph had been drawn with a vertical scale of $0 - 6\%$ instead of $4.60 - 5.10\%$, how would its appearance change? Would this new scale change your impression of the interest rate behavior during these months?

With the new scale, the interest rates would appear to change less over these months, since the vertical changes would be diminished with the broader vertical scale.

8. The graph that follows shows the unemployment rate in a city for many recent years.

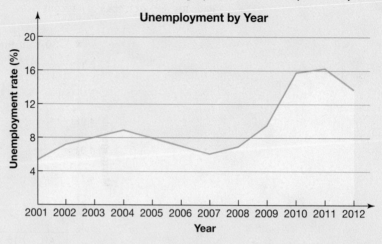

Unemployment by Year

a. Describe the general trend in the unemployment rate suggested by this graph.

The unemployment rate increased from 2001 to 2004, then decreased from 2004 to 2007. The unemployment rate rose sharply until 2010, when it began to level off and then decrease.

b. In which year (of those shown) was the unemployment rate the highest? What was the rate that year (to the nearest whole percent)? Illustrate this information with an appropriate ordered pair from the graph.

2011; 16%; (2011, 16)

c. In which year (of those shown) was the unemployment rate the lowest? What was the rate that year? Illustrate this information with an appropriate ordered pair from the graph.

2001; approximately 5%; (2001, 5)

d. If the city has a population of around 150,000 people (for all the years shown in the graph), what is the difference in the number of unemployed individuals between the lowest unemployment rate and the highest unemployment rate?

16,500 people

e. Could this information have been displayed with a bar graph instead? Why do you think the author chose this particular style of graph to display the data?

A bar graph could have been used instead, but the line graph smooths out the changes and illustrates a trend in a way that bar graphs do not.

9. Looking Forward, Looking Back Plot the following points:

A. $(-2.75, 1)$

B. $(0, -0.8)$

C. $\left(1, -\dfrac{5}{2}\right)$

D. $(2.3, 0)$

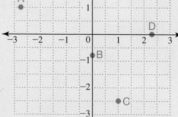

1.10 Two by Two: Scatterplots

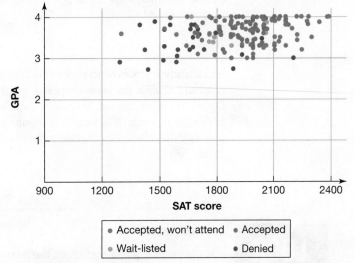

SECTION OBJECTIVES:
- Create and interpret scatterplots
- Sketch a curve to best fit a scatterplot

Explore

1. The following graph shows high school GPAs and SAT scores for students who applied to a particular university. Each dot represents a student and is color coded to indicate the student's admission status.

Data from "Trinity University GPA, SAT, and ACT Data," Allen Grove, *About Education*

a. What do the blue dots represent and where do they seem to be clustered on the graph? Explain why that makes sense.

The blue dots represent students who were denied admission. They tend to be clustered in the lower left corner of the cluster of points, representing students with lower test scores and lower GPAs.

b. What do the green and purple dots represent and where do they seem to be clustered on the graph? Explain why that makes sense. Are there any green or purple dots that seem to contradict the trend?

The green and purple dots represent students who were accepted, whether they attended or not. They tend to be clustered in the upper right corner of the graph, representing students with high test scores and high GPAs. There are some green and purple dots that contradict this trend. For example, there is a green dot which represents a lower test score of approximately 1,300.

c. Notice that there are blue and orange dots representing students who were waitlisted or denied admission behind some of the green and purple dots. Explain what this means.

Some students were denied admission even though their scores were as high as those of other students who were admitted. Clearly this school considers other criteria in the admission process, such as interviews, letters of recommendation, and extracurricular activities.

Discover

This section will focus on making and interpreting scatterplots, the type of graph shown in the *Explore*.

As you progress through this section, think about these questions:

1 When is a scatterplot the best graph to create?

2 Do you know how to set up a scale to illustrate data on a scatterplot?

3 Can you interpret a scatterplot that someone else has created and draw reasonable conclusions?

You already know how to make and interpret bar and pie graphs, but not all data can be represented with these two types of graphs. Sometimes a situation, like the one in the *Explore*, involves related or paired data such as height/weight, study time/grades, or education/income. If a situation involves paired data and you wish to investigate or illustrate the relationship between the two variables, then a scatterplot is a useful graph to make.

LOOK IT UP

Scatterplot

A **scatterplot** is a graph that shows the relationship between two quantitative variables. The horizontal axis is used to represent one variable, and the vertical axis is used to represent the other variable. The data values are plotted as ordered pairs, but the points are not connected as they are in a line graph.

FOR EXAMPLE, the following scatterplot shows the relationship between horsepower and highway mileage for a selection of vehicles. It appears that vehicles with more horse-power get lower mileage on the highway.

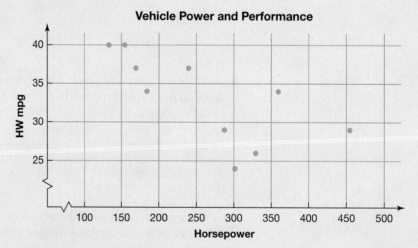

Vehicle Power and Performance

Notice that neither axis starts at zero. This is indicated by a jagged start to each axis, which shows a break in the scale.

To make a scatterplot, you first need to decide which variable will be plotted on the horizontal axis and which will be plotted on the vertical axis for the given data. Sometimes this will be decided for you. If it's not, think about which variable depends on the other. In the mileage and horsepower example shown in the *Look It Up*, it makes sense to graph the mileage on the vertical axis, since it is affected by, or depends on, horsepower.

Independent and Dependent Variables

When two variables are paired, the one that is thought to depend on the other is called the **dependent variable**. The other is called the **independent variable**. This distinction is often a matter of perspective.

FOR EXAMPLE, suppose you are driving 60 mph. If you travel for 1 hour, then you will go 60 miles. If you travel for 2 hours, then you will go 120 miles. In this sense, your distance depends on the time. You might consider distance to be the dependent variable.

However, you could also say that in order to go 60 miles, you need to drive for 1 hour. From this perspective, travel time could be considered the dependent variable.

In a scatterplot, the independent variable is plotted on the horizontal axis and the dependent variable is plotted on the vertical axis. It makes sense to plot the value of the independent variable (x) first and then the value of the dependent variable (y).

INSTRUCTOR NOTE: Consider asking students to identify the independent and dependent variables as well as the scale used in the *Explore* graph.

After choosing which variable is independent and which is dependent, a **scale** needs to be determined for each axis. To determine a scale, consider the smallest and largest numbers (approximately) that need to be graphed for each variable, and then choose a consistent **increment**, or step, to cover the distance between them. For example, you might need to show numbers from 100 to 250 on an axis and choose to use an increment of 25 to cover that range. To show this scale, you draw tick marks at 100, 125, 150, and so on, until you reach 250.

Once the axes and scales are determined, plot each pair of values as an ordered pair on the graph. The value of the independent variable is the first number in the ordered pair, and the value of the dependent variable is the second number in the ordered pair.

EXAMPLE

Create a scatterplot to represent the following data. GDP stands for gross domestic product and is given in international dollars.

Country	GDP per Capita	Percent Using Social Media
Israel	30,000	53
United States	48,000	50
Britain	36,000	43
Russia	17,000	43
Poland	20,000	39
Germany	38,000	35
China	9,000	32
Ukraine	7,000	30
Egypt	6,000	28
Japan	35,000	25
Mexico	15,000	22
India	4,000	12
Pakistan	3,000	2

Data from www.pewglobal.org

SOLUTION

It makes sense to think that the percent of the population on social media depends, at least in part, on the wealth of the country. So we can decide to make GDP the independent variable and place it on the horizontal axis, leaving percent using social media for the vertical axis. The following choices for the minimum, maximum, and increment are not the only possible choices, but they are reasonable for the numbers in the table. We want the minimum to be at or below the smallest value shown in the table and the maximum to be at or above the largest value. Since our choices will affect the look of the scatterplot, it is important to make reasonable choices that will display the data well.

Variable on the x-axis:	GDP per Capita	Variable on the y-axis:	Percent Using Social Media
Minimum value:	0	Minimum value:	0
Maximum value:	60,000	Maximum value:	60
Increment:	5,000	Increment:	5

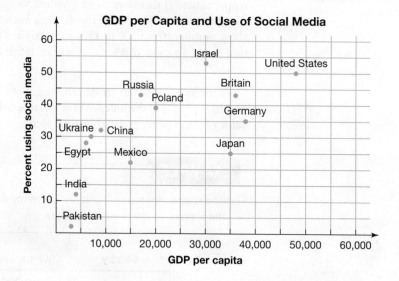

We can see from the scatterplot that countries with higher GDPs tend to have a larger percent of the population using social media.

Next, you will use the procedure shown in the example to create a scatterplot with a new data set.

2. Consider the following data on years of experience and salaries for the employees of a small business:

Experience (years)	0	1	3	4	5	5	11	15	16	23
Salary (dollars)	48,000	67,000	54,000	58,000	62,000	64,000	74,000	81,000	85,000	92,000

Identify the following details to prepare for making a scatterplot. Keep in mind that there is no single correct answer for the minimum/maximum values and increment.

Since your choices will affect the look of your scatterplot, it is important to make reasonable choices that will display the data clearly.

Independent variable:	Experience	Dependent variable:	Salary
Axis:	Horizontal	Axis:	Vertical
Minimum value:	0	Minimum value:	40,000
Maximum value:	25	Maximum value:	100,000
Increment:	5	Increment:	10,000

Now draw the scatterplot. What conclusions can be reached from this scatterplot?

Possible scatterplot:

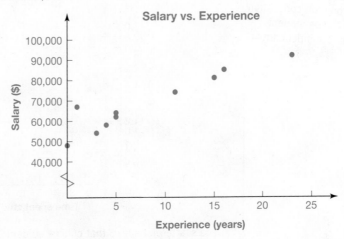

Salaries tend to increase as experience increases.

Tech TIP

A graphing calculator can be used to make a scatterplot. Access the STAT menu and choose EDIT. Enter the data into lists. Then use the STAT PLOT menu to make the graph.

HOW IT WORKS

To make a scatterplot:

1. Decide which variable is the independent variable and which is the dependent variable in the paired data.
2. Draw axes and label each axis with the appropriate variable name. Title the graph.
3. Choose a minimum and maximum value to be graphed on each axis by considering the range of data values. If an axis does not start at zero, indicate this with a jagged start.
4. Choose a convenient increment for each axis.
5. Plot the data as ordered pairs, with the value of the independent variable first.
6. Do not connect the points. The dots will not necessarily fall in a line or other pattern.

EXAMPLE: Make a scatterplot to represent the following data:

Time Spent Studying per Week (hours)	College GPA
2	1.8
4	2.2
7	2.4
10	3.2
14	3.8

(continued)

HOW IT WORKS

It makes sense to consider the time spent studying as the independent variable and graph it on the horizontal axis. College GPA is then the dependent variable and graphed on the vertical axis.

For the time spent studying, the scale can go from 0 to 15 hours with an increment of 1 hour. For college GPA, the scale can go from 0 to 4, since that represents the full range of possibilities, or it can be shortened to show just the data in the table. A reasonable increment for this scale might be 0.2.

Remember to title your graph. A common convention is to state the dependent variable vs. the independent variable.

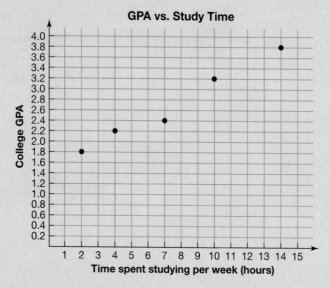

The scatterplot shows that college students who spend more hours per week studying tend to have higher GPA's.

Mathematicians often look at scatterplots to try to find a general pattern that the data points seem to follow, even if it is not a perfect fit. This process of choosing an appropriate graph shape to represent the trend in the scatterplot is one type of mathematical **modeling**.

3. For each of the following scatterplots, sketch a smooth curve through the points to indicate the **model** you think the data points seem to follow. The line or curve should look like it passes through the center of the cloud of points; it does not necessarily need to go through each, or any, point on the graph.

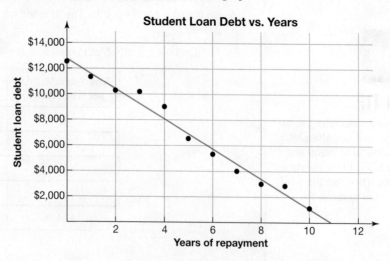

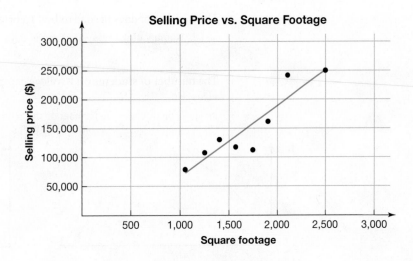

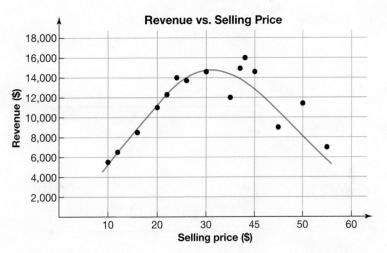

It's also useful to be able to predict the shape of a graph from a description of the situation when you do not have any points to plot. To do this, first think about which variable depends on the other, and then consider how it depends on the other variable. Does that variable increase as the other increases? Decrease? Does that variable increase steadily or level off?

4. Match each description to the graph that you think best illustrates it. Identify the independent and dependent variables in each case.

a. You save the same amount of money each month in a non-interest-bearing account until you reach your goal. Then you leave the account at that balance.

vi;   I: Time in months   D: Account balance

b. Average monthly temperatures in a given location

v;   I: Time in months   D: Temperature

c. A population of bacteria doubles every hour.

iii;   I: Time in hours   D: Population

d. The height of a ball each second after it's shot from a catapult

i;   I: Time in seconds   D: Height

e. The number of days needed to build a bridge compared to the number of workers

iv;   I: Number of workers   D: Time in days

f. The number of students on a college campus who contract a highly contagious flu virus

ii;   I: Time in days   D: Number of students

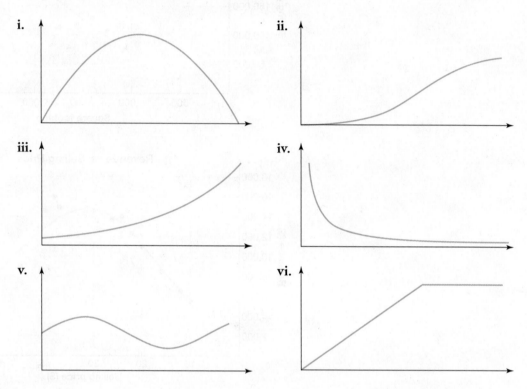

i.   **ii.**

iii.   **iv.**

v.   **vi.**

Connect

The ratio of elderly people to working-age people in the United States has been increasing for many years. This is a potentially serious problem, since working people tend to pay taxes that support services for the elderly. The ratio of elderly people to working-age people is affected by both increasing longevity and decreasing birth rates.

5. Consider the following scatterplots, which show the number of kids per woman versus the number of elderly per 100 working-age adults for various countries in 1960 and 2015. The orange dot represents the United States.

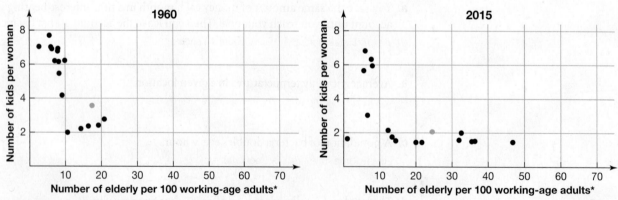

1960 — Number of kids per woman vs. Number of elderly per 100 working-age adults*

2015 — Number of kids per woman vs. Number of elderly per 100 working-age adults*

*Elderly refers to 65 and over. Working-age refers to people age 20–64.

Data from www.npr.org

a. How did the United States compare to the other countries shown in 1960?

Compared to the other countries in 1960, the United States had a relatively high number of elderly per 100 working-age adults, about 17.5. The number of kids per woman was in the lower half, at 3.6.

b. How had these measures changed for the United States by 2015?

By 2015, the number of elderly per 100 working-age adults in the United States had increased to about 25, which is closer to the average for the countries shown. The number of kids per woman had dropped to 2.

c. Describe the change in the scatterplot from 1960 to 2015.

The cloud of points is less vertical and more horizontal in 2015. The 2015 graph shows that many countries had a higher ratio of elderly to working-age adults and a lower number of kids per woman than in 1960.

Reflect

WRAP-UP

What's the point?

Scatterplots are often seen in the news because they are so useful for showing the relationship between two variables. You should be able to draw a scatterplot, as well as interpret one and identify the general shape or trend shown in a graph.

What did you learn?

How to create and interpret scatterplots
How to sketch a curve to best fit a scatterplot

INSTRUCTOR NOTE: Remind students to read the *Getting Ready* article for the next section.

1.10 Homework

Skills MyMathLab

First complete the MyMathLab homework online. Then work the two exercises to check your understanding.

1. From each verbal description, identify the independent and dependent variables.

a. A tree grows approximately 6 inches per year and is 10 feet tall when planted.

Independent variable: Time in years

Dependent variable: Tree height in feet

b. When a swimmer gets 8 hours of sleep the previous night, she is able to complete a 200-meter swim in 3 minutes. When she gets only 6 hours of sleep, it takes her 3 minutes and 16 seconds to complete the same swim.

Independent variable: Amount of sleep in hours

Dependent variable: Swim time in minutes

2. Which of the following graphs best shows how the number of hours of daylight changes over the days of a year in a midwestern state in the United States?   C

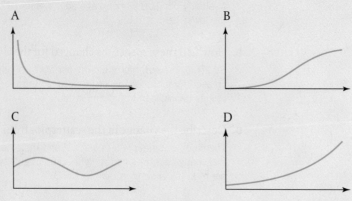

Identify the independent and dependent variables in the graphs.

Independent: Day of the year

Dependent: Number of hours of daylight

Concepts and Applications

Complete the problems to practice applying the skills and concepts learned in the section.

3. The number of times per day that students check Facebook seems to have a negative effect on the amount of time they spend preparing for class. Consider the following data:

Number of Times Facebook Checked per Day	Time Spent Preparing for College Classes per Day (minutes)
0	120
3	95
5	90
6	80
6	60
7	70
8	40
9	20
10	10

Provide the following information to prepare for making a scatterplot. Keep in mind that there is no single correct answer for the minimum/maximum values and increment. Since your choices will affect the look of your scatterplot, it is important to make reasonable choices that will display the data clearly.

Independent variable: Number of Facebook checks   Dependent variable: Time spent preparing for class

Axis: Horizontal   Axis: Vertical

Minimum value: 0   Minimum value: 0

Maximum value: 12   Maximum value: 120

Increment: 2   Increment: 10

Draw the scatterplot.

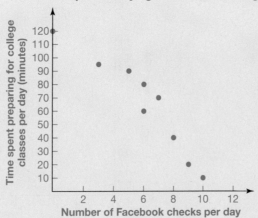

Time Spent Studying vs. Facebook Usage

What conclusion can you draw from this scatterplot?

Preparation time tends to decrease as the number of Facebook checks increases.

4. High school GPA is well known to be a good predictor of college GPA. Consider the following data:

HS GPA	College GPA
3.9	3.7
3.6	3.5
3.5	3.6
3.1	3.0
2.5	2.0
2.3	2.2
2.1	1.8
1.9	1.5
1.5	1.0

a. Provide the following information to prepare for making a scatterplot. Keep in mind that there is no single correct answer for the minimum/maximum values and increment. Since your choices will affect the look of your scatterplot, it is important to make reasonable choices that will display the data clearly.

Independent variable:	HS GPA		Dependent variable:	College GPA
Axis:	Horizontal		Axis:	Vertical
Minimum value:	0		Minimum value:	0
Maximum value:	4		Maximum value:	4
Increment:	0.2		Increment:	0.2

b. Draw the scatterplot.

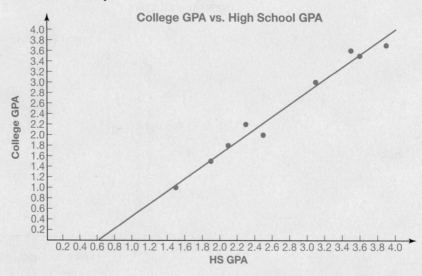

c. Draw a line or curve on the scatterplot to indicate the trend you think the data points follow.

d. What conclusion can you draw from this scatterplot?

The higher your HS GPA, the higher your college GPA is likely to be.

5. Consider the following data:

Age (years)	Weight (pounds)
21	100
22	112
24	110
25	130
28	145
35	146
37	148

What is wrong with the following scatterplot that was made to represent these data? How could you fix it?

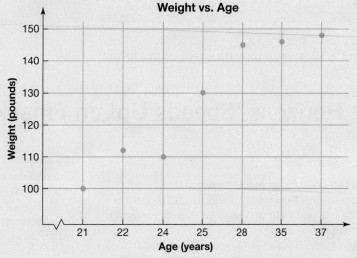

Weight vs. Age

The vertical axis should show a jagged start, since it begins at 100 and not 0. The scale on the horizontal axis is not consistent. The first increment between tick marks is 1, while the second is 2. One way to fix the scale on the horizontal axis would be to show a jagged start and begin at 20 with an increment of 2.

6. Use the predicted data on the ratio of elderly to 100 working-age adults and number of kids per woman in 2060 to make a scatterplot. Compare the scatterplot to the one shown in the *Connect* for 2015.

Number of Elderly per 100 Working-Age Adults	Number of Kids per Woman
6.8	3.94
7.6	3.53
8.6	3.51
9.8	2.42
22.6	1.95
41.1	2.09
42.3	1.75
48.4	2.07
47.5	1.76
56.7	1.72
57.5	1.95
56.5	1.86
61.4	1.94
64.2	1.94
68.7	1.83
70.5	1.91
75.6	1.92

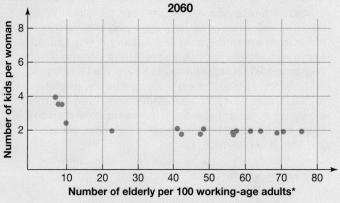

2060

*Elderly refers to 65 and over. Working-age refers to people age 20–64.

In 2060, the predictions show that many more countries will have high ratios of elderly to working-age adults and low ratios of children per woman.

7. **Looking Forward, Looking Back** Scale 60 mph to miles per minute.

1 mile per minute

Getting Ready for Section 1.11

Read the article below and answer the questions that follow.

New Horizons 'Speeds Up' on Final Approach to Pluto

Last Updated: July 30, 2015
Editor: Lillian Gipson

With just two weeks to go before its historic July 14 flight past Pluto, NASA's New Horizons spacecraft tapped the accelerator late last night and tweaked its path toward the Pluto system.

The 23-second thruster burst was the third and final planned targeting maneuver of New Horizons' approach phase to Pluto; it was also the smallest of the nine course corrections since New Horizons launched in January 2006. It bumped the spacecraft's velocity by just 27 centimeters per second—about one-half mile per hour—slightly adjusting its arrival time and position at a flyby close-approach target point approximately 7,750 miles (12,500 kilometers) above Pluto's surface.

While it may appear to be a minute adjustment for a spacecraft moving 32,500 miles per hour, the impact is significant. New Horizons Mission Design Lead Yanping Guo, of the Johns Hopkins Applied Physics Laboratory in Laurel, Maryland, says without the adjustment, New Horizons would have arrived 20 seconds late and 114 miles (184 kilometers) off-target from the spot where it will measure the properties of Pluto's atmosphere. Those measurements depend on radio signals being sent from Earth to New Horizons at precise times as the spacecraft flies through the shadows of Pluto and Pluto's largest moon, Charon.

In fact, timing and accuracy are critical for *all* New Horizons flyby observations, since those commands are stored in the spacecraft's computers and programmed to "execute" at exact times.

This latest shift was based on radio-tracking data on the spacecraft and range-to-Pluto measurements made by optical-navigation imaging of the Pluto system taken by New Horizons in recent weeks. Using commands transmitted to the spacecraft on June 28, the thrusters began firing at 11:01 p.m. EDT on June 29 and stopped 23 seconds later. Telemetry indicating the spacecraft was healthy and that the maneuver went as designed began reaching the New Horizons Mission Operations Center at APL, through NASA's Deep Space Network at 5:30 a.m. EDT on June 30.

"We are really on the final path," said New Horizons Project Manager Glen Fountain, of APL. "It just gets better and more exciting every day."

"This maneuver was perfectly performed by the spacecraft and its operations team," added mission principal investigator Alan Stern, of Southwest Research Institute, Boulder, Colorado. "Now we're set to fly right down the middle of the optimal approach corridor."

New Horizons is now about 10 million miles (16 million kilometers) from the Pluto system—some 2.95 billion miles (4.75 billion kilometers) from Earth.

Source: www.nasa.gov

Questions

1. The article mentions that New Horizons's thruster burst increased its velocity by 27 centimeters per second or one-half mile per hour. Convert each of these speeds to feet per minute. Why are they not exactly the same?

   About 53 feet per minute; about 44 feet per minute; because the values stated in the article have probably been rounded

2. Assuming New Horizons encountered Pluto as planned after 3,462 days, what was its average speed (in miles per hour) to cover the approximately 3 billion miles?

   36,106 mph

1.11 Multiply vs. Divide: Converting Units

Explore

SECTION OBJECTIVE:

- Convert units by multiplying or dividing

You might need to Google a conversion fact if you don't know how many MB are in a GB.

1. A cell phone carrier offers a family plan that includes 10 GB of data to be shared on four lines for $100/month. If the family goes over the 10-GB limit, they are charged 2.5 cents/MB. If last month the family used 10.5 GB of data, would they have been better off with a plan for 12 GB at $110/month?

The family went over by 0.5 GB. Since 1 GB is 1,024 MB, they went over by 512 MB. This will cost them $12.80. They would have been better off paying $110 for 12 GB, since their bill is now $112.80.

Discover

INSTRUCTOR NOTE: This section will focus on multiplying and dividing to convert units. Dimensional analysis will be briefly mentioned in the *Connect*, and will be developed in detail in Section 4.2.

The *Explore* involves a conversion between units. There is more than one way to convert units, but the most common method used in daily life involves multiplying or dividing using a conversion fact. This method can be challenging if there are several conversions to be made or if the context makes it difficult to know which operation to use. The problems we will solve in this cycle have a manageable number of conversions and difficulty level. So for now, we will focus on understanding when we should multiply or divide to convert.

As you work through this section, think about these questions:

1 Do you know where or how to find a conversion fact when you need one?

2 How do you decide whether to multiply or divide when converting units?

We will begin by finding a relevant conversion fact and trying to understand the relationship it states, using a picture if necessary. We will then determine the operation needed for the conversion and calculate the answer. Finally, we will check to make sure the answer is reasonable.

EXAMPLE 1

John has decided to run a 10K race. How many miles will he be running?

SOLUTION

This conversion of 10 kilometers to miles can be done in different ways, depending on which conversion fact you use. We will show two possible solutions, one in each column.

1. **Find a relationship between the units:**

$$1 \text{ km} \approx 0.62 \text{ mi} \qquad\qquad\qquad 1 \text{ mi} \approx 1.61 \text{ km}$$

2. **Understand the relationship:**

Draw a picture and label it with the conversion fact.	Draw a picture and label it with the conversion fact.

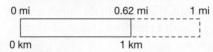

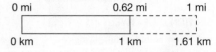

Notice from the picture that 1 mile is larger than 1 kilometer. Since the units in the original problem are kilometers, let's look at it from that perspective. The fact says for every 1 km, we get 0.62 mi.	The fact says that for every 1 mile, we get 1.61 kilometers. Since the units in the original problem are kilometers, let's look at it from that perspective. The fact says that for every 1.61 km, we get 1 mi.
Since we have 10 km, we need 10 groups of 0.62 mi.	We need to determine how many 1.61 km are in the 10 km.

3. **Decide whether to multiply or divide.**

Since each km is 0.62 mi, we need to multiply 10 by 0.62.	Since we need to know how many times 1.61 goes into 10, we need to divide 10 by 1.61.

4. **Calculate the answer.**

$$10(0.62) = 6.2 \text{ mi}$$	$$\dfrac{10}{1.61} \approx 6.2 \text{ mi}$$

5. **Check if the answer is reasonable.**

Since miles are longer than kilometers, it takes fewer of them to cover the same distance.

HOW IT WORKS

To convert units:

1. Find a relationship between the units. This may involve one or more conversion facts.
2. Understand the relationship. Use a picture if needed.
3. Determine whether to multiply or divide.
4. Calculate the answer.
5. Check to ensure that the answer is reasonable.

EXAMPLE: Convert 120 pounds to kilograms.

Begin by finding a conversion fact: 1 kg ≈ 2.2 lb. So, for every kilogram, there are 2.2 pounds. Or for every 2.2 pounds, there is 1 kilogram. Determine how many 2.2's are in 120. That indicates division.

$$\frac{120}{2.2} \approx 54.5 \text{ kg}$$

Kilograms are a larger unit of weight than pounds, so it makes sense that there are fewer of them.

Need more practice?

1. Convert 80 ounces to pounds.
   Answer: 5 pounds

2. Convert 15 pounds to ounces.
   Answer: 240 ounces

The following list shows some commonly used conversion facts that will be useful in this section and the homework. However, you should not hesitate to simply Google a conversion fact when you need one.

1 ft = 12 in.	1 in. = 2.54 cm	1 L = 1,000 mL	1 min = 60 sec
1 mi = 5,280 ft	1 km = 1,000 m	1 lb = 16 oz	1 hr = 60 min
1 yd = 3 ft	1 m = 100 cm	1 kg = 1,000 g	1 yr = 52 weeks
1 km ≈ 0.62 mi	1 cm = 10 mm	1 g = 1,000 mg	1 week = 7 days
1 mi ≈ 1.61 km		1 ton = 2,000 lb	
		1 gal = 4 qt	
		1 kg ≈ 2.2 lb	

2. If you are 5′6″, are you 5.6 feet tall? How tall are you in inches?

No, 5′6″ is not 5.6 feet. 6 inches = 0.5 foot, so you are 5.5 feet or 66 inches tall.

INSTRUCTOR NOTE: Tell students that the difference in minutes is easier to find if they convert both times to minutes first instead of hours.

3. You are going to a movie that is 189 minutes long. Your friend is seeing a movie that starts at the same time, but is 2 hours, 10 minutes long. Who will be finished first? By how many minutes?

Your friend's movie is 130 minutes long. Your friend will be finished 59 minutes sooner.

In the next example, you will see how scaling can be used as a way to convert units.

EXAMPLE 2

Convert 300 millimeters to meters.

SOLUTION

If we do not have a conversion fact relating millimeters and meters, we can convert millimeters to centimeters first. We can then use scaling to change the conversion fact.

10 mm = 1 cm	Start with a conversion fact you know.
1,000 mm = 100 cm	Multiply both sides by 100 to get 100 cm.
1,000 mm = 1 m	100 cm is the same as 1 m.
1 mm = 0.001 m	Since it is not easy to scale 1,000 down to 300, divide by 1,000 to scale to 1.
300 mm = 0.3 m	Multiply both quantities by 300 to get 300 mm.

Since metric units differ by powers of ten, the decimal point placement is the only thing that changes when you convert metric units. Another way to convert metric units is to memorize the order of the metric prefixes and the direction that the decimal point should move.

4. a. Convert 4,650 milliliters to liters.

4,650 mL = 4.65 L

b. Convert 5 kg to g and then to mg.

5 kg = 5,000 g = 5,000,000 mg

c. Convert 500 mg/hour to mg/day and then to g/day.

500 mg/hour = 12,000 mg/day = 12 g/day

You won't always be told exactly which conversion you need to make as you were in the last problem. Sometimes the conversion that needs to be done is not obvious but a little more subtle, as is the case in the next example. In this example, pay particular attention to the thought process and explanation, not just the calculations that are done. This type of thinking will help you solve the focus problem for the cycle.

INSTRUCTOR NOTE: Consider having students work the problem in Example 3 a different way by converting the range of 75–150 mg/kg per day to different units.

EXAMPLE 3

A doctor orders 200 mg of an antibiotic for an infant who weighs 15.4 lb. The medication should be taken every 8 hours. The medication label shows that 75–150 mg/kg per day is the appropriate dosage range. Determine if the dosage ordered is within the appropriate range.

SOLUTION

Since the dose is given every 8 hours, the infant will receive 3 doses in a 24-hour period. Three 200-milligram doses will amount to 600 mg of the antibiotic per day.

$$3 \text{ doses} \cdot \frac{200 \text{ mg}}{1 \text{ dose}} = 600 \text{ mg}$$

The infant weighs 15.4 lb and will get 600 mg per day. Since the dosage information on the label is shown for kilograms, not pounds, we next need to convert the infant's weight to kilograms.

For every 2.2 pounds, there is 1 kilogram. Since we need to know how many kilograms are in 15.4 pounds, we need to determine how many 2.2's are in 15.4. So we should divide 15.4 by 2.2.

$$15.4 \text{ lb} \cdot \frac{1 \text{ kg}}{2.2 \text{ lb}} = 7 \text{ kg}$$

To check the dosing against the label, we need to know how many milligrams the infant is getting per kilogram of weight. So we need the ratio of mg to kg.

$$\frac{600 \text{ mg}}{7 \text{ kg}} \approx 85.7 \text{ mg/kg}$$

Since 85.7 is in the range 75–150 mg/kg, the infant has been appropriately dosed.

It is more comfortable to be in a class where the content is not challenging, but then little learning takes place. Learning something new is often not easy and can be uncomfortable. This is normal and to be expected. Think about a flower that, when it tries to grow, has to push its roots through the soil. To do that, the soil has to be upturned and disrupted.

Connect

So far this section has stressed using logic and unit sense to decide whether to multiply or divide when doing a unit conversion. Although this is a great way to begin and is extremely practical, it does have its limitations as conversions get more complicated. In Example 3, the conversions were also shown with a new technique called **dimensional analysis**, which involves multiplying by conversion facts written as fractions. The unit you start with and don't want divides off with the denominator of the fraction, and you are left with the new unit that you want. You can use a string of such conversion factors to do a multistep conversion. Try this technique on the next problem.

Need more practice?

1. Convert 8 meters to centimeters and then inches.
   Answer: 800 cm; about 315 in.

2. Convert 50 km/hr to ft/min.
   Answer: 2,728 ft/min

5. Convert 2.5 miles to yards by changing the miles to feet and then the feet to yards.

$$2.5 \text{ mi} \cdot \frac{5{,}280 \text{ ft}}{1 \text{ mi}} \cdot \frac{1 \text{ yd}}{3 \text{ ft}} = 4{,}400 \text{ yd}$$

Reflect

WRAP-UP

What's the point?

When people do simple unit conversions, they don't usually write out a long process. Instead, they typically focus on whether they should multiply or divide to get an answer that makes sense. You can always look up a conversion fact if you need one and then use this logic.

What did you learn?

How to convert units by multiplying or dividing

1.11 Homework

Skills MyMathLab

First complete the MyMathLab homework online. Then work the two exercises to check your understanding.

1. A pill contains 1,257 mg of an active ingredient. How many grams is that?
   1.257 grams

2. Convert 12.5 miles to inches.
   792,000 inches

Concepts and Applications

Complete the problems to practice applying the skills and concepts learned in the section.

3. If you are converting from a large unit to a smaller unit, will you always divide? Explain and include an example.

   No; it depends on the conversion fact you are using. For example, to convert 3 miles to kilometers using the fact 1 km ≈ 0.62 mi, you divide by 0.62. However, to convert 3 miles to kilometers using the fact 1 mi ≈ 1.61 km, you multiply by 1.61.

4. a. Convert 30 days to seconds by first converting days to hours, then hours to minutes, then minutes to seconds.

30 days = 30 days (24 hours/day) = 720 hours

720 hours = 720 hours (60 minutes/hour) = 43,200 minutes

43,200 minutes = 43,200 minutes (60 seconds/minute) = 2,592,000 seconds

b. Convert 30 days directly to seconds using the fact that each day contains 86,400 seconds.

30 days = 30 days (86,400 seconds/day) = 2,592,000 seconds

c. What do you notice about your answers to parts a and b?   They are the same.

5. a. If you are converting from a smaller unit to a larger unit, the number will get _____smaller_____.

b. If you are converting from a larger unit to a smaller unit, the number will get _____larger_____.

6. a. If there are exactly 2.54 centimeters in 1 inch, how many inches are in 1 centimeter? Round to two decimal places.   0.39 inch

b. If 1 mile is 5,280 feet, how many miles are in 1 foot? Round to six decimal places.
0.000189 mile

7. One desk requires 84 inches of a type of trim. Find the length of trim needed in inches for 5 desks. Then convert your answer to a potentially more useful unit.
420 inches; 35 feet

8. If you are driving 70 mph and you look down for 3 seconds, how far have you driven in that time?   308 feet

9. If a hose is filling a swimming pool at a rate of 20 liters per minute, how long will it take to fill the 3,600-gallon pool?   Approximately 11 hours and 21 minutes

10. Convert New Horizons's speed when it left our atmosphere, 10 miles per second, to miles per hour to get a better sense of how fast the spacecraft was traveling.
36,000 mph

11. Looking Forward, Looking Back Use scaling to change 35% to a simplified fraction.
$35\% = \dfrac{35}{100} = \dfrac{7}{20}$

1.12 Up and Down: Percent Change

Explore

SECTION OBJECTIVES:
- Apply a percent change
- Find percent change
- Interpret percent change

1. The ticket price for a one-day admission has steadily increased each year at many theme parks, including Disneyland in Anaheim, California. Use the line graph below to answer the following questions.

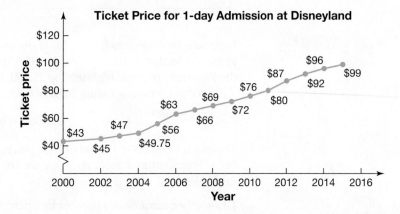

Ticket Price for 1-day Admission at Disneyland

Data from www.ocregister.com; this source also sourced Disney.com for ticket prices

  a. By what percent did the ticket price increase from 2014 to 2015?   3%

  b. Use your answer from part a to forecast the ticket price in 2016 to the nearest dollar.
  $102

  c. Between 2000 and 2015, by what percent did the ticket price increase?   130%

Discover

INSTRUCTOR NOTE: Percent change is shown in this section using a one-step and a two-step method. Encourage students to get comfortable with the one-step approach as it will be useful when they learn how to generalize sequences and write functions.

There are many situations in which we need to put a change into perspective. Percents give us one way to do this and allow us to make convenient comparisons. However, since it is easy to misinterpret percents, care must be taken. This section will explore several aspects of percent change.

As you work through this section, think about these questions:

1 Can you increase or decrease a number by a percent in one step?

2 Is a 200% increase the same as doubling or tripling a quantity?

3 Do you know why changes are often stated as percent change?

Let's begin by looking at ways to increase a number by a percent. If we want to increase 25,000 by 5% in two steps, we take 5% of the base quantity, 25,000, and then add the result to 25,000. To find 5% percent of 25,000, we multiply 25,000 by 0.05.

Step 1: $25{,}000(0.05) = 1{,}250$

Step 2: $25{,}000 + 1{,}250 = 26{,}250$

The following picture verifies that our answer is reasonable. The whole rectangle (100%) represents 25,000, since 25,000 is the whole for which we are finding a part. We can see that 1,250 is a reasonable result for 5% of 25,000. The new amount after the increase is 26,250.

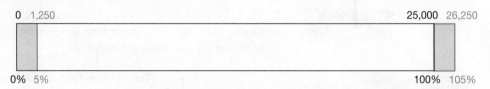

Increasing by 5% means finding 5% of the original whole (100%) and then adding this 5% to the whole; that is, 100% + 5% = 105%. In the picture, this is shown by the addition of the 5% to the rectangle representing 25,000. Finding the percent and adding it to the base are two steps, whereas finding 105% of the base is equivalent but requires only one step.

$$25{,}000(1.05) = 26{,}250$$

Therefore, to increase a number by a percent in one step, we can multiply the number by (1 + percent as a decimal). Since we are *multiplying* by this number, it is known as a **multiplier**.

Likewise, we can *decrease* a number by a percent using either two steps or one step.

EXAMPLE 1

Decrease $100 by 5% by using two steps and by using one step.

SOLUTION

To use the two-step approach, we begin by finding 5% of $100 and then subtracting.

$$\$100(0.05) = \$5 \quad \text{Find the amount of decrease}$$
$$\$100 - \$5 = \$95 \quad \text{Subtract the amount}$$

We began with a whole (100%) and decreased it by 5%, which is the same as subtracting 5%. So, 100% − 5% = 95%. To find the new amount after the decrease in one step, we find 95% of the whole. The multiplier is 0.95.

$$\$100(0.95) = \$95$$

In general, to decrease a number by a percent in one step, multiply by (1-percent as a decimal).

> For a percent increase, the multiplier is greater than 1. For a percent decrease, the multiplier is less than 1.

2. Find the multiplier for each percent change.

   a. Increase by 22%   1.22

   b. Decrease by 22%   0.78

HOW IT WORKS

Need more practice?

Olivia will receive a 2.5% cost-of-living increase in her salary. What will she make after the raise if she makes $41,000 now?

Answer: $42,025

To apply a percent increase:

Option 1: 1. Find the percent of the number.   **Option 2:** Multiply the number by
2. Add this amount to the number.   (1 + percent as a decimal)

EXAMPLE: A company's executives expect a 2% increase in sales in the upcoming year. If the company had $62,000 in sales last year, what can the executives expect in the following year?

Find 2% of $62,000 and then add it:	(0.02)($62,000) = $1,240	$62,000 + $1,240 = $63,240
Alternatively, find 102% of $62,000:	(1.02)($62,000) = $63,240	

To apply a percent decrease:

Option 1: 1. Find the percent of the number.   **Option 2:** Multiply the number by
2. Subtract this amount from the number.   (1 − percent as a decimal)

EXAMPLE: The average price of a gallon of milk is $2.79, and it is expected to drop 4% in the next year. What will the average price be in one year?

Find 4% of $2.79 and then subtract it:	(0.04)($2.79) = $0.11	$2.79 − $0.11 = $2.68
Alternatively, find 96% of $2.79:	(0.96)($2.97) = $2.68	

While you might have the choice between using the one-step or two-step method, there will be times when the one-step method is preferable or even necessary. The sooner you can get comfortable with the one-step method, the easier it will be to work with percent change.

Even if you are not in a labor union, you might be presented with more than one pay structure when offered a job. Many employees negotiate their salaries on an individual basis. Being able to determine which structure is more beneficial is a valuable skill.

3. At your job, there are two proposed pay raise structures being offered by management to the labor union.

> **Option 1:**   5% increase
>
> **Option 2:**   3.5% increase and then a $1,000 bonus

The labor union has to vote on which to use, but different members of the union have different opinions based on the effect the options will have on their salary. The union's leadership has to determine which option is best for whom.

Decide who benefits most from each pay raise option by using the following numeric approach that involves looking at many salaries—high, low, and in between—and determining how they will change under each option.

INSTRUCTOR NOTE: The table calculations work well with student groups. They can divide the work and check their calculations as they go.

While students are working, consider loading Excel using the screenshot on the next page as a guide.

a. Complete the table.

Name	Current Salary ($)	New Salary ($) Option 1 5% increase	New Salary ($) Option 2 3.5% increase + $1,000	Difference ($) Option 1 − Option 2
John	15,000	15,750	16,525	−775
Marissa	18,000	18,900	19,630	−730
Jay	20,000	21,000	21,700	−700
Carrie	38,000	39,900	40,330	−430
Bob	40,000	42,000	42,400	−400
Paul	46,000	48,300	48,610	−310
Leah	55,000	57,750	57,925	−175
Juan	60,000	63,000	63,100	−100
Evelyn	80,000	84,000	83,800	200
Emily	95,000	99,750	99,325	425

b. List the names of the employees who benefit more from Option 1 and those who benefit more from Option 2.

Option 1: Evelyn, Emily

Option 2: John, Marissa, Jay, Carrie, Bob, Paul, Leah, Juan

c. What does the result in the difference column mean for John? For Emily?

John will make $775 less under Option 1 than Option 2.

Emily will make $425 more under Option 1 than Option 2.

d. In general, which salaries benefit more under each option?

Lower salaries do better with Option 2.

Higher salaries do better with Option 1.

e. Is there a salary for which the two options would be the same? Explain how you can logically find that salary.

The options are the same, giving a value of zero in the difference column, for the salary $66,666.66. To find this, we can see that somewhere between Juan's and Evelyn's salaries will be the difference of zero. The differences of their salaries are −$100 and $200, which are $300 apart. Juan's difference of −$100 is one-third of the $300 distance away from a difference of zero. So the salary for which the options give the same result will be one-third of the way between Juan's salary of $60,000 and Evelyn's salary of $80,000. The distance is $20,000, and one-third of it is $6,666.66. Adding that value to Juan's salary, we get $66,666.66.

f. State the salary range of employees who would benefit more from each pay raise option.

Option 1:   Higher than $66,666.66

Option 2:   Lower than $66,666.66

If we can generalize the calculations being done, we can eventually use algebra to find the value for which the options are equivalent. To do this, let's start with a specific case and then move to a general one using a variable. We will use the one-step approach to percent change, since it is simpler to write.

To increase a salary of $10,000 by 5%, multiply by 1.05: $10,000(1.05) = $10,500

Likewise, to increase a salary, S, by 5%, multiply by 1.05: $S(1.05)$ or $1.05S$

In 3e, you wanted to know the exact salary at which the options give the *same* new salary. We can write a formula for each salary option using the variable S for the original salary.

Salary under Option 1 $= 1.05S$

Salary under Option 2 $= 1.035S + 1,000$

To find the original salary S where the options result in the same amount, we need to write and solve the equation $1.05S = 1.035S + 1,000$. We will not solve this using algebra now, but we will in future cycles.

The calculations to complete the table by hand can be tedious. Generalizing the calculations also allows us to make use of Excel to create the table of values faster. The following screenshot from Excel illustrates how we can use the calculation features of Excel. We used the variable S for salaries, but within Excel, the salaries are denoted by their location, such as B3. The *Fill Down* command can be used to quickly replicate calculations. Specific instructions for these and other commands in Excel are provided in the appendix.

	SUM	▼	× ✓ ƒx	=1.05*B3		
	A	B	C	D	E	F	G
1			Option 1		Option 2		Difference
2	Name	Current Salary ($)	1.05S	0.05S + S	1.035S + 1000	0.035S + S + 1000	Opt 1 - Opt 2
3	John	15,000	=1.05*B3				
4	Marissa	18,000					
5	Jay	20,000					
6	Carrie	38,000					
7	Bob	40,000					
8	Paul	46,000					
9	Leah	55,000					
10	Juan	60,000					
11	Evelyn	80,000					
12	Emily	95,000					

Screenshot of Microsoft Excel. Used by permission of Microsoft Corporation.

Although we can guess and check by adding more original salaries to see which salary produces a difference of zero, it can be time consuming to do so. Also, that method doesn't guarantee that we can get the answer. We still need good logic or algebra to solve the problem in 3e.

The last problem involved applying a percent change to a number, but how do we find the percent change when we know the initial and final numbers?

EXAMPLE 2

A salesman starts the year with 200 clients and ends with 260. By what percent did his client base increase?

SOLUTION

We can begin by finding the number of clients he gained: $260 - 200 = 60$ new clients.

If he had 10 clients at the start of the year, an increase of 60 would be a lot. If he had 1,000, it would not be. The quantity 60 is relative to his initial number of clients. Using percents allows us to put the number 60 in perspective. We can compare 60 to his number of clients at the start of the year to find the percent of clients added. The fraction scales easily to a percent without the use of a calculator.

$$\frac{60}{200} = \frac{30}{100} = 30\%$$

Summarizing, we can find percent change by finding the amount of change and then comparing that against our starting point by writing a fraction before converting to a percent. Thinking of percent change in this way is easier than remembering rules.

Remember that when we were increasing or decreasing a number by a percent we could use one or two steps. The same is true with finding the percent change. In this previous example, we subtracted and then divided (two steps). Another option is to divide the final amount by the starting amount and then interpret the result, which is the multiplier required to get from the original amount to the final amount.

$$\frac{260}{200} = 1.3$$

Since the multiplier is $1.3 = 130\% = 100\% + 30\%$, his client base increased by 30%.

Need more practice?

The number of downloads of Beth's program has increased from 22,000 to 28,560. By what percent did the number of downloads increase?

Answer: 30%

To compute percent change:

Option 1: 1. Subtract the initial value from the final value.
2. Compare that difference against the initial value by writing a fraction.
3. Multiply by 100 to convert the fraction to a percent.

Written as a formula,

$$\text{Percent change} = \frac{\text{final} - \text{initial}}{\text{initial value}} \cdot 100.$$

Option 2: 1. Divide the final value by the initial value.
2. Interpret the result. Multipliers greater than 1 imply a percent increase; multipliers less than 1 imply a percent decrease.

Written as a formula,

$$\text{Multiplier} = \frac{\text{final value}}{\text{initial value}}.$$

EXAMPLE: The population of a town increased from 40,000 people to 45,000 people. Find the percent change.

The change in population is $45,000 - 40,000 = 5,000$ people. Comparing this against the starting population of 40,000, the result is $\frac{5,000}{40,000} = 0.125 = 12.5\%$.

Alternatively, divide and interpret: $\frac{45,000}{40,000} = 1.125$. The multiplier 1.125 is greater than 1 by 0.125, which means there has been a 12.5% increase.

EXAMPLE: Sales decreased from 80,000 units to 68,000 units. Find the percent change.

The change in sales is $68,000 - 80,000 = -12,000$ units. Comparing this against the starting value of 80,000, the result is $\frac{-12,000}{80,000} = -0.15 = -15\%$.

Alternatively, divide and interpret: $\frac{68{,}000}{80{,}000} = 0.85$. The multiplier 0.85 is less than 1 by 0.15, which means there was a 15% loss of units.

Note: When finding the percent decrease using the two-step method, you can subtract the smaller number from the larger number to avoid negative numbers as long as you are careful with your result. Clarify the percent decrease either with a negative sign on the answer or with words, as shown in the previous example.

4. Interpret the multiplier for each percent change.

   a. 1.37   Increase of 37%

   b. 0.51   Decrease of 49%

5. A husband and wife are working together to lose weight. After the first week, they weigh in and compare their results.

   a. Fill in the table. When completing the percent column, use your calculator for the husband's information and scaling for the wife's.

	Starting Weight	Ending Weight	Pounds Lost	Percent of Weight Lost
Husband	380	366	14	3.7
Wife	200	190	10	5

   b. Using the information from the table, determine which of the following statements are true:

   a. The husband lost more pounds.

   b. The wife lost a higher percent of her weight.

   c. The wife's new weight is 95% of what it was originally.

   d. The wife's starting weight was 5% more than her ending weight.

   a, b, and c are correct.

INSTRUCTOR NOTE: If students are struggling with why parts a and b of #6 are the same, consider this explanation: If there is a 100% increase, then you have the original 100% plus the 100% increase, for a total of 200% or twice the initial price.

6. The price of an item in an online auction goes from $200 to $400. Determine which of the following statements are true.

   a. There was a 100% increase in the price.

   b. The new price is double the initial price.

   c. There was a 200% increase in the price.

   d. The new price is 200% of the original price.

   a, b, and d are correct.

Connect

You may leave class some days with some ideas not completely clear to you. If that happens, reread your notes until they make sense. Seek out your instructor for help right away. If you have a student group, compare notes with the others and ask questions.

7. A college raises its tuition each year by the percent noted in the table:

2012	8%
2013	5%
2014	9%
2015	13%
2016	7%

a. Find the overall percent increase in tuition from 2012 through 2016.

49.5%

b. Adding the percent change for each year will not result in the correct percent change. Why is that?

You cannot simply add the percents, since the base changes each year.

c. Is there ever a time when we can add percents? If so, when?

Yes; if we are finding the percent of the same number each time.

Reflect

WRAP-UP

What's the point?

Change, especially percent change, is common in many real-life contexts. To find a percent change or change a number by a percent, more than one option is available. Choose the process that makes sense to you, but knowing how to do both will prove useful. The one-step approach might require a greater understanding of percents, but it can make calculations faster and simpler.

What did you learn?

How to apply a percent change
How to find percent change
How to interpret percent change

INSTRUCTOR NOTE: Consider providing the focus problem grading rubric and writing template to students at this point to help them form their written solution. Both are available in MyMathLab.

Focus Problem Check-In
Use what you have learned to finish solving the focus problem. Write a rough draft to explain your solution, showing and explaining all mathematical work. Edit your draft into a final write-up by making any needed corrections and proofreading for grammar, spelling, and mathematical errors. Have someone verify the mathematical work and proofread the final draft for mistakes.

1.12 Homework

Skills MyMathLab

First complete the MyMathLab homework online. Then work the two exercises to check your understanding.

1. At dinner, the meal's total bill is $42. Find the amount of the dinner with a 20% tip added to the total.

   $42(1.20) = $50.40

2. Review the bar graph to answer the questions that follow.

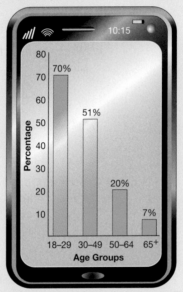

Use of Cellphones to Avoid Boredom

Source: Pew Research Center

   a. The percent of 30–49-year-olds who use their phone to avoid boredom is how many percentage points lower than the percent of 18–29-year-olds who do the same?   19 percentage points

   b. By what percent does the percent of people who use cell phones to avoid boredom decrease from the first to second age group shown in the graph?   37.3%

Concepts and Applications

Complete the problems to practice applying the skills and concepts learned in the section.

3. In the section, we explored two pay structures. It was stated that in Option 2, the order should be that the 3.5% increase is applied first and then the $1,000 bonus is added. Does the order of these two increases matter?

a. Pick five salaries and apply the increase both ways.

Salaries chosen are for illustration purposes.

Salary	Add $1,000, then Increase by 3.5%	Increase by 3.5%, then Add $1,000
$10,000	$11,385	$11,350
$20,000	$21,735	$21,700
$30,000	$32,085	$32,050
$40,000	$42,435	$42,400
$50,000	$52,785	$52,750

b. Do the results in your table indicate that the order of the two increases matters? Defend your position.

Order matters. Increasing by $1,000 first and then adding 3.5% always produces a higher salary than using the other order because we apply the percent increase to a larger amount.

c. Write a formula for computing the new salary under each order using S for the old salary.

New salary $= 1.035(S + 1,000)$

New salary $= 1.035S + 1,000$

d. Use the formulas from part c to confirm your conclusion from part b.

If the $1,000 increment is applied first, then the 3.5% is applied to that as well. So the final amount with the increment applied first is always larger by 3.5% of $1,000, or $35.

e. If these are possible pay structures for a company, which order would the management prefer if they are concerned primarily with keeping employee salaries low? Which would the employees prefer?

Management would prefer the second way, which adds the percent first. Employees would prefer the first method, which applies the percent second.

4. Suppose an online electronics retailer is running two different specials. One deal will get you 10% off your total purchase, and the other deal is worth $20 off a purchase of $100 or more. Assuming you spend at least $100, if you are allowed to use both deals, in which order do you want to apply them? Try the calculation both ways with a few different prices, or write a formula to represent each option.

Option 1: $0.90P - 20$

Option 2: $0.90(P - 20)$

Option 1 always results in a lower price and is better for the customer.

5. Ava scored 80 on the first exam in her history course and then scored 100 on the second exam. She calculated her percent increase as follows:

$$\frac{100 - 80}{100} = \frac{20}{100} = 20\%$$

Is she correct? If not, identify her mistake.

She is not correct. She needs to divide by the first score of 80 instead of the second score of 100. Her percent increase is really 25%.

6. Which of the following statements is equivalent to saying that there is a 50% increase in the cost of a movie? Circle all that apply.

    a. The new cost of the movie is twice the old cost.

    b. The new cost of the movie is one and a half times the old cost.

    c. The increase in the cost of the movie is half the original cost.

    d. The increase in the cost of the movie is half the new cost.

    e. The cost of the movie doubled.

    Statements b and c are correct.

7. You have an investment of $1,000 in a stock. Consider using the two-step method to find the percent change when answering the problems, since it might be more illuminating.

    a. In one year, the stock loses 10% of its value. What is it worth now? $900

    b. In the following year, the stock's value increases by 10% over its previous ending value. What is it worth now? $990

    c. Is it true that a 10% loss followed by a 10% gain has no net effect? Why or why not?

    No; the amount the percent is applied to is not the same in each case, so the two percents do not just "cancel each other out."

Tech TIP

If you need to apply a percent change using the same percent repeatedly, apply the percent change once using the one-step approach. Then press the multiplication key, the multiplier you used before, and Enter. Doing so changes the previous result by the percent desired. Continue pressing Enter until the calculation is complete.

8. A college has a 30% completion rate, meaning that 30% of all students who start at the college complete their studies. The president of the college sets a goal of increasing this number by 50%. What will the completion rate goal be as a percent?

The goal should be 45% of students completing their goal.

9. An elementary school is in a growing suburb. The school's principal estimates that the current enrollment of 750 students will increase 5% each year for the next 10 years. By what percent will the enrollment increase over 10 years?

Approximately 63%

10. Looking Forward, Looking Back Write the next four terms of the sequence.

$-22, -6, 2, 6, 8,$ _____, _____, _____, _____

9, 9.5, 9.75, 9.875

1.13 The X Factor: Algebraic Terminology

Explore

SECTION OBJECTIVES:

- Differentiate between variables and constants
- Differentiate between expressions and equations
- Differentiate between factors and terms

1. Evan is working on math homework with his friend Ryan. Evan says that he can cancel the 2's in $\frac{2}{3} + \frac{7}{2}$, but he isn't sure what will be left. Ryan says that he could divide off the 2's if the computation used multiplication like $\frac{2}{3} \cdot \frac{7}{2}$ but not with addition as it is now. Who is right and why?

Ryan is correct. Since we multiply the fractions to create one fraction with a factor of 2 in its numerator and denominator, we can divide off that factor of 2 before or after the multiplying occurs. The 2's in the addition version of the computation are not common factors of the same fraction, whereas the 2's in the multiplication version of the computation are.

Discover

We need terminology in mathematics to clarify what objects we're working with and the differences that exist between those objects. While some differences seem slight, they can have a large impact. Understanding the subtleties can make many procedures in arithmetic and algebra easier and help us to know whether a shortcut is valid or not.

As you work through this section, think about these questions:

1 Can you use algebraic terminology correctly?

2 Do you know which operations separate terms and which operations separate factors?

3 Is π a variable or a constant?

Representing a quantity with a letter or word is commonplace in real life and has already been done in this cycle. When we move from a specific numeric situation to a generalized situation that uses variables, we move from arithmetic to algebra. Here is some key vocabulary we will use.

LOOK IT UP

Algebraic Terminology

Algebra is a branch of mathematics in which variables are used to represent numbers and numeric operations are generalized.

A **constant** is a value that does not change—that is, a number.

A **term** is a constant or a variable or the product or quotient of constants and/or variables.

FOR EXAMPLE, 5, M, and $5M$ are all terms. If a term consists of only a number, it is known as a **constant term**. In the term $5M$, the 5 and the M are being multiplied, which makes them **factors** of $5M$. Since the 5 is the numeric factor of the term, it is known as the **coefficient** of the term.

An **expression** is a mathematical phrase that contains one term or more than one term with the terms separated by addition or subtraction signs.

FOR EXAMPLE, $1.035S + 1,000$ is an expression with two terms.

An **equation** is a mathematical statement that two expressions are equal. The two expressions are known as the sides of the equation.

FOR EXAMPLE, $1.05S = 1.035S + 1,000$ is an equation.

It is helpful to think of an expression as a phrase and an equation as a complete sentence. Also, we *simplify* expressions but *solve* equations.

An example of a constant is the number pi, which is 3.14159. . . . Its decimal expansion continues forever and never forms a pattern that repeats. We have a symbol, π, to allow us to write the number exactly and succinctly. Even though the number π is represented by a symbol, it is still a constant and not a variable. Numerals, like 4, function in the same way as the symbol π does to symbolize a number.

One way to make sense of the terminology is to compare and contrast the terms using a Venn diagram.

2. Draw a Venn diagram with one circle labeled "Constant" and the other labeled "Variable." Decide if the circles will overlap or not before you draw them. If you can think of some characteristic that constants and variables share, then the circles should overlap. In each section of the diagram, write examples and attributes.

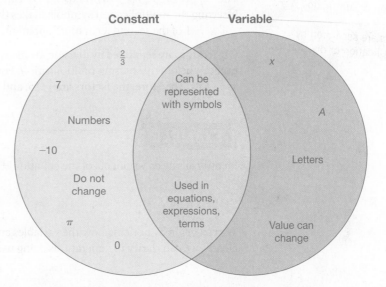

3. Draw a Venn diagram with one circle labeled "Expression" and the other labeled "Equation." Decide if the circles will overlap or not before you draw them. In each section of the diagram, write examples and attributes.

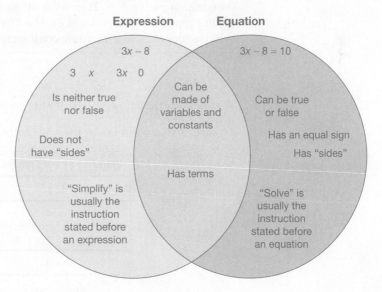

Understanding the differences between factors and terms is essential for the upcoming work we will do with algebra. One way to differentiate between factors and terms is to notice how variables and constants are connected to one another.

For example, consider the expression $3x - 12y$. There are two terms separated by a subtraction sign. The constant 3 and the variable x are being multiplied, which makes them factors of the term $3x$. Likewise, the constant 12 and the variable y are being multiplied, which makes them factors of the term $12y$. To make it easier to recognize the factors of a term, you can insert multiplication dots.

$$\underset{\text{Term}}{\underbrace{\overset{\overset{\text{Factors}}{\downarrow\ \searrow}}{3 \cdot x}}} - \underset{\text{Term}}{\underbrace{\overset{\overset{\text{Factors}}{\downarrow\ \searrow}}{12 \cdot y}}}$$

The expression $3x - 12y$ can be written as $3x + (-12y)$. So the second term in the expression can be thought of as $12y$ or $-12y$.

The expression $7(x - 1) + x(x - 1)$ can be written as $7 \cdot (x - 1) + x \cdot (x - 1)$. Inserting the multiplication symbols makes the factors 7 and $x - 1$ of the first term and x and $x - 1$ of the second term more apparent.

Objects that are separated by division are also considered factors since any division calculation can be rewritten using multiplication. For example, the expression $\frac{15w}{4}$ can be written as $\frac{1}{4} \cdot 15w$. Therefore, the factors are $\frac{1}{4}$, 15, and w.

> **Terms** are separated by addition or subtraction signs.
>
> **Factors** are separated by multiplication or division signs.

EXAMPLE

Identify all the components of the equation $-8x + 1 = 2x + 6$.

SOLUTION

We can begin by noticing how the variables and constants are connected. To see that more clearly, we can clarify the operations being used.

$$\underset{\text{Expression}}{\underbrace{\underset{\text{Term}}{\underbrace{\overset{\overset{\text{Factors}}{\downarrow\ \searrow}}{-8 \cdot x}}} + \underset{\text{Term}}{\underbrace{1}}}} = \underset{\text{Expression}}{\underbrace{\underset{\text{Term}}{\underbrace{\overset{\overset{\text{Factors}}{\downarrow\ \searrow}}{2 \cdot x}}} + \underset{\text{Term}}{\underbrace{6}}}}$$

The equation $-8x + 1 = 2x + 6$ has the variable x and constants -8, 1, 2, and 6. Its sides are the expressions $-8x + 1$ and $2x + 6$. The term $-8x$ has factors of -8 and x. The term $2x$ has factors of 2 and x. There are constant terms 1 and 6.

4. Complete the following table for the expression $\dfrac{2z(x - 1)}{3y + 4}$. Classify each set of variables and/or constants by their role in the expression.

Variables and/or Constants	Role
3, y	Factors
$3y$, 4	Terms
2, 3	Coefficients or factors
x, -1	Terms
2, z, $x - 1$	Factors

Identifying factors and terms can help us with fraction operations. We know that to simplify a fraction, we divide both the numerator and the denominator by a common factor. Thinking of this process as division instead of canceling, which is not a defined operation, makes it clear whether a 0 or 1 is left.

For example, in the fraction $\frac{3 + (-3)}{9}$, we don't cancel the 3 and -3 but instead add them to get 0. So $\frac{3 + (-3)}{9} = \frac{0}{9} = 0$.

In the fraction $\frac{5}{5x}$, we don't cancel the 5's but instead divide off a factor of 5 from the numerator and denominator, leaving 1's. So $\frac{\overset{1}{\cancel{5}}}{\underset{1}{\cancel{5}x}} = \frac{1}{x}$.

Since we can only divide off common factors that are multiplied together, we cannot divide off a 3 from the fraction $\frac{3}{5 + 3}$. The 3 in the denominator is a term, not a factor. It is attached to the 5 by addition, not multiplication. "Canceling" the 3's could leave $\frac{1}{6}$ or $0.1\overline{6}$, which is not the correct value. Also, it's not clear whether a 0 or 1 is left after "canceling." Instead, we simplify the denominator to start. Since there are no common factors between the numerator and the denominator, the result is $\frac{3}{5 + 3} = \frac{3}{8} = 0.375$.

5. If possible, simplify before performing the operation indicated in each part.

   a. $\frac{4}{9} \cdot \frac{3}{4}$ $\frac{1}{3}$

   b. $\frac{4}{9} \div \frac{3}{4}$ $\frac{16}{27}$

   c. $\frac{4 + 3}{9 + 4}$ $\frac{7}{13}$

Connect

Recall that exponents can be used to express repeated multiplication. For example, instead of writing $2 \cdot 2 \cdot 2 \cdot 2$, we can write 2^4. The factor being repeated is 2, and the number of factors is 4. In this example, 2 is called the **base**, and 4 is called the **exponent**.

When an expression has only whole-number exponents on the variables, as in $-27x^3 + 12.5x + \frac{5}{4}$, we call the expression a polynomial. Polynomials have no variables in a denominator and no variables under square roots.

LOOK IT UP

Polynomial

A **polynomial** is an expression that has one or more terms that are constants or products of variables and constants. Any exponents on the variables must be whole numbers. There are some specific types of polynomials:

- A **monomial** is a polynomial with one term.
- A **binomial** is a polynomial with two terms.
- A **trinomial** is a polynomial with three terms.

FOR EXAMPLE, $4x^2 + 3x + 5$ is a polynomial. Since it has three terms, it is a trinomial.

The **degree** of a polynomial in one variable is the highest exponent that appears on that variable in the polynomial. The polynomial $4x^2 + 3x + 5$ has only one variable, x. The highest exponent on x is 2, so the degree of the polynomial is 2.

When the exponent on a variable is written, the degree is easier to determine. But what about when the exponent on a variable isn't written or there is no variable? In the case of a term like $-7x$, there is one factor of x. So the degree of the term is 1. In the case of a constant like 14.5, there are zero factors of a variable present in the term. So the degree of a constant term is 0.

6. If the expression is a polynomial, identify it as a monomial, binomial, trinomial, or none of these and identify its degree. If the expression is not a polynomial, state this and give a reason why.

   a. $5 - 3y^2$ binomial; 2

   b. $5n^2 - 9n + 2$ trinomial; 2

   c. 4 monomial; 0

   d. $x^3 + x^2 - x - 1$ none of these; 3

   e. $3x^{1.5}$ not a polynomial, since the exponent is not a whole number

   f. $\frac{1}{4}x^3$ monomial; 3

   g. $2x^3 - 2x^4$ binomial; 4

   h. $\dfrac{7}{x + 8}$ not a polynomial, since there is a variable in the denominator

Reflect

WRAP-UP

What's the point?

The terminology in the section is very important as we move forward in the course. Often, we need to know what is varying and what is constant in a situation. We also need to know the difference between an expression and an equation and between a factor and a term so that we know how to proceed in a problem. Vocabulary matters in every discipline, including mathematics.

What did you learn?

How to differentiate between variables and constants
How to differentiate between expressions and equations
How to differentiate between factors and terms

1.13 Homework

Skills MyMathLab

First complete the MyMathLab homework online. Then work the two exercises to check your understanding.

1. Identify all specified components of the statement $3x - 7 = -18$. There may be more than one answer for each. Keep a negative sign with the term directly after it.

   a. Expression(s): $3x - 7, -18$

   b. Equation(s): $3x - 7 = -18$

   c. Variable(s): x

   d. Constant(s): $3, -7, -18$

   e. Term(s): $3x, -7, -18$

   f. Factor(s): $3, x$

   g. Coefficient(s): 3

2. Identify each of the following as either an *expression* or an *equation*.

   a. $3(x - 2)$ expression

   b. $3(x - 2) = 0$ equation

   c. $3 - 2$ expression

   d. $3 - 2x$ expression

   e. $3 = 2x$ equation

   f. 3 expression

Concepts and Applications

Complete the problems to practice applying the skills and concepts learned in the section.

3. The acceleration due to gravity is $g = 9.8\,\text{m/s}^2$. One student argues that acceleration is a variable, since it is represented by a letter. Another student claims it's a constant, since it doesn't change. Who is right?

   The second student is correct. It is a constant, since its value is 9.8. Although it has a letter to represent it, the quantity is considered a constant.

4. When we begin solving equations, one allowed operation we will use is to add the same quantity to both sides of the equation. Find the mistake in the student's work.

$$5x - 9$$
$$\underline{+9 \quad +9}$$
$$5x + 9 = 0$$
$$\underline{-9 \quad -9}$$
$$\frac{5x}{5} = \frac{-9}{5}$$
$$x = \frac{-9}{5}$$

$5x - 9$ is an expression, not an equation. It should not be solved. It does not have sides, so adding 9 twice increases the quantity by 18.

5. **a.** Evaluate the expression $5x - 20$ when $x = 4$ by replacing the variable with the number 4 and computing the result.   0

   b. Check that the equation $5x - 20 = 0$ is true when $x = 4$ by replacing the variable with the number 4. Compute the left side of the equation. Determine if it is equal to the right side of the equation. $5(4) - 20 = 0$. This is a true statement.

6. Identify the terms and factors for each expression or equation.

Expression or Equation	Term(s)	Factor(s)
$5x$	$5x$	5 and x
$5x - 8 = 0$	$5x, -8, 0$	5 and x
$x(5x - 8) = 0$	$5x, -8$	x and $5x - 8$, 5 and x
$\dfrac{x}{(5x - 8)(y + 7)}$	$x, 5x, -8, y, 7$	5 and x, $5x - 8$, and $y + 7$

7. **a.** Write an expression that has at least three terms. One of the terms should have more than two factors.   Answers will vary; possible answer: $2xyz + 3 + 5w$

   b. Write an equation that has at least two terms on each side. At least one term should have a negative coefficient.   Answers will vary; possible answer: $-2x + 4 = 3x - 1$

8. The volume formula for a right circular cylinder is $V = \pi r^2 h$, where V is the volume of the cylinder, r is the radius of the base, and h is the height of the cylinder.

   a. What is constant and what can vary in the formula?
   Pi is constant. Volume, radius, and height can vary.

   b. What happens to the volume as the radius and height increase?   The volume increases as the radius and height of the cylinder increase.

   c. If the radius and height are each doubled, by what factor does the volume change?
   The volume increases by a factor of 8.

In this book, we strive to use meaningful variables and we do not use x unless there is not a context to the problem. Using meaningful variables serves as a reminder of the quantities in a problem.

9. **Looking Forward, Looking Back** Your gym is increasing its fees by 15% and then adding on a $10 fee for a parking pass. Write an expression for your new gym fees if your old fees are represented by the variable F. $1.15F + 10$

1.14 General Number: Recognizing Patterns

Explore

SECTION OBJECTIVES:

- Make conjectures and generalize patterns
- Identify and use arithmetic and geometric sequences

1. Consider this sequence of numbers: 2, −4, _____, _____, _____

Fill in the next three numbers in the sequence based on a pattern you observe. Write the rule you used to generate the numbers. Repeat this at least two more times by considering a different pattern in the numbers.

Rule used

a. 2, −4, _____, _____, _____   _____

b. 2, −4, _____, _____, _____   _____

c. 2, −4, _____, _____, _____   _____

Possible answers: 2, −4, −10, −16, −22 (subtract 6 to get the next number)

2, −4, 8, −16, 32 (multiply by −2 to get the next number)

2, −4, 2, −4, 2 (repeat the pattern alternating 2's and −4's)

2, −4, −16, −256, −65,536 (take the opposite of the square of the previous number)

Discover

INSTRUCTOR NOTE: The primary goal of this section is to generalize arithmetic and geometric sequences. This skill will be essential for linear and exponential functions in Section 1.15.

It is human nature to generalize patterns that we see. This is an especially useful skill in mathematics. It will often be necessary in this course for you to write an expression to represent a pattern as you try to model a situation. This section will focus on a couple of particularly useful types of patterns and on how to generalize patterns and write expressions to represent them.

As you work through this section, think about these questions:

1 Can you look at a sequence of numbers and explain the pattern?

2 Are you able to express a pattern mathematically?

3 What are two different types of reasoning that are required in mathematics?

We will often need to look at a list of numbers and make a guess about the pattern that exists. This skill can help us create models, graphs, and projections for the future. The process of making an educated guess or conclusion based on the evidence provided is known as forming a **conjecture**. You will need to make a conjecture in the next problem in order to write more terms of a sequence.

2. The Fibonacci sequence is a famous sequence of numbers that follows an interesting pattern. Write the next six numbers in the sequence and describe the rule you are using to generate those values.

1, 1, 2, 3, 5, __8__, __13__, __21__, __34__, __55__, __89__, …

Each number (after the first two) is the sum of the preceding two numbers.

A sequence begins with the first term, followed by the second term, and so on. The *n*th term refers to the term in the "*n*th" position and is a generic expression that shows the general form of each term.

Need more practice?

Write an expression for the *n*th term of the sequence:

1, 8, 27, 64, . . .

Answer: n^3

3. Determine the pattern in the following sequence and write an expression for the *n*th term. It may help to label the term number for each term to see the relationship between the term number and the values in the term.

$$\frac{3}{2}, \frac{4}{3}, \frac{5}{4}, \frac{6}{5}, \frac{7}{6}, \ldots \qquad \frac{n+2}{n+1}$$

A particular type of reasoning, inductive reasoning, was needed to generalize the pattern in the sequence in #3.

Types of Reasoning

When we generalize and move from a specific case to the general one, the process is known as **inductive reasoning**. When we move from the general rule to a specific case, we are using **deductive reasoning**.

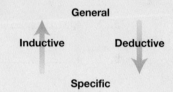

Since we use both types of reasoning so often in our daily lives, we often have trouble distinguishing them. Inductive reasoning is used to form **hypotheses** or **conjectures**. Deductive reasoning is used to prove conjectures.

We can make conjectures based on what we have seen, but the conjectures may not be accurate for a variety of reasons. A conjecture can be proven true using deductive reasoning or proven false with a counterexample. A counterexample is simply an example that does not follow the conjecture and therefore shows that the conjecture is not true.

FOR EXAMPLE, you might conjecture that squaring a number always produces a positive result, since $1^2 = 1$, $2^2 = 4$, and so on. However, $0^2 = 0$, which is not positive. This counterexample disproves the conjecture because it shows a case in which squaring a number does not produce a positive result.

It takes only one counterexample to disprove a conjecture. However, a conjecture cannot be proven even with multiple examples. Instead, a conjecture can be proven only with deductive logic.

EXAMPLE

A student knows that a number is even if the last digit in the number is even. So she wonders if a number must be divisible by 3 if the last digit is divisible by 3. Find a counterexample that proves her conjecture is false.

SOLUTION

The student's conjecture is that if the last digit is divisible by 3, then the number is also divisible by 3. So a counterexample would be a number with a last digit that is divisible by 3 but that is not itself divisible by 3. Try the first two-digit number that could possibly work as a counterexample: 13. This number has a last digit (3) that is divisible by 3, but the number 13 is not divisible by 3. We do not get a whole-number answer when we divide 13 by 3, so 13 serves as a counterexample.

Using inductive reasoning to make conjectures is helpful, but it does have its limits, since some conjectures turn out to be false, as we saw in the last example. If we are making conjectures to determine the pattern in a sequence, our job will be a lot easier if we can identify when we have particular types of sequences. There are two types of sequences that are common and are important for other topics in this course.

LOOK IT UP

INSTRUCTOR NOTE: Consider asking students to look at the sequences generated for the *Explore* and note if any are arithmetic or geometric.

Sequences

In an **arithmetic sequence**, the next term is always found by adding the same number to the previous term.

FOR EXAMPLE, $-8, -6, -4, -2, 0, 2, \ldots$ is an arithmetic sequence because each term is found by adding 2 to the previous term.

In a **geometric sequence**, the next term is always found by multiplying the previous term by the same number.

FOR EXAMPLE, $-8, 16, -32, 64, -128, \ldots$ is a geometric sequence because each term is found by multiplying the previous term by -2.

Consider the sequence $-17, -12, -7, -2, \ldots$. Notice that each term can be found by adding 5 to the previous term, which makes the sequence an arithmetic sequence. Each term requires one more 5 than the previous term. Specifically,

$$
\begin{array}{ll}
 & \text{1st} \\
 & \text{Term} \\
\text{2nd term:} & -12 = -17 + 5 = -17 + 1(5) \\
\text{3rd term:} & -7 \;\; = -17 + 5 + 5 = -17 + 2(5) \\
\text{4th term:} & -2 \;\; = -17 + 5 + 5 + 5 = -17 + 3(5)
\end{array}
$$

Each term can be found by starting with the first term, -17, and adding a certain number of 5's. The number of 5's added to get each term is always *one fewer* than the term number. So, if the term number is called n, then the number of 5's is $n - 1$. This is a key insight if we want to write an expression for a generic nth term in the sequence. The nth term can be found using the formula

$$n\text{th term:} \quad -17 + (n - 1)5$$

Remember?

Division can always be rewritten as multiplication.

4. Consider this sequence: $100, 50, 25, 12.5, \ldots$

   a. Describe how each term is generated from the previous term. Is this sequence arithmetic, geometric, or neither?

   Each number is half of the previous number. Since you can multiply the previous term by one-half to get the next term, it is a geometric sequence.

   b. Complete the following table:

Term #	Term	Calculation to get term from the first term
1	100	
2	50	$100\left(\dfrac{1}{2}\right)$
3	25	$100\left(\dfrac{1}{2}\right)\left(\dfrac{1}{2}\right) = 100\left(\dfrac{1}{2}\right)^2$
4	12.5	$100\left(\dfrac{1}{2}\right)\left(\dfrac{1}{2}\right)\left(\dfrac{1}{2}\right) = 100\left(\dfrac{1}{2}\right)^3$
5	6.25	$100\left(\dfrac{1}{2}\right)\left(\dfrac{1}{2}\right)\left(\dfrac{1}{2}\right)\left(\dfrac{1}{2}\right) = 100\left(\dfrac{1}{2}\right)^4$
6	3.125	$100\left(\dfrac{1}{2}\right)\left(\dfrac{1}{2}\right)\left(\dfrac{1}{2}\right)\left(\dfrac{1}{2}\right)\left(\dfrac{1}{2}\right) = 100\left(\dfrac{1}{2}\right)^5$

c. Write an expression for the nth term.

nth term: $100\left(\dfrac{1}{2}\right)^{n-1}$

5. Identify each of the following sequences as arithmetic, geometric, or neither. If the sequence is arithmetic or geometric, write an expression for the nth term.

a. $6, 18, 54, 162, \ldots$ geometric; nth term: $6(3)^{n-1}$

b. $1, 2, 4, 7, 11, \ldots$ neither

c. $-16, -20, -24, -28, \ldots$ arithmetic; nth term: $-16 + (n-1)(-4)$

Sometimes it is easier to write an expression for the nth term of a sequence if you use the "0th term" instead of the 1st term that's shown in the sequence. That is, you can use the term that would come before the first term.

Consider the arithmetic sequence that begins $1, 3, 5, 7, \ldots$. If we write a formula for the nth term based on the first term of 1, we get $1 + (n-1)2$.

Another option is to work backward and determine what number should come before the 1 in the pattern. If each number is found by adding 2 to the previous number, then we can work the sequence backward by subtracting 2. So the "0th term" that would come before the 1st term is $1 - 2 = -1$.

We can then use this "0th term" to write an expression for the nth term in the sequence. Notice that, if we start with -1, the number of 2's needed is always the same as the term number.

$$
\begin{array}{ll}
 & \text{0th} \\
 & \text{Term} \\
\text{1st term:} & 1 = -1 + 2 = -1 + 1(2) \\
\text{2nd term:} & 3 = -1 + 2 + 2 = -1 + 2(2) \\
\text{3rd term:} & 5 = -1 + 2 + 2 + 2 = -1 + 3(2) \\
\text{4th term:} & 7 = -1 + 2 + 2 + 2 + 2 = -1 + 4(2)
\end{array}
$$

INSTRUCTOR NOTE: Consider showing students that the two expressions for the nth term are algebraically equivalent. This provides a nice preview of skills the students will learn in Cycle 2.

If we write a formula for the nth term based on the first term of 1, we get $-1 + 2n$. This formula is equivalent to the first formula we found and will generate the same terms, but it looks simpler.

Connect

You can't always identify a sequence from the first couple of terms. It might be possible to generate more than one sequence from the first two terms.

INSTRUCTOR NOTE: For more practice on writing 0th terms, consider having students write each arithmetic and geometric sequence in #5 using the 0th term.

6. Consider the following start of a sequence: $2, 10, \ldots$

a. Write the next five terms of the sequence if it's arithmetic. Write an expression for the nth term using the 1st term of the sequence and another expression using the 0th term.

2, 10, 18, 26, 34, 42, 50

nth term: $2 + (n-1)8$; nth term: $-6 + n \cdot 8$

b. Write the next five terms of the sequence if it's geometric. Write an expression for the nth term using the 1st term of the sequence and another expression using the 0th term.

2, 10, 50, 250, 1250, 6250, 31250

nth term: $2 \cdot (5)^{n-1}$; nth term: $\frac{2}{5} \cdot (5)^n$

Reflect

WRAP-UP

What's the point?

Identifying and generalizing patterns are powerful skills that can require the use of both inductive and deductive reasoning. Learning particular patterns, like arithmetic and geometric, that occur in life and nature can be particularly useful.

What did you learn?

How to make conjectures and generalize patterns
How to identify and use arithmetic and geometric sequences

1.14 Homework

Skills MyMathLab

First complete the MyMathLab homework online. Then work the two exercises to check your understanding.

It is fine to use the MyMathLab help aids *View an Example* and *Help Me Solve This*, but make sure you are not dependent on them. To be successful in the course, you need to know how to start and finish a problem.

1. Find the pattern in each sequence and use it to list the next two terms.

a. 5, 17, 29, 41, 53, 65

b. 18, 14, 10, 6, 2, −2

c. −9, 4, −8, 5, −7, 6, −6, 7

2. What is the next term of the sequence?
$3x, 4x + 1, 5x + 2, 6x + 3, 7x + 4$

Concepts and Applications

Complete the problems to practice applying the skills and concepts learned in the section.

3. a. Write the first five terms of the arithmetic sequence if the nth term is given by $25 + (n - 1)(7)$.

25, 32, 39, 46, 53

b. Write the first five terms of the geometric sequence if the nth term is given by $36\left(\frac{1}{3}\right)^{n-1}$.   36, 12, 4, $\frac{4}{3}$, $\frac{4}{9}$

4. a. Look at the following sequence of numbers. Study the sequence and find the next numbers in the fraction row. Under each fraction, write the decimal.

Term #:	1	2	3	4	5	10	50	100	150	200	n
Fraction:	1,	$\frac{1}{2}$,	$\frac{1}{3}$,	$\frac{1}{4}$,	$\frac{1}{5}$,	$\frac{1}{10}$,	$\frac{1}{50}$,	$\frac{1}{100}$,	$\frac{1}{150}$,	$\frac{1}{200}$,	$\frac{1}{n}$
Decimal:	1.0,	0.5,	0.33,	0.25,	0.2,	0.1,	0.02,	0.01,	0.007,	0.005	

b. Look at the sequence of numbers in the decimal row. Make a conjecture about the numbers and what value they are approaching.

The numbers appear to be getting smaller and approaching zero.

5. a. Look at the following sequence of numbers. Study the sequence and find the next numbers in the fraction row. Under each fraction, write the decimal.

Term #:	1	2	3	4	5	6	7	8	9	200	n
Fraction:	$\frac{2}{3}$,	$\frac{4}{5}$,	$\frac{6}{7}$,	$\frac{8}{9}$,	$\frac{10}{11}$,	$\frac{12}{13}$,	$\frac{14}{15}$,	$\frac{16}{17}$,	$\frac{18}{19}$,	$\frac{400}{401}$,	$\frac{2n}{2n + 1}$
Decimal:	0.67.	0.8.	0.86.	0.89.	0.91.	0.92.	0.93.	0.94.	0.95	1	

b. One student conjectures that the numbers in the decimal row are always getting bigger and will continue to increase forever, eventually approaching the number 1 million. Another student disagrees but cannot come up with a counterexample. Instead, he tries to argue that the numbers will never get close to being that large. How can he defend his position?

Each number's denominator will always be 1 greater than the numerator. So the fraction will always be proper and therefore never greater than 1.

6. A student conjectures that the square of a number is always greater than the number itself. Find a counterexample for this conjecture.

If you square a positive number less than 1, the result will be less than the original number. Also, if 0 or 1 is squared, the result is equal to the number but not greater.

7. Suppose you have an investment fund that began with $200 and is increasing by 5% each month. Write a sequence that represents how much the fund is worth each month for the first six months (starting at month 0) and then write an expression for the value after n months.

$200, $210, $220.50, $231.53, $243.10, $255.26; $200(1.05)^n$ or $210(1.05)^{n-1}$

8. Consider the following sequence of figures and how each one is generated from the previous figure. Count the number of squares in each figure and determine the pattern in this sequence. Write an expression for the number of squares in the nth figure.

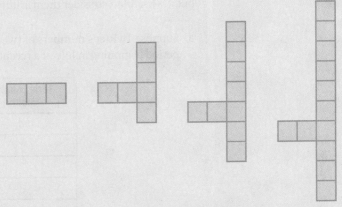

Each figure has three more squares than the previous figure. The sequence for the number of squares is 3, 6, 9, 12, ...

Since the number of squares is increasing by 3 each time, the sequence is an arithmetic sequence.

nth term: $3 + (n - 1)3$ or $3n$

9. Looking Forward, Looking Back

a. State the multiplier needed to increase a number by 15%.   1.15

b. State the multiplier needed to decrease a number by 15%.   0.85

1.15 Social Network: Linear and Exponential Change

SECTION OBJECTIVE:
- Model change with linear and exponential functions

Explore

Many social media companies experience rapid changes in their use as their popularity increases or wanes. Understanding the changes, both positive and negative, can help companies estimate future growth or decline.

1. For each part of the problem, organize the sequence of numbers into a table by pairing the term with the term's number. Treat the first number in the sequence as the 0th term, since it is the initial quantity before any change occurs. Determine if the sequence is arithmetic, geometric, or neither. If it is arithmetic or geometric, write an expression for the nth term. Write a sentence explaining how the sequence of values is changing within the context given. With real data, sequences are rarely perfectly arithmetic or geometric, but it is useful to consider them arithmetic or geometric if they are approximately so.

a. Suppose Twitter's numbers of tweets, in billions per month, over a six-month period beginning in July of a recent year were 8.00, 8.40, 8.82, 9.26, 9.72, and 10.21.

Term Number (month)	Number of Tweets (in billions)
0	8.00
1	8.40
2	8.82
3	9.26
4	9.72
5	10.21

Geometric; nth term: $8 \cdot (1.05)^n$
The number of tweets increased by approximately 5% each month.

INSTRUCTOR NOTE: The examples in this section are long. They allow you to focus on the connections between the representations of functions instead of the creation of tables and graphs.

b. Suppose the numbers of active Google+ users, in millions per month, over a six-month period beginning in July of a recent year were 3.8, 3.68, 3.56, 3.44, 3.32, 3.2.

Term Number (month)	Number of Active Users (in millions)
0	3.8
1	3.68
2	3.56
3	3.44
4	3.32
5	3.2

Arithmetic; nth term: $3.8 - 0.12n$
The number of active users for Google+ decreased by approximately 0.12 million, or 120,000 users each month.

Discover

The problems in the *Explore* began with arithmetic and geometric sequences that were written in lists. The table made the process of relating the terms to the term numbers easier. This section will look more closely at the representations of sequences like these in tables as well as graphs and equations.

As you work through this section, think about these questions:

1 Can you determine if a function is linear, exponential, or neither from a graph, a table, or a scenario?

2 How do arithmetic and geometric sequences relate to linear and exponential functions?

3 If a graph appears to show linear or exponential change, is that enough to conclude that it does?

Section 1.14 developed concepts related to recognizing change in a sequence of numbers so that the sequence can be continued and generalized with an expression for the *n*th term of the sequence. This section will now extend those ideas from a single sequence of numbers to a specific type of pairing of numbers known as a function.

Function

A **function** is a rule that assigns a value of the independent variable, known as an **input**, to exactly one value of the dependent variable, known as an **output** or a function value. If a relationship pairs an input with more than one output, then the relationship is not a function.

We say that the dependent variable is **a function of** the independent variable. Functions can be displayed as graphs, tables, and equations.

FOR EXAMPLE, the weekly salary, *S*, of a worker who makes $15.75/hour is a function of the number of hours worked per week, *n*. Each number of hours worked pairs with exactly one salary. Since the salary depends on how many hours are worked, *n* is the independent variable and *S* is the dependent variable. The numbers of hours worked each week are the inputs, and the weekly salaries are the outputs from the function.

Table

Number of hours worked in a week, *n*	Weekly salary, *S*
15	$236.25
20	$315.00
25	$393.75
30	$472.50
35	$551.25
40	$630.00

Equation

$S = \$15.75n$

Graph

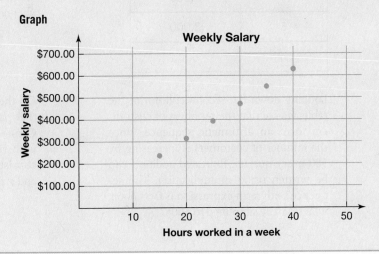

Weekly Salary

We will often be presented with a situation or problem that we want to understand better. To do so, we determine the variables involved and their relationship to each other using a process known as **modeling**. In this book, we will often model a situation by looking at changes we see in a table or graph and finding patterns in those changes. Based on those patterns, we will write an equation to describe the situation and use the table, graph, or equation to analyze the situation further.

Although many kinds of functions can be used to model data, the examples in the *Explore* provide two common types: linear functions and exponential functions. Let's look more closely at each of these types of function.

EXAMPLE 1

Suppose the owner of a new company has the goal of gaining 500 new customers every fiscal quarter. Write a function to model the relationship between the number of fiscal quarters and the total number of customers. Use a table, graph, or equation to determine how many customers the company should expect to have after 5 years and how long it will take to reach 100,000 total customers.

SOLUTION

We can begin this problem by constructing a table, graph, or equation. In this example, we look at all three to see each representation of the function.

We start with no time having passed for the company and no customers. From there, in each additional quarter the total number of customers will increase by 500. Using this information, we get the following table of values.

Quarter	Total Customers
0	0
1	500
2	1,000
3	1,500
4	2,000
5	2,500
6	3,000
n	$500n$

(The left side shows +1 between each quarter; the right side shows +500 between each total customers value.)

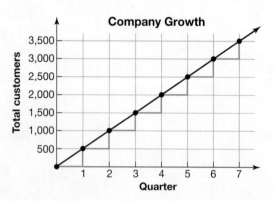

Looking down the second column of the table, we see that the numbers of customers form an arithmetic sequence, since the number of customers is increasing by 500 every quarter. Repeated addition can be written using multiplication. The sequence's nth term expression is $0 + 500n$, which can be simplified to $500n$.

On the graph, we can see that for each quarter that passes, the company gains 500 customers. Connecting consecutive points on the graph with right triangles shows that this change is consistent, since each triangle is the same shape and size.

We can label the expression of the nth term of the sequence as "Total customers." The equation formed is another way to represent the function.

$$\text{Total customers after } n \text{ quarters} = 500n$$

Notice in the table that each output is 500 times its corresponding input, confirming our result.

The equation can be used to find the number of total customers after 5 years, which is 20 quarters. Substituting 20 for n in the function, we find there will be $500(20) = 10,000$ total customers after 5 years. The table or graph could be extended using a calculator or Excel to find this result, but the equation provides the fastest approach in this situation.

Likewise, to find out how long it will take to reach 100,000 total customers, use logic and the function. Since the total number of customers is 500 times that of the number of quarters, divide 100,000 by 500 to find the number of quarters. In this case, it will be 200 quarters. Every four quarters is a year. So 200 quarters or $\frac{200}{4} = 50$ years need to pass to reach 100,000 total customers. This is a lengthy and unrealistic amount of time for a new company, which might lead the company to other methods to gain customers more quickly.

The type of consistent change shown in the last example is known as linear change, since the graph through the points forms a line. The function used to model this change is known as a linear function.

Linear Function

A **linear function** is a function whose outputs increase or decrease by a constant amount when the inputs increase by 1. Therefore, the function values form an arithmetic sequence. If the function values are increasing, the function represents **linear growth**. If the values are decreasing, the function represents **linear decay**.

FOR EXAMPLE, if Facebook gains 1 million new users each month, then the number of new users is a linear function of the number of months.

To use a table to determine if a function is linear, find the amount of change in the function values when the inputs increase by 1. The amount of change should be a constant value if the function is linear. The points can also be plotted to see if they fall in a line. Conversely, if we know the data has a linear pattern, we can connect the dots with a line, as we did in the previous example.

Equations of linear functions have the form
Dependent variable = starting value + rate of change · independent variable,
where the starting value is the 0th term of the corresponding arithmetic sequence.

The nth term of an arithmetic sequence is given by the expression
Starting value + rate of change · x.

The generalized form of a linear function is given by the equation
$y = starting\ value + rate\ of\ change \cdot x.$

Linear Growth

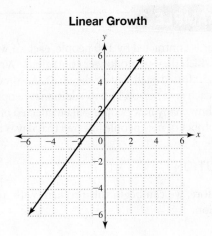

Linear Decay

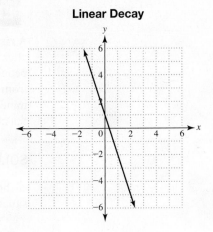

Linear functions follow directly from arithmetic sequences. If a sequence is arithmetic, then the terms are a linear function of the term number.

2. Use the following data to determine if y is a linear function of x. If it is, explain why and write a linear function that models the data.

x	y
0	−18
1	−12
2	−6
3	0
4	6
5	12

Since the y-values increase by 6 as the x-values increase by 1, y is a linear function of x.
Function: $y = -18 + 6x$

3. You have an interest-free loan of $800 and are paying it off at a rate of $50 a month. Model the situation with a table, graph, and equation. Is the function a linear function? How many months will it take to repay the loan?

Month, m	Amount left to pay, A
0	$800
1	$750
2	$700
3	$650
4	$600
5	$550

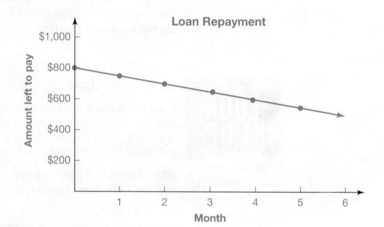

$A = 800 - 50m$
Yes, it is a linear function. It will take 16 months to repay the loan.

Linear functions are useful for modeling situations with changes that are consistent amounts. Some situations involve change that is not a consistent *amount*, but rather a consistent *percent*. Let's look at this kind of change and a type of function used to represent it.

EXAMPLE 2

A virus is contracted by 100 people, each of whom infects two more people the first day, after which they are no longer contagious. Write a function to model the number of new people infected each day by the virus. Assume that the infection will continue to spread from each sick person to two additional people. Use a table, graph, or equation to determine how many new people will be infected on the 14th day. Determine the day on which the number of new people infected that day is at least 100,000.

SOLUTION

As before, we can begin this problem with a table, graph, or equation to represent the situation.

Let's explore this in a small setting, with just one initial person infected, known as patient zero, before moving to the 100 people given in the problem. This process, known as solving a simpler problem, is a useful tool in mathematics.

We know that each person will pass the virus to two more people. For ease of drawing, let's represent a person with a box. Each time a person infects someone, we draw lines from that square to two more squares. Notice that on the 0th day, there is 1 person infected, which can be considered the 0th term. On day 1, there are 2 new people infected. On day 2, there are 4 new people infected. On day 3, there are 8. The number of new people infected is doubling each day, forming a geometric sequence.

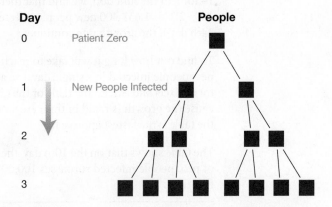

Using this idea, let's return to the given number of initial infected people and change the one person infected to 100 people. We continue the same pattern, doubling the number of new people infected each day. Using this information, we get the following table of values and graph:

	Day	Number of New People Infected That Day	
+1	0	100	· 2
+1	1	200	· 2
+1	2	400	· 2
+1	3	800	· 2
+1	4	1,600	· 2
+1	5	3,200	· 2
+1	6	6,400	· 2
	n	$100 \cdot 2^n$	

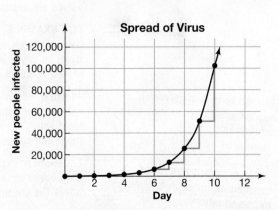

Looking down the second column of the table, we see that the numbers of new people infected form a geometric sequence, since the number infected is doubling each day. Repeated multiplication can be written using exponents. The expression for the sequence's nth term is $100 \cdot 2^n$.

Notice that the relationship between the day and the number of new infections is not linear. The graph reflects this with triangles connecting consecutive points that are not the same shape or size. And clearly the points do not fall in a line.

We can label the expression for the nth term of the sequence as "Number of new people infected." The equation formed is another way to represent the function.

$$\text{Number of new people infected on the } n\text{th day} = 100 \cdot 2^n$$

Notice in the table that each output is 100 times the value of 2 raised to its corresponding input, which confirms the result.

The equation can be used to find the number of new people infected on the 14th day. Substituting 14 for n in the function, we find that there will be $100 \cdot 2^{14} = 1{,}638{,}400$ new people infected on the 14th day if the growth rate continues.

To find out how long it will take to reach 100,000 new people infected in a single day, use a calculator or Excel and extend the table or the graph. Because the growth is rapid in this case, continuing the table is the fastest approach.

The table shows that on the 10th day, the number of new people infected surpasses 100,000 for the first time.

Day	Number of New People Infected that Day
0	100
1	200
2	400
3	800
4	1,600
5	3,200
6	6,400
7	12,800
8	25,600
9	51,200
10	102,400

The type of change shown in the previous scenario, table, and graph is known as exponential change, since the independent variable becomes an exponent in the function. The type of function used to model this change is known as an exponential function.

Exponential Function

An **exponential function** is a function whose outputs experience a constant multiplier as the inputs increase by 1. Therefore, the function values form a geometric sequence. If the function values are increasing, the function represents **exponential growth**. If the values are decreasing, the function represents **exponential decay**.

FOR EXAMPLE, if the number of new Facebook users triples every day, the number of new Facebook users is an exponential function of the number of days. The number of new users on day 2 is three times as large as the number of new users on day 1. The number of new users on day 3 is three times as large as the number of new users on day 2.

The nth term of a geometric sequence is given by the expression
$Starting\ value \cdot (multiplier)^x$.

The generalized form of an exponential function is given by the equation
$y = starting\ value \cdot (multiplier)^x$.

To use a table to determine if there is an exponential relationship, find the **multiplier** between the function values as the inputs change by 1. The multiplier should be a constant value. If there is an exponential relationship in the data, a curve can be drawn on the graph connecting the points, as we did in the previous example.

Equations of exponential functions have the form
$Dependent\ variable = starting\ value \times (multiplier)^{independent\ variable}$,
where the starting value is the 0th term of the corresponding geometric sequence.

Exponential functions follow directly from geometric sequences. If a sequence is geometric, then the terms are an exponential function of the term number.

To see if there is an exponential relationship, the points can also be plotted to see if they fall in an exponential curve. The graph could curve up with exponential growth, or it could curve down in the case of exponential decay. However, although a curve can be used to *support* a conjecture that an exponential relationship exists, it is not enough to *determine* if there is one. Other functions that are not exponential can have a similarly shaped graph. The scenario, table, or equation can be used to definitively determine if a relationship is exponential or not.

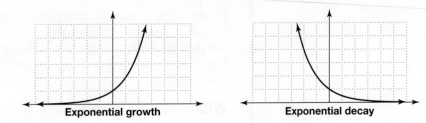

Exponential growth   **Exponential decay**

4. A company wants to project growth using both linear and exponential change to see the long-term effects. The company begins with 150 clients. The CEO wants to project growth assuming 15 new clients per month versus a 10% growth in the client list each month. Model each situation with a function. How many months will it take the company to reach 500 clients with each approach?

   First approach: Number of clients $= 150 + 15n$, where n is the number of months
   $23\frac{1}{3}$ months to reach 500 clients

   Second approach: Number of clients $= 150 \cdot (1.10)^n$, where n is the number of months
   13 months to reach 500 clients

Notice that each scenario in #4 begins with the points (0, 150) and (1, 165) but grows differently afterward. Two points are not necessarily enough to determine a type of change without additional information.

5. City planners for a city with a current population of 100,000 residents are modeling the population in various ways to help them make decisions related to resources. Experts disagree on the growth rate and model, since they are using different data, but they do know that they have resources for only 200,000 residents.

   For each scenario, write a function and use it to determine when the city will be at its maximum capacity.

   a. Population increases by 5,000 residents per year.
   Population $= 100,000 + 5,000n$, where n is the number of years; 20 years

   b. Population increases by 8,500 residents per year.
   Population $= 100,000 + 8,500n$, where n is the number of years; 12 years

   c. Population increases by 2.5% each year.
   Population $= 100,000(1.025)^n$, where n is the number of years; 29 years

   d. Population increases by 3% each year.
   Population $= 100,000(1.03)^n$, where n is the number of years; 24 years

Reflect

WRAP-UP

What's the point?

Change presents itself in many patterns, two of which are linear and exponential. It is useful to be able to recognize change in sequences and situations and then model the changes with tables, graphs, and equations that represent functions. Other types of change are possible and will be addressed later in the book.

What did you learn?

How to model change with linear and exponential functions

1.15 Homework

Skills MyMathLab

First complete the MyMathLab homework online. Then work the two exercises to check your understanding.

1. **a.** In the following table, compute a y-value by multiplying each previous y-value by 5. What type of growth is this? Write a function to model the data.

You may think that multiplying the inputs by the same value to get the outputs is exponential growth, but that is not the case. Instead, exponential change means that each output can be found from the previous output by using the same multiplier.

x	y
0	1
1	5
2	25
3	125
4	625
5	3,125

Exponential; $y = 5^x$

b. In the following table, compute a y-value for a given x-value by multiplying the x-value by 5. What type of growth is this? Write a function to model the data.

x	y
0	0
1	5
2	10
3	15
4	20
5	25

Linear; $y = 5x$

2. Determine if the change shown in the table is linear, exponential, or neither. If it is linear or exponential, explain why and write a function that models the data.

x	y
0	−16
1	−8
2	−4
3	−2
4	−1
5	$-\frac{1}{2}$

Each time the y-values are multiplied by $\frac{1}{2}$ as the x-values increase by 1. Therefore there is an exponential relationship in the data.

Function: $y = -16\left(\frac{1}{2}\right)^x$

Concepts and Applications

Complete the problems to practice applying the skills and concepts learned in the section.

3. Consider the following data. Is y a function x? Explain.

x	y
0	48,000
1	67,000
3	54,000
4	58,000
5	62,000
5	64,000
11	74,000
15	81,000
16	85,000
23	92,000

No; there is not one unique y-value for each x-value. Specifically, 5 corresponds to two different y-values, 62,000 and 64,000.

4. Give at least one example of something that you think grows exponentially and something that you think grows linearly.

Answers will vary.

National debt, human populations, and bacteria populations can all grow exponentially. Hair growth and tree growth are examples of linear growth.

5. Explore the effect of increasing the price of a gallon of gas by 10 cents each week.

a. Complete the table.

Week, w	Gas Price per Gallon ($)
0	3.50
1	3.60
2	3.70
3	3.80
4	3.90
5	4.00

b. Make a graph.

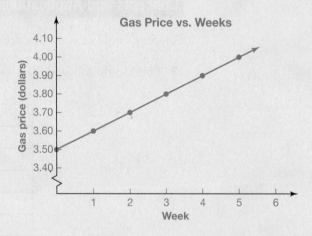

c. What type of growth is this? Why?   Linear, because each output is $0.10 more than the previous output

d. Write a formula to give the price per gallon after w weeks:
Gas price per gallon $= 3.50 + 0.10w$

e. If the gas price keeps increasing in this same pattern, when will it reach $6 per gallon?   Week 25

f. Is it realistic to assume that this growth will continue?
It's unlikely. At some point, the growth will probably slow or the price may begin to fall due to other economic factors.

6. A company has 1,500 customers and wants to gain 10% more customers each quarter. The executives want to project when the company will reach certain milestones to see if this approach will enable them to meet their goals.

   a. Use your calculator to complete the table. Round values to the nearest whole number before entering them in the table, but use the exact values to continue the calculation on your calculator. This method provides more exact results.

Quarter, q	Total Customers, T
0	1,500
1	1,650
2	1,815
3	1,997
4	2,196
5	2,416

   b. What kind of change is this? Why?
   Exponential, because each output is 1.1 times the previous output

   c. Write a function to model the data.
   $T = 1,500 \cdot (1.10)^q$

   d. How many years will it take to surpass the total customer mark of 5,000?
   It will take 13 quarters, which is 3.25 years.

   e. Find the number of total customers after 100 quarters.
   Approximately 20,670,919

7. The value of your neighbor's house was $200,000, but it dropped to $180,000 over the last year. If the value keeps falling by the same dollar amount each year, what kind of change is that? If the value keeps falling by the same percent each year, what kind of change is that?
   Falling by $20,000 each year is a linear change. Falling by the same percent each year is an exponential change.

8. Assume the tuition at a college is currently $100 per credit hour. Create a table of tuition values for each year of the next decade, assuming some kind of linear growth in the tuition. Create a table of tuition values for each year of the next decade, assuming some kind of exponential growth in the tuition.

Answers will vary.

9. Looking Forward, Looking Back On average, a major league baseball game lasts 3 hours. If the game goes through the bottom of the ninth inning, there will be three outs for each team in each inning. Approximately how long does each out take?

Each out takes around $3.\overline{3}$ minutes.

1.16 Infinity and Beyond: Perimeter and Area

Explore

SECTION OBJECTIVE:

• Calculate perimeter and area

1. a. Draw and label the dimensions of two different rectangles that have an area of 48 square inches. Find the perimeter of each rectangle. How many rectangles are possible with this area if the sides have whole-number lengths?

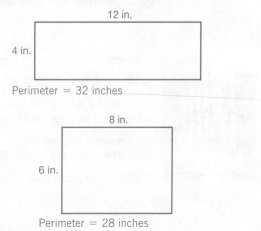

Perimeter = 32 inches

Perimeter = 28 inches

There are five such rectangles, with the following dimensions in inches: 1×48, 2×24, 3×16, 4×12, 6×8.

b. Draw and label the dimensions of two different rectangles that have a perimeter of 48 inches. Find the area of each rectangle. How many rectangles are possible with this perimeter if the sides have whole-number lengths?

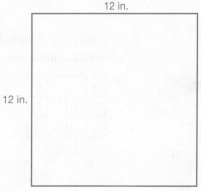

Area = 144 square inches

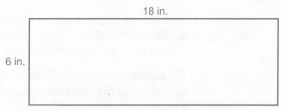

Area = 108 square inches

There are 12 such rectangles, with the following dimensions in inches: 1×23, 2×22, 3×21, 4×20, 5×19, 6×18, 7×17, 8×16, 9×15, 10×14, 11×13, 12×12.

Discover

There are many situations that require us to calculate the distance around a two-dimensional object or the amount of space contained inside it. First, we will make sure that the definitions of these two measurements are clear. Then we will tie together concepts already learned in this cycle with the new ideas of this section.

As you learn about perimeter and area in this section, keep these questions in mind:

1 Do you understand the difference between perimeter and area?

2 Do you know how to determine the units in a perimeter or area?

Area and Perimeter

The **area** of a two-dimensional figure is the size of the region enclosed by the figure. Area is measured in square units, such as square inches or square meters.

The **perimeter** of a two-dimensional figure is the distance around the figure. Perimeter is measured in units of length, such as inches or meters.

EXAMPLE: Find the area and perimeter of the rectangle shown here.

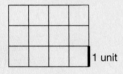

1 unit

SOLUTION: By counting the individual square units, we can determine that the area inside the rectangle is 12 square units. By counting the line segments around the exterior of the rectangle, we can determine that the perimeter of the rectangle is 14 units.

The best way to read a quantity like 25 in.^{2} is "25 square inches" not "25 inches squared." If you say "25 inches squared," then it sounds like you mean (25 in.)^2.

2. Find the areas of the following shapes (shown in orange) by counting the number of square units inside each one.

a.

4.5 square units

b.

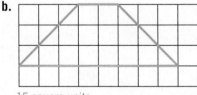

15 square units

How did you find the area of the triangle in the last problem? One good technique is to double the triangle and form a rectangle. The area of the rectangle should be easy to find, and the triangle's area is half of that.

The perimeters of the two figures in the last problem are not as easy to find as the areas, since some of the edges are not on the grid lines and therefore you need to calculate their lengths.

3. For the following shape, find the perimeter by determining the total length of its sides, and find the area by counting the number of squares inside it.

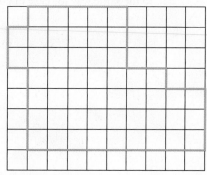

P = 34 units; *A* = 51 square units

4. Find the perimeter and area of the following figure if the slanted lines each have a length of 5 units.

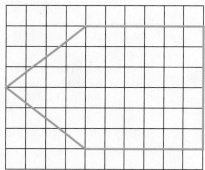

P = 28 units; *A* = 48 square units

Two figures might have different perimeters and areas but the same general shape. There's a name for this type of relationship and a way to confirm it mathematically.

Similarity

When two figures are **similar**, they have the same shape but not necessarily the same size. Numerically, the ratios of their corresponding sides are always the same. Corresponding sides are the sides in the same relative locations when the objects are oriented in the same way.

FOR EXAMPLE, the following trapezoids are similar, although this is easier to see when the orientations are the same, as shown in the second pair of figures.

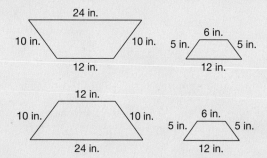

All three ratios formed by comparing a side of the large trapezoid to a corresponding side of the small trapezoid simplify to 2:1. So we say that the sides are **in proportion**.

$$\frac{12}{6} = \frac{24}{12} = \frac{10}{5} = \frac{2}{1}$$

Each side of the large trapezoid is twice the length of the corresponding side of the small trapezoid.

The triangles shown below look similar because they have the same shape. To verify this, we need to check that the corresponding sides are proportional. They have been oriented in the same way for ease of comparison.

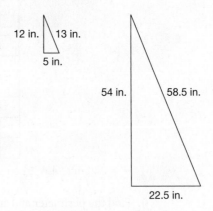

First, we write the ratio of the left side of the large triangle to the left side of the small triangle.

$$54:12 \quad \text{or} \quad 4.5:1$$

Then we can write the ratio of the bottom sides, being careful to write the ratio in the same order as the first one, large triangle side to small triangle side.

$$22.5:5 \quad \text{or} \quad 4.5:1$$

Finally, we can calculate the ratio of the right sides, in the same order.

$$58.5:13 \quad \text{or} \quad 4.5:1$$

Notice that all three ratios are the same when simplified, which confirms that the triangles are similar. The simplified ratio tells us that each side of the large triangle is four and a half times longer than the corresponding side of the small triangle.

5. The following two triangles are similar. Find the lengths of the remaining two sides of the larger triangle.

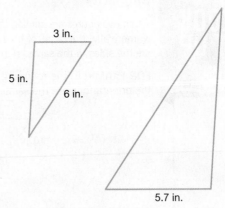

9.5 inches, 11.4 inches

Figures with the same shape but not necessarily the same size are similar. If figures are the same shape *and* size, they are **congruent**.

Connect

Look at the following sequence of figures called **Sierpinski triangles**. All the included triangles, of every size, are equilateral, which means their sides are all the same length.

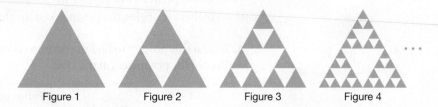

Figure 1   Figure 2   Figure 3   Figure 4

Notice how each triangle is formed from the previous triangle. Each shaded triangle is split evenly into four triangles in the next figure. Three of the four smaller triangles are shaded and the center one is not. This process continues with each new figure in the sequence.

One interesting thing about these figures is that as the pattern is continued, the perimeters of the shaded portions continue to increase without bound even though the areas decrease. The perimeters are approaching infinity, but the areas are approaching zero.

Sierpinski triangles are also **self-similar**, which means that any triangle in one of the figures has the same shape as one of the smaller triangles inside it.

6. Assume the first figure has an area of 1 square unit. Then each of the triangles in the second figure has an area of $\frac{1}{4}$. Since the shaded area in the second figure is composed of three of these triangles, its area is $\frac{3}{4}$.

   a. Repeat this process to find the shaded area in the third figure, fourth figure, and so on. Record the areas in the table.

Figure Number	Area of Shaded Triangles (in square units)
1	1
2	$\frac{3}{4}$
3	$\frac{9}{16}$
4	$\frac{27}{64}$
5	$\frac{81}{256}$
n	$\left(\frac{3}{4}\right)^{n-1}$

Tech TIP

You can use Excel or a calculator to continue the pattern in the areas and perimeters to help you see what they are approaching.

   b. If more figures are produced following this same pattern, what will the shaded area be in the nth figure? Consider the pattern in the shaded areas in the table. What kind of sequence do they form?

   geometric

   c. Confirm, using your results in the table, that the shaded area does get smaller as more figures are produced in this same pattern.

   The areas in the table as decimals are 1, 0.75, 0.56, 0.42, 0.32. The areas start at 1 and get smaller. This should also be evident from the figures, since the shaded areas are getting smaller and the white area is more predominant.

7. Assume the first figure has a perimeter of 3 units. Then each side of the triangle in the first figure has a length of 1 unit. So each triangle in the second figure has a length of $\frac{1}{2}$ and a total perimeter of $\frac{3}{2}$. Since the shaded area in the second figure is composed of three of these triangles, the perimeter of the shaded area is $3 \cdot \frac{3}{2} = \frac{9}{2}$.

a. Repeat this process to find the perimeters of the shaded areas in the figures. Record the perimeters in the table.

Figure Number	Perimeter of Shaded Triangles (in units)
1	3
2	$\frac{9}{2}$
3	$\frac{27}{4}$
4	$\frac{81}{8}$
5	$\frac{243}{16}$
n	$3\left(\frac{3}{2}\right)^{n-1}$

b. If more figures are produced following this same pattern, what will the perimeter of the shaded areas be in the nth figure? Consider the pattern in the shaded perimeters in the table. What kind of sequence do they form?

geometric

c. Confirm, using your results in the table, that the perimeters of the shaded portions do get larger as more figures are produced in this same pattern.

The perimeters in the table as decimals are 3, 4.5, 6.75, 10.125, 15.1875. The perimeters get larger each time.

Reflect

INSTRUCTOR NOTE: Consider discussing the solution to the focus problem. You might want to show the sample solution, available in MyMathLab, or a correct solution from a student or student group in the class. See the Instructor Guide for more debriefing ideas.

WRAP-UP

What's the point?

Area is a measure of the space enclosed within a two-dimensional figure, and perimeter is a measure of the distance around the figure. It's important to be able to not just calculate them but also know when to use each measure.

What did you learn?

How to calculate perimeter and area

Focus Problem Check-In
Do you have a complete solution to the focus problem? If your instructor has provided you with a rubric, use it to grade your solution. Are there areas you can improve?

1.16 Homework

Skills MyMathLab

First complete the MyMathLab homework online. Then work the two exercises to check your understanding.

Use the following figure for #1 and #2.

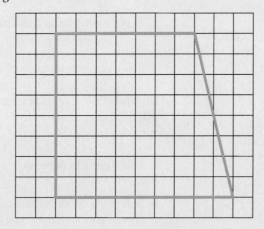

1. Find the area of the figure in square units.   64 square units

2. Find the perimeter of the figure in units if the slanted side is 8.2 units long.
   32.2 units

Concepts and Applications

Complete the problems to practice applying the skills and concepts learned in the section.

3. A student looks at a figure and finds that the perimeter is 10 inches and the area is 10 square inches. He then claims that the perimeter and area are equal. Explain his mistake.

   He is ignoring the units. The perimeter and area are not the same, since the units are different. They are both 10, but they are 10 of different things.

4. Explain what is wrong with each of the following statements.

   a. Seen in a tourist guide: "This state has the largest area of shoreline of any state in the region."

   The relevant shoreline measure should be perimeter or length, not area.

   b. Seen in a house description: "This house has over 3,200 feet."

   The measurement is most likely square feet.

   c. Seen in an advertisement: "Spring Special: Mulch is only $39/yard."

   Mulch is sold by volume, usually cubic yards.

5. A recipe for lemon bars calls for a 9″ × 13″ pan. How many 1″ × 1″ bars can be cut? How many 2″ × 2″ bars can be cut? How much is wasted if you serve only the 2″ × 2″ square bars?

   117 1″ × 1″ bars; 24 2″ × 2″ bars with 21 square inches wasted

6. A rectangular garden has dimensions of 12 feet by 20 feet. The homeowner wants to increase each dimension by 20%.

    a. Find the original perimeter and area of the garden.
      64 feet, 240 square feet

    b. Find the perimeter and area of the garden with a 20% increase in each dimension.
      76.8 feet, 345.6 square feet

    c. By what percent will the perimeter increase? By what percent will the area increase?
      20%, 44%

    d. Explain why the area and perimeter will not increase by the same percentage.
      The new perimeter is found by adding the increased quantities, whereas the area is found by multiplying them. When the 20% increase is multiplied by another 20% increase, the result is a 44% increase.

7. Do you believe that you can take a standard sheet of notebook paper and cut a hole in it that is large enough to walk through? Follow the directions below to try it for yourself.

> This exercise illustrates the idea that you can have a fixed area (the sheet of paper) but an infinite length (the cut line inside the sheet of paper).

    a. Fold a piece of paper in half the long way.

    b. Make cuts on all the vertical lines as indicated in the following picture. The lines should alternate sides, one from the top, one from the bottom, and so on. The closer the lines are to each other, the larger the hole will be.

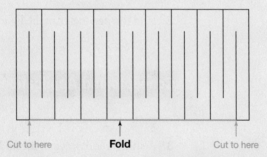

Cut to here   **Fold**   Cut to here

    c. Cut the fold line, but cut only between the first and last vertical lines. Do not cut all the way to the edge of the paper. Unfolded, the paper should look like this:

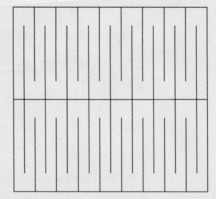

    d. See if you can actually walk through the hole you cut in the paper! Try the exercise again, making the cuts closer together, and see if you can make an even larger hole.

8. **Looking Forward, Looking Back**   The following figure is called **Pascal's triangle**. Determine the pattern in each row's numbers, and use it to complete the next row.

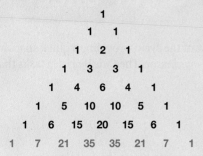

```
                    1
                 1     1
              1     2     1
           1     3     3     1
        1     4     6     4     1
     1     5    10    10     5     1
  1     6    15    20    15     6     1
1     7    21    35    35    21     7     1
```

Cycle 1   Study Sheet

This page is an overview of the cycle, providing a quick snapshot of the material covered to help you as you begin studying for a test. *However, it is not comprehensive.* The *Cycle Wrap-Up* tasks that follow will help you review the cycle in more detail.

SKILLS

Before adding or subtracting fractions, find a common denominator.

Sign patterns for integer operations

+	Positive	Negative
Positive	+	depends
Negative	depends	−

×	Positive	Negative
Positive	+	−
Negative	−	+

To increase 15 by 22%:

1 step: $15(1.22) = 18.3$

2 steps: $15(0.22) = 3.3 \rightarrow 15 + 3.3 = 18.3$

To convert 5 yards to inches:

Multiply by 36 because there are 36 inches in a yard. $5(36) = 180$ in.

Or, use dimensional analysis: $5 \text{ yd} \cdot \frac{36 \text{ in.}}{1 \text{ yd.}} = 180$ in.

ALGEBRAIC VOCABULARY

$$\overbrace{10y}^{\text{Term}} - \overbrace{15z}^{\text{Term}}$$
$$\underbrace{\qquad\qquad}_{\text{Expression}}$$

10 and y are factors of the term 10y

$$\underset{\underset{\text{Coefficient}}{\nearrow} \ \underset{\text{Variable}}{\nwarrow}}{10y}$$

GRAPHS

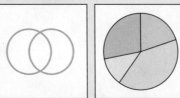

Venn diagram   Pie chart   Bar graph

Scatterplot   Linear function   Exponential Function

TYPES OF CHANGE

Arithmetic sequence

14, 22, 30, . . .

14 is the first term. Use the change (+8) to work backward to find the 0th term, 6.

nth term: $6 + 8n$

Linear function

x	y
0	6
1	14
2	22
3	30

$y = 6 + 8x$

Geometric sequence

15, 45, 135, . . .

15 is the first term. Use the change (·3) to work backward to find the 0th term, 5.

nth term: $5 \cdot (3)^n$

Exponential function

x	y
0	5
1	15
2	45
3	135

$y = 5 \cdot (3)^x$

SELF-ASSESSMENT: **REVIEW**

For each objective, use the boxes provided to indicate your current level of expertise. If you cannot indicate a high level of expertise for a skill or concept, continue working on it and seek help if necessary. Section numbers, like 1.2, are provided for reference.

INSTRUCTOR NOTE: Consider discussing this page before a test. The objectives are the same ones that are provided at the beginning of the cycle, giving students two points of comparison on their learning. Completing the page can help students see where they need to focus their studying for a test.

SKILL or CONCEPT

Low   High

1. Simplify, add, subtract, multiply, and divide fractions. *(1.2)*

2. Solve applied problems involving fractions. *(1.2)*

3. Plot ordered pairs. *(1.3)*

4. Determine the coordinates of a point. *(1.3)*

5. Interpret ratios. *(1.4)*

6. Scale ratios to produce equivalent ratios. *(1.4)*

7. Determine if quantities are proportional. *(1.4)*

8. Find and interpret experimental probabilities. *(1.5)*

9. Interpret signed numbers in situations. *(1.6)*

10. Add, subtract, multiply, and divide signed numbers. *(1.7)*

11. Find and interpret the mean of a set of numbers. *(1.8)*

12. Create and interpret pie, bar, and line graphs. *(1.9)*

13. Create and interpret scatterplots. *(1.10)*

14. Sketch a curve to best fit a scatterplot. *(1.10)*

15. Convert units by multiplying or dividing. *(1.11)*

16. Apply, find, and interpret percent change. *(1.12)*

17. Differentiate between variables/constants, expressions/equations, and factors/terms. *(1.13)*

18. Make conjectures and generalize patterns. *(1.14)*

19. Model change with linear and exponential functions. *(1.15)*

20. Calculate perimeter and area. *(1.16)*

WRAP-UP

> If you can find a path with no obstacles, it probably doesn't lead anywhere.
>
> —Frank A. Clark

INSTRUCTOR NOTE: Discuss the steps included in the *Cycle Wrap-Up* in detail with students so they understand the actions that lead to success on a test.

The *Cycle Wrap-Up* will help you bring together the skills and concepts you have learned in the cycle so that you can apply them successfully on a test. Since studying is an active process, the *Wrap-Up* will provide you with activities to improve your understanding and your ability to show it in a test environment.

STEP 1

GOAL Revisit the cycle content

Action Skim your cycle notes

Action Read the Cycle Study Sheet

Action Complete the Self-Assessment: Review

Don't wait until the last minute to get help on problem areas. Also, **study at least 2–3 hours**. If this turns out to be too much time spent studying, you can adjust for the next test. But if you do not study enough, you won't be able to compensate for that during the test.

Are you having anxiety about the test? If so, consider yourself normal. Most students are somewhat anxious before any test, but especially the first one. Follow the plan outlined in this *Wrap-Up* so that you have an effective approach for a test that will have a variety of types of problems. Being prepared will reduce your anxiety.

STEP 2

GOAL Practice cycle skills

Action Complete the MyMathLab assignment

Action Address your areas of difficulty

MyMathLab

Complete the MyMathLab assignment. Reference your notes if you are having trouble with a skill. For more practice, you can rework Skills problems from your homework. You can also access your MyMathLab homework problems, even if the assignment is past due, by using the Review mode available through the Gradebook.

STEP 3

GOAL Review and learn cycle vocabulary

Action Use your notes to complete the Vocabulary Check

Action Memorize the answers

Vocabulary Check

Cycle 1 Word Bank

absolute value	degree	mean	ratio
algebra	dependent variable	monomial	rate
area	experimental probability	negative number	real numbers
arithmetic	exponential function	opposite	scale
axes	expression	ordered pair	scatterplot
bar graph	equation	origin	similar
binomial	factor	percent change	term
conjecture	function	perimeter	trinomial
constant	geometric	pie graph	variable
counterexample	independent variable	polynomial	Venn diagram
Cartesian coordinate system	inductive reasoning	proportion	
	integers	proportional	
deductive reasoning	linear function	quadrants	

Choose a word from the word bank to complete each statement.

1. A(n) _____Venn diagram_____ is a picture that can be used to compare and contrast sets.

2. The _____Cartesian coordinate system_____ is composed of two number lines that intersect at a right angle. The two number lines, called _____axes_____, intersect at the _____origin_____. The number lines divide the plane into four areas called _____quadrants_____.

3. In the _____ordered pair_____ (5, 4), the first number tells how far to move to the right and the second number tells how far to move up.

4. A(n) _____ratio_____ is a comparison of two quantities and can be written as a fraction. When the two quantities have different units, it is also known as a(n) _____rate_____.

5. When two rates are equal, they form a(n) _____proportion_____, and the quantities are said to be _____proportional_____.

6. To _____scale_____ a fraction, multiply the numerator and denominator by the same nonzero number.

7. A relative frequency that tells the percent of the time something did happen can also be considered a(n) _____experimental probability_____.

8. To indicate a quantity that is less than zero, we write a(n) _____negative number_____.

9. The whole numbers and their opposites form the set of numbers called the _____integers_____.

10. One way to read -5 is as the _____opposite_____ of 5.

11. A number's distance from zero is known as its _____absolute value_____.

12. The _____real numbers_____ include all the numbers that can be represented on a number line.

13. To find the _____mean_____ of a set of numbers, add the numbers and divide the sum by how many numbers are in the set.

14. The best graph to use to make part-to-part comparisons is often a(n) _____bar graph_____. The best graph to use to make part-to-whole comparisons is often a(n) _____pie graph_____.

15. A(n) _____scatterplot_____ can be drawn to show the relationship between two variables.

16. When two variables are related, the one that affects the other is called the _____independent variable_____, while the one that is affected is called the _____dependent variable_____.

17. The branch of mathematics in which letters are used to represent numbers is called _____algebra_____.

18. A value that does not change is called a(n) _____constant_____.

19. A(n) _____variable_____ is a letter used to represent an unknown quantity.

20. A(n) _____term_____ is a constant or variable or the product or quotient of constants and/or variables that are _____factors_____.

21. A(n) _____expression_____ is a mathematical phrase that has one or more terms separated by addition or subtraction signs, while a(n) _____equation_____ is a statement that two of these mathematical phrases are equal.

22. The expression $3x^3 - 2x^2 - x + 3$ is an example of a(n) _____polynomial_____.

23. A(n) _____monomial_____ has one term, a(n) _____binomial_____ has two terms, and a(n) _____trinomial_____ has three terms.

24. The _____degree_____ of the expression $6x^3 - 2x^2 - x + 5$ is 3.

25. When specific examples are used to form a general rule, _____inductive reasoning_____ is used. When a general rule is used to give specific examples, _____deductive reasoning_____ is used.

26. A hypothesis is also called a(n) _____conjecture_____. Only one _____counterexample_____ is needed to disprove a hypothesis.

27. To calculate _____percent change_____, divide the amount of change by the initial value.

28. A(n) _____function_____ is a rule that assigns each input to exactly one output.

29. If the output values are multiplied by the same factor repeatedly as the input values change by 1, there is a(n) _____exponential function_____. The function values form a(n) _____geometric_____ sequence.

30. If the same value is repeatedly added or subtracted to the output values as the input values change by 1, there is a(n) _____linear function_____. The function values form a(n) _____arithmetic_____ sequence.

31. The area of a two-dimensional figure is a measure of the region it encloses. The _____perimeter_____ of a two-dimensional figure is the distance around it.

32. Geometric shapes that have the same shape but not necessarily the same size are _____similar_____.

STEP 4

GOAL Practice cycle applications

Action Complete the Concepts and Applications Review

Action Address your areas of difficulty

Work all the applications problems until you can get the correct answers on your own. This may require looking up the topic in the notes and working on more problems of a similar nature.

Concepts and Applications Review

Complete the problems to practice applying the skills and concepts learned in the cycle.

> Don't just go through the motions when working the applications problems. These problems resemble test questions in complexity and format and deserve your full attention.

1. Use this image from a computer download to answer the questions that follow.

Screenshot from Mozilla Firefox. Copyright © by Mozilla Firefox. Used by permission of Mozilla Corporation. Firefox is a registered trademark of the Mozilla Foundation.

a. What percent of the Firefox Setup file has been downloaded? Round to the nearest percent. $1.1/5.8 \cdot 100 \approx 19\%$

b. How many MB are left to download in the setupeng.exe file?
$23.1 - 1.5 = 21.6$ MB

c. What percent of the setupeng.exe file is left to download? Round to the nearest percent. $21.6/23.1 \cdot 100 \approx 94\%$

d. State the Firefox Setup file download rate.   60.8 KB/sec

e. Interpret the download rate from part d in words.

60.8 kilobytes per second are being downloaded.

f. The header of the box says "5% of 3 files." Verify that this percent is correct. Show a calculation to support your work.

Total downloaded so far: 3.9 MB

Total to download: 76 MB

$3.9/76 \cdot 100 \approx 5\%$ (rounded to the nearest percent)

2. Complete each of the following signed-number calculations.

$-8 + (-8) = -16$

$-8 - (-8) = 0$

$-8 \cdot (-8) = 64$

$-8 \div (-8) = 1$

> A characteristic that differentiates problems from exercises is that you don't immediately know how to start solving a problem. That can be frustrating, but it is necessary to work through that frustration. Persisting through challenges, also known as having grit, will help you pass a math class as well as be successful in college in general.

3. Suppose you haven't paid attention to the balance in your checking account and you begin to overdraw the account. Assume you had $200 in your account on Friday morning and make the following charges over the weekend:

Friday	Saturday	Sunday
$100 ComEd	$10 IHOP	$85 Woodman's
$150 Book Store	$35 Mobil	
$15 Pizza Hut		
$20 ATM		
$20 AMC Theater		

Your bank charges a $5 overdraft fee for each item (up to four per day) if your account is overdrawn. There is also a consecutive-day overdraft fee of $5 per day for each day your account is overdrawn after the first day.

a. If the charges go through in the order listed, calculate your balance at the end of each day.

Friday:

$200 - 100 - 150 - 15 - 20 - 20 = -105; -105 - 20$ (overdraft fees) $= -\$125$

Saturday:

$-125 - 10 - 35 = -170; -170 - 10$ (overdraft fees) $- 5$ (overdraft fee) $= -\$185$

Sunday:

$-185 - 85 = -270; -270 - 5$ (overdraft fee) $- 5$ (overdraft fee) $= -\$280$

b. If you could decide on the order for the charges to clear, in which order would you like them to be processed?

It would be better if the smaller charges cleared first on Friday. Then there would be fewer overdraft charges.

c. Suppose you notice the problem first thing on Saturday and transfer $150 from your savings to your checking account. Will you avoid overdrawing the account on Saturday?

No; transferring $150 on Saturday morning would avoid one of the overdraft fees on Saturday, though.

4. Suppose you scored 72, 78, and 75 on the first three exams in your math course. Imagine your instructor enters your next exam score incorrectly as 8 instead of 82.

a. What is your average with the correct exam score?   76.75

b. What is your average with the mistake?   58.25

c. What percent change in your average is caused by the instructor's mistake?

$$\frac{58.25 - 76.75}{76.25} \cdot 100 \approx -24\%$$

5. a. Suppose the following data show the percent of students in a college class who use various forms of social media. Decide if a bar or pie graph is more appropriate for the data and explain why.

A pie graph is not appropriate to represent this data, since some students in the class are in more than one social media category. A pie graph can be used only when the categories do not overlap and all the categories make up the whole. So a bar graph would be more appropriate.

Social Media	Percent
Twitter	45
Instagram	20
Pinterest	8
LinkedIn	4
Google+	23

b. Make the appropriate graph from the data in part a.

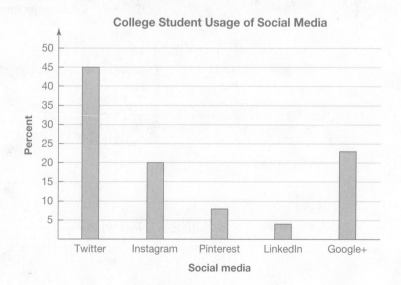

6. Use this part of the nutrition label to determine if the recommended amount of dietary fiber is proportional to the amount of total fat in a diet. Support your work with calculations.

25 g/65 g ≈ 0.38

30 g/80 g ≈ 0.38

*Percent Daily Values are based on a 2,000 calorie diet. Your daily values may be higher or lower depending on your calorie needs.			
	Calories	2,000	2,500
Total Fat	Less than	65g	80g
Saturated Fat	Less than	20g	25g
Cholesterol	Less than	300mg	300mg
Sodium	Less than	2,400mg	2,400mg
Total Carbohydrate		300g	375g
Dietary Fiber		25g	30g
Protein		50g	65g
Calories per gram:			
Fat 9 • Carbohydrate 4 • Protein 4			

If the answers are rounded, they look proportional. The decimals are not identical, though. For the purposes of the nutrition label, the results are close enough to claim proportionality.

7. Imagine that you have a sinus infection, and your doctor prescribes a Zpack. A Zpack is the drug Zithromax given over five days in the following way: 500 mg on day 1 and 250 mg each day for the remaining four days. The higher dose given on the first day is known as a "loading dose."

a. Complete this table:

Day	Cumulative Amount of Zithromax Taken (mg)
1	500
2	750
3	1,000
4	1,250
5	1,500

b. Make a line graph by plotting points. Label the axes.

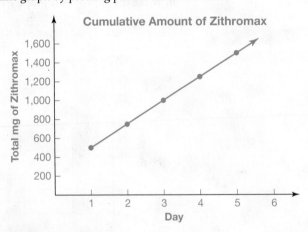

c. Is the data linear? Why or why not?

Yes; when the day increases by 1, the total amount of Zithromax increases by 250 mg.

d. Why is this function for the total amount, using *D* to represent the number of days, incorrect?

$$\text{Total amount} = 500 + 250D$$

The answer will be off by 250 mg each time. On day 2, you should calculate 500 + 250, not 500 + 250(2).

e. Write a function for the total amount after *D* days.

Total amount = 500 + 250(*D* − 1) or Total amount = 250 + 250*D*

8. You purchase a car for $20,000 and find out it will lose approximately $1,500 in value each year for the next 5 years. Make a table to compare the age of the car with its value. Use the table to answer the questions that follow.

Age (years)	Value ($)
0	20,000
1	18,500
2	17,000
3	15,500
4	14,000
5	12,500

a. Write a function to model the data.

Value = $20,000 − $1,500*n*, where *n* is the age of the car in years

b. How much will the car be worth in 5 years?

$12,500

c. Approximate when the car will be worthless if this depreciation rate continues.

Between 13 and 14 years

9. Create a scatterplot of waist circumference vs. neck circumference, and draw in a line or curve that models the data.

Neck Circumference (in.)	Waist Circumference (in.)
13.5	27.5
14	30
14.5	34
15	31
15.5	35
16	33
14.2	29
15.4	33
14.9	33
15	29

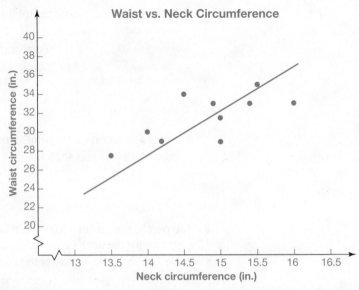

Waist vs. Neck Circumference

10. An employee has been offered a 3.5% raise and a $15,000 bonus on top of the raise for the following year.

    a. Write an expression for the employee's new salary if his old salary is *S*.
    1.035S + 15,000

    b. Write an equation that indicates the employee's new salary will be twice his old salary.
    1.035S + 15,000 = 2S

11. For each sequence, fill in the next two numbers in the pattern. Then write an expression for the *n*th term in the sequence.

    a. 2, 5, 10, 17, 26, 37, _____, _____, ... , _____
    50, 65, *n*th term: $1 + n^2$

    b. −22, −16, −10, _____, _____, ... , _____
    −4, 2, *n*th term: $-22 + 6(n - 1)$ or $-28 + 6n$

12. The leading edge of some lava flows can travel as fast as 10 km/hr on steep slopes. How fast is this rate in miles/day? One mile is approximately 1.61 km.
    149 miles/day

13. **a.** Draw a square with an area of 36 square units. What is its perimeter?

    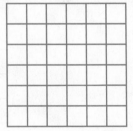

    24 units

Sometimes students will study for a test and feel confident, but find the test problems look unfamiliar. To avoid this, challenge yourself to work new problems without help aids and without knowing which section they come from. Continue working problems in this way until you can do them confidently and correctly. Step 5 in the *Cycle Wrap-Up* gives some specific ideas for studying with this technique.

b. Draw a square with a perimeter of 36 units. What is its area?

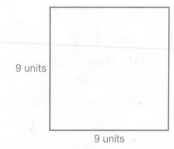

9 units

9 units

81 square units

14. Decide if each statement is true or false. If it is true, explain why. If it is false, give a counterexample.

a. A 50% increase, followed by a 50% decrease, results in no net change.

The statement is false. If the cost of a $100 item is increased 50%, the cost rises to $150. If the cost is then decreased 50%, the cost falls to $75, not to the original cost of $100.

b. A 200% increase is the same as tripling a quantity.

The statement is true. The original quantity, 100%, is increased by 200%, resulting in 300% or three times the original quantity.

15. Use the following diagram to answer the questions.

Children's Acetaminophen Liquid (Generic)

Use weight to determine the appropriate dose. If the weight is not available, then use age. Dose may be repeated every 4 hours as needed, to a maximum of 5 doses in 24 hours. Medicine may be given a maximum of 5 days unless directed by a doctor.

Weight	Age	Children's Acetaminophen
Under 24 lb	Under 2 years	Seek medical advice
24–35 lb	2–3 years	$\frac{1}{2}$ teaspoon or 2.5 mL
36–47 lb	4–5 years	$\frac{3}{4}$ teaspoon or 3.75 mL
48–59 lb	6–8 years	1 teaspoon or 5 mL
60–71 lb	9–10 years	$1\frac{1}{4}$ teaspoons or 6.25 mL
72–95 lb	11 years	1.5 teaspoons or 7.5 mL

Acetaminophen 320 mg in each 5 mL = 1 teaspoon

a. If a 5-year-old is given the maximum recommended dose of the medicine for 24 hours, how many grams of acetaminophen has she consumed?

The medicine can be administered at most five times a day. The dose for a 5-year-old is $\frac{3}{4}$ teaspoon. 0.75 tsp · 5 times per day = 3.75 tsp

Each teaspoon contains 320 mg of acetaminophen.
3.75 tsp · 320 mg acet./tsp = 1,200 mg acetaminophen

One thousand milligrams is 1 gram. $\frac{1,200}{1,000}$ = 1.2 grams. So the child consumes 1.2 grams of acetaminophen in a 24-hour period.

b. If that child had been dosed by her weight, 33 pounds, what is the maximum she would have received in 24 hours?

By her weight, she should have been given only $\frac{1}{2}$ teaspoon per dose. This would have resulted in $2\frac{1}{2}$ teaspoons over the 24-hour period, or 0.8 gram of acetaminophen.

c. How much extra acetaminophen has she been given? Give your answer as an amount and a percent.

0.4 gram extra. $\frac{0.4}{0.8} \cdot 100 = 50\%$

16. The circumference of a circle is its perimeter. The formula for circumference is $C = 2\pi r$ or $C = \pi d$, where r is the circle's radius and d is the circle's diameter.

   a. In these formulas, what is constant and what is variable?

   2 and π are constant; C, r, and d vary.

   b. What are the independent variables? What are the dependent variables?

   Circumference depends on the radius or diameter of the circle, so circumference is dependent and radius and diameter are independent.

   c. Suppose a test question states the radius of a circle as $\frac{2}{3}$ inch. You decide to calculate the circumference using $r = 0.67$ and 3.14 for π. What will you get? Round to the nearest hundredth. $C \approx 4.21$ inches

   d. Your instructor wants a more precise answer and calculates the circumference with $r = \frac{2}{3}$ and the π button on the calculator. What will she get? Round to the nearest hundredth. $C \approx 4.19$ inches

   e. What is the percent error of your calculation?

   The error is 0.02 inch, which is 0.48% of 4.19 inches.

STEP 5

GOAL Simulate a test environment

Action Construct a mock test and take it while being timed

Creating a Mock Test

To pass a math test, you must be able to recall vocabulary, skills, and concepts and apply them to new problems in a fixed amount of time.

Up to this point, you have worked on all the components of the cycle, but they were separated and identified by their topic or type. Although helpful, this is not a test situation. You need to create a test situation to help overcome anxiety and determine where problem areas still exist. Think about a basketball player. She must practice her skills, but until she engages in a practice game, she doesn't really have a feel for a game situation.

The key is to practice problems without assistance under time pressure.

Use the following steps to create a mock test.

- Pick a sample of 5 to 10 problems from your notes and homework and write them on note cards. Choose both skills and applications. On the back of the note card, write the page number where the problem was found.
- Shuffle the cards.
- Set a timer for 20 minutes.
- Work as many problems as you can without outside help.
- Check your work by consulting the page number on the card.

This test will show you where your weaknesses are and what needs more work. It will also give you a feel for your pace. If you completed only a few problems, you are not ready for a test.

Continue to work on the areas that need improvement.

Now is the time to make sure that your calculator batteries are working and that you have pencils for the test. If you still have issues with using your calculator, seek help.

Get plenty of sleep the night before the test. Don't go into the test not having slept or eaten.

When you get your test, take a deep breath and exhale. This will reduce anxiety. Then start with the easiest problems on the test to build your confidence.

CYCLE **TWO**

HOW DOES THAT WORK?

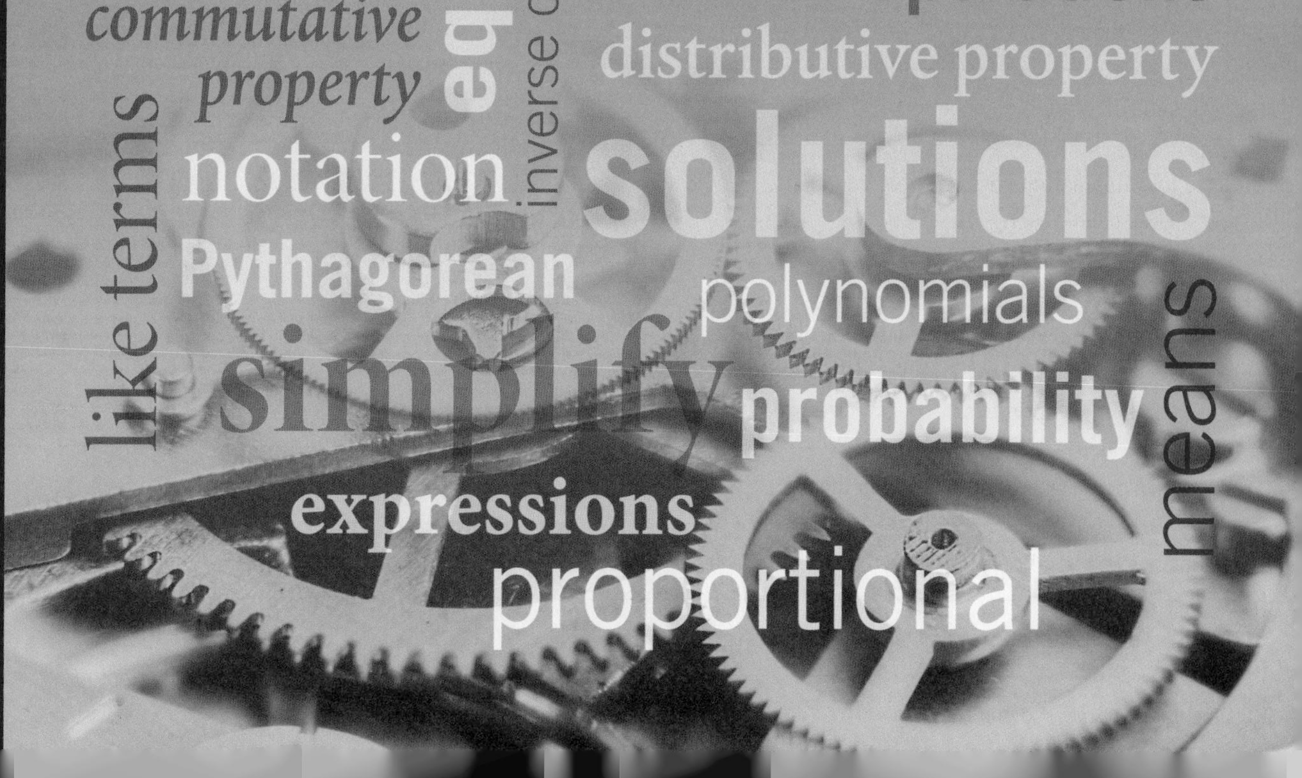

mode
equations
inverse operations
volume
order of operations
commutative property
distributive property
like terms
notation
solutions
Pythagorean
polynomials
simplify
probability
means
expressions
proportional

SELF-ASSESSMENT: PREVIEW

Below is a list of objectives for this cycle. For each objective, use the boxes provided to indicate your current level of expertise.

INSTRUCTOR NOTE: Consider discussing this page when you begin the cycle. The same list is presented at the end of the cycle so that students can note their progress and identify areas that they still need to work on before the test.

SKILL or CONCEPT	Low → High
1. Find and interpret weighted means.	
2. Find the median and mode of a data set.	
3. Apply basic exponent rules.	
4. Use geometric formulas.	
5. Identify and add like terms.	
6. Use the order of operations to simplify expressions.	
7. Apply the commutative and associative properties.	
8. Use and apply the distributive property.	
9. Determine if two expressions are equivalent by using the commutative, associative, and distributive properties.	
10. Recognize when the distributive property can be applied.	
11. Verify a solution to an equation.	
12. Write and solve one-step equations.	
13. Solve two-step and multi-step linear equations.	
14. Write an equation to model a situation.	
15. Solve a problem numerically and algebraically.	
16. Write and solve proportions.	
17. Solve problems using the Pythagorean theorem.	
18. Calculate theoretical probabilities.	
19. Compare theoretical and experimental probabilities.	
20. Calculate volume and surface area.	

One thing is sure. We have to do something. We have to do the best we know how at the moment . . . If it doesn't turn out right, we can modify it as we go along.

—Franklin D. Roosevelt

2.1 Predicting a Child's Height
Focus Problem

Parents often wonder, usually out of sheer curiosity, how tall their child will be as an adult. When a child has a medical condition or appears to be growing abnormally, the need to understand his or her growth can be important. One early method involved simply doubling a child's height at age 2. This gave very inaccurate results, especially for girls. As a result, many better methods have been developed to try to accurately predict a child's eventual adult height from various factors.

One simple method to predict adult height is to average the parents' heights and then add 2.5 inches if the child is a boy or subtract 2.5 inches if the child is a girl. This method can be fairly accurate, but can also be as much as 5 inches above or below the child's eventual height. Write a formula or formulas to express this calculation, and clearly define any variables that you use.

Another method is to add the parents' heights and then add an additional 5 inches if the child is a boy or subtract 5 inches if the child is a girl. This total is then divided by 2 to predict the child's ultimate height. Write a formula or formulas for this method, using variables you've already defined. How do these formulas compare to the first ones described? Are they equivalent? Show how they are equivalent or explain why they are not.

A third method, referred to as the "mid-parent rule," uses a weighted average of the parents' heights in which the height of the opposite-gender parent is multiplied by a factor of 12/13 before it is averaged with the other parent's height. Write a formula to express this calculation. Is it equivalent to the first two formulas? Explain fully.

These early, simple methods gave pediatricians and parents quick ways to estimate a child's adult height, but they are not terribly accurate. A new statistical model was eventually developed based on age, weight, and parental height data from a large, longitudinal study. This method, called the Khamis-Roche method, can be applied only to healthy Caucasian children over the age of 4. To use this method, coefficients must be obtained from a table, multiplied by the corresponding measurement, and then added together with the constant to generate a formula to predict a child's height. The partial tables that follow give the coefficients for girls and boys aged 4 to 7. Write a formula for this method to predict the height of a 5-year-old girl, clearly defining any variables you use.

	Age	Constant Term	Coefficient of Current Height (in.)	Coefficient of Current Weight (lb)	Coefficient of Average Parent Height (in.)
Girls	4	−8.13250	1.24768	−0.019435	0.44774
	5	−5.13582	1.19932	−0.017530	0.38467
	6	−3.51039	1.15866	−0.015400	0.34105
	7	−2.87645	1.11342	−0.013184	0.31748

	Age	Constant Term	Coefficient of Current Height (in.)	Coefficient of Current Weight (lb)	Coefficient of Average Parent Height (in.)
Boys	4	−10.2567	1.23812	−0.0087235	0.50286
	5	−11.0213	1.10674	−0.0064778	0.53919
	6	−11.1138	1.05923	−0.0052947	0.52513
	7	−10.9984	1.05877	−0.0048144	0.48538

Data from: "Predicting Adult Stature Without Using Skeletal Age: The Khamis-Roche Method," Harry J. Khamis and Alex F. Roche.

Use all four methods described to predict the ultimate adult height of a 5-year-old girl if her current height is 3'6" and her weight is 43 pounds. Assume her mother is 5'5" and her father is 6'1" tall. Then use all four methods again to predict the height for you or someone you know. Which of these methods would have most accurately predicted the height?

Maybe you've seen or used an online calculator to predict height and you've wondered . . .

HOW DOES THAT
WORK?

How did the calculator use your data to determine an estimate? How much error can you expect in that estimate? Would a different calculator give you the same result?

INSTRUCTOR NOTE: Additional focus problems are available in MyMathLab. Each focus problem is accompanied by a writing template to help students form their solution. A grading rubric is also available. For more information on ways to use the focus problems, see the Instructor Guide at the beginning of the worktext.

If you are using groups and would like more detailed information on using them effectively, please consult the Instructor Guide at the beginning of the worktext. It contains techniques and resources for group work.

This focus problem is an example of a real-world challenge for which we continue to improve our response. We are much better at predicting a child's adult height now than we were decades ago. But in order to improve on a method, you need to first understand how it works.

This cycle will focus on algebraic details that will help you write and simplify expressions and write and solve equations. These skills can help you build models to do things like estimate height, as required in this problem. Understanding the details of how something works and the rules that apply is important in every discipline. This cycle will lay important foundations for the rest of your work in algebra.

As you learn skills and concepts in this cycle, return to the focus problem and see if you have any additional insights to help you solve it. You'll see sticky notes throughout the cycle to encourage and remind you to work on the focus problem.

INSTRUCTOR NOTE: Remind students to read the *Getting Ready* article for the next section.

Getting Ready for Section 2.2

Read the article and answer the questions that follow.

Here's Why Real-World MPG Doesn't Match EPA Ratings

Has this happened to you? You go shopping for a new car and great fuel economy is high on the list of things you want. You buy a car that's rated 30 mpg on the highway and 28 mpg overall by the Environmental Protection Agency (EPA). But after a month or so of driving around, you find that the best fuel efficiency you can get is a measly 24 mpg average. You might acknowledge that you drive a little faster than the speed limit, but you're no hot-rodder.

So why doesn't your fuel economy match the EPA rating?

This question comes up most often when gasoline prices are rising. It is especially pertinent now as a fuel-efficiency ratings scandal sullies Hyundai Motor Corp.'s image and leaves millions of consumers wondering if they can trust the EPA window sticker the next time they shop for a new car or truck.

Hyundai has had to roll back its fuel-efficiency claims for eight models, along with an additional five rollbacks for cars and crossovers sold by its sister company, Kia Motors. Hyundai and Kia issued a joint statement saying they will give customers a personalized debit card that will reimburse them for their difference in the EPA combined fuel economy rating, based on the fuel price in their area and their own actual miles driven. The carmakers explained that the issue was rooted in an error by its South Korean test crew. The problem has been corrected and won't happen again, Kia and Hyundai say.

That's good, and the fact that the inflated Hyundai and Kia fuel economy figures were caught and ordered corrected by the EPA is even better. It proves that the agency's system works.

But it took the EPA two years to act after fielding hundreds of complaints from Hyundai customers that the mileage they were experiencing wasn't even coming close to the official EPA estimates posted on the vehicles when they were on the dealers' lots.

For many, the explanation of what happened was an eye-opening look at how the system works: The EPA establishes the tests that yield the fuel economy figures, but for the most part it doesn't conduct the tests itself. It doesn't have the budget, equipment or manpower to test the hundreds of individual models with unique engine and transmission combinations that automakers produce each year.

Instead, the agency gives its test protocols to the auto companies and lets each test its own cars and trucks. It accepts as true the "EPA estimated" fuel-efficiency numbers each car company submits. To keep the industry honest, the agency runs scores of spot checks each year. Until the Hyundai/Kia mess, only two rollbacks had been ordered in the past decade, each for a single model. The 13 Hyundai and Kia models with overstated fuel efficiency represent an unprecedented breach that has some consumer advocates calling for the agency to conduct an industry-wide verification of fuel-efficiency claims.

Even if they don't own Hyundai or Kia vehicles, many car and truck owners are stewing over the gap between the fuel efficiency they've been promised in advertising and window stickers and what they're actually getting on the road. Honda, for example, continues to struggle with customer dissatisfaction with the fuel economy of older model gas-electric Civic Hybrid sedans.

Events may prove that there have been deliberate efforts by one or more automakers to game the system and achieve higher testing scores than warranted under the very specific, federally mandated testing cycles. But for the most part, the fuel-economy gap exists for a more mundane reason: Real people drive real cars in the real world. There are so many variables that the idea of an absolutely accurate rating of average mpg is laughable. But to new-car buyers, it often feels as if the joke is on them.

City vs. Highway

A key element in assessing the EPA rating for a vehicle's average fuel economy (EPA combined) is the split between highway and city driving. Almost all cars and trucks deliver better fuel economy while cruising at 55 mph on the open highway than they do while stopping and starting at low speed on city streets.

The EPA rating for combined mpg presumes that we drive 55 percent of the time in the city and 45 percent of the time on the highway. Most people simply assume that's the case in their own driving. But many motorists—especially those in urban regions with lots of traffic congestion—spend far more time driving in city conditions than they do on the open road.

It's also useful to remember that even when a lot of driving time is spent on a highway or freeway, it only counts as highway driving when the average speed is 50 mph or so. Crawling toward Manhattan on the Staten Island Expressway at 10 mph is actually city driving, even though it's technically done on a highway.

(continued)

Here's Why Real-World MPG Doesn't Match EPA Ratings (*continued*)

If you diligently keep track of the number of miles that you are really driving at highway speeds each day for a month or so, you might find that you really are doing most of your driving at less efficient city speeds. If you have a newer vehicle with a trip meter that displays your average speed, keeping track is easy. Just set it to zero each morning, and then check each evening for your actual average speed for the day.

No Bias

Fuel-efficiency researchers David Greene and Zhenhong Lin at Oak Ridge National Laboratory in Tennessee recently compared EPA estimates and "Your MPG" reports and found no evidence of over-performance bias in the government's numbers. But what the researchers did identify were six factors they said could cause a vehicle's real-world fuel economy to vary significantly from its EPA rating.

At the top of the list is how much city driving the car does. The proportion of stop-and-go driving could reduce the EPA combined city/highway fuel efficiency for a particular vehicle by as much as 27 percent, they say. Individual driving style can cause the second biggest variation, lowering fuel efficiency by up to 18 percent. Edmunds testing supports these conclusions. "Calm" drivers, those motorists who don't accelerate constantly and who avoid unnecessary lane changes, get 35 percent better fuel economy than other drivers.

Other factors in play are air-conditioner use (up to 14 percent reduction in fuel efficiency), vehicle size (up to 15 percent reduction) and the region in which the vehicle is primarily operated (up to 12 percent less fuel efficiency, because hot weather and mountainous conditions take a toll on fuel economy).

Fuel type also affects mileage. Most gasoline in the U.S. today is 8-10 percent ethanol, but the EPA does its tests with 100 percent gasoline in the tank. The use of ethanol to increase the amount of oxygen in gasoline for better combustion can reduce fuel efficiency by around 2 percent all by itself.

Questions

1. According to the article, "calm" drivers get 35% higher gas mileage than other drivers. If a driver who accelerates and changes lanes often gets 21 mpg, what gas mileage should a "calm" driver get?

   28.35 mpg

2. The EPA rating for combined gas mileage assumes a driver spends 55% of driving time in the city and 45% on the highway. Assume your car gets 22 mpg in the city and 32 mpg on the highway. Multiply each gas mileage by the percent of time the EPA expects a driver to be at that rate to find the combined gas mileage.

   26.5 mpg

3. If you spend 35% of your driving time in the city and the rest of the time on the highway, find your combined gas mileage using the miles per gallon provided in #2. How does your answer compare to your answer for #2? Explain.

   28.5 mpg; the mileage is higher since a greater percentage of time was spent on the highway.

2.2 Rule of Thumb: Weighted Means

SECTION OBJECTIVES:
- Find and interpret weighted means
- Find the median and mode of a data set

Explore

One rule of thumb in college is that a student should spend 2 hours studying outside of class for every hour spent in class. For example, it is recommended that a student study 6 hours outside of class if the student is taking a 3-credit-hour course.

1. Here are study logs for two students for one week.

Ben

Class	Credit Hours	Total Hours Studied
History	3	6
Statistics	3	6
Psychology	3	4
English	3	5
Total	**12**	**21**

Kate

Class	Credit Hours	Total Hours Studied
History	3	6
Math	6	6
Chemistry	4	4
English	3	3
Total	**16**	**19**

a. Find the overall average number of study hours per credit hour for each student. Are they meeting the college rule of thumb?

Ben: $\frac{21}{12} = 1.75$ study hr/credit hr; Kate: $\frac{19}{16} = 1.19$ study hr/credit hr; no

b. Find the hours studied per credit hour for each course Ben takes. Average the four rates. Does this result match Ben's overall average in part a? Why or why not? Repeat this process with Kate's information.

Ben

Class	Hours Studied Per Credit Hour
History	2
Statistics	2
Psychology	1.33
English	1.67
Average	1.75

Kate

Class	Hours Studied Per Credit Hour
History	2
Math	1
Chemistry	1
English	1
Average	1.25

Ben's average of 1.75 study hours/credit hour matches the result in part a.
Kate's average of 1.25 study hours/credit hour does not match the result in part a.

Ben's averages match because he has the same number of credit hours for each subject; Kate does not.

Discover

In previous sections, we worked with the mean, or average, as a measure of the center of a data set. Many times the data values do not have the same importance, which affects how the mean is computed. In the *Explore*, because the number of credit hours varied for Kate, the average of her individual course study rates was not the same as the overall average. Thus, we need a way to find the mean and factor in the credit hours, or weight, of each course.

As you work through this section, think about these questions:

1 Can you recognize when to calculate a weighted mean?

2 How is the median of a data set different from its mean?

Weighted Mean

A **weighted mean** is an average in which some values carry more importance, or more weight, than others. To find the weighted mean, multiply each data value by its weight. Add these values and divide the total by the sum of the weights.

Weighted mean =

$$\frac{(\text{1st value} \cdot \text{1st weight}) + (\text{2nd value} \cdot \text{2nd weight}) + \cdots + (\text{last value} \cdot \text{last weight})}{\text{sum of weights}}$$

An abbreviated way of writing the formula uses the summation symbol, $\sum$. The formula can be written as

$$\text{Weighted mean} = \frac{\sum xw}{\sum w}$$

where x indicates a data value and w its corresponding weight.

EXAMPLE: Let 3, 4, and 5 be the data values. Suppose the 3 carries twice as much weight as the 4 and 5. Find the weighted mean.

SOLUTION: For the data values of 3, 4, and 5, the weights are 2, 1, and 1, respectively.

$$\text{Weighted mean} = \frac{3 \cdot 2 + 4 \cdot 1 + 5 \cdot 1}{2 + 1 + 1} = \frac{15}{4} = 3.75$$

This is the same mean we would get if we simply listed the 3 twice and found the mean of 3, 3, 4, and 5.

Notice that if we weighted the numbers equally and found the mean of 3, 4, and 5, we would get 4. The weighted mean is lower, since the lowest value (3) carries a greater weight.

2. Find the weighted mean of 15, 25, 100, and 250 if 15 carries four times as much weight as each of the other values.

62.1

Weights can come in many forms. It is not always obvious when a mean is weighted or what the weights are. When finding any average, look for information that indicates whether or not the numbers being averaged are equally important. Let's look at a grading situation where a weighted average is used.

EXAMPLE

Suppose a student has earned the points given in four categories. The teacher uses the percent weights listed in the table that follows. Find the student's percent and letter grade in the class. Assume the instructor uses a 90-80-70-60 letter grade scale.

Student's points

Homework: $\dfrac{85}{90} \approx 94.4\%$ Midterm: $\dfrac{66}{100} = 66\%$

Quizzes: $\dfrac{70}{80} = 87.5\%$ Final Exam: $\dfrac{63}{100} = 63\%$

Grade Percent by Category	
Homework	10%
Quizzes	10%
Midterm	40%
Final Exam	40%

SOLUTION

The table with the grade percents indicates that each category does not have the same importance. Because of that, we cannot simply average the student's grades in each category.

Instead, we need to find a weighted mean. It is helpful to use a table to organize the work involved with a weighted mean calculation. To do so, we have to identify the data and how it is being weighted. In this situation, the student's grades in each category are the data, and they are weighted according to the grade percents by category.

Enter the student's percent for each category into the table in the Grade column. The weights can be listed as decimals, as shown, or as given in the original table: 10%, 10%, 40%, and 40%. Multiply each grade by its weight. Add the Grade · Weight values. Divide this sum by the sum of the weights.

Category	Grade	Weight	Grade · Weight
Homework	94.4	0.10	9.44
Quizzes	87.5	0.10	8.75
Midterm	66	0.40	26.4
Final Exam	63	0.40	25.2
		Sum = 1	Sum = 69.79

The final grade is $\frac{69.79}{1} = 69.79$. If the teacher rounds to the nearest whole number, the student's grade in the course is a 70 or C. If the teacher does not round, the student's grade is a D. If the student had assumed the category percents could simply be averaged, he would have incorrectly concluded that he had a 77.7% in the class.

To find a weighted mean:

$$\text{Weighted mean} = \frac{\sum xw}{\sum w}$$

where x indicates a data value and w its corresponding weight.

1. Identify the data and weights.
2. Multiply each data value by its weight.
3. Find the sum of the products.
4. Divide by the sum of the weights.

EXAMPLE: At a department store, there are 20 employees who earn $18.75 per hour, 5 employees who earn $22.00 per hour, and 100 employees who earn $9.00 per hour. Find the average hourly wage per employee.

To begin, notice that there are different numbers of employees at each hourly rate. Because of that, the hourly wages cannot be averaged by themselves. Instead, the numbers of employees weight the wages.

$$\text{Average hourly wage per employee} = \frac{\$18.75 \cdot 20 + \$22.00 \cdot 5 + \$9.00 \cdot 100}{20 + 5 + 100}$$

$$= \frac{\$1385.00}{125}$$

$$= \$11.08 \text{ per hour per employee}$$

Tech TIP

A graphing calculator can be used to find the weighted mean of a data set. Choose STAT, then EDIT. Enter the data values into L_1. Enter the weights into the second list, L_2. Then find one variable statistics by choosing STAT, then CALC, then #1 (1-Var Stats). Type L_1, L_2 after 1-Var Stats and press ENTER.

3. The following data represents the ages in years of cats at a shelter. Find the average age of the cats. Instead of entering the values into your calculator separately, use the frequencies of the ages to find a weighted mean. The table is provided to organize your work.

1, 1, 1, 1, 1, 1, 1, 2, 2, 2, 3, 3, 3, 3, 3, 3, 3, 3, 3, 3, 4, 4, 4, 4, 5, 5, 5, 5, 5, 5, 5, 5, 5, 5, 5, 6, 6, 8, 11, 15

Age	Frequency	Age · Frequency
1	7	7
2	3	6
3	10	30
4	4	16
5	11	55
6	2	12
8	1	8
11	1	11
15	1	15
	Sum = 40	Sum = 160

Average age of the cats is 4 years.

In addition to means and weighted means, there are other measures of the center of a data set. Two such measures are the median and the mode.

Median and Mode

The **median** of a set of data is the middle value when the data values are arranged in order.

EXAMPLES:

Data	Median
72 74 (76) 78 80	76

If there is an even number of values, the median is found by averaging the two middle values.

Data	Median
72 74 76 78 80 88 Average = 77	77

The **mode** is the most common value or values in a set of data. Data sets can have no mode, one mode, or more than one mode.

EXAMPLES:

Data	Mode	Explanation
72 74 76 76 78 80	76	76 appears twice.
72 74 74 76 76 76 78 80 80 80	76 and 80	76 and 80 each appear three times.
72 74 76 78 80	No mode	Each value has the same frequency.

Tech TIP

A graphing calculator can be used to find the median of a data set. Choose STAT, then EDIT. Enter the data values as one list. Then find one variable statistics on the list with the data.

---

4. Find the mean and median for the data set of ages of college students in a small study group. If a student who is 55 years old joins the group, how will that data value affect each measure of center?

$$18, 22, 25, 24, 19, 20, 19, 17$$

Original data: mean = 20.5 years old, median = 19.5 years old

Data with additional student: mean = 24.3 years old, median = 20 years old

The addition of the 55-year-old student increases the mean age of the group because the value is much larger than the other values. The median does not change much with the addition of the 55-year-old student.

The mean is affected by values that are much larger or smaller than the rest of the data and is pulled in the direction of those larger or smaller values. However, the median is not as affected by extreme values. Thus, the median is the preferred measure of the center when there are extreme values in a data set.

Tech TIP

If your calculator cannot handle all the digits when finding the mean and gives you an overflow error, reduce the number of digits entered by using each salary in thousands. For example, enter 31.2 in your calculator instead of 31,200. Once you find the mean of the data, write "thousands" for the units or adjust the mean value accordingly.

5. Use this data of annual salaries for nurses in a clinic to answer the questions that follow.

$31,200	$31,200	$31,200	$34,000	$35,000
$44,000	$46,000	$46,000	$50,000	$51,000
$54,000	$54,000	$63,000	$65,000	$70,000
$72,000	$74,000	$75,000	$100,000	$125,000

a. What is the mean salary?

$57,580

b. What is the median salary?

$52,500

c. Which is higher: the mean or the median? Why?

The mean is higher because some large values pull the mean up.

d. If the nurses were trying to make a case for raising salaries, which measure of center should they use?

Median

e. What is the most common salary? What could account for this?

$31,200. The clinic could have hired three nurses at the same level of education and experience.

f. It is cumbersome to average all the data values at one time. Could you average each column and then average those results? Why or why not?

Yes; each column is evenly weighted with the same number of values.

6. Grade point averages (GPAs) are computed as weighted averages in which each course grade is weighted by the number of credit hours for the course. Use the information in the tables to calculate the student's GPA for each semester. Also, calculate the cumulative GPA for each semester, starting with the second semester. Convert the letter grades into points as follows:

A = 4 points,   B = 3 points,   C = 2 points, D = 1 point,   F = 0 points

a. Fall semester, freshman year

Course	Credit Hours	Grade
ENG 101	3	C
MTH 096	6	D
BIO 104	5	B
PSY 121	3	C

Semester GPA = 1.94

b. Spring semester, freshman year

Course	Credit Hours	Grade
ENG 102	3	C
MTH 096	3	C
CHM 120	5	B
SOC 131	3	C

Semester GPA: 2.36; cumulative GPA: 2.13

c. Fall semester, sophomore year

Course	Credit Hours	Grade
HIS 120	3	A
MTH 115	3	A
CHM 125	4	B
BIO 105	3	A

Semester GPA: 3.69; cumulative GPA: 2.59

d. Why don't the strong grades in the fall semester of the student's sophomore year raise the overall GPA much?

There was a greater number of credit hours, or weight, at the lower grades, which keeps the overall GPA lower.

Reflect

WRAP-UP

What's the point?

There are multiple ways to describe the center of a data set. We can use a mean when the values are weighted equally or a weighted mean when they are not. A median is a good measure of the center for data sets that have extreme values. The mode indicates the most common data value or values.

What did you learn?

How to find and interpret weighted means
How to find the median and mode of a data set

2.2 Homework

Skills MyMathLab

Often, students earn 100% on MyMathLab homework but not on quizzes and tests. If this happens to you, take an honest look at *how* you are studying. Are you using *View an Example* to start and finish most problems? If so, then you are not yet ready for a test and need to do additional work.

First complete the MyMathLab homework online. Then work the two exercises to check your understanding.

1. **a.** Find the mean of these values assuming they are each weighted equally: 2, 6, 10, 14, 20, 32.

   14

   b. Find the mean of these values assuming the first five values are each worth 10% and the last value is worth 50%:   2, 6, 10, 14, 20, 32

   21.2

2. Find the median and mode of the following data: 5   10   5   15   5   20

   mode = 5; median = 7.5

Concepts and Applications

Complete the problems to practice applying the skills and concepts learned in the section.

3. Suppose you spend 2 hours between Monday and Saturday studying for your 5-credit-hour chemistry course. To meet the college rule of thumb, how many hours do you need to study on Sunday? Is it possible? Is it effective?

   2 hours for each credit hour (5) means 10 hours per week. So you need to study 8 additional hours on Sunday. It's possible but very difficult to do and not an effective way to study.

4. **a.** Estimate how many hours you spent studying for each of your classes last week. List your classes, credit hours, and study times in the following table.

   Answers will vary.

Class	Credit Hours	Study Hours

   b. What is your overall average number of study hours per credit hour?

   Answers will vary.

   c. Does your average meet the college rule of thumb?

   Answers will vary.

5. A college dean has received the following results from a department chair about MTH 123, which is a math course with six sections. The dean wants to compute the overall pass rate using a spreadsheet and is trying to determine if she can average the values in the Pass Rate column.

Course & Section	A	B	C	D	F	Incomplete	Withdraw	Total	Pass Rate
MTH 123-1	1	9	6	5	9	1	1	32	0.500
MTH 123-2	10	6	5	1	3	0	4	29	0.724
MTH 123-3	3	5	3	3	9	0	6	29	0.379
MTH 123-4	2	5	5	0	10	0	1	23	0.522
MTH 123-5	5	9	7	1	4	0	1	27	0.778
MTH 123-6	7	9	2	3	5	1	3	30	0.600
Total	28	43	28	13	40	2	16	170	

a. Compute the overall pass rate for this course using the summary data in the last row. In other words, find the total number of students who passed—that is, earned an A, B, or C—and divide that number by the total number of students.

58.2% of students in MTH 123 are passing.

b. Compute the average of the pass rates in the last column. Is it correct to average the pass rates this way? Why or why not?

58.4%; no, it is not correct to average the pass rates this way. The section sizes are not the same.

c. Find the average of the pass rates using the class sizes as the weights. Do you get the same answer as you did when you used the raw data in part a?

58.2%; yes

6. Many employers screen by GPA, in addition to seeking the skills mentioned in the sticky note. Suppose you are in your last semester of college and have a 2.75 GPA after 105 credit hours. If you are taking 15 credit hours in your last semester and get a perfect 4.0 GPA, what will your overall GPA be? Is it possible to get your GPA up to 3.0 for graduation?

Your GPA will be only 2.91; no.

7. Recall these student grades used earlier in the section:

Student's points

Homework: $\dfrac{85}{90} \approx 94.4\%$ Midterm: $\dfrac{66}{100} = 66\%$

Quizzes: $\dfrac{70}{80} = 87.5\%$ Final Exam: $\dfrac{63}{100} = 63\%$

Find the student's grade assuming these weights: homework (5%), quizzes (5%), midterm (30%), final exam (60%). What effect does the heavily weighted final exam have in this case?

Category	Grade	Weight	Grade · Weight
Homework	94.4	0.05	4.72
Quizzes	87.5	0.05	4.38
Midterm	66	0.30	19.8
Final exam	63	0.60	37.8
		Sum = 1	Sum = 66.7

The final grade is 66.7, which is a D. Since the final exam is weighted so heavily, the overall grade is pulled toward the final exam's grade of D.

8. You have seen the mean used as an average or balancing point of data, and the median used as the center of a list of data. The physical balancing point of a geometric shape is actually located at the intersections of its medians, although the word has a different meaning in this context than the way we used it in this section.

The **median** of a triangle is a line segment that connects a vertex (the intersection point of two sides) with the midpoint of the side opposite it.

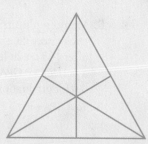

a. For each of the following objects, find the balancing point. The balancing point is the point under which you could place a pencil tip and the shape would be perfectly horizontal and balanced.

Once you find the balancing point of the object, give a general description of its location so that someone else could locate the balancing point on a similar shape.

Line segment   Midpoint

Rectangle   Intersection of diagonals or intersection of
                lines connecting midpoints of opposite sides

Triangle   Intersection of medians

b. For each of the following quadrilaterals (four-sided shapes), find the balancing point and generalize its location. Trace and cut out each shape to start.

Parallelogram (a four-sided shape with opposite sides parallel)

Balancing point is at the intersection of the diagonals.

Rhombus (a parallelogram with all sides the same length)

Balancing point is at the intersection of the diagonals.

Square (a rhombus with four right angles)

Balancing point is at the intersection of the diagonals. It is also at the intersection of the line segments connecting the midpoints of opposite sides.

9. **Looking Forward, Looking back** For each statement, incorrect units are given. Correct the statement and explain why it was wrong.

a. The average height of the males in our class is 69 square inches.

It should be 69 inches. Height is a linear measurement, not an area.

b. The average square footage of homes in this town is 1,500 feet.

It should be 1,500 square feet. Square footage is a measurement of area, which uses square units.

2.3 Measure Up: Basic Exponent Rules

SECTION OBJECTIVES:
- Apply basic exponent rules
- Use geometric formulas

1. Ana is making brownies and debating between an 8-inch round pan and an 8-inch square pan.

   a. If Ana likes thick brownies, which pan should she use? Explain why and use an appropriate calculation to support your answer.

   She should use the round pan because the bottom has a smaller area, which makes the brownies thicker. The area of the round pan is approximately 50.3 square inches, whereas the area of the square pan is 64 square inches.

   b. If Ana wants as much crispy edge as possible, which pan should she use? Explain why and use an appropriate calculation to support your answer.

   She should use the square pan because that pan has a larger perimeter, which produces more crust on the brownies. The perimeter of the square pan is 32 inches, whereas the perimeter of the round pan is approximately 25 inches.

Discover

INSTRUCTOR NOTE: Only whole-number exponents are used in this section. Integer exponents will be used in Section 4.4.

The problems in the *Explore* required basic geometric formulas for area and perimeter along with an understanding of units and exponents. This section will help you develop the skills to solve such problems.

As you progress through this section, keep the following questions in mind:

1 Do you know how to determine when the answer to a problem is in square units vs. cubic units?

2 When do you add exponents and when do you multiply them?

3 Do you have techniques to use when you forget an exponent rule?

The two formulas needed for the *Explore*, along with some other commonly used geometric formulas, are given here.

Rectangle	$A = lw$	
Parallelogram	$A = bh$	
Triangle	$A = \frac{1}{2}bh$	
Trapezoid	$A = \frac{1}{2}h(B + b)$	

Circle	$d = 2r,\ A = \pi r^2,\ C = 2\pi r = \pi d$	
Cylinder	$V = \pi r^2 h,\ S = 2\pi r^2 + 2\pi rh$	
Cone	$V = \frac{1}{3}\pi r^2 h$	
Sphere	$V = \frac{4}{3}\pi r^3,\ S = 4\pi r^2$	
Rectangular prism	$V = lwh$	

INSTRUCTOR NOTE: Consider having students label the figures with the appropriate variables as they are defining them.

2. The formulas are useful only if you know what the variables represent. What do each of the following variables represent when used in these geometric formulas?

A area

V volume

S surface area

B longer base of a trapezoid

l length

w width

b base

r radius

d diameter

h height

When we use geometric formulas to find perimeter, area, volume, or surface area, it is common to use only numbers in the formulas, without units, and then state the units with the answer. This method of leaving the units until last can sometimes be problematic, since it is easy to forget to include the units at the end. And even if we do remember to state the units, we may forget which type of unit goes with the calculation. Including units in the calculation can solve both of these problems but requires that we know some exponent rules.

Recall that exponents are used to condense repeated multiplication. Instead of writing $2 \cdot 2 \cdot 2 \cdot 2$, we can write 2^4. The factor being repeated is 2, which is called the **base**. The number of times the base appears as a factor is 4, which is the **exponent**.

3. For each row in the following table, complete the third and fourth columns and then write a conjecture by comparing the second and fourth columns. For the fourth column, find a new way to write the expression using exponents that differs from the form given in the second column. The first row has been completed as an example.

1	2 Expression	3 Rewrite expression without exponents	4 Rewrite result from the third column	5 Conjecture
a.	$5^3 \cdot 5^4$	$5 \cdot 5 \cdot 5 \cdot 5 \cdot 5 \cdot 5 \cdot 5$	5^7	When multiplying numbers with like bases, keep the base and add the exponents.
b.	$(3 \cdot 5)^2$	$(3 \cdot 5) \cdot (3 \cdot 5)$	$3^2 \cdot 5^2$	When raising a product to a power, apply the exponent to each factor.
c.	$\left(\dfrac{2}{3}\right)^2$	$\left(\dfrac{2}{3}\right) \cdot \left(\dfrac{2}{3}\right)$	$\dfrac{2^2}{3^2}$	When raising a quotient to a power, apply the exponent to the numerator and the denominator.
d.	$(7^3)^2$	$(7 \cdot 7 \cdot 7)^2 = (7 \cdot 7 \cdot 7) \cdot (7 \cdot 7 \cdot 7)$	7^6	When raising a power to a power, multiply the exponents.
e.	$\dfrac{11^5}{11^2}$	$\dfrac{11 \cdot 11 \cdot 11 \cdot 11 \cdot 11}{11 \cdot 11} = \dfrac{\cancel{11} \cdot \cancel{11} \cdot 11 \cdot 11 \cdot 11}{\cancel{11} \cdot \cancel{11}}$	11^3	When dividing numbers with like bases, keep the base and subtract the exponents (top − bottom).

The patterns in the table will be summarized as exponent rules after we add one more to the list. What does a zero exponent mean?

Consider powers of four and the pattern formed when we repeatedly decrease the exponent by 1.

$$4^4 = 256$$
$$4^3 = 64$$
$$4^2 = 16$$
$$4^1 = 4$$

To continue this pattern, what should 4^0 equal? Each result was divided by 4 to get the next result. According to this pattern, $4^0 = 4 \div 4 = 1$. It turns out that any number to the zero power is 1, with one exception that will be discussed after the summary.

We have investigated some important properties of whole-number exponents. Let's summarize them. Since we are generalizing these number properties, we will use variables to represent numbers.

HOW IT WORKS

If you forget an exponent rule, write out the expression without exponents. For example, should you add or multiply the exponents for x^4x^2? Writing it out, you get $x \cdot x \cdot x \cdot x \cdot x \cdot x$ which is x^6. We can see the rule is to keep the base and *add* the exponents.

Need more practice?

Simplify each expression.

1. $\dfrac{a^4 b^5}{ab^2}$

   Answer: $a^3 b^3$

2. $\left(\dfrac{4xy^3}{8y}\right)^2$

   Answer: $\dfrac{x^2 y^4}{4}$

3. $(5xy^5)^3$

   Answer: $125x^3 y^{15}$

To simplify expressions with exponents:

Assume a and b are whole numbers and x and y are real numbers.

1. $x^a x^b = x^{a+b}$

   When multiplying two numbers with the same base, keep the base and add the exponents.

2. $\dfrac{x^a}{x^b} = x^{a-b}$

   When dividing two numbers with the same base, keep the base and subtract the bottom exponent from the top exponent.

3. $x^0 = 1$

   A number to the zero power is 1. This is true for all numbers except zero.

4. $(x^a)^b = x^{ab}$

   When raising a power to a power, multiply the exponents.

5. $(xy)^a = x^a y^a$

   When raising a product to a power, apply the exponent to each factor.

6. $\left(\dfrac{x}{y}\right)^a = \dfrac{x^a}{y^a}$

   When raising a quotient to a power, apply the exponent to the numerator and the denominator.

EXAMPLES:

1. $(-3x^2 x^5)(-6x^2) = (-3x^7)(-6x^2) = 18x^9$ — Work with the coefficients and the exponents separately.

2. $(-2x^4)^4 = (-2)^4(x^4)^4 = 16x^{16}$ — Be careful to apply the exponent of 4 to -2, not just to 2.

3. $\dfrac{x^{15}}{(x^2)^3} = \dfrac{x^{15}}{x^6} = x^9$ — Simplify the numerator and denominator separately first.

There is an exception to the zero exponent rule that we need to consider. Let's look at powers of zero and the pattern formed when we repeatedly decrease the exponent by 1.

$$0^4 = 0$$
$$0^3 = 0$$
$$0^2 = 0$$
$$0^1 = 0$$

To continue this pattern, what should 0^0 equal? Every power of zero so far gave a result of zero. According to this pattern, $0^0 = 0$. However, it also makes sense that 0^0 should equal 1, since every other number to the zero power equals 1. Since there is not a unique solution to 0^0, it is called indeterminate. You can try entering 0^0 on your calculator. You should get an error message because it is not defined.

Now that we have a list of basic exponent rules, you need to practice using them and deciding when to use each particular rule.

4. Simplify each expression.

a. $17x^2 \cdot 2x^5$ $34x^7$

b. $(4x^4)^2$ $16x^8$

c. $(-5x^3)^2$ $25x^6$

d. $(-8)^0$ 1

e. -8^0 -1

f. $\left(\dfrac{3x^4}{6x^3}\right)^2$ $\dfrac{x^2}{4}$

g. $\dfrac{(3xy^2)^2}{3y^3}$ $3x^2y$

h. $(8xyz)(4x^2y^2z)^2$ $128x^5y^5z^3$

Now that we've seen the basic exponent rules, we can revisit using a geometry formula with the units included.

EXAMPLE

Find the volume of a cylinder with a radius of 4 cm and a height of 8 cm. Include units in your work.

SOLUTION

Using the formula shown earlier in the section, we can substitute the measurements with their units into the formula.

$V = \pi r^2 h$

$= \pi(4\text{ cm})^2(8\text{ cm})$

$= \pi(16\text{ cm}^2)(8\text{ cm})$ The exponent is applied to the number and the unit.

$= \pi(128\text{ cm}^3)$ When multiplying with the same base, add exponents.

$= 128\pi\text{ cm}^3$

Now practice using geometric formulas with exponents.

5. a. Find the area of a triangle whose base is 5 inches and height is 20 inches. Include units in your work.

$A = \dfrac{1}{2}bh = \dfrac{1}{2}(5\text{ in.})(20\text{ in.}) = 50\text{ in.}^2$

b. Find the volume of a cylinder whose radius is 2 centimeters and height is 1 centimeter. Find an approximate answer by using 3.14 for π and round to the nearest tenth. Include units in your work.

$V = \pi r^2 h = \pi(2\text{ cm})^2(1\text{ cm}) = \pi(4\text{ cm}^2)(1\text{ cm}) \approx 12.6\text{ cm}^3$

c. For a sphere, find the ratio of its volume to its surface area.

$\dfrac{V}{S} = \dfrac{\frac{4}{3}\pi r^3}{4\pi r^2} = \dfrac{\frac{4}{3}r}{4} = \dfrac{4}{3}r \div 4 = \dfrac{4}{3}r \cdot \dfrac{1}{4} = \dfrac{r}{3} \text{ or } \dfrac{1}{3}r$

Tech TIP

You can find exponents and roots on a scientific calculator. Depending on your calculator, you might have a generic exponent button y^x or a button like $\wedge$ to indicate an exponent. You should also have a root button like $\sqrt[x]{y}$. Experiment with your calculator now so you know how to evaluate an expression with exponents.

6. a. Find the area of a square whose sides have lengths of 3 centimeters. Why do you think 3^2 is sometimes read as "three squared"?

9 cm^2; the quantity 3^2 gives the area of a square with sides of length 3.

b. Find the volume of a cube whose sides have lengths of 5 feet. Why do you think 5^3 is sometimes read as "five cubed"?

125 feet^3; the quantity 5^3 gives the volume of a cube with sides of length 5.

Connect

All the exponents shown so far in this section have been whole numbers. In fact, fractions can also be used as exponents.

LOOK IT UP

Fractional Exponents

A **fractional exponent** is simply another way to indicate a root. For example, $x^{1/2} = \sqrt{x}$ and $x^{1/3} = \sqrt[3]{x}$.

In general, $x^{1/n} = \sqrt[n]{x}$.

FOR EXAMPLE, $16^{1/2} = \sqrt[2]{16} = \sqrt{16} = 4$.

You can always check that you've found the correct root by working backward. For example, the cube root of 125 is 5, and 5 cubed is 125.

$$125^{1/3} = \sqrt[3]{125} = 5 \text{ since } 5^3 = 125$$

The cube root, or 1/3 power of 8 can be found by taking one of the three equal factors of 8. $8^{1/3} = \sqrt[3]{8} = \sqrt[3]{2 \cdot 2 \cdot 2} = 2$

7. Evaluate each of the following. Check your answers by working backward and on your calculator.

a. $64^{1/3}$ 4

b. $(-64)^{1/3}$ -4

c. $16^{1/4}$ 2

d. $144^{1/2}$ 12

Reflect

WRAP-UP

What's the point?

Exponent rules are important not only for simplifying algebraic expressions but also for determining units in geometric formulas. The rules need to be memorized, and they should also make sense to you so that you can recreate them if needed.

What did you learn?

How to apply basic exponent rules
How to use geometric formulas

2.3 Homework

Skills MyMathLab

First complete the MyMathLab homework online. Then work the two exercises to check your understanding.

Each time you learn a new skill, ask yourself if you know *when* to use it, not just *how* to use it. If you practice homework in a rote way by mimicking help aids or problems you have already seen, you will find yourself frustrated when you encounter new problems.

1. Simplify.

   a. $(14ab)^3(ab)$ $2{,}744a^4b^4$

   b. $15^0(3x^2)^3$ $27x^6$

   c. $\dfrac{4m^9}{8m^3n}$ $\dfrac{m^6}{2n}$

2. **a.** An interior designer is designing a round table with a radius of 24 inches. She would like to use a special kind of trim, which is quite expensive, for the edge of the table. Instead of building the table and measuring for the length of trim, she is working on paper in case the trim is too expensive and she needs to make the table smaller. How much trim will she need? Round up to the nearest whole inch to ensure that she has enough.

      151 inches

   b. If she wants to cover the top of the table with a special crackle paint, how many square feet of surface will the paint need to cover? Round up to the nearest whole square foot to ensure she has enough.

      13 ft^2

Concepts and Applications

Complete the problems to practice applying the skills and concepts learned in the section.

3. The volume of a substance is related to its mass and density by this formula: $V = \frac{m}{d}$. If the mass is in grams and the density is in grams per cubic centimeter, what are the units for volume? Show how to determine this algebraically.   cubic centimeters

4. The distance a projectile will travel when it is fired at an initial angle of 30° can be found using the formula $D = \frac{v^2}{g} \cdot 0.87$. If velocity is in meters/second and g is in meters/second^2, what are the units for distance? Show how to determine this algebraically.   meters

5. The formula $T = 2\pi\sqrt{\dfrac{L}{g}}$ gives the time T for a full cycle of a pendulum's swing, where the pendulum's length in feet is L, and g is the acceleration due to gravity, 32.2 ft/sec^2. When T is calculated from this formula, what are the units? Show how to determine this algebraically.   seconds

6. The following formula calculates the gravitational force between two objects based on their masses (m_1 and m_2) and the distance between them (r):

$$F = G\frac{m_1 m_2}{r^2}$$

a. If the masses increase, what happens to the force between them?
It increases as well.

b. If the distance between the objects increases, what happens to the force between them?   It decreases.

c. The variables m_1 and m_2 have units of kilograms, and r has units of meters. What units must the constant G have in order for the units on F to be newtons (N)?
$\dfrac{Nm^2}{kg^2}$

7. Three students were asked to simplify $(-3x^3)^2 x^5$. Find the mistake in each student's work.

a. $(-3x^3)^2 x^5 = -3x^6 x^5$
$= -3x^{11}$

b. $(-3x^3)^2 x^5 = -9x^6 x^5$
$= -9x^{11}$

c. $(-3x^3)^2 x^5 = 9x^6 x^5$
$= 9x^{30}$

a. The student did not square the coefficient −3.
b. The student squared 3 but not −3. The new coefficient should be 9, not −9.
c. The student multiplied the exponents in the last step instead of adding them.

8. **Looking Forward, Looking Back** Write an expression for the nth term of the sequence.

$\dfrac{5}{2}, \dfrac{5}{4}, \dfrac{5}{8},$ _____, _____, _____, ... $\dfrac{5}{2^n}$

2.4 Count Up: Adding Polynomials

Explore

1. a. Complete the following table. When finding each sum, state the answer first using the given units. If possible, convert to a common unit and combine further.

SECTION OBJECTIVE:
• Identify and add like terms

Expression	Objects Being Combined	Sum
2 cm + 7 cm + 8 m	centimeters, meters	9 cm + 8 m 8.09 m or 809 cm
$\frac{2}{5} + \frac{1}{5}$	fifths	$\frac{3}{5}$
0.03 + 0.05	hundredths	0.08
5 quarters + 10 dimes + 20 quarters	quarters, dimes	25 quarters + 10 dimes 35 coins or 725 cents
$\frac{2}{3} + \frac{5}{3} + \frac{1}{2}$	thirds, halves	$\frac{7}{3} + \frac{1}{2}$ $\frac{17}{6}$
5 hours, 2 minutes + 7 hours, 54 minutes	hours, minutes	12 hours, 56 minutes 776 minutes or 12.93 hours
0.05 + 0.004 + 0.0123	hundredths, thousandths, ten-thousandths	0.05 + 0.004 + 0.0123 0.0663

b. When are we able to add objects? When are we not?
We can add only like objects.

Discover

As you work through this section, think about these questions:

1 Can you read an expression and determine the number and type of objects being added?

2 Do you know when to add exponents vs. when to add coefficients when you are simplifying an expression?

We can see from the *Explore* that we are able to add only like objects. If objects are unlike, we can either find something common about them and combine further or leave the objects as they are.

For example, if we have 5 quarters + 6 dimes, we do not have 11 quarterdimes. We have 11 coins (a trait they share in common), or we have 5 quarters + 6 dimes. Keep this example in mind when we move to the more abstract cases with variables.

Before we look at adding objects, we need to understand the notation used to indicate objects that are being added and the number of them.

There are several ways to indicate multiplication of two numbers.

$$3 \times 4 \qquad (3)(4) \qquad 3(4) \qquad 3 \cdot 4$$

Using an $\times$ for multiplication is common in arithmetic but not in algebra, since it can be mistaken for a variable instead of an operation. We will typically use a dot for multiplication between numbers.

Likewise, there are several ways to indicate multiplication of a number and a variable.

$$3n \qquad (3)(n) \qquad 3(n) \qquad (3)n \qquad 3 \cdot n$$

Since multiplication is implied when a variable is next to a number, we will often use that option, like $3n$, since it is the simplest way to write this type of product.

The factors in a product indicate the number of objects being counted and the type of object. The expression $2(8)$ means $2 \cdot 8$ or two eights. The fraction $\frac{2}{8}$ means $2 \cdot \frac{1}{8}$ or two eighths. Likewise, $2n$ means two n's. In an algebraic expression, the coefficient tells how many of the objects there are. The variable factor that follows is the object being recorded. If the same type of objects are being added, also known as like terms, they can be combined further to make a simpler expression.

LOOK IT UP

Like Terms

Like terms have the same variables raised to the same exponents. They need not have the same coefficients.

FOR EXAMPLE, $3xy$ and $4xy$ are like terms, but $4xy^3$ and $4x^3y$ are not.

Often, an expression will be given with a brief direction: simplify. This one word can have many meanings depending on the context. In general, to simplify an expression is to perform the given operations to shorten or condense the expression. To simplify or combine like terms, the coefficients are added but the object does not change. This is true in numeric expressions like 2 meters + 7 meters, which is 9 meters. It is also the case in algebraic expressions that contain variable factors.

EXAMPLE 1

Simplify $7x + 2y - 3x - 4y$.

SOLUTION

To simplify the expression, we combine the like terms. Notice there are four terms, two that have x as a variable factor and two that have y as a variable factor. We can use color or symbols, such as boxes and circles, to make the like terms more noticeable. Keep the term with its sign. $7x$ and $-3x$ are like terms, as shown with circles. Also, $2y$ and $-4y$ are like terms, as indicated with squares.

$$\boxed{7x} \quad +2y \quad \boxed{-3x} \quad -4y$$
$$7x - 3x + 2y - 4y$$
$$(7 - 3)x + (2 - 4)y$$
$$4x - 2y$$

Once we have identified the like terms, we can combine them—that is, add them. Reading the expression in words helps determine how to simplify it: seven x's plus two y's minus three x's minus four y's. Add the coefficients and leave the variable factors alone. We get $4x - 2y$ or $-2y + 4x$, depending on which variable is listed first.

To add like terms:

1. Identify the like terms in an expression.
2. Add the coefficients of the like terms. Keep the variable factors the same.

EXAMPLE: Simplify $-5xy + 7y^2 + 10xy + 2$.

Notice the like terms are $-5xy$ and $10xy$. Adding them, the result is $5xy$. The expression becomes $5xy + 7y^2 + 2$. Notice that the terms cannot be combined further because they are unlike.

Need more practice?

Simplify the expression
$-10x - 8y - y + x$.

Answer: $-9x - 9y$

Using abbreviations such as LOL, BTW, and OMG when you text saves time, but too many abbreviations can make a message difficult to read. The same issue applies to algebra. Notation can save time, but it's also easy to lose the bigger picture. It's important to ask yourself regularly, "Do I understand the notation?" and not just, "Can I use it?"

Since adding like terms depends on using the coefficient, it is important to be able to recognize the coefficient even when it is not written. When no coefficient is written, it is assumed to be 1. For example, x means we have 1 factor of x. So x is the same as $1x$. In the term $-x$, the coefficient is -1, so $-x$ is the same as $-1x$.

If you understand how to combine like terms, you can transfer that skill to a variety of new situations. Each part of the following problem has an expression with some like terms to add. If you have trouble simplifying any of them, read them out loud or silently to yourself. If you can read them, it will help you add them.

2. In each problem, identify the like terms and then simplify the expression by combining like terms.

a. $a - 9a + 23$
    $-8a + 23$

b. $8x^2 + 4y + 8y + 16x^2 - 12x$
    $24x^2 + 12y - 12x$

c. $17w^2 - \dfrac{3}{2}w^2 + \dfrac{1}{4}w - w^2 - w - 1$
    $\dfrac{29}{2}w^2 - \dfrac{3}{4}w - 1$

   d. $3\sqrt{x} + 5\sqrt{x}$
      $8\sqrt{x}$

   e. $9xy - 9x + 17xy + 8x$
      $26xy - x$

   f. $(3 + 4i) + (-7 - 9i)$
      $-4 - 5i$

In Section 2.3, you learned about exponent rules for multiplying and dividing. This section explores the ideas of addition and subtraction of terms, some of which have exponents. Let's put these ideas together.

3. Simplify each expression.

   a. $4x^3 \cdot 4x^3$ $16x^6$

   b. $4x^3 + 4x^3$ $8x^3$

   c. $4x^3 \cdot 4x$ $16x^4$

   d. $4x^3 - 4x$ $4x^3 - 4x$

Connect

We saw in Section 2.3 that using units while simplifying geometric formulas can make it easier to correctly identify and include the units in the final answer. Some geometric formulas also require the combining of like terms in their simplification.

EXAMPLE 2

Find the area of a trapezoid whose height is 3 feet, long base is 6 feet, and short base is 4 feet. Include units in your work.

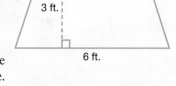

SOLUTION

The area formula for a trapezoid is $A = \frac{1}{2}h(B + b)$, where h is the height, B is the long base, and b is the short base.
We begin by replacing the variables in the formula with the values and units from the problem statement. Next, we simplify the expression, treating the unit like a variable factor; that is, 3 ft is like $3x$.

$$A = \frac{1}{2}h(B + b)$$

$$= \frac{1}{2}(3 \text{ ft})(6 \text{ ft} + 4 \text{ ft}) \quad \text{Combine like terms.}$$

$$= \frac{1}{2}(3 \text{ ft})(10 \text{ ft}) \quad\quad \text{Use exponent rules to simplify.}$$

$$= 15 \text{ ft}^2$$

4. Solve each problem. Include units in your work.

a. Find the amount of edging needed for a rectangular desk that is 76 centimeters by 46 centimeters.

76 cm + 46 cm + 76 cm + 46 cm = 244 cm

INSTRUCTOR NOTE: An option is to think of the objects being combined as π in.^{2}, even though technically π is part of the coefficient.

b. Find the surface area of an aluminum can that is 6 inches tall and has a diameter of 2.5 inches. Keep π in the calculation until the last step, and then use a calculator at the end to estimate the answer to the nearest square inch.

$$S = 2\pi r^2 + 2\pi rh$$
$$= 2\pi(1.25 \text{ in.})^2 + 2\pi(1.25 \text{ in.})(6 \text{ in.})$$
$$= 3.125\pi \text{ in.}^2 + 15\pi \text{ in.}^2$$
$$= 18.125\pi \text{ in.}^2 \approx 57 \text{ in.}^2$$

Reflect

WRAP-UP

What's the point?
Understanding notation can help you simplify expressions that have like objects.

What did you learn?
How to identify and add like terms

2.4 Homework

Skills MyMathLab

First complete the MyMathLab homework online. Then work the two exercises to check your understanding.

1. Are $10xy^2$ and $10x^2y$ like terms? Why or why not?

No; they do not have the same variables raised to exactly the same exponents.

2. Simplify each expression.

a. $-11x^2 + 11xy + 22xy - 12x^2 + 2$

$-23x^2 + 33xy + 2$

b. $(-11x^2)(11xy)(22xy)(-12x^2)(2)$

$63{,}888x^6y^2$

Concepts and Applications

Complete the problems to practice applying the skills and concepts learned in the section.

3. A student was asked to simplify two expressions by combining like terms. Explain the mistake in each problem and then find the correct answer.

   a. $-3xy + 5x + 6y - 8xy$

      $= -3xy + 11xy - 8xy$

      $= 8xy - 8xy$

      $= 0$

      The student added $5x$ and $6y$ to get $11xy$, but x and y are not like terms. They should not be combined. The correct answer is $-11xy + 5x + 6y$.

   b. $15x^2 + 10x - 21x - 18x^2$

      $= 15x^2 - 11x - 18x^2$

      The expression still has like terms ($15x^2$ and $-18x^2$) that should be combined. The correct answer is $-3x^2 - 11x$.

4. Give an example of an expression that contains like terms and another that does not.

   Answers will vary.

5. Simplify each expression.

   a. $\dfrac{10}{x+1} - \dfrac{8}{x+1}$ $\dfrac{2}{x+1}$

   b. $-3x\sqrt{5yw} + 3xyw + 11x\sqrt{5yw}$ $8x\sqrt{5yw} + 3xyw$

   c. $12(x+2)^2 + 15(x+2) + 18(x+2)^2$ $30(x+2)^2 + 15(x+2)$

6. The formula for the surface area of a right circular cone is $S = \pi r^2 + \pi rl$, where r is the radius of the base and l is the slant height of the cone. Find the surface area of a cone with $r = 2$ inches and $l = 4$ inches. Include units in your work. Give an exact answer in terms of π as well as an approximate answer rounded to the nearest tenth of a square inch. Use the π button on your calculator.

   $S = \pi(2 \text{ in.})^2 + \pi(2 \text{ in.})(4 \text{ in.}) = \pi(4 \text{ in.}^2) + \pi(8 \text{ in.}^2) = 12\pi \text{ in.}^2 \approx 37.7 \text{ in.}^2$

7. **Looking Forward, Looking Back** Classify the polynomial $-17x^3 - 17x + 17$ by the number of terms. State the degree.

   Trinomial; 3

2.5 A Winning Formula: Applying Order of Operations

Explore

There are several ways to measure an NFL quarterback's skills. One numeric value that is highly correlated with success is the adjusted net passing yards per attempt, or ANY/A for short. It summarizes a quarterback's passing performance in one number by rewarding touchdowns and deducting for interceptions and sacks. The formula is

[passing yards + (number of passing touchdowns) · 20 − (number of interceptions thrown) · 45 − sack yards lost]/(number of passing attempts + number of sacks)

SECTION OBJECTIVES:
- Use the order of operations to simplify expressions
- Evaluate formulas and expressions

1. a. Rewrite the formula as a vertical fraction and remove any unnecessary parentheses. Define and use meaningful variables.

$$\frac{PY + 20PT - 45I - SYL}{PA + S}$$, where PY = number of passing yards, PT = number of passing touchdowns, I = number of interceptions thrown, SYL = sack yards lost, PA = number of passing attempts, and S = number of sacks

b. Find the ANY/A value for the player who has the values given in the table.

Completed Passes	Passing Attempts	Passing Yards	Passing Touchdowns	Interceptions	Sacks	Sack Yards Lost	ANY/A
450	659	5477	55	10	18	120	8.87

c. If a player wants to increase his ANY/A value, what should he do?

Pass fewer times for long distances, increase the number of passing touchdowns, and reduce the number of interceptions and sacks.

Discover

The formula in the *Explore* uses multiple sets of parentheses to clarify which operations to do first. While this is an option when writing expressions, it can make a formula difficult to read. Instead of using many parentheses to indicate the order of operations, a set of rules has been established in mathematics to specify the order. These rules, which are similar to the rules of grammar in English, dictate the order in which we perform operations to maintain clarity and consistency. Without them, it is unclear where to begin when simplifying an expression.

As you work through this section, think about these questions:

1 Can you evaluate formulas using the correct order of operations?

2 Do you know how your calculator handles order of operations?

3 Do you know what the "P" in PEMDAS means?

Tech TIP

Not all calculators and computer programs use the same order of operations. You should experiment to see how your calculator works.

To guarantee that there is only one correct answer for a calculation, operations are done in the following order:

1. Operations inside brackets, parentheses, or other grouping symbols are done first.
2. Exponents and roots are evaluated next.
3. Multiplications and divisions are done from left to right in the order in which they appear.
4. Additions and subtractions are done from left to right in the order in which they appear.

This process is often remembered with the acronym PEMDAS.

Parentheses
Exponents
Multiplication
Division
Addition
Subtraction

The rationale behind this ordering comes from the hierarchy of operations and how they relate.

Multiplication is repeated **addition**.
Division is repeated **subtraction**.
Exponents are repeated **multiplication**.
Roots are repeated **division**.

If you use the memory aid PEMDAS to remember the order of operations, be sure to understand that multiplication and division have the same priority level even though M is listed before D. Likewise, addition and subtraction have the same priority level even though A is listed before S.

An expression like (3)(4) is not what "parentheses" implies. It implies the simplifying of operations within parentheses, such as $-5(8 - 17)$. GEMDAS may be a more appropriate acronym, since we begin by simplifying within any *grouping* symbol, which includes several possibilities like square roots, fractions, absolute value, and brackets, not just parentheses.

A common memory aid for remembering the order of operations is Please Excuse My Dear Aunt Sally.

EXAMPLE

Simplify $\dfrac{18 - 8 \cdot 4^2}{3}$ using the order of operations.

SOLUTION

To begin, notice that the division bar is a grouping symbol. It groups the numerator as one expression and the denominator as another expression. Each of these expressions is simplified separately before dividing.

$\dfrac{18 - 8 \cdot 4^2}{3}$ Apply the exponent.

$\dfrac{18 - 8 \cdot 16}{3}$ Next, perform the multiplication.

$\dfrac{18 - 128}{3}$ Then subtract in the numerator.

$-\dfrac{110}{3}$ Since there are no common factors between 110 and 3 other than 1, the fraction is simplified.

2. Simplify each expression, using the correct order of operations.

a. $4 - (-3)$ 7

b. $4 - |-3|$ 1

c. $4 - (-3)^2$ -5

d. $[4 - (-3)]^2$ 49

Let's summarize the process of simplifying expressions using the order of operations.

HOW IT WORKS

To simplify an expression, operations should be performed in the following order:

1. Start with expressions within grouping symbols such as parentheses, brackets, absolute value, fraction bars, or root symbols.
2. Simplify exponents and roots.
3. Perform multiplications and divisions from left to right.
4. Perform additions and subtractions from left to right.

EXAMPLE: Simplify $2 + 8^2 - (3 - 6)$.

Begin by performing the subtraction in parentheses. Then apply the exponent. Next, the opposite of negative 3 is positive 3. Once the expression contains only addition and subtraction, compute from left to right. The result is 69.

$$2 + 8^2 - (3 - 6)$$
$$2 + 8^2 - (-3)$$
$$2 + 64 - (-3)$$
$$2 + 64 + 3$$
$$69$$

3. Simplify each expression, using the correct order of operations.

a. $3 + 5^2$ 28

b. $7 - 3(2 - 6)^2$ -41

c. $4^3 - \dfrac{\sqrt{16 \cdot 2^2}}{4}$ 62

d. $\dfrac{-4 + \sqrt{5^2 - 4(-3)(2)}}{2}$ $\dfrac{3}{2}$

> The quantities $(-6)^2$ and -6^2 may not look very different, but their meanings and results are. In the expression $(-6)^2$, we must take the opposite of 6 before squaring. The base being squared is -6, so the result is 36. In the expression -6^2, the order of operations indicates to square the positive 6 and then take the opposite of that answer, which is -36.

4. Evaluate the expression $\dfrac{2 - 3x^2}{2}$ when $x = -2$ by replacing x with -2 and simplifying. -5

One common situation in which the order of operations rules are used is the evaluation of formulas. We have already seen several geometry formulas in this cycle. Now we will work with some formulas from science, finance, and statistics.

To evaluate a formula:

1. Replace the variables in the formula with the appropriate numerical values.

   a. When replacing a variable with a negative number, use parentheses around the number.

   b. State the units with the numerical values if appropriate.

2. Use the order of operations rules to simplify the expression.

3. State the result, with units if appropriate.

EXAMPLE: Use the formula $F = \dfrac{9}{5}C + 32°$ to convert 100° Celsius to Fahrenheit.

Begin by replacing C with 100° and then simplify.

$$F = \frac{9}{5}(100°) + 32°$$

$$F = 180° + 32°$$

$$F = 212°$$

So 100° Celsius is the same as 212° Fahrenheit.

Tech TIP

You may need two sets of parentheses to enter the formula in #5 into your calculator in one step. One set is shown in the formula, and the other set goes around the exponent if it is written as a product.

5. The formula $A = P\left(1 + \dfrac{r}{n}\right)^{nt}$ calculates the future value of an investment if you invest P dollars at an interest rate of r (as a decimal). The variable n is the number of times per year the interest is compounded. The variable t is the number of years the money is invested. Assume that you have \$1,000 to invest at an interest rate of $3\frac{5}{8}$% compounded monthly. How much will you have after 20 years?   \$2,062.48

6. **a.** In statistics, there is a formula for the lower fence in a box-and-whisker plot:
   $$LF = Q_1 - 1.5IQR$$

   Find the lower fence when $IQR = 33$ and $Q_1 = 60$. $LF = 10.5$

   b. Another formula in statistics is for z-scores: $z = \dfrac{x - \mu}{\sigma}$

   Find the z-score associated with $x = 85, \mu = 100, \sigma = 15$. $z = -1$

Connect

Consider the following formula, which calculates the monthly payment for a mortgage:

$$M = P\frac{\left(1 + \dfrac{r}{n}\right)^{nt} \cdot \dfrac{r}{n}}{\left(1 + \dfrac{r}{n}\right)^{nt} - 1}$$

M is the monthly payment, P is the principal or amount borrowed, n is the number of payments per year, t is the number of years, and r is the annual interest rate as a decimal.

7. Find the monthly payment for a loan of \$150,000 at 4.1% for 30 years. $724.80

Reflect

Focus Problem Check-In
Use what you have learned so far this cycle to work toward a solution on the focus problem. If you are working in a group, work with your group members to create a list of any remaining tasks that need to be completed. Determine who will do each task, what can be done now, and what will have to wait until you have learned more.

WRAP-UP

What's the point?

We need a set of ground rules for performing calculations with consistency and accuracy. The order in which mathematical operations are applied is known as the *order of operations*.

What did you learn?

How to use the order of operations to simplify expressions
How to evaluate formulas and expressions

2.5 Homework

Skills MyMathLab

First complete the MyMathLab homework online. Then work the two exercises to check your understanding.

1. a. Simplify: $\dfrac{5}{9}(14 - 32)$ −10

b. Simplify: $20 \cdot \dfrac{9}{5} + 32$ 68

2. a. Evaluate $\dfrac{5x}{y^2}$ for each of the following combinations:

$x = -2, y = -4$ $-\dfrac{5}{8}$

$x = -0, y = -4$ 0

$x = -4, y = 0$ undefined

b. Use the following body mass index (BMI) formula to find the BMI for a person who weighs 150 pounds and is 6 feet tall: $\text{BMI} = \dfrac{(\text{weight in pounds})(703)}{(\text{height in inches})^2}$.
BMI = 20.34 pounds per square inch

Concepts and Applications

Complete the problems to practice applying the skills and concepts learned in the section.

3. We have seen that $(-6)^2 = 36$ and $-6^2 = -36$. Determine the order necessary to enter these expressions on your calculator to achieve these answers.
Answers will vary based on the calculator used.

4. Height is sometimes used to determine medicinal or nutritional needs. When a patient is bedridden, it can be difficult to determine height accurately. To deal with this issue, formulas have been developed that are based on a measurement that can be easily obtained on a prone patient.

Height in cm (females) = 84.88 − (0.24 × age) + (1.83 × knee height)
Height in cm (males)   = 64.19 − (0.04 × age) + (2.02 × knee height)

To find a patient's knee height, the knee and ankle of the patient are held at a 90° angle and the length from the heel to the knee is measured in centimeters. Age is in years.

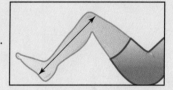

Based on: www.rxkinetics.com

a. Find the height of an 80-year-old female patient whose knee height is 16 inches. Give the final result in inches.   Knee height is 40.64 cm; height in cm = 140.05 cm; height in inches = 55.14 inches

b. Test the formula on yourself to see how accurately your knee height and age predict your height.   Answers will vary.

5. A cupcake recipe has the following instructions: Mix the dry ingredients in a bowl. Cream the wet ingredients in another bowl. Combine the dry and wet mixtures. Pour the batter into muffin cups. Yield: 24 cupcakes

Which of the following expressions best describes the process used to make the cupcakes?

a. $\big[(\text{dry} + \text{dry}) + (\text{wet} + \text{wet})\big] \cdot 24$ **b.** $\dfrac{(\text{dry} + \text{wet}) + (\text{dry} + \text{wet})}{24}$

c. $\dfrac{(\text{dry} + \text{dry})(\text{wet} + \text{wet})}{24}$ **d.** $\dfrac{(\text{dry} + \text{dry}) + (\text{wet} + \text{wet})}{24}$

d

6. When previously earned interest is added to an investment and then allowed to earn more interest, we say that the interest is compounded. Compounding interest increases the effective interest rate of an investment compared to simple interest, where the interest earned is not added to the investment to earn additional interest.

   The annual percentage yield, or APY, gives the effective interest rate on an investment based on the number of times the interest is compounded in a year, n, and the interest rate, r, as a decimal.

   $$APY = \left(1 + \frac{r}{n}\right)^n - 1$$

   Find the annual percentage yield if the interest rate is 4% and interest is compounded monthly. State the answer as a percent rounded to the nearest hundredth.
   4.07%

7. The following formula is used to calculate the interest rate per period of a loan when the compounding period does not match the payment period.

   $$r = \left(1 + \frac{i}{n}\right)^{n/p} - 1$$

   In this formula, $r =$ the rate per payment period, $i =$ the annual rate, $n =$ the number of compounding periods per year, and $p =$ the number of payments per year.

   a. List the operations involved in evaluating this formula in the order they are completed.
   Divide i by n
   Add 1
   Raise to the $\frac{n}{p}$ power
   Subtract 1

   b. Use the formula to calculate r when $i = 7.5\%$, $n = 2$, and $p = 12$. Write down intermediate calculations as accurately as possible, or use parentheses to enter the entire expression into your calculator at once. Remember that percents must be written as decimals in these types of formulas. Write your final answer as a percent rounded to four decimal places. $r \approx 0.6155\%$

   c. Use the formula to calculate the interest rate per period for a loan if there are 24 payments per year, the annual rate is 8.25%, and there are 12 compounding periods per year. Write your final answer as a percent rounded to four decimal places. $r \approx 0.3432\%$

8. **Looking Forward, Looking Back** Solve each problem using only one operation.

   a. Apply a 9.75% sales tax to a $45 restaurant bill to find the total after taxes. State the multiplier and the result.
   Multiply by 1.0975; $49.39

   b. Apply a 20% discount to a $75 jacket to find the total after the discount. State the multiplier and the result.
   Multiply by 0.80; $60

2.6 Does Order Matter?: Rewriting Expressions

SECTION OBJECTIVE:
- Apply the commutative and associative properties

Explore

1. Imagine that you are making purchases at a department store and you see that there is a 10%-off sale. You also have an online coupon for 5% off your purchase.

   a. Is the final price affected by the order in which the discounts are applied? Why or why not?

   No; for example, if the purchase amount is $100, then after the 10% discount is applied, the new total is 0.90($100) = $90. Taking an additional 5% off gives a final amount of 0.95($90) = $85.50.

   If the discounts are applied in the reverse order, the new total after the 5% discount is 0.95($100) = $95. Taking an additional 10% off gives a final amount of 0.90($95) = $85.50.

   The final price is the same either way. Regardless of the order in which the discounts are applied, the multiplier to determine the final amount is (0.95)(0.90) = 0.855.

   b. Will applying the 10% discount and then the 5% discount result in a 15% discount? Explain why or why not.

   No; if the discounts combine to a 15% discount, then the final amount would be 85% of $100, or $85. The second discount is not applied to the $100 original purchase but to the reduced amount.

Discover

As we saw with order of operations in Section 2.5, the order in which multiple operations are done usually makes a difference. However, there are mathematical calculations in which order does not matter. This section will help you consider when and where order might not matter in mathematics.

By the end of the section, you should be able to answer the following questions:

① For which operations does order not matter?

② How can mathematical properties be used to make mental math easier to do?

In the *Explore*, the order of the discounts does not make a difference because we found the amounts by applying two multipliers. When you multiply numbers, the order of the factors does not affect the final product. Similarly, when you add numbers, the order of those numbers does not affect the final sum.

The commutative property states that, for some operations, the order of the numbers involved does not matter. For example, imagine your commute to work. You travel from home to work and then from work to home. The order changes, but the trip is the same length (provided you use the same route).

Commutative Properties

The **commutative properties** of addition and multiplication say that we can change the order in which we multiply or add numbers without changing the result. The word "commute" means to exchange or interchange two things.

Suppose *a* and *b* are real numbers.

$$a + b = b + a \quad \text{(commutative property of addition)}$$
$$ab = ba \qquad \text{(commutative property of multiplication)}$$

EXAMPLES:

$$4 + 5 = 5 + 4$$
$$4(5) = 5(4)$$
$$x + 2y = 2y + x$$
$$xy = yx$$

The commutative properties are often used when you first learn the addition and multiplication facts in order to reduce the number of facts that you have to memorize. Knowing that 3×4 is the same as 4×3 reduces two facts to one. Likewise, $2 + 9$ is the same as $9 + 2$, and it is easier to count up 2 from 9 than to count up 9 from 2.

2. Use the commutative properties to rewrite each expression and then combine like terms.

a. $ab + ba$ This is the same as $ab + ab$, which is $2ab$.

b. $ab - ba$ This is the same as $ab - ab$, which is 0.

c. $3x + 4y - 5x + 8y$ This is the same as $3x - 5x + 4y + 8y$, which is $-2x + 12y$.

The commutative properties tell us that the order does not matter when we are performing addition and multiplication. Are there operations for which order *does* matter? When we are performing subtraction and division, the order of the numbers does affect the answer. For example, $12 \div 6 = 2$ but $6 \div 12 = \frac{1}{2}$.

3. We know that $4 - 11$ and $11 - 4$ are not equivalent expressions, but sometimes we need to reverse the order of the numbers in an expression involving subtraction. For the following expressions, write the terms in a different order, using only the commutative property for addition.

a. $x + 2$ $2 + x$

b. $4 - 11$ $4 + (-11)$, so $-11 + 4$

c. $x - 2$ $x + (-2)$, so $-2 + x$

d. $2 - x$ $2 + (-x)$, so $-x + 2$

To summarize, if we are performing addition or multiplication, we can reverse numbers without affecting the result. If we are performing subtraction, we can still reverse the numbers, but we need to keep the appropriate sign with each number.

The commutative property says that we can change the order when adding or multiplying. Another property, the associative property, says that terms can be regrouped when we add or multiply.

Associative Properties

The **associative properties** of addition and multiplication say that we can change the grouping when we multiply or add three or more numbers without changing the result.

Suppose a, b, and c are real numbers.

$$a + (b + c) = (a + b) + c \quad \text{(associative property of addition)}$$
$$a(bc) = (ab)c \qquad \text{(associative property of multiplication)}$$

Sometimes the associative property is applied in order to group compatible numbers together.

FOR EXAMPLE, instead of $15 + (5 + 17)$, we can apply the associative property to get $(15 + 5) + 17$. This allows the addition of 15 and 5 to be performed first, since it is easy to add those numbers mentally. We get $20 + 17$, which is 37.

Similarly, the expression $(26 \cdot 5) \cdot 2$ can be written as $26 \cdot (5 \cdot 2)$ to make the calculations easier to perform. We get $26(10)$, which is 260.

Remember that the commutative property involves changing order, and the associative property involves regrouping.

Sometimes we combine the commutative property with the associative property to make it easier to perform calculations mentally.

4. Use the stated property to rewrite the computation to make it easier to compute mentally, and then find the result.

   a. $17 + (13 + 68)$ associative property $(17 + 13) + 68 = 98$

   b. $(8 \cdot 15)(2)$ associative property $8(15 \cdot 2) = 240$

   c. $19 + (72 + 11)$ commutative property and then associative property
   $(19 + 11) + 72 = 102$

Connect

To add and subtract polynomials, you can first use the commutative and associative properties to rearrange the terms. Suppose we want to add the following trinomials. We need to reorder and regroup the terms to get like terms together.

$$(5x^4 + 8x^3 - 2x) + (x^4 - 2x^3 - 5x)$$
$$= (5x^4 + x^4) + (8x^3 - 2x^3) + (-2x - 5x)$$
$$= 6x^4 + 6x^3 - 7x$$

5. Add or subtract as indicated.

   a. $(3x^2 - 5x + 6) + (4x^2 - 7x - 5)$ $7x^2 - 12x + 1$

   b. $(5y - 14) + (7y^2 - 8y + 3)$ $7y^2 - 3y - 11$

   c. $(9x^2 + 6x + 15) - 4x^2$ $5x^2 + 6x + 15$

WRAP-UP

What's the point?

Understanding properties of numbers can make mental math calculations easier. It will often be necessary to apply these properties when simplifying algebraic expressions.

What did you learn?

How to apply the commutative and associative properties

2.6 Homework

Skills MyMathLab

First complete the MyMathLab homework online. Then work the two exercises to check your understanding.

1. Rewrite each expression, using the stated property.

   a. $(x + y) + z$ commutative property of addition $(y + x) + z$ or $z + (x + y)$

   b. $(x + y) + z$ associative property of addition $x + (y + z)$

2. Rewrite each expression, using the stated property.

   a. $(xy)z$ commutative property of multiplication $(yx)z$ or $z(xy)$

   b. $(xy)z$ associative property of multiplication $x(yz)$

Concepts and Applications

Complete the problems to practice applying the skills and concepts learned in the section.

3. **a.** The commutative property does not apply to division. Give an example that shows why it does not apply to this operation.

   Example: $6 \div 2 = 3$, but $2 \div 6 = \dfrac{1}{3}$

   b. A friend in class is trying to do the following division problem: $5\overline{)25}$.

   She says, "Divide 5 by 25." Why is that not a correct way to say it? Explain and then give two correct ways to verbalize the problem.

   "Divide 5 by 25" implies $5 \div 25$, but $25 \div 5$ is the intended problem. You could say, "Divide 25 by 5" or "Divide 5 into 25." How you verbalize the problem matters, since it indicates the order of the numbers in the division problem.

4. Children are taught to think of multiplication as repeated addition, so a child will first think of 3×4 as 3 groups of 4 or $4 + 4 + 4$ and then use addition facts to learn the multiplication facts.

   a. How would a child use addition to find 8×2?
   $2 + 2 + 2 + 2 + 2 + 2 + 2 + 2 = 16$

   b. How could the commutative property be used to make this easier?
   Rewrite the problem as $2 \times 8 = 8 + 8 = 16$.

   c. Why is it easier?
   It is easier because there are fewer numbers to keep track of in your head.

5. a. A friend says that his grade on the test is 5 points lower than your grade. If your grade is represented by x, how can you represent his grade? Should you use $x - 5$ or $5 - x$? $x - 5$

   b. What if x represents his grade instead of yours? Then how can you represent your grade? Should you use $x - 5$ or $5 - x$ or $x + 5$ or $5 + x$? $x + 5$ or $5 + x$

6. Explain how the commutative and associative properties can be used to make the following computation easier to perform:

$$-5 + 8 - 2 + 10 - 3 + 15 - 7 + 19 - 2$$

Group all the positive numbers together and all the negative numbers together and then combine.

7. Which property should be used *first* in each of the following expressions if the goal is to combine like terms?

   a. $2x + (3y - 4x)$ Commutative

   b. $(3y + 2x) - 4x$ Associative

8. a. Multiply and then simplify: $\dfrac{2}{3} \cdot \dfrac{15}{14}$ $\dfrac{5}{7}$

   b. Simplify first and then multiply: $\dfrac{2}{3} \cdot \dfrac{15}{14}$ $\dfrac{5}{7}$

   c. Did you get the same result for parts a and b? What does this tell you about the operations of multiplying and simplifying?
   The two operations commute.

9. Looking Forward, Looking Back Find the error in the following work and then show how to simplify the expression correctly.

$$\frac{5 + (3 - 8)^2}{5} = \frac{\cancel{5} + (3 - 8)^2}{\cancel{5}} \qquad \frac{5 + (3 - 8)^2}{5} = \frac{5 + (-5)^2}{5}$$
$$= 1 + (3 - 8)^2 \qquad\qquad = \frac{5 + 25}{5}$$
$$= 1 + (-5)^2 \qquad\qquad$$
$$= 1 + 25 \qquad\qquad = \frac{30}{5}$$
$$= 26 \qquad\qquad\qquad = 6$$

The 5's cannot be "canceled" in the first step. The 5 in the numerator cannot be divided off, since it is a term and not a factor.

2.7 Fair Share: Distributive Property

- Apply the distributive property to expressions
- Use the distributive property in applied contexts

Explore

1. At a department store, you purchase items with the following prices: $13, $73, and $27. An 8% sales tax in your state is applied to the subtotal. Will the final cost be affected by whether the tax is added to each item separately or the tax is applied to the subtotal?

No; the total, $122.04, will be the same regardless of when the sales tax is added.

Pre-tax total: $13 + $73 + $27 = $113
Final cost = 1.08($113) = $122.04

1.08($13) = $14.04   1.08($73) = $78.84   1.08($27) = $29.16
Final cost = $14.04 + $78.84 + $29.16 = $122.04

Discover

In Section 2.6 you learned about the commutative and associative properties for addition and multiplication. In this section, you will learn about another useful property that relates these two operations.

As you work through this section, think about these questions:

1 Do you know how to use the distributive property with mental math computations?

2 Can you give an example of when the distributive property can be used and when it cannot?

We saw in the *Explore* that we can apply the tax to each item and then add or we can add the items and then apply the tax to the subtotal. As long as the same tax rate is applied to each item, we will get the same total regardless of when it is applied. In this situation, we are distributing a multiplier throughout a quantity. Many situations from arithmetic, algebra, and real life use this idea. For example, to triple a recipe for chocolate frosting, we need to triple each ingredient.

Total and then triple   **Triple each and then total**

3(1 c cream + 8 oz chocolate + 1 tbsp butter) = 3(1 c cream) + 3(8 oz chocolate) + 3(1 tbsp butter)

The act of tripling, or multiplying by 3, is applied to each of the terms in the expression in the parentheses. This example illustrates a mathematical property known as the distributive property.

Distributive Property

The **distributive property** is a property of numbers that allows multiplication to be distributed over addition or subtraction. If a, b, and c are real numbers, it is written as follows:

$$a(b + c) = ab + ac \quad \text{(distributive property of multiplication over addition)}$$
$$a(b - c) = ab - ac \quad \text{(distributive property of multiplication over subtraction)}$$

In words, this means we can find a sum or difference and then multiply it by a number. Or we can multiply each term in the sum or difference by that number and then find the sum or difference.

EXAMPLES: $3(x - 4) = 3x - 3(4)$ Multiply each term by a factor of 3.
$$= 3x - 12$$

$-(2x + 5) = -1(2x + 5)$ Write the negative sign as -1 and distribute that factor.
$$= -1(2x) + (-1)(5) \quad \text{Multiply each term by } -1.$$
$$= -2x - 5$$

2. Apply the distributive property in each expression. Simplify.

a. $-3(x^2 - 4x + 1)$ $-3x^2 + 12x - 3$

b. $-(-4 - 7)$ $4 + 7 = 11$

c. $10x(x^2 - 5x + 6)$ $10x^3 - 50x^2 + 60x$

INSTRUCTOR NOTE: If students are struggling with #2d, encourage them to use the commutative property to write the 2 in front of the parentheses or distribute from the back.

d. $(8 + x) \cdot 2$ $16 + 2x$

e. $4(3 - 2i) + 6(8 - i)$ $60 - 14i$

f. $2x(6x^3 - 4x + 11) - 5(x + 9)$ $12x^4 - 8x^2 + 17x - 45$

Although the distributive property is useful, there are times when an expression can be simplified without using it. In #2b, you could simplify inside the parentheses to get -11 and then take the opposite to get 11. If you have the choice, pick the method to simplify an expression that is the fastest or makes most sense to you.

The distributive property is an important property that is used often in mathematics. We need to understand it and be careful not to apply it carelessly. Understand that the distributive property applies only when you have a sum or difference in parentheses. Since the expression $3(5h)$ has only one term in the parentheses, the distributive property does not apply. So $3(5h)$ simplifies to $15h$, not $15 + 3h$.

When you can apply the distributive property, be careful to distribute the factor to all terms in the sum or difference, not just the first term.

For example, in the following expression we distribute the factor of 3 to the 4 and 5:

$$
\begin{aligned}
3(4 + 5) &= 3 \cdot 4 + 3 \cdot 5 \\
&= 12 + 15 \\
&= 27
\end{aligned}
$$

If we distribute the factor of 3 only to the 4, we get 17, which is not correct.

The distributive property is useful in many situations in mathematics, including mental mathematics. For example, you might decide to buy 4 shirts priced at $19 each. If you want to do a mental calculation of the cost of the shirts, 4×19 done the traditional way is not the best option, since most people cannot do it quickly and accurately. Instead, we can use the distributive property.

$$4 \times \$19 = 4(\$20 - \$1)$$

Rewrite the right side of the equation using the distributive property.

$$4(\$20 - \$1) = 4(\$20) - 4(\$1)$$

Complete the calculation. This can be done more quickly than the original calculation.

$$\$80 - \$4 = \$76$$

Rewriting 19 as $20 - 1$ is just one option. 19 can also be thought of as $9 + 10$. Distributing the 4 gives 4 nines $+$ 4 tens or $\$36 + \$40 = \$76$.

3. Suppose you are planning a birthday party at a water park that charges $17 per child. 11 children will be attending. Show how the distributive property can help you calculate the cost mentally.

Answers will vary. One possibility: $\$17 \cdot 11 = \$17(10 + 1) = \$170 + \$17 = \$187$

We have seen that the distributive property can be used with mental math and with a multiplier over an expression that has two or more terms. We can extend the distributive property to polynomial multiplication as follows.

EXAMPLE

Multiply $(x + 2)(x + 3)$.

SOLUTION

There are a few ways to find this product of two binomials.

The first method involves distributing the quantity $(x + 2)$ to both the x and the 3 in the second set of parentheses.

$\boxed{(x + 2)}\ (x + 3)$	Distribute the quantity $(x + 2)$ to both the x and the 3.
$(x + 2)x + (x + 2)3$	Distribute or use the commutative property to rewrite the order of the multiplication in each term before distributing.
$x^2 + 2x + 3x + 6$	
$x^2 + 5x + 6$	Combine like terms.

Notice that there were two terms in each expression and four terms after all the distributing took place. Each of the two terms in the first factor times each of the two terms in the second factor gives four terms. This will be true every time you multiply two binomials. Sometimes there will be like terms to combine, reducing the number of terms in the final answer.

We can use these ideas of finding all possible products and counting the number of products to get methods that make it easier to keep track of all the products to be found.

The following method works only in the case of a binomial multiplied by a binomial. It is often known as **FOIL,** which stands for FIRST, OUTER, INNER, and LAST. This method organizes the four products by their locations. Again we have four possible products resulting in four terms before like terms are collected.

F = product of first terms of each binomial $= x \cdot x = x^2$
O = product of outer terms of the binomials $= x \cdot 3 = 3x$
I = product of inner terms of the binomials $= 2 \cdot x = 2x$
L = product of last terms of each binomial $= 2 \cdot 3 = 6$

Write the four products and find their sum by combining any like terms.

$$x^2 + 2x + 3x + 6$$
$$x^2 + 5x + 6$$

A third method uses a box to multiply $(x + 2)(x + 3)$. The product $(x + 2)(x + 3)$ describes the area of the box with dimensions $x + 2$ and $x + 3$.

The number of terms determines the number of boxes high and wide that are needed. If we have a binomial multiplied by a binomial, then the box is 2 high by 2 wide.

We write the terms of each binomial outside the box. We then find the corresponding products, which indicate the areas of the small boxes, and write them inside the boxes.

	x	3
x	x^2	$3x$
2	$2x$	6

We add the products in the boxes and collect like terms.

The result is $x^2 + 5x + 6$, which is another expression for the total area of the box. Therefore the area of the box can be represented with the product $(x + 2)(x + 3)$ or the sum $x^2 + 5x + 6$. So these two expressions are equal.

It is common to write a polynomial with the terms in decreasing order of degree. The previous result had the term x^2 with degree 2 first, the term $5x$ with degree 1 next, and finally the constant term 6 with degree 0.

The box method organizes the work well so that all products are included. This method has an advantage over FOIL because it can be used with any polynomial multiplication, not just multiplication of binomials. Notice that the products in the boxes are the same products we found using the previous methods.

4. Find each product using whichever method you prefer.

a. $(x - 2)(x - 8)$
$x^2 - 10x + 16$

b. $(x + 1)(x^2 - 2x + 1)$
$x^3 - x^2 - x + 1$

Connect

5. Each of the following problems uses the box method of multiplying polynomials. Fill in the missing terms outside or inside the box. Then write the area of the box as a product and sum.

a.

	$2x^2$	$-x$	5
$3x$	$6x^3$	$-3x^2$	$15x$

$3x(2x^2 - x + 5) = 6x^3 - 3x^2 + 15x$

b.

	x	3
x	x^2	$3x$
4	$4x$	12

$(x + 3)(x + 4) = x^2 + 7x + 12$

c.

	x^2	$2x$	5
x	x^3	$2x^2$	$5x$
1	x^2	$2x$	5

$(x + 1)(x^2 + 2x + 5) = x^3 + 3x^2 + 7x + 5$

Reflect

WRAP-UP

What's the point?

We have now seen three number properties: commutative, associative, and distributive. Each is useful when simplifying algebraic expressions, but they can also be used to improve mental math calculations.

What did you learn?

How to apply the distributive property to expressions
How to use the distributive property in applied contexts

2.7 Homework

Skills MyMathLab

First complete the MyMathLab homework online. Then work the two exercises to check your understanding.

1. Use the distributive property to rewrite the expression.

   a. $-4(x - 16x^2)$ $-4x + 64x^2$

   b. $(2x - 1)(x + 1)$ $2x^2 + x - 1$

2. Use mental math and the distributive property to find this product: 7×22
   $7(20 + 2) = 140 + 14 = 154$

Concepts and Applications

Complete the problems to practice applying the skills and concepts learned in the section.

3. A student is trying to multiply two mixed numbers and shows the following work:

$$2\frac{1}{2} \cdot 3\frac{2}{3} = 2 \cdot 3 + \frac{1}{2} \cdot \frac{2}{3} = 6 + \frac{1}{3} = 6\frac{1}{3}$$

Unfortunately, both the work and result are incorrect.

   a. A way to avoid this particular mistake is to rewrite each mixed number in the original calculation as an improper fraction and then multiply the fractions as we have done in the past. Rewrite the calculation in this way, calculate, and simplify your result. Write the answer as a mixed number.
   $2\frac{1}{2} \cdot 3\frac{2}{3} = \frac{5}{2} \cdot \frac{11}{3} = \frac{55}{6} = 9\frac{1}{6}$

   b. There is a way to perform the multiplication and keep the numbers in mixed-number form. Let's rewrite the calculation in a way that shows what each mixed number really represents.

$$2\frac{1}{2} \cdot 3\frac{2}{3} = \left(2 + \frac{1}{2}\right)\left(3 + \frac{2}{3}\right)$$

Tech TIP

Use the fraction button on your calculator to multiply the fractions as mixed numbers: $2\frac{1}{2} \cdot 3\frac{2}{3}$. Another option is to convert them to improper fractions, and use division to enter the fraction bar: $5/2 \cdot 11/3$.

Use the distributive property to finish the calculation.
$2\frac{1}{2} \cdot 3\frac{2}{3} = \left(2 + \frac{1}{2}\right)\left(3 + \frac{2}{3}\right) = 2\left(3 + \frac{2}{3}\right) + \frac{1}{2}\left(3 + \frac{2}{3}\right) = 6 + \frac{4}{3} + \frac{3}{2} + \frac{1}{3} = 9\frac{1}{6}$

Did you arrive at the same result as you did in part a?   Yes

   c. Use your calculator to find the answer to the original calculation and verify your result from part b.

4. A student claims to have a shortcut for subtracting two mixed numbers that allows him to avoid turning them both into improper fractions. Here is his work.

$$6\frac{2}{3} - 3\frac{1}{3} = \left(6 + \frac{2}{3}\right) - \left(3 + \frac{1}{3}\right)$$

$$= 6 + \frac{2}{3} - 3 - \frac{1}{3}$$

$$= 6 - 3 + \frac{2}{3} - \frac{1}{3}$$

$$= 3\frac{1}{3}$$

His method is valid and sometimes a faster way to do the computation.

He starts by rewriting each mixed number according to the definition; that is,

$$6\frac{2}{3} = 6 + \frac{2}{3} \quad \text{and} \quad 3\frac{1}{3} = 3 + \frac{1}{3}.$$

a. Which property allows him to move from this notation to the second line in the calculation?   Distributive property

b. Which property allows him to move from the second to the third line in the calculation?   Commutative property

c. Show that you get the same result if you convert both mixed numbers to improper fractions first.

$$\frac{20}{3} - \frac{10}{3} = \frac{10}{3} = 3\frac{1}{3}$$

You should decide which method you prefer. Although the student's method can be used, it can get fairly complicated if the numbers do not have common denominators as they do in this situation.

d. Try both methods on this problem: $2\frac{1}{4} - 1\frac{3}{4}$

Method 1 (student's nontraditional way):

$$\left(2 + \frac{1}{4}\right) - \left(1 + \frac{3}{4}\right) = 2 + \frac{1}{4} - 1 - \frac{3}{4} = 1 - \frac{2}{4} = \frac{1}{2}$$

Method 2 (traditional approach using improper fractions): $\frac{9}{4} - \frac{7}{4} = \frac{2}{4} = \frac{1}{2}$

5. An application of the distributive property shows that $(a - b)(a + b) = a^2 - b^2$.

    a. Verify that this is a true statement by applying the distributive property to the left side of the equation.

    $(a - b)(a + b) = (a - b)a + (a - b)b = a^2 - ba + ab - b^2 = a^2 - b^2$

    This formula can be useful for multiplying two numbers mentally. For example, $37 \times 43 = (40 - 3)(40 + 3) = 40^2 - 3^2 = 1,600 - 9 = 1,591$.

    Show how you can use this mental strategy to find the following products without using a calculator or written computations.

    b. 49×51

    $49 \times 51 = (50 - 1)(50 + 1) = 50^2 - 1^2 = 2,500 - 1 = 2,499$

    c. 72×68

    $72 \times 68 = 68 \times 72 = (70 - 2)(70 + 2) = 70^2 - 2^2 = 4,900 - 4 = 4,896$

    Which property must be applied to the expression in part c before you can use the formula?

    Commutative property of multiplication

6. Each problem uses the box method of multiplying polynomials. Fill in the missing terms outside or inside the box. Then write the area of the box as a product and sum.

    a.

	$6x^2$	$-x$	10
$5x$	$30x^3$	$-5x^2$	$50x$

    $5x(6x^2 - x + 10) = 30x^3 - 5x^2 + 50x$

    b.

	x	-5
x	x^2	$-5x$
-8	$-8x$	40

    $(x - 5)(x - 8) = x^2 - 13x + 40$

7. Looking Forward, Looking Back Determine the pattern and write an expression for the output when the input is n.

Input	Output
0	-1
1	2
2	5
3	8
4	11
n	$3n - 1$

2.8 Seat Yourself: Equivalent Expressions

Explore

1. Consider the following sequence of figures. How many squares will there be in the nth figure? Explain how you wrote your expression based on the physical pattern in the figures.

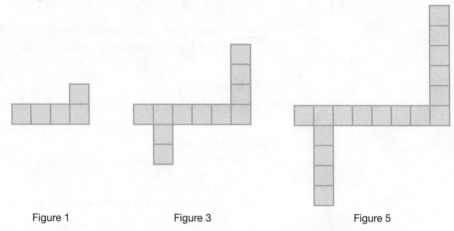

Figure 1   Figure 3   Figure 5

There are $3n + 2$ squares in the nth figure. One way to determine this is to realize that there are $n + 3$ squares in the middle long row, n squares going vertically up at one end, and $n - 1$ squares going vertically down.

Discover

If you compare answers in the *Explore* with other students, you'll likely realize that they saw the situation differently than you did. Even if they saw a different pattern in the squares, their final expression for the number of squares should match yours if you're both correct. In order to compare your expressions, you need to be able to simplify them. In this section, you will get more practice using the commutative, associative, and distributive properties to simplify expressions.

By the end of the section, you should be able to answer the following questions:

1 How can you tell if expressions are equivalent?

2 How can you decide if your way of modeling a situation is equivalent to someone else's?

Suppose you are planning an event that will require renting tables and chairs. The tables are to be arranged in one line with the sides of the tables touching and chairs placed on all available sides. We will assume the tables are square and can fit one chair on each side. The first few arrangements are shown here.

INSTRUCTOR NOTE: While the context of tables and chairs may appear somewhat contrived, the problem in #2 is a simpler version of a carbon chain problem students will solve in Section 3.7.

2. Complete the following table by determining how many chairs can be accommodated for each number of tables and writing an expression for n tables.

Number of Tables	Number of Chairs
1	4
2	6
3	8
4	10
20	42
200	402
n	$2n + 2$

There are many ways of looking at this physical situation to generalize the number of chairs needed, but are those different approaches equivalent? We can model each physical description with an expression for the number of chairs at n tables. Then we can simplify the expressions to determine if they are equivalent. The following table describes four different ways to count the number of chairs and an algebraic expression to model each of them.

	Physical Description	Algebraic Expression
a.	There are 4 seats per table, but you need to subtract the seats you lose when you put the tables against each other. You lose 2 seats at each juncture, and there are $n - 1$ junctures.	$4n - 2(n - 1)$
b.	This arrangement of tables forms a long rectangle. One long side has n seats, and the other long side has n seats. But there are 2 more seats on the ends.	$n + n + 2$
c.	Each table has 2 seats except for the tables on the end, which have 3. So there are $n - 2$ tables that have 2 seats and 2 tables that have 3 seats.	$2(n - 2) + 2 \cdot 3$
d.	The first table has 4 seats, but each table after that adds only 2 more seats to the total. There are $n - 1$ additional tables after the first.	$4 + (n - 1) \cdot 2$

Each expression in the table can be simplified to help us determine if they are equivalent.

Here are some common phrases and the corresponding expressions:

Two more than a number	$n + 2$
All but two	$n - 2$
Two less than a number	$n - 2$
Two less a number	$2 - n$
Twice a number	$2n$
Half a number	$\frac{n}{2}$

3. Simplify each expression. Are they equivalent?

a. $4n - 2(n - 1)$ $4n - 2(n - 1) = 4n - 2n + 2 = 2n + 2$

b. $n + n + 2$ $n + n + 2 = 2n + 2$

c. $2(n - 2) + 2 \cdot 3$ $2(n - 2) + 2 \cdot 3 = 2n - 4 + 6 = 2n + 2$

d. $4 + (n - 1) \cdot 2$ $4 + (n - 1) \cdot 2 = 4 + 2n - 2 = 2n + 2$

All the expressions are equivalent.

It's interesting to note that there were several correct ways to count the number of chairs, but each method resulted in the same expression. We can try this same approach with a different configuration of tables and chairs.

INSTRUCTOR NOTE: Students will likely notice that they obtained the same $2n + 2$ expression as they did for the linear arrangement. Consider mentioning that this is just a rearranged configuration in which some tables are attached in a line vertically instead of horizontally without affecting the number of seats. Be sure students realize that not every situation will simplify to $2n + 2$.

4. Let's consider another way of arranging the tables and chairs we just discussed. Assume the tables are arranged in a plus-sign shape as follows:

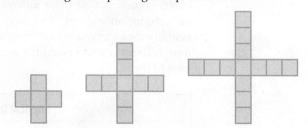

Use the following table to record how many tables there are of each type. Then use that information to write an expression for the number of chairs.

Number of tables with 0 chairs	Number of tables with 1 chair	Number of tables with 2 chairs	Number of tables with 3 chairs	Unsimplified expression for the total number of chairs
1	0	$n - 5$	4	$2(n - 5) + 4(3)$

Simplify the expression. $2n + 2$

EXAMPLE

Your neighbor is building onto his square patio. He plans to add 5 linear feet to both the width and the length of the patio. He claims that the patio will then be 25 square feet larger. How do you know he's incorrect?

SOLUTION

Since we don't know the current size of the patio, we can assume it's x by x feet. If he adds 5 feet in each direction, the patio would be $(x + 5)$ feet by $(x + 5)$ feet with an area of $(x + 5)^2$ square feet. Is this equivalent to $x^2 + 25$ as he claims? We can simplify $(x + 5)^2$ to find out.

$$(x + 5)^2 = (x + 5)(x + 5)$$
$$= x^2 + 5x + 5x + 25$$
$$= x^2 + 10x + 25$$

We can see that $(x + 5)^2$ is not equivalent to $x^2 + 25$, since it includes the extra $10x$ term. These expressions would be equal only if $x = 0$. So your neighbor is wrong about the new patio being 25 square feet larger than the original. The patio area will increase by more than 25 square feet.

Connect

5. Which of the following expressions are equivalent? If an expression has no match, write two equivalent expressions for it.

   a. $2x^2 - 4x + 7$

   b. $2(2x^2 - 2x - 1) - 2(x^2 - 2x + 3)$

   c. $2(2x - 1) + 3(x^2 - 2x) - (x^2 - 2x + 6)$

   d. $2(x^2 - 4x + 7)$

   e. $(x^2 - 2x - 2) + (x^2 - 2x + 9)$

   f. $(x^2 - x + 1) - (-x^2 + 3x - 6)$

   a, e, and f are equivalent; b and c are equivalent; d is equivalent to $2x^2 - 8x + 14$ and $(4x^2 - 5x + 10) - (2x^2 + 3x - 4)$

Reflect

WRAP-UP

What's the point?

A physical situation can often be represented by many expressions that might appear different. To determine whether a proposed expression is accurate, you must be able to simplify expressions and compare them.

What did you learn?

How to write an expression to represent a scenario
How to determine if two expressions are equivalent by using the commutative, associative, and distributive properties

2.8 Homework

Skills MyMathLab

First complete the MyMathLab homework online. Then work the two exercises to check your understanding.

1. Simplify the expression: $5(x + 8) - 4(x - 1) + 6$ $x + 50$

2. Simplify the expression: $8 - 3(2 - n) - 17$ $-15 + 3n$ or $3n - 15$

Concepts and Applications

Complete the problems to practice applying the skills and concepts learned in the section.

3. Complete the table to write an expression for the number of chairs at n tables for each of the following arrangements:

Shape	Number of tables with 0 chairs	Number of tables with 1 chair	Number of tables with 2 chairs	Number of tables with 3 chairs	Unsimplified expression for the total number of chairs	Simplified expression
L-shaped	0	0	$n - 2$	2	$2(n - 2) + 3(2)$	$2n + 2$
T-shaped	0	1	$n - 4$	3	$1(1) + 2(n - 4) + 3(3)$	$2n + 2$

L-shaped   **T-shaped**

4. Suppose you decide to arrange tables in a square configuration with only a single row of tables on the edge of the square. Chairs will be set on only the outside of the square. As the arrangements increase in size, there will be open space in the shape of a square on the inside. Chairs will not be set in this space.

   a. Study the first two possible arrangements and then draw the next two in the sequence.

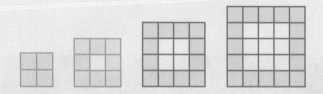

   b. Generalize the number of chairs needed for n tables. Consider using an approach like the one shown in the section in which we determined how many chairs are needed for tables of a certain type. In this problem, there are corner tables and the interior tables are between the corner tables. Write an expression for the number of chairs and simplify it.

   One way to think about this is that there are 4 corner tables that each seat 2 people. All the other tables $(n - 4)$ seat one person. So there are $4 \cdot 2 + (n - 4) \cdot 1 = n + 4$ seats.

c. Use the figures in part a to complete the following chart. Generalize the calculation and list the result in the last row, looking left to right in the chart.

Number of Tables	Number of Chairs
4	8
8	12
12	16
16	20
n	$n + 4$

What do you notice?

You get the same general result, $n + 4$, as you found by using the physical situation in part b.

d. Would it be possible to have 20 tables and make a square? Why or why not? If so, how many people could be seated?

Yes; the number of tables in this arrangement is always a multiple of 4, and 20 is a multiple of 4. 24 people could be seated.

e. If we have 36 people at a party, how many tables would be needed for the arrangement?   32

5. You and a friend are hiking on a trail that is labeled with mile markers. You have started at opposite ends of the trail and decide to call each other to check in and meet for lunch. Your friend is at mile marker 19, and you are at marker 3.

a. Draw a picture of this situation.

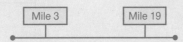

b. While on the phone, you decide to meet halfway between your current locations. At which mile marker will you meet?   11

c. If we generalize this calculation, we can find the middle point between any two locations. Call the first location A and the second B. Write an expression for the distance between them if B is the larger number. $B - A$

d. Because we want to split that distance evenly, rewrite the expression to show that we're taking half of it.

$\frac{1}{2}(B - A)$

e. Now we can add that distance to the smaller of the locations, A. Write this and simplify.

$$A + \frac{1}{2}(B - A) = \frac{1}{2}A + \frac{1}{2}B$$

f. You can also subtract half the distance from the larger of the mile markers, B. Show that you get the same result as in part e if you do this.

$$B - \frac{1}{2}(B - A) = \frac{1}{2}A + \frac{1}{2}B$$

g. Your friend tells how she found the middle point. Her method was shorter, just adding the two numbers and dividing by 2. Write this expression. Is it equivalent to the result in part f?

$$\frac{A + B}{2} = \frac{1}{2}(A + B) = \frac{1}{2}A + \frac{1}{2}B; \text{ yes, they are equivalent.}$$

6. Follow the steps and fill in the blanks as you go. Do this process three times, each time recording your starting number and ending number in the table.

Step 1: Pick a number.   _____

Step 2: Quadruple it.   _____

Step 3: Add 4 to your result. _____

Step 4: Subtract 2.   _____

Number from Step 1	Number from Step 4
0	2
4	18
10	42
n	$4n + 2$

Make a conjecture that will allow you to get from the left column to the right column in one step. If you can see the general rule, write it in the last row of the table.

To help you conjecture the general rule, do this process again but with n as the number in Step 1. Write your work in the blanks as you go. Simplify the expression.

Step 1: Pick a number. n

Step 2: Quadruple it. $4n$

Step 3: Add 4 to your result. $4n + 4$

Step 4: Subtract 2. $4n + 2$

7. Looking Forward, Looking Back Find the error in the following student work:

$$(3x - 4)(2x - 5) = 3x(2x - 5) - 4(2x - 5)$$
$$= 6x^2 - 15x - 8x - 20$$
$$= 6x^2 - 23x - 20$$

The student did not distribute the negative sign to the second term. The student should have multiplied -4 by -5 to get 20 for the constant term.

2.9 Parts of Speech: Using Operations Correctly

Explore

SECTION OBJECTIVES:

- Distinguish between an operator and an object
- Determine the object on which an operator is acting
- Recognize when the distributive property can be applied

1. Complete this table by filing in the missing boxes. Write the operations in words, not symbols. The first line has been completed as an example.

	Expression	Object(s)	Operation
a.	$\sqrt{17}$	17	Square root
b.	$4\frac{2}{5}$	$4, \frac{2}{5}$	Addition
c.	$\frac{8}{9}$	8, 9	Division
d.	$\lvert-10\rvert$	-10	Absolute value
e.	$(-8)^2$	(-8)	Square
f.	-8^2	8	Square
g.	$2(x + y)$	$2, (x + y)$	Multiplication
h.	$\dfrac{a + b}{2}$	a, b	Addition
i.	$\dfrac{a + b}{2}$	$a + b, 2$	Division
j.	$\dfrac{a}{2} + \dfrac{b}{2}$	$\dfrac{a}{2}, \dfrac{b}{2}$	Addition
k.	$\dfrac{a}{2} + \dfrac{b}{2}$	a and 2, b and 2	Division

Discover

Mathematics uses operators and objects, similar to how language uses verbs and nouns. Comparing mathematics and language can improve your understanding of both. In mathematics, we often interpret operations as actions that are applied to numbers or expressions. Operations are active; they do something. Numbers and expressions can be seen as objects to which the operations, or actions, are applied. Understanding what operation is being applied to which object is essential as we continue our work in algebra.

As you work through this section, think about these questions:

1 Can you use deductive reasoning to prove a claim or find a counterexample to show that the claim is false?

2 Can you determine the operations being applied and the order in which they are applied in an expression?

Parentheses affect the value of an expression in much the same way that punctuation affects the interpretation of setnence. The meanings of these sentences are very different with the addition of commas.

"I like cooking my family and my dog."

"I like cooking, my family, and my dog."

Likewise, the meanings of these expressions are very different with the addition of parentheses:

$$-8 - 8$$
$$(-8)(-8)$$

The first expression is a difference while the second expression is a product.

It is necessary to distinguish between the operations of subtraction and multiplication in the above expressions. Similarly, the distinction between operations and objects is essential for using the distributive property correctly. In the following example, the operation of multiplying by 3 is applied to each of the objects or terms in parentheses.

$$3(2x + 9) = 6x + 27 \quad \text{Distribute the factor of 3 to both terms.}$$

In contrast, we cannot "distribute" the factor of 3 in the expression $3(2x \cdot 9)$ because there is only one term that the factor of 3 is applied to. Instead, we multiply the constants.

$$3(2x \cdot 9) = 54x$$

Let's look at other situations where the distributive property may or may not be used.

Suppose you want to halve a recipe that requires $4\frac{1}{2}$ cups flour. We can take half of the 4 and half of the $\frac{1}{2}$ and then combine them. This method is valid, since it is an application of the distributive property.

$$\frac{4\frac{1}{2}}{2} = \frac{1}{2}\left(4 + \frac{1}{2}\right) = \frac{1}{2}(4) + \frac{1}{2}\left(\frac{1}{2}\right) = 2 + \frac{1}{4} = 2\frac{1}{4}$$

Notice that the operation is halving, and it is applied to the object $4\frac{1}{2}$ cups. This example is a specific case of the general idea that $\dfrac{a + b}{2} = \dfrac{a}{2} + \dfrac{b}{2}$. We can see that the expressions are equivalent using the distributive property.

$$\frac{a + b}{2} = \frac{1}{2}(a + b) = \frac{1}{2}a + \frac{1}{2}b = \frac{a}{2} + \frac{b}{2}$$

Since division is multiplication by the reciprocal, division can also be distributed over a sum or difference.

EXAMPLE

Simplify $\dfrac{4x + 1}{4}$.

SOLUTION

Notice that the operation of division by 4 is being applied to the sum $4x + 1$. We can use the fact that division can be distributed over a sum to simplify. Resist the urge to divide off the 4's as a first step.

$$\frac{4x + 1}{4} = \frac{4x}{4} + \frac{1}{4} = \frac{\overset{1}{\cancel{4}}x}{\underset{1}{\cancel{4}}} + \frac{1}{4} = x + \frac{1}{4}$$

2. Simplify each expression.

a. $\dfrac{7x + 7}{7}$ $x + 1$

b. $\dfrac{x - 7}{7}$ $\dfrac{x}{7} - 1$

c. $-\dfrac{x - 7}{7}$ $-\dfrac{x}{7} + 1$

As we have seen, we can distribute division over a sum, but can we distribute a sum over division? When answering questions like this, remember that only deductive reasoning using the general case can prove a statement.

Remember?

Examples are not enough to prove a statement, but one counterexample is enough to disprove a claim.

3. Are the following two expressions always equal: $\dfrac{2}{a + b}$ and $\dfrac{2}{a} + \dfrac{2}{b}$? If they are, show why. If not, provide a counterexample.

No; if a and b are each 1, the expressions are not equal.

4. Can an exponent be distributed over addition? In other words, is $(a + b)^2$ equivalent to $a^2 + b^2$? If not, provide a counterexample and show how $(a + b)^2$ can be rewritten.

No; this is not valid. A counterexample can be shown with $a = 2$ and $b = 4$.
We need to use the distributive property after rewriting the expression.
$(a + b)^2 = (a + b)(a + b) = a^2 + 2ab + b^2$

Connect

Operations that undo each other, like multiplication and division, are called inverse operations. Addition and subtraction are another pair of inverse operations. Understanding the process needed to undo a set of operations is necessary when you begin solving equations later in this cycle. For example, these steps are needed to create a sealed envelope with a piece of folded paper inside:

1. Fold the paper.

2. Place the paper in the envelope.

3. Seal the envelope.

To undo this process and get the contents out of the envelope, you need to undo each task working backward from the last task to the first.

1. Unseal the envelope.

2. Take out the paper.

3. Unfold the paper.

Use these ideas to complete the next problem.

5. Evaluate each expression when $n = 5$. List the operations that you applied to the value of n in the order they were applied. Then provide a second list of operations that contains the inverse operations in the reverse order of the first list. These operations undo the original set of operations done to n.

a. $3n^2 - 2$ 73

Square, multiply by 3, subtract 2

Add 2, divide by 3, square root

b. $\dfrac{2(n - 1)}{3}$ $\dfrac{8}{3}$

Subtract 1, multiply by 2, divide by 3

Multiply by 3, divide by 2, add 1

Reflect

WRAP-UP

What's the point?

It is important to understand the role of operators and objects in mathematics. This knowledge will pay off as you move into the second half of the cycle and learn to solve equations.

What did you learn?

How to distinguish between an operator and an object
How to determine on which object an operator is acting
How to recognize when the distributive property can be applied

2.9 Homework

Skills MyMathLab

First complete the MyMathLab homework online. Then work the two exercises to check your understanding.

1. For the expression $-15(x + 7)$, which objects are being multiplied? Which objects are being added?

-15 and $x + 7$; x and 7

2. For the expression -10^2, which object is being squared?

10

Concepts and Applications

Complete the problems to practice applying the skills and concepts learned in the section.

3. In each expression, the square root is one of several operations. Identify the object to which the square root is being applied. This object is known as the radicand.

a. $\sqrt{4}(x + y)$ 4

b. $\sqrt{4(x + y)}$ $4(x + y)$

c. $\dfrac{\sqrt{x}}{4}$ x

d. $\sqrt{\dfrac{x}{4}}$ $\dfrac{x}{4}$

4. Determine whether $\sqrt{x + y}$ is equivalent to $\sqrt{x} + \sqrt{y}$. If they are equivalent, use deductive reasoning to support your conclusion. Otherwise, provide a counterexample.

They are not equivalent. If $x = 25$ and $y = 9$, then $\sqrt{25 + 9} = \sqrt{34} \approx 5.83 \neq 8$.

5. Some students were asked to simplify $\dfrac{8(2x + 6)}{2}$. Find the mistake in each student's work. Then find the correct answer.

a. $\dfrac{8(2x + 6)}{2} = 8(x + 6)$

$= 8x + 48$

b. $\dfrac{8(2x + 6)}{2} = \dfrac{16x + 6}{2}$

$= 8x + 3$

c. $\dfrac{8(2x + 6)}{2} = \dfrac{16x + 8x + 48}{2}$

$= 8x + 4x + 24$

$= 12x + 24$

a. The student divided off the 2 from the first term in the parentheses instead of from both terms.

b. The student distributed the factor of 8 to the first term in the parentheses but not to the second.

c. The student distributed the factor of 8 to both the 2 and the x in the term 2x.

The correct answer is $8x + 24$.

6. Can an exponent be distributed over subtraction? Is $(a - b)^2$ equivalent to $a^2 - b^2$? If not, how can $(a - b)^2$ be rewritten?

No; the expressions are not equivalent. We need to use the distributive property after rewriting the problem as $(a - b)(a - b)$. Thus $(a - b)^2 = a^2 - 2ab + b^2$.

7. For each expression, list the operations being applied and the order in which they are applied to the variable. Then provide a second list of operations that contains the inverse operations in the reverse order of the first list.

   a. $6(x - 3)$

   Subtract 3, multiply by 6.
   Divide by 6, add 3.

   b. $\dfrac{8x + 11}{7}$

   Multiply by 8, add 11, divide by 7
   Multiply by 7, subtract 11, divide by 8

   c. $5x^2 + 1$

   Square, multiply by 5, add 1
   Subtract 1, divide by 5, square root

   d. $\sqrt{x^2 - 8}$

   Square, subtract 8, square root
   Square, add 8, square root

   e. $\dfrac{1}{x - 5}$

   Subtract 5, take reciprocal
   Take reciprocal, add 5

8. **Looking Forward, Looking Back** Simplify the polynomial expression. State the degree of the result.

$$2 - 8(2 - 8x) - 2$$

   $64x - 16$, 1st degree

Mid-Cycle Recap

INSTRUCTOR NOTE: The Mid-Cycle Recap provides students a chance to see how they are accomplishing the goals of the cycle so far. A quiz is available in MyMathLab as an option for assessing students' progress on the objectives at this point. See the Instructor Guide for more assessment ideas.

Complete the problems to practice applying the skills and concepts learned in the first half of the cycle.

Skills

1. Simplify each expression.

   a. $(5x^5)^2$ $25x^{10}$

   b. $\dfrac{50x^{50}}{5x^5}$ $10x^{45}$

   c. $5x^4 + 5x^4$ $10x^4$

   d. $5x^4 \cdot 5x^4$ $25x^8$

   e. -5^0 -1

   f. $(-5)^0$ 1

Concepts and Applications

2. A cylindrical water tank has a height of 10 feet and a radius of 2 feet. Use the following formula to calculate its surface area. Write units in every step of your calculation and round your final answer to the nearest tenth.

   $$\text{Cylinder:}\quad S = 2\pi r^2 + 2\pi rh$$

   150.8 square feet

3. Show how to make the following computation easier to do mentally by using the commutative, associative, or distributive properties: $99(25) =$
   $(100 - 1)(25) = 2{,}500 - 25 = 2{,}475$

4. A forester is charged with the task of determining the average tree diameter in a section of woods. He is to select 10 trees randomly and use the measurements to determine the average. Since he is not going to cut the trees down, the diameter cannot be measured directly. He uses a flexible metal measuring tape and finds the circumference of each. Following are his measurements in inches:

   57   65.5   72   71   64.25   69.5   58.75   57.75   69.25   64

   a. Find the average of the circumferences.

   64.9 inches

   b. Use the average circumference to find the average diameter by using the formula $C = 2\pi r$.

   20.7 inches

   c. Now find the average diameter in a different way. Find the diameter of each tree that was measured. Then average these diameters. Do you get the same result?
   The result should be the same if individual diameters were not rounded.

2.10   A Fine Balance: Verifying Solutions

Explore

• Verify a solution to an equation

> Vocabulary and notation are important. Understanding the vocabulary is essential to working in any context.

Chemical equations are similar to algebraic equations in many ways, but they are "balanced" instead of "solved." The process of balancing a chemical equation involves finding coefficients so that the equation has the same number of atoms of each element on both sides of the equation.

Let's explore a chemical equation that is balanced to understand the notation and process:

$$Ra + 2\,H_2O \rightarrow Ra(OH)_2 + H_2$$

In this balanced chemical equation, there are two different molecules on each side [Ra and H_2O on the left and $Ra(OH)_2$ and H_2 on the right] that are comparable to terms in an algebraic equation.

The subscript on an element tells how many atoms of the element are in that compound. If a subscript is not written, it is understood to be 1. Coefficients are multiplied by subscripts to give the total number of atoms of each element in a particular molecule or compound. For example, there are 2 hydrogen atoms and 1 oxygen atom in H_2O, and the coefficient of 2 on that molecule in the equation makes 4 hydrogen atoms and 2 oxygen atoms. On the right side of the equation, the subscript in the molecule $Ra(OH)_2$ is applied to each element in parentheses, just as we would apply an exponent to each factor in parentheses in algebra.

The following table shows the total element counts for each side of the balanced chemical equation. Notice that the element counts are the same for each side, which indicates that the equation is balanced.

	Left Side	Right Side
Ra	1	1
H	4	4
O	2	2

1. Confirm that the following chemical equations are balanced by showing that each element has the same count on both sides of the equation.

   a. $2\,H_2O_2 \rightarrow 2\,H_2O + O_2$

	Left Side	Right Side
H	4	4
O	4	4

   b. $2\,N_2H_4 + N_2O_4 \rightarrow 3\,N_2 + 4\,H_2O$

	Left Side	Right Side
N	6	6
H	8	8
O	4	4

c. $2\,H_3PO_4 \ + \ 3\,Ca(OH)_2 \rightarrow Ca_3(PO_4)_2 \ + \ 6\,H_2O$

	Left Side	Right Side
H	12	12
P	2	2
O	14	14
Ca	3	3

Discover

In chemistry, you must understand what it means to have an equation balanced. Similarly, in algebra, you must understand what it means to have an equation solved.

As you work through this section, think about these questions:

1 How can you tell if a number is a solution to an equation?

2 Could you write an equation with no solution?

3 Are there equations that have more than one solution?

Solution

The **solution** of an equation is a number that results in a true statement when it is substituted for the variable in the equation.

FOR EXAMPLE, $x = 3$ is a solution to $2x - 5 = x - 2$, since a true statement results when $x = 3$ is substituted into the equation.

$$2x - 5 = x - 2$$
$$2 \cdot 3 - 5 \stackrel{?}{=} 3 - 2$$
$$6 - 5 \stackrel{?}{=} 3 - 2$$
$$1 \stackrel{\checkmark}{=} 1$$

2. For each of the following equations, verify that the given value is a solution to the equation by substituting it into the equation and checking that a true statement results.

a. $2 - 3x = -7; x = 3$

$$2 - 3(3) \stackrel{?}{=} -7$$
$$2 - 9 \stackrel{?}{=} -7$$
$$-7 \stackrel{\checkmark}{=} -7$$

b. $\dfrac{14}{x - 1} - 4 = 10; x = 2$

$$\frac{14}{2 - 1} - 4 \stackrel{?}{=} 10$$
$$\frac{14}{1} - 4 \stackrel{?}{=} 10$$
$$14 - 4 \stackrel{?}{=} 10$$
$$10 \stackrel{\checkmark}{=} 10$$

c. $\sqrt{2x - 1} = 8 - x; x = 5$

$\sqrt{2(5) - 1} \overset{?}{=} 8 - 5$

$\sqrt{10 - 1} \overset{?}{=} 3$

$\sqrt{9} \overset{?}{=} 3$

$3 \overset{\checkmark}{=} 3$

d. $-x^2 + 5x + 14 = 0; x = -2$

$-(-2)^2 + 5(-2) + 14 \overset{?}{=} 0$

$-4 + (-10) + 14 \overset{?}{=} 0$

$-14 + 14 \overset{?}{=} 0$

$0 \overset{\checkmark}{=} 0$

e. $2^{x-1} - 8 = 0; x = 4$

$2^{4-1} - 8 \overset{?}{=} 0$

$2^3 - 8 \overset{?}{=} 0$

$8 - 8 \overset{?}{=} 0$

$0 \overset{\checkmark}{=} 0$

HOW IT WORKS

To verify a solution to an equation:

1. Substitute the value of the variable into the equation.
2. Simplify the left and right sides of the equation.

- If a true statement results, the value is a solution.
- If a false statement results, the value is not a solution to the equation.

EXAMPLE: Verify that $x = -5$ is a solution to the equation $3x - 2 = -17$.

When $x = -5$ is substituted into the equation, a true statement results.

$$3(-5) - 2 \overset{?}{=} -17$$
$$-15 - 2 \overset{?}{=} -17$$
$$-17 \overset{\checkmark}{=} -17 \quad \text{True statement}$$

EXAMPLE: Verify that $x = 0$ is not a solution to the equation $3x - 2 = -17$.

When $x = 0$ is substituted into the equation, a false statement results.

$$3(0) - 2 \overset{?}{=} -17$$
$$0 - 2 \overset{?}{=} -17$$
$$-2 \overset{\checkmark}{\ne} -17 \quad \text{False statement}$$

Some equations have one solution. Others may have no solutions or many solutions.

3. Use the following algebraic equation to answer the questions:

$$-3(2x - 1) - x + 8 = -7(x + 1) + 18$$

a. Simplify each side of the equation. What do you notice?

$-7x + 11 = -7x + 11$; both sides are the same expression.

b. Confirm that $x = 1, x = 0$, and $x = -4$ are solutions to the equation. Without using algebra, state one more solution to this equation.

Answers will vary.

c. Describe the solutions to the equation. Defend your response.

The solutions are all real numbers. Since the left side is identical to the right side and both sides are defined for all real numbers, every number is a solution.

The equation in the last problem is a special type of equation known as an **identity**. Each side contained the exact same terms when it was simplified. Balanced chemical equations are, in a sense, also a type of identity, since in a balanced equation, each side has the same number of atoms of each element.

Connect

INSTRUCTOR NOTE: You might show students how to create this table on Excel. Fill in the first column with consecutive integers, and then use the left side and right side of the equation as formulas to fill in the second and third columns.

4. Use the following table to identify the solution(s) to the equation $x^2 = 5x + 24$.

x	Left Side	Right Side
−4	16	4
−3	9	9
−2	4	14
−1	1	19
0	0	24
1	1	29
2	4	34
3	9	39
4	16	44
5	25	49
6	36	54
7	49	59
8	64	64
9	81	69
10	100	74

$x = -3, 8$

Reflect

WRAP-UP

What's the point?

Verifying that a chemical equation is balanced is similar to verifying that an algebraic equation is solved. Understanding what it means to have a solution to an equation is the first step toward solving equations, which we will begin to do in the next section.

What did you learn?

How to verify a solution to an equation

2.10 Homework

Skills MyMathLab

First complete the MyMathLab homework online. Then work the two exercises to check your understanding.

1. Verify that the given value is a solution to the equation.

$$3x - 7 = -5(x + 3); x = -1$$

$$3(-1) - 7 \stackrel{?}{=} -5(-1 + 3)$$

$$-3 - 7 \stackrel{?}{=} -5(2)$$

$$-10 \stackrel{\checkmark}{=} -10$$

2. Verify that the given value is a solution to the equation.

$$2x^2 = 4x - \frac{3}{2}; x = \frac{3}{2}$$

$$2\left(\frac{3}{2}\right)^2 \stackrel{?}{=} 4\left(\frac{3}{2}\right) - \frac{3}{2}$$

$$2\left(\frac{9}{4}\right) \stackrel{?}{=} \frac{12}{2} - \frac{3}{2}$$

$$\frac{9}{2} \stackrel{\checkmark}{=} \frac{9}{2}$$

Concepts and Applications

Complete the problems to practice applying the skills and concepts learned in the section.

3. Fill in each blank.

a. A person is to a family as an atom is to a(n) _____molecule_____.

b. A subscript in a chemical equation is like a(n) _____exponent_____ in an algebraic expression.

c. The _____coefficient_____ of the expression $15x^2$ is 15.

d. If a chemical equation is _____balanced_____, both sides contain the same number of atoms of each element.

4. a. Is CH_4 the same as $(CH)_4$? How many carbon atoms are in each? How many hydrogen atoms?

No; CH_4 has 1 carbon and 4 hydrogen atoms.

$(CH)_4$ has 4 carbon and 4 hydrogen atoms.

b. Is xy^4 the same as $(xy)^4$? How many factors of x and y are in each expression?

No; xy^4 has one x and four y factors.

$(xy)^4$ has four x and four y factors.

c. Find $CH_4 + CH_4$.

$2CH_4$

d. Find $xy^4 + xy^4$.

$2xy^4$

5. Explain how the notation for three water molecules, $3H_2O$, is like the expression $3x^2y$.

In $3H_2O$, the coefficient of 3 indicates how many water molecules are present.

In $3x^2y$, the coefficient of 3 indicates how many x^2y terms are present.

The subscript in the water molecule is like the exponent on the x-variable.

6. A recipe includes the following ingredients and yields two dozen cookies:

2 eggs + 4 cups of flour + 4 tablespoons of butter + 1 cup of sugar $\rightarrow$ 24 cookies

Rewrite the recipe but scale it to use 4 eggs. Now how many cookies does the recipe yield?

4 eggs + 8 cups of flour + 8 tablespoons of butter + 2 cups of sugar $\rightarrow$ 48 cookies

The recipe yields 48 cookies.

7. Consider an applesauce recipe that says

$$4 \text{ cup apples} + \frac{1}{2} \text{ cup sugar} + \frac{1}{4} \text{ cup water} = 4 \text{ cups applesauce.}$$

a. What is the ratio of cups of apples to cups of sugar? 8:1

b. What is the ratio of cups of sugar to cups of water? 2:1

c. What is the ratio of cups of water to cups of sugar? 1:2

d. How would you "balance" this equation if you wanted to have only whole numbers of cups? Rewrite the recipe using only whole numbers.

You would quadruple the recipe:

16 cups apples + 2 cups sugar + 1 cup water = 16 cups applesauce.

8. Consider the equation for burning methane gas in oxygen to produce carbon dioxide and water: $CH_4 + 2O_2 \rightarrow CO_2 + 2H_2O$.

a. In every molecule of water, there are ___2___ atoms of hydrogen for one atom of oxygen.

b. In every molecule of methane, there are ___4___ atoms of hydrogen for one atom of carbon.

    c. In the balanced equation, what is the ratio of methane to water molecules?

         1:2

    d. In the balanced equation, what is the ratio of water to methane molecules?

         2:1

9. Balance each chemical equation by determining coefficients that will produce the same total number of atoms for each element on each side of the equation.

    a. $Zn + 2\,HCl \rightarrow ZnCl_2 + H_2$

    b. $2Fe + 3\,H_2SO_4 \rightarrow Fe_2(SO_4)_3 + 3\,H_2$

    c. $3\,Ca(OH)_2 + 2\,H_3PO_4 \rightarrow Ca_3(PO_4)_2 + 6\,H_2O$

    d. $2\,C_2H_6 + 7\,O_2 \rightarrow 6\,H_2O + 4\,CO_2$

10. Looking Forward, Looking Back Find the error in this student's work.

$$\frac{16}{2x + 2y} = \frac{16}{2x} + \frac{16}{2y}$$

$$= \frac{8}{x} + \frac{8}{y}$$

A numerator with multiple terms can be split into multiple fractions over the same original denominator, but a denominator cannot be split this way.

2.11 Separate but Equal: Solving Simple Equations

Explore

SECTION OBJECTIVE:

- Write and solve one-step equations

1. Erin runs a concession stand for a high school football game. She wants the prices to be whole-dollar amounts that already include a 6% sales tax. For each item in the table, the price for the customer is given with the sales tax included. What is the actual price for each item before the tax is added?

Regular hot dog	$2.00	$1.89
Jumbo hot dog	$4.00	$3.77
Soda	$1.00	$0.94
Pork chop sandwich	$3.00	$2.83
Combo deal (jumbo hot dog, chips, soda)	$5.00	$4.72

Discover

When solving a problem like the one in the *Explore,* you can use numeric or algebraic methods. Numeric methods rely on numbers alone to solve problems while algebraic methods also include the use of variables. Numeric methods don't require knowledge of algebra and can sometimes be simpler. Algebraic methods become very helpful as a situation grows more complicated, since they organize the calculations done in the problem. In this section you will learn how to write and solve one-step equations to solve problems for which algebra may be preferable to numeric approaches.

As you work through this section, think about these questions:

1 Can you write a simple equation to model a given situation?

2 Do you know the inverse operation for a given operation?

3 When is a problem easier for you to solve with algebra than with a numeric approach?

You may be able to solve some or all of the problems in this section without algebra. However, we will use both algebraic and numeric approaches to help you learn the processes and notation needed for solving algebraic equations. When using an algebraic approach, we need to define a variable, write an equation, and solve it. With either approach we get a result that should be checked to see if it satisfies the problem. We should also be careful to answer the original question, which may or may not be the initial result from a numeric or algebraic approach.

In practice, you should choose the method, numeric or algebraic, that makes the most sense to you and that you can execute accurately.

EXAMPLE 1

Eli says that he needs only $225 more in his account to reach the $1,875 necessary for a down payment on a car. How much is in his account now?

SOLUTION

a. Solve the problem numerically.	b. Write an algebraic equation.
We will use numbers to solve the problem, showing work and including units with the answer. Eli's account is $225 away from the $1,875 necessary. Subtracting $225 from $1,875 gives his current account balance. $$\$1,875 - \$225 = \$1,650$$ Notice that we began with his goal, $1,875, and *subtracted* $225 to get the current balance.	We need to choose a meaningful variable and define it specifically. To write the equation, we pretend to know the unknown value and write a true statement about it. Let B = current account balance. If Eli has $225 in addition to the current balance, B, he will reach his goal of $1,875. $$B + \$225 = \$1,875$$ Notice we began with the unknown, B, and *added* $225 to get the goal of $1,875. The operation of addition is implied in the problem statement, since Eli needs *more* money to reach his goal.
c. Solve the equation.	d. Check your work. State the answer to the question.
The goal when solving an algebraic equation is to isolate the variable so that we can find the value of the variable that makes the equation true. To begin, we identify the operation being applied to the variable. Isolate the variable by applying the inverse operation to both sides of the equation. To maintain a balanced equation, any operation that is applied to one side must also be applied to the other side. Since $225 is added to B, we need to perform the inverse operation, subtraction, to isolate B. Subtract $225 from both sides. $$B + \$225 = \$1,875$$ $$\underline{-\$225 \quad -\$225}$$ $$B = \$1,650$$ Subtracting $225 from the left side isolates the variable. Subtracting $225 from the right side solves the problem and is the identical calculation that was done in part a.	Replace the variable with the solution and simplify. The right and left sides of the equation should be the same. $$B + \$225 = \$1,875$$ $$\$1,650 + \$225 \overset{?}{=} \$1,875$$ $$\$1,875 \overset{\checkmark}{=} \$1,875$$ We need to make sure that we have answered the question that was asked in the problem statement and not a related question. Eli's current account balance is $1,650.

When you solve a problem numerically, you work backward from the result to the unknown using the inverse of the operation implied in the problem. When you write an equation to solve a problem algebraically, you work forward from the unknown using the operation implied in the problem.

2. You watch a show in which people compete against each other to lose weight. The trainer exclaims, "Down 7 pounds in a week!" On a screen behind the contestant is his current weight, 168 pounds. What was his weight last week?

a. Solve the problem numerically.	**b. Write an algebraic equation.**
168 lbs + 7 lbs = 175 lbs	Let w = his weight last week.
	$w - 7 = 168$
Which operation was used?	
Addition	

c. Solve the equation.	**d. Check your work. State the answer to the question.**
Since the operation done to w is the subtraction of 7, we need to add 7 to both sides.	$175 - 7 = 168$
$$w - 7 = 168$$ $$\underline{+7 \quad +7}$$ $$w = 175$$	His weight last week was 175 pounds.

HOW IT WORKS

To solve a one-step equation:

1. Identify the operation performed on the variable in the equation.
2. Isolate the variable by performing the inverse operation on both sides of the equation.
3. State the solution and check.

EXAMPLES:

A number is added to x.

$$x + 5 = 12$$
$$\underline{-5 = -5} \quad \text{Subtract 5 from both sides.}$$
$$x = 7$$

Check:

$$7 + 5 \overset{?}{=} 12$$
$$12 \overset{\checkmark}{=} 12$$

(continued)

HOW IT WORKS

Need more practice?

Solve.

1. $x + 4\frac{2}{3} = 8$

2. $x - 8.7 = -10.15$

3. $-15x = 3$

4. $\frac{x}{-21} = 2.8$

Answers:

1. $\frac{10}{3}$

2. -1.45

3. $-\frac{1}{5}$

4. -58.8

(*continued*)

EXAMPLES:

A number is subtracted from x.

$$x - 5 = 12$$
$$\underline{+5 \quad +5} \quad \text{Add 5 to both sides.}$$
$$x = 17$$

Check:

$$17 - 5 \overset{?}{=} 12$$
$$12 \overset{\checkmark}{=} 12$$

A number is multiplied by x.

$$5x = 20$$
$$\frac{5x}{5} = \frac{20}{5} \quad \text{Divide both sides by 5.}$$
$$x = 4$$

Check:

$$5(4) \overset{?}{=} 20$$
$$20 \overset{\checkmark}{=} 20$$

x is divided by a number.

$$\frac{x}{5} = -20$$
$$5 \cdot \frac{x}{5} = -20 \cdot 5 \quad \text{Multiply both sides by 5.}$$
$$x = -100$$

Check:

$$\frac{-100}{5} \overset{?}{=} -20$$
$$-20 \overset{\checkmark}{=} -20$$

---

3. Your e-mail account states that 3.36 GB (21%) of storage space has been used. How much total storage is allotted to your e-mail account?

a. Solve the problem numerically.	b. Write an algebraic equation.
$$\frac{3.36 \text{ GB}}{0.21} = 16 \text{ GB}$$ What operation was used? Division	Let T = total storage. $0.21T = 3.36$
c. Solve the equation. Since the equation is $0.21T = 3.36$ and the operation done to T is multiplication by 0.21, we need to divide both sides by 0.21. $$\frac{0.21T}{0.21} = \frac{3.36}{0.21}$$ $$T = 16$$	**d. Check your work. State the answer to the question.** $$0.21(16) = 3.36$$ The total storage provided is 16 GB.

INSTRUCTOR NOTE: If students are struggling with part a, suggest using a pie graph with three sectors. The whole pie represents the true the value of the house. Label one sector $42,340. This visual approach illuminates the operation needed to solve the problem numerically.

4. You're reading the newspaper and see home value assessments. Your house is listed at $42,340. Assessments are one-third of the true value of the home. What is your home actually worth?

a. **Solve the problem numerically.**	b. **Write an algebraic equation.**
$3 \cdot 42{,}340 = \$127{,}020$	Let $v =$ true value of the house. $$\frac{1}{3}v = 42{,}340 \quad \text{or} \quad \frac{v}{3} = 42{,}340$$
Which operation was used? Multiplication	

c. **Solve the equation.**	d. **Check your work. State the answer to the question.**
Since the equation is $\frac{v}{3} = 42{,}320$ and the operation done to v is division by 3, we need to multiply both sides by 3. $$3 \cdot \frac{v}{3} = 42{,}340 \cdot 3$$ $$v = 127{,}020$$	$$\frac{127{,}020}{3} = 42{,}340$$ Your home is actually worth \$127,020.

If you have a one-step equation in which a variable is multiplied by a fraction, as was the case in #4, you can undo the operation by dividing both sides by the fraction. This is equivalent to multiplying both sides by the reciprocal, as seen in the following example.

EXAMPLE 2

Only $\frac{2}{3}$ of the class was present on Friday. If 16 students were present, how many students are in the class?

SOLUTION

a. Solve the problem numerically.

Consider using proportional reasoning.

$\frac{2}{3}$ of the class is 16, so $\frac{1}{3}$ of the class is 8.

$\frac{2}{3} + \frac{1}{3} = 1$, so $16 + 8 = 24$ students.

Another approach is to use a picture. In the pie graph shown here, the green area is $\frac{2}{3}$ of the pie and represents 16 students. Since each sector is the same size, divide the 16 by 2 to get the size of one sector. It is 8 students. Totaling the value of the three sections, we get 24 students.

16 students

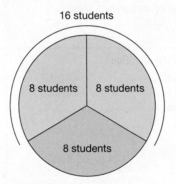

b. Write an algebraic equation.

Let n = number of students in the class.

Since $\frac{2}{3}$ of the class size, n, is 16 students, we get the equation $\frac{2}{3}n = 16$.

c. Solve the equation.

We can divide both sides by $\frac{2}{3}$ to solve or, equivalently, multiply both sides by the reciprocal of $\frac{2}{3}$, which is $\frac{3}{2}$.

Divide both sides by $\frac{2}{3}$.

$$\frac{2}{3}n = 16$$

$$\frac{\frac{2}{3}n}{\frac{2}{3}} = \frac{16}{\frac{2}{3}}$$

or

$$\frac{\frac{\cancel{2}}{\cancel{3}}n}{\cancel{\frac{2}{3}}} = 16 \cdot \frac{3}{2}$$

$$n = 24$$

Multiply both sides by the reciprocal of $\frac{2}{3}$.

$$\frac{2}{3}n = 16$$

$$\frac{3}{2} \cdot \frac{2}{3}n = 16 \cdot \frac{3}{2}$$

$$n = 24$$

d. Check your work. State the answer to the question.

Replace n in the equation with the solution, 24.

$$\frac{2}{3}n = 16$$

$$\frac{2}{3}(24) \overset{?}{=} 16$$

$$16 \overset{\checkmark}{=} 16$$

There are 24 students in the class.

Connect

Throughout this section you have solved one-step **linear equations**, which are polynomial equations of degree 1. One-step equations can also contain a polynomial with a higher degree, such as $x^3 = 64$. Whenever you choose to use algebra to solve a problem, you need to determine the operation done to the variable. Doing so allows you to determine the inverse operation that should be performed on both sides to isolate the variable.

5. a. A homeowner has enough patio pavers to make a 200-square-foot patio. He wants the patio to be square. What will the dimensions of the patio be if he wants to use all the pavers? State your answer in feet and inches, such as 5 feet 6 inches.

Each dimension will be approximately 14 ft, 2 in.

b. After building the patio, the homeowner is planning to build a cube-shaped bin to hold compost. It needs to hold 5 cubic yards of compost. What will its dimensions be in feet and inches?

Each dimension will be approximately 5 ft, 2 in.

Remember?

Section 2.3 has a list of geometric formulas.

Reflect

WRAP-UP

What's the point?
Most problems can be solved in more than one way. Numeric and algebraic methods both help to solve a problem and answer the question. Both methods require that you understand the operations involved and how to undo them.

What did you learn?
How to write and solve one-step equations

2.11 Homework

Skills MyMathLab

First complete the MyMathLab homework online. Then work the two exercises to check your understanding.

1. Solve $x - 3.2 = -9.8$.   −6.6

Check your answer:

$-6.6 - 3.2 \stackrel{?}{=} -9.8$

$-9.8 \stackrel{?}{=} -9.8$

2. Solve $-\dfrac{4}{5}x = -20$.   25

Check your answer:

$-\dfrac{4}{5}(25) \stackrel{?}{=} -20$

$-20 \stackrel{?}{=} -20$

Concepts and Applications

Complete the problems to practice applying the skills and concepts learned in the section.

3. During his morning forecast, a weatherman predicted the day's high would be 17° Fahrenheit. In the 5 p.m. news, he comments that his prediction was 3 degrees higher than the actual high temperature. What was the actual high temperature for the day? Solve both numerically and algebraically.

a. Solve the problem numerically.	b. Write an algebraic equation.
$17 - 3 = 14$ 14° Fahrenheit	$H =$ actual high temperature for the day $17 = H + 3$
c. Solve the equation.	d. Check your work. State the answer to the question.
Subtract 3 from each side of the equation. $\begin{array}{r} 17 = H + 3 \\ -3 \quad -3 \\ \hline 14 = H \end{array}$ $H = 14$	$14 + 3 = 17$ The actual high temperature for the day was 14° Fahrenheit.

4. Suppose a particular city estimates that only $\frac{2}{5}$ of its eligible voters went to the polls. If 30,000 voters were at the polls, how many eligible voters does the city have?

a. Solve the problem numerically.	b. Write an algebraic equation.
$\frac{2}{5}$ of the voters is 30,000, so $\frac{1}{5}$ of the voters is 15,000. Then $\frac{5}{5}$ of the voters is 75,000.	$V =$ number of eligible voters in the city $\frac{2}{5}V = 30,000$
c. Solve the equation.	d. Check your work. State the answer to the question.
$\frac{2}{5}V = 30,000$ $\dfrac{\frac{2}{5}V}{\frac{2}{5}} = \dfrac{30,000}{\frac{2}{5}}$ $V = 30,000 \cdot \dfrac{5}{2}$ $V = 75,000$	$\frac{2}{5}(75,000) = 30,000$ There are 75,000 eligible voters in this city.

5. Assume food prices have risen about 4% per year for the last five years.

    a. If a gallon of milk cost \$2 five years ago, what does it cost today?

        $2(1.04)(1.04)(1.04)(1.04)(1.04) = \$2.43$

    b. If a gallon of milk cost \$$x$ five years ago, write an expression for its cost today.

        $x(1.04)(1.04)(1.04)(1.04)(1.04)$ or $(1.04)^5 x$

    c. If a gallon of milk costs \$3.50 now, what did it cost five years ago? Write and solve an equation to answer the question.

        Let $x = $ cost of a gallon of milk five years ago

        $x(1.04)(1.04)(1.04)(1.04)(1.04) = 3.50$

                              $x = \$2.88$

        \$2.88 per gallon

    d. If milk prices increase 4% per year for five years, what is the overall percent increase?   Approximately 22%

6. a. Evaluate the expression for $x = 2$.

        $4x - 8$ 0

    b. Solve the equation by first adding 8 to both sides. Then divide both sides by 4.

        $4x - 8 = 0$ $x = 2$

7. a. If a one-step equation involves addition, which operation do you use to solve it?

        Subtraction

    b. If a one-step equation involves subtraction, which operation do you use to solve it?   Addition

    c. If a one-step equation involves multiplication, which operation do you use to solve it?   Division

    d. If a one-step equation involves division, which operation do you use to solve it?

        Multiplication

8. Identify the student error in each of the following incorrect equation solutions:

a.
$$x - 3 = 3$$
$$\underline{-3 \quad -3}$$
$$x = 0$$

b.
$$\frac{x}{-2} = -3$$
$$\frac{x}{-2} \cdot -2 = -3 \cdot -2$$
$$x = -5$$

a. The student should have added 3 to both sides instead of subtracting 3.

b. The student should have multiplied −3 and −2 instead of adding them.

9. **Looking Forward, Looking Back** Simplify each polynomial expression. State the degree of the resulting polynomial.

a. $-9(2 - 3x) + 15$ $27x - 3$, 1st degree

b. $(-9x + 15)(2 - 3x)$ $27x^2 - 63x + 30$, 2nd degree

c. $3x(-9x + 15)(2 - 3x)$ $81x^3 - 189x^2 + 90x$, 3rd degree

2.12 A State of Equality: More Equation Solving

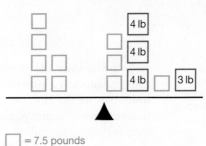

Explore

1. Determine the weight of a small box if the scale shown here is balanced.

□ = 7.5 pounds

SECTION OBJECTIVE:

- Solve two-step and multi-step linear equations

Discover

In Section 2.11, you learned how to solve one-step equations using algebra. However, not all equations can be solved with one step. Some contexts and situations are more involved and so are the resulting algebraic equations to model them.

As you work through this section, think about these questions:

① Do you know when to perform an operation on both sides of an equation?

② Can you tell if an equation is an identity, a contradiction, or neither?

③ Do you have a strategy for solving equations?

Throughout the book numeric methods and reasoning are used to solve equations. This section will show additional algebraic techniques to help you solve more complicated linear equations.

LOOK IT UP

Linear Equations

A **linear equation** is an equation that can be written in the form $ax + b = c$, where a, b, and c are real numbers and a is not zero. A linear equation is a first-degree polynomial equation in which the highest exponent on a variable is 1.

FOR EXAMPLE, $15x - 11 = 2$ and $7x = 9$ are linear equations. Since $7x^2 = 9$ contains a second-degree polynomial, it is not a linear equation.

INSTRUCTOR NOTE: Consider showing students how to solve each equation in #2 in more than one way.

2. Solve each one-step linear equation. Simplify first if that seems helpful.

a. $-x = 7$ $x = -7$

b. $-\dfrac{5}{7}x = 9$ $x = -\dfrac{63}{5}$

c. $15 + x = -8$ $x = -23$ **d.** $x - (-6) = 20$ $x = 14$

INSTRUCTOR NOTE: In this section, balances are used along with algebraic methods to solve equations. Doing so promotes a deeper understanding of the operations used to solve an equation.

To help you learn the process of solving more involved equations, we provide pictures using balances as a visual aid. The balances show why we perform the operations in the order that we do. However, problems with fractions or decimals are not well suited to modeling with balance pictures. For that reason, we will use the balance pictures only as long as they are helpful.

A balance has two sides, a left side and a right side, separated by the fulcrum, or balancing point, of the balance. In an equation, the equal sign serves as the fulcrum.

The following picture models the equation $x + 3 = 8$. The small square represents an unknown x, and the larger squares with numbers inside them represent the constant terms in the equation. The operation of addition is implied when two squares are next to each other on one side of the fulcrum.

$x + 3 = 8$

Algebraically, our goal is to have one x on one side and one number on the other so that the variable is alone or isolated. To maintain balance, whatever operation is done to one side must also be done to the other side. In this case, we isolate the x by removing 3 from each side.

$$\begin{array}{r} x + 3 = 8 \\ -3 \quad -3 \\ \hline x = 5 \end{array}$$

Numerically, we are trying to find the solution(s) to the equation, that is, the number or numbers that make the left side equal to the right side. Physically, we are trying to make the two sides balance. The number 5 is the solution because it makes the equation true and balances the scale.

INSTRUCTOR NOTE: Consider having students write an equation for the balance shown in the *Explore*.

$$\begin{array}{c} x + 3 = 8 \\ 5 + 3 \overset{?}{=} 8 \\ 8 \overset{\checkmark}{=} 8 \end{array}$$

The equation $x + 3 = 8$ is a one-step equation because only one operation, adding 3, is done to the variable. Only one step, subtracting 3, is then needed to isolate x. Let's look at equations that have two operations performed on the variable. Such equations need two steps to isolate x.

3. Solve the following equation algebraically after reading through the solution using balances.

Using balances:	**Algebraically:**

Model the equation $4x + 1 = 9$ using a balance.

The goal is to get down to one x. Start by taking away the number, 1, on the left. Do the same from the right side.

Four of the same number sum to 8. So the 8 is divided equally 4 ways.

x is isolated and is equal to 2.

Check the answer using the balance. The sides are balanced, since each is equal to 9.

Check:

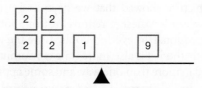

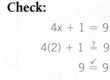

$$4x + 1 = 9$$
$$\underline{-1 \quad -1}$$
$$4x = 8$$
$$\frac{4x}{4} = \frac{8}{4}$$
$$x = 2$$

$$4x + 1 = 9$$
$$4(2) + 1 \overset{?}{=} 9$$
$$9 \overset{\checkmark}{=} 9$$

Complete the table that follows for the equation $4x + 1 = 9$. The second list should have the inverse operations in the reverse order of the first list.

Operations Done to x	**Steps to Solve for x**
1. Multiply by 4	1. Subtract 1
2. Add 1	2. Divide by 4

Need more practice?

Solve $\frac{x}{4} - 3 = 8$.

Answer: $x = 44$

Notice that the steps to solve for x in the table in the last problem were the same steps used in the algebraic solution to the equation. Making a list of operations done to the variable will help you determine the operations needed to solve for the variable. To solve for the variable, you want to undo the operations in the reverse order. While the balance pictures are helpful, we ultimately need to be able to solve equations like the one in the previous problem without them. This technique of undoing the operations in the reverse order can be used to solve an equation once the variable is on only one side.

Now let's use a balance picture to look at equations that have variables on both sides. The picture will provide guidance on how to begin.

EXAMPLE

Solve $7 + 3x = 4x$.

SOLUTION

Using balances:

We model the equation $7 + 3x = 4x$ using a balance.

The goal is to get the variable on one side and a number on the other. To achieve that, we remove $3x$ from both sides.

x is isolated and is equal to 7. The variable can end up on either the right or left side.

We check the answer using the balance. The sides are balanced, since each is equal to 28.

Algebraically:

$$\begin{array}{rr} 7 + 3x = & 4x \\ -\ 3x & -3x \\ \hline 7 = & x \end{array}$$

Check:

$$7 + 3x = 4x$$
$$7 + 3(7) \overset{?}{=} 4(7)$$
$$7 + 21 \overset{?}{=} 28$$
$$28 \overset{\checkmark}{=} 28$$

The picture showed that we need to get the variables on one side eventually. In this example, it is easiest to remove $3x$ from both sides and get the x term on the right side of the equation. However, we could move the variables to the left to begin. Although this is a valid method, it is also more time consuming than the approach shown. Equations can be solved in more than one way, and some approaches are faster than others.

INSTRUCTOR NOTE: Discuss alternate ways to solve the equation in #4. Students can clear the denominators using the lowest common denominator (LCD). Another option is to get the x terms on the right side first. In that case, it is easier to multiply by the reciprocal, 8, instead of dividing both sides by $\frac{1}{8}$.

4. Solve $\dfrac{1}{2}x + \dfrac{3}{2} = \dfrac{5}{8}x$.

12

It is possible for an equation to have more than one or two operations on each side. Let's use a balance picture one more time to see how to approach equations like this.

Often, math teachers ask students to show their work when solving an equation. If you feel this isn't always necessary, consider someone else's point of view. Other people don't know your thought process unless you explain it to them. It's like when you get a bill at a restaurant. If the waiter gave you only the total for your bill, the first question you would ask is "Where did he get that number?"

5. Solve the following equation algebraically after reading through the solution using balances.

Using balances:

Model the equation $4x = 2(x + 1) + 6$ with a balance.

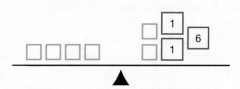

Although the right side represents $2(x + 1) + 6$, the picture shows that we have 2 x's and 8. To get the variables on one side, we can take 2 x's from each side.

Two of the same number sum to 8. Therefore, we are dividing 8 into two equal pieces.

x is isolated and is equal to 4.

Check the answer using the balance. The sides are balanced, since each is equal to 16.

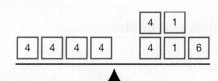

Algebraically:

$$4x = 2(x + 1) + 6$$
$$4x = 2x + 2 + 6$$
$$4x = 2x + 8$$
$$\underline{-2x \quad -2x}$$
$$2x = 8$$
$$\frac{2x}{2} = \frac{8}{2}$$
$$x = 4$$

Check:

$$4x = 2(x + 1) + 6$$
$$4(4) \stackrel{?}{=} 2(4 + 1) + 6$$
$$16 \stackrel{?}{=} 2(5) + 6$$
$$16 \stackrel{?}{=} 10 + 6$$
$$16 \stackrel{\checkmark}{=} 16$$

From this problem, we can see that we needed to simplify each side before doing any operations on both sides. Let's summarize the guidelines that will help you solve linear equations.

HOW IT WORKS

To solve a linear equation:

1. Simplify each side of the equation before doing anything to both sides. Use the distributive property and collect like terms, if necessary.

2. If the equation has variables on both sides, add or subtract a variable term from both sides to get all the variables on one side. If the equation has numbers on both sides, add or subtract a number from both sides to get all the numbers on the opposite side as the variable term.

   Determine what operations have been done to the variable. Undo them in the reverse order that they were applied.

   Whatever operations are done to one side of the equation must also be done to the other side.

3. Check your answer by replacing the variable in the original equation with the number. The left side of the equation must simplify to the same value as the right side for your answer to be a true solution.

Need more practice?

Solve $-3(2x - 5) = \frac{1}{2}(4x - 8)$.

Answer: $x = \frac{19}{8}$

EXAMPLE:

$2(-x + 4) - (-5) = 2x + 5$ Distribute on the left side.

$-2x + 8 + 5 = 2x + 5$ Collect like terms on the left side.

$-2x + 13 = 2x + 5$ Add $2x$ to both sides.

$\underline{+2x \qquad +2x}$

$13 = 4x + 5$ Subtract 5 from both sides.

$\underline{-5 \qquad -5}$

$\dfrac{8}{4} = \dfrac{4x}{4}$ Divide both sides by 4.

$2 = x$

Check:

$2(-x + 4) - (-5) = 2x + 5$

$2(-2 + 4) - (-5) \overset{?}{=} 2(2) + 5$ Replace the variable with the value of the solution.

$2(2) + 5 \overset{?}{=} 4 + 5$ Simplify.

$9 \overset{\checkmark}{=} 9$ Since the left side of the equation equals the right side, the solution is correct.

The equations that we have solved so far have all had one solution. However, some linear equations do not have a solution, and some have every number as a solution.

LOOK IT UP

Identities and Contradictions

When a linear equation results in a statement that is always true, the equation is known as an **identity**. The solution is all real numbers.

When a linear equation results in a statement that is always false, the equation is known as a **contradiction**. There is no solution to the equation.

EXAMPLE: Solve $x - 1 = x$.

SOLUTION: To solve the equation, isolate x.

$x - 1 = x$

$\underline{-x \quad -x}$ Subtract x from both sides.

$-1 = 0$

Since $-1 = 0$ is a false statement, the original equation is a contradiction. There is no solution, and the equation will never check.

There is a difference between verifying a solution of an equation and getting $0 = 0$, and solving an equation that is an identity and getting $0 = 0$. The first tells you that the number you are checking is a solution. The second tells you that any number x is a solution.

EXAMPLE: Solve $-2(x - 1) = -2x + 2$.

SOLUTION: To solve the equation, begin by simplifying the left side.

$$-2(x - 1) = -2x + 2$$

$$-2x + 2 = -2x + 2 \quad \text{Distribute.}$$

$$\underline{+2x \qquad\quad +2x} \quad \text{Add } 2x \text{ to each side.}$$

$$2 = 2$$

Since $2 = 2$ is always a true statement, the original equation is an identity. The two expressions that form the equation are equivalent, and every real number is a solution. This type of equation will check for any value of the variable.

INSTRUCTOR NOTE: Consider drawing a picture using the balance method to help students understand the solution sets to #6.

6. Solve the following equations:

   a. $3x - x = 2(x + 1) + 4$
   No solution

   b. $2(x + 1) = 2x + 2$
   All real numbers

Connect

7. For each equation, state if it is sometimes, never, or always true. Explain your response.

   a. $5 - 7x = 9x + 24$

   When $x = -\frac{19}{16}$, the equation is true, but for other values of x, it is false. So the equation is sometimes true.

   b. $-2(5x - 15x + 6) = -8 - 4 + 20x$

   If the left side is simplified, it is the same as the right side. Since the equation is an identity, it is always true.

   c. $3x - 2 = 3x - 5$

   The number negative 2 isn't the same as the number negative 5. The equation has no solution, so it is never true.

Reflect

WRAP-UP

What's the point?

Learning how to solve equations is a major goal of this cycle. Undoing a process that has been done is one key strategy used when solving equations. It is important that you can apply other techniques, such as simplifying each side before performing operations on both sides, as needed.

What did you learn?

How to solve two-step and multi-step linear equations

2.12 Homework

Skills MyMathLab

First complete the MyMathLab homework online. Then work the two exercises to check your understanding.

1. Solve $3x + (-1) = -10 + 2x$ -9 **2.** Solve $2 - 3(x - 4) = 17$ -1

Concepts and Applications

Complete the problems to practice applying the skills and concepts learned in the section.

3. Suppose you are tutoring a friend on solving basic equations. She wants a procedure or a list of steps to follow as a guideline. What should you tell her? Write at least three steps.

Answers will vary.

1. Always simplify both sides of the equation as much as possible first.

2. Undo operations in the reverse order.

3. Check your answer.

4. Suppose there is an equation in which the following operations have been performed on x, in this order:

　　1. Multiply by 8

　　2. Square both sides

　　3. Subtract 2

List the operations you would need to do to solve the equation for x, in the order you would do them.

1. Add 2

2. Square root both sides

3. Divide by 8.

5. Write at least two equations that can be modeled with this balance.

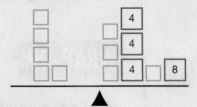

Possible equations:

$$4x + x = 3(x + 4) + x + 8$$
$$4x + x = 3x + 4 + 4 + 4 + x + 8$$
$$4x + x = x + x + x + 4 + 4 + 4 + x + 8$$
$$x + x + x + x + x = x + x + x + 4 + 4 + 4 + x + 8$$
$$5x = 4x + 20$$

6. The following solutions are examples of student work. For each example, state whether or not the student's method is correct and whether or not it is efficient. If the student's method is not correct or not efficient, show a better solution.

a.
$$2x - 5 + 3 = 18$$
$$\underline{+\ 5 \qquad\quad +5}$$
$$2x \quad + 3 = 23$$
$$\underline{\qquad - 3 \quad -3}$$
$$2x = 20$$
$$x = 10$$

Correct but inefficient

Combine the -5 and 3 first.

b.
$$2x - 5 + 3 = 18$$
$$\underline{+\ 5 + 5 \quad +5}$$
$$2x \qquad + 8 = 23$$
$$\underline{\qquad -8 \quad -8}$$
$$2x = 15$$
$$x = 7.5$$

Incorrect, 5 is added to all constants instead of to each side.

Simplify to get $2x - 2 = 18$, add 2, divide by 2, $x = 10$

7. a. Check that this chemical equation is balanced: $2\,H_2 + O_2 \rightarrow 2\,H_2O$.

There are 4 H's and 2 O's on each side of the equation.

b. Check that $x = 2$ is a solution to the following equation:
$$2(x - 4) + 10 = x - (-4)$$

$$2(2 - 4) + 10 \overset{?}{=} 2 - (-4)$$
$$2(-2) + 10 \overset{?}{=} 2 + 4$$
$$-4 + 10 \overset{?}{=} 6$$
$$6 \overset{\checkmark}{=} 6$$

8. Consider the linear equation $3(x - 1) + 12 = 4x - 3 + 2x$.

a. Solve the equation by first distributing on the left side and collecting like terms.

$x = 4$

b. Solve the equation by first subtracting 12 from each side and eventually getting x on the left side.

$x = 4$

c. Solve the equation by first adding 3 to each side and eventually getting x on the right side.

$x = 4$

9. a. Complete the equation so that $x = 0$ is a solution.

$2x - 1 = -1$

b. Complete the equation so that there are no solutions.

$2x - 1 =$ Answers will vary; $2x - 1 = 2x$ is one possibility.

10. a. Write a linear equation that is an identity.   Answers will vary.

b. Write a linear equation that is a contradiction.   Answers will vary.

11. Use the following table to make a conjecture about the solution(s) to the equation $18(x - 2) + 5 = 9x - 21 + 9x - 10$. Verify your answer by solving the equation algebraically.

x	Left Side	Right Side
−4	−103	−103
−3	−85	−85
−2	−67	−67
−1	−49	−49
0	−31	−31
1	−13	−13
2	5	5
3	23	23
4	41	41
5	59	59
6	77	77
7	95	95
8	113	113
9	131	131
10	149	149

All real numbers

12. Looking Forward, Looking Back Calculate the total after tax on a $20 bill at a restaurant if the tax is 8%. Do this with one step and show your calculation.

$1.08(20) = 21.6$

The total with tax is $21.60.

2.13 Quarter Wing Night: Writing and Solving Equations

Explore

SECTION OBJECTIVES:
- Write an equation to model a situation
- Solve a problem numerically and algebraically

Remember?

To increase a number by 20%, multiply by 1.20.

1. Write an algebraic expression to represent each of the following calculations. Use *x* to represent the number.

   a. Increase a number by 5 and then by 35%. $1.35(x + 5)$

   b. Increase a number by 35% and then by 5. $1.35x + 5$

   c. Quadruple a number, decrease that by 10, and then take 80%. $0.80(4x - 10)$

   d. Take a quarter of a number, add 15, then increase the total by 10%, and then 25%.
   $1.25 \cdot 1.10\left(\dfrac{x}{4} + 15\right)$

Discover

Now that you've learned how to solve basic equations by undoing operations, we can focus on using that skill to solve problems. In this section, we will concentrate on writing an equation to represent a situation and then solving it using skills you learned in the previous sections. To do this, you'll need the skill you just practiced in the *Explore* to write an equation.

As you work through this section, think about the following:

1 Is there more than one way to write an equation to represent a situation?

2 How does writing an equation compare to solving that equation?

3 When is it worth using algebra to answer a question?

We will begin by considering a simple situation and then add details to make it more realistic.

It is important to use a meaningful variable and define it specifically. Letting *w* stand for wings is not specific enough. Is *w* the cost of the wings or the number of them?

2. A local bar and grill is having a quarter wing night during which chicken wings cost a quarter each. As you leave home to join your friends there, all you grab is a $20 bill along with your ID, phone, and keys. Determine how many wings you can purchase if you also want to buy a pitcher of your favorite beverage for $5.

a. Solve numerically.	b. Write an equation where *w* = number of wings. Solve it and answer the question.
$20 - 5 = 15$ $\dfrac{15}{0.25} = 60$ wings	Let w = number of wings $5 + 0.25w = 20$ 60 wings

Notice that the numeric approach involved dividing 15 by 0.25. This was the same calculation used to solve the algebraic equation. For this first simple situation, both of these approaches are reasonable and fairly fast. However, many important considerations were omitted from the problem, and the number of wings you could purchase is quite large and unrealistic as a result. Next, we will start to include some of those details, starting simply and building to a more realistic situation.

3. Suppose you would like to buy a pitcher of a beverage for $5, buy blue cheese dressing for $1.50, and use the remaining money to buy wings. How many wings can you buy now?

a. Solve numerically.	b. Write an equation where w = number of wings. Solve it and answer the question.
$20 - 5 - 1.50 = 13.50$ $\dfrac{13.50}{0.25} = 54$ wings	$1.5 + 5 + 0.25w = 20$ 54 wings

4. Suppose you want to buy wings, the pitcher for $5, the dressing for $1.50, and also add a 20% tip. A friend says, "Just figure a $4 tip, since $4 is 20% of $20. Then figure out how many wings you can get." Determine how many wings you can buy and the percent tip you'd be leaving if you followed his advice.

a. Solve numerically.	b. Write an equation where w = number of wings. Solve it and answer the question.
$20 - 5 - 1.50 - 4 = 9.5$ $\dfrac{9.5}{0.25} = 38$ wings	$1.5 + 5 + 4 + 0.25w = 20$ 38 wings

Percent tip:

$38(0.25) + 5 + 1.5 = \$16$

$\dfrac{4}{16} = 0.25$ or 25% tip

5. Notice that your friend's method did not result in a 20% tip. Solve the problem again with the goal of leaving a 20% tip after buying a pitcher for $5, dressing for $1.50, and as many wings as possible.

a. Solve numerically.	b. Write an equation where w = number of wings. Solve it and answer the question.
HINT: What is the last thing you would do if you were purchasing food and then adding a tip? The last operation done will be the first operation undone. $$Purchases = \frac{20}{1.2} = 16.67$$ $$16.67 - 5 - 1.5 = 10.17$$ $$\frac{10.17}{0.25} = 40.67$$ 40 wings	$1.2(5 + 1.5 + 0.25w) = 20$ 40 wings

6. For the most realistic result, we also need to incorporate sales tax. Assuming you will need to pay 7% sales tax and you want to buy a pitcher, dressing, as many wings as possible, and leave a 20% tip on the total (which includes the tax), how many wings can you order?

a. Solve numerically.	b. Write an equation where w = number of wings. Solve it and answer the question.
Remove tip: $\frac{20}{1.20} = 16.67$ Remove tax: $\frac{16.67}{1.07} = 15.58$ Purchase = 15.58 $15.58 - 5 - 1.50 = 9.08$ $\frac{9.08}{0.25} = 36.32$ wings 36 wings	$1.2[1.07(5 + 1.5 + 0.25w)] = 20$ 36 wings

Notice in the last version of the problem that applying a 7% tax and then a 20% tip does not result in a 27% increase. If you multiply 1.07 by 1.20, you can see that the combined effect of these two percents is actually a 28.4% increase.

You have to decide for yourself at what point the situation became complicated enough that you found the algebraic solution easier than the numeric solution. This is a personal preference. Most people eventually prefer the algebraic approach when a situation becomes too complicated to think through numerically.

7. If the price per wing changes to $0.40 and the price of a pitcher to $5.50, how many wings can you now afford for $20? Assume you are still paying a 7% tax and leaving a 20% tip. Which method do you prefer to use to solve this problem?

21 wings; answers will vary.

Connect

When equations contain many decimal numbers or fractions, it can be helpful to eliminate them in the first step so that the equation can be written with only integers. If an equation contains decimal numbers, they can be eliminated by multiplying each side of the equation by some power of ten depending on the number of decimal places. Since the only decimal numbers involved in the following equation have two decimal places (hundredths position), multiplying each side of the equation by 100 will eliminate or clear the decimals. Once the factor of 100 is distributed to each term on both sides of the equation, the equation can be solved more easily with integer values.

$$0.52(x - 2) + 0.6 = 1.29$$
$$100 \cdot 0.52(x - 2) + 100 \cdot 0.6 = 100 \cdot 1.29$$
$$52(x - 2) + 60 = 129$$

A similar technique can be employed when an equation contains fractions. We can multiply both sides of the equation by the least common denominator to eliminate the fractions. In the following equation, we could multiply both sides by the LCD of 5, 3, and 9, which is 45, before proceeding. This factor of 45 is distributed to each term on both sides of the equation.

$$\frac{2}{5}x - \frac{1}{3}(x + 1) = \frac{2}{9}x - 5$$

$$45 \cdot \frac{2}{5}x - 45\frac{1}{3}(x + 1) = 45 \cdot \frac{2}{9}x - 45 \cdot 5$$

$$18x - 15(x + 1) = 10x - 225$$

Focus Problem Check-In
Use what you have learned to finish solving the focus problem. Write a rough draft to explain your solution, showing and explaining all mathematical work. Edit your draft into a final write-up by making any needed corrections and proofreading for grammar, spelling, and mathematical errors. Have someone verify the mathematical work and proofread the final draft for mistakes.

8. Solve the following equations by first eliminating the fractions or decimal numbers:

a. $\dfrac{x - 1}{6} + 3x = \dfrac{1}{4}x - 1$

$x = -\dfrac{2}{7}$

b. $1.359(x + 2) - 0.25(x - 2) = 2.5$

$x \approx -0.647$

Reflect

INSTRUCTOR NOTE: Consider providing the focus problem grading rubric and writing template to students at this point to help them form their written solution. Both are available in MyMathLab.

WRAP-UP

What's the point?

Problems can be solved numerically or algebraically; neither method is superior. You should choose the method that makes the most sense for the given situation and for you personally.

What did you learn?

How to write an equation to model a situation
How to solve a problem numerically and algebraically

2.13 Homework

Skills MyMathLab

First complete the MyMathLab homework online. Then work the two exercises to check your understanding.

1. Solve $3.2(1 + 0.5x) = -8$.   −7

2. Solve $1.08(1.28)(8 - x) = 10$. Round your answer to the nearest tenth.   0.8

Concepts and Applications

Complete the problems to practice applying the skills and concepts learned in the section.

3. An online retailer is selling used books for $5.99, and you have $50 to spend. How many books can you buy if you must pay 7.5% sales tax and an additional 10% of the total purchase price (with tax) for shipping?

7 books

4. Suppose you want to treat your office team to lattes at a local coffee shop. The drinks are $4.25 each, plus you plan to buy a $2 scone for yourself. You also tip the barista 10% and have to pay 8.25% sales tax. How many lattes can you afford if you have $60 to spend?

11 lattes

5. Two numbers sum to 40. One number is 8 more than the other number. Find the numbers by each method.

   a. Numerically: Guess and check or reason numerically to find the numbers.   16, 24

   b. Algebraically: Call the first number x. Write an expression for the second number, which is 8 more than the first. Write an equation that says these two numbers add to 40. Solve it.
   x and $x + 8$
   $x + x + 8 = 40$
   $x = 16, x + 8 = 24$

6. The sum of the three interior angles of a triangle is 180°. Suppose one angle is 70° and the remaining two angles are the same measure. Find their measures by both methods.

   a. Numerically: Reason numerically to find the measures of the remaining angles.
   55°, 55°

   b. Algebraically: Call one of the unknown angles x. Write an equation that says the three angles sum to 180°. Solve it.
   $x + x + 70 = 180$
   $x = 55$
   55°, 55°

7. Suppose you get a $100 iTunes gift card. Each $1.29 song has an additional sales tax charge of 7% in your state. How many songs can you buy with the gift card?
   72 songs

8. **Looking Forward, Looking Back** If your company expects salaries to increase by 3% each year for the next four years, what type of growth will the salaries show?
   Exponential

2.14 Outwit and Outlast: Using Proportions

SECTION OBJECTIVE:
• Write and solve proportions

Explore

Five players on a television reality show are stranded on an island as a part of the competition. The group is given 50 cups of rice with a one-cup scoop plus bottled water for nourishment, and they are dependent on their surroundings for shelter. Since the rice is so important to their survival, they need to figure out how to divide the rations fairly.

Evan suggests dividing the rice equally by giving 10 cups to each person. Alan says this is unfair because he weighs more than the other contestants, yet his portion of food is the same. Becca agrees with Alan, suggesting that they find a more fair way to split the rice. They discuss ideas as a group and agree that everyone will get at least 1 cup. They decide that rice should be given in proportion to weight. Since they have no other measuring tools besides the scoop, they decide that only whole cups of rice will be distributed.

Here are the five players and their weights:

Alan	Becca	Chris	David	Evan
297 lb	195 lb	161 lb	122 lb	105 lb

1. a. Find a distribution of rice that is consistent with the players' agreement.

17 cups, 11 cups, 9 cups, 7 cups, 6 cups

INSTRUCTOR NOTE: Some students will use trial and error to find the distribution, while others will use algebraic methods or percents.

b. Using your results from part a, find the ratio $\frac{\text{cups of rice}}{\text{weight of player}}$ for each player and write each as a decimal number. Is the distribution fair? How are we judging "fair?"

Alan: $\frac{17 \text{ cups}}{297 \text{ lbs}} \approx 0.057$, Becca: $\frac{11 \text{ cups}}{195 \text{ lbs}} \approx 0.056$, Chris: $\frac{9 \text{ cups}}{161 \text{ lbs}} \approx 0.056$,

David: $\frac{7 \text{ cups}}{122 \text{ lbs}} \approx 0.057$, Evan: $\frac{6 \text{ cups}}{105 \text{ lbs}} \approx 0.057$

Answers will vary. In this case, "fair" means that the amount of rice is proportional to weight.

Discover

While guess and check can be used to solve the problem in the *Explore,* that can be tedious. Proportionality provides a more direct way to find a fair distribution of rice. This section will explore problems that involve proportional relationships and using algebraic methods to solve them.

As you work through this section, think about these questions:

1 Can you set up a proportion in multiple ways for the same situation?

2 Do you know when to set cross products equal vs. when to scale to solve a proportion?

EXAMPLE

One method in statistics for finding a sample from a larger population is to divide the population into distinct groups called strata. A sample can be found by choosing proportionally from each stratum. This is known as stratified sampling. Suppose a stratified sample of 250 students is needed from a population of students with various degrees as shown in the table. How many students should be chosen from each degree type if the number of students with a particular degree in the sample is to be proportional to the number of students who hold that degree in the population?

Degree	Number of Students in the Population
A.A.	1,575
A.S.	565
A.A.S.	415

SOLUTION

One way to find the number in the sample from each stratum is to find the percent each degree group represents of the total population (1,575 + 565 + 415 = 2,555 students) and then apply that percent to the sample size, 250.

For example, the students with A.A. degrees represent $\frac{1,575}{2,555} \approx 61.6\%$ of the population. Finding 61.6% of 250, we get $0.616(250) \approx 154$ students with A.A. degrees in the sample. We can perform a similar set of calculations for the other degrees.

Another option is to use the concept of proportionality to set up one equation to solve for each degree instead of using a series of related calculations as we did with the percent approach. The goal for the sample is that the relationship between the number of students with each degree and the total, 250, is the same as the relationship between the number of students in the population with each degree and the total number of students, 2,555.

Set $\frac{\text{students with degree}}{\text{total number of students}}$ equal for the population and the sample. Start with the A.A. degree. Let x be the number of students with an A.A. degree in the sample.

$$\underset{\substack{\text{Students with A.A.}\\ \text{degree in population}}}{\underset{\substack{\text{Total students in}\\ \text{population}}}{\frac{1,575}{2,555}}} = \underset{\substack{\text{Students with A.A.}\\ \text{degree in sample}}}{\underset{\substack{\text{Total students in}\\ \text{sample}}}{\frac{x}{250}}}$$

To solve this proportion, isolate the variable by multiplying both sides by 250. This is preferred to scaling in this case, since 250 is not a factor of 2,555 and so cannot be easily scaled to that number.

$$250 \cdot \frac{1,575}{2,555} = \frac{x}{250} \cdot 250$$

$$154 \approx x$$

Rounding to the nearest whole number, we need 154 students with A.A. degrees in the sample. Numerically, the calculation ends up being the same as the one done before with percents, but the thought process is different. Use the method that makes sense to you.

Similarly, we can find the number of students to include in the sample with A.S. and A.A.S. degrees.

A.S.: $\frac{565}{2,555} = \frac{y}{250}$ $y \approx 55$ students

A.A.S.: $\frac{415}{2,555} = \frac{z}{250}$ $z \approx 41$ students

2. A paper company has been reducing the size of its paper towel rolls in an effort to address rising costs. By keeping the price the same but providing less of the product, the company can recoup the increasing costs to manufacture paper towels without customers paying more per roll.

    a. Before the size reduction, 138 sheets of the paper towels cost $3. What was the cost per sheet?

      2.17 cents per sheet

    b. If the CEO of the company still wants to charge $3 for the roll, how many sheets need to be in a roll to bring the price per sheet up to 3 cents per sheet?

      100 sheets

    c. Is this a significant difference in the number of sheets? Determine the percent change to quantify your answer.

      Yes; this is a decrease of about 28% in the number of sheets.

We can refine how we solve proportions when the ratios are not easily scaled up or down, such as in this proportion: $\dfrac{x}{8} = \dfrac{12}{17}$.

Notice that this proportion is true: $\dfrac{2}{3} = \dfrac{8}{12}$

In the equation, 2 and 12 are diagonally across from each other, as are 3 and 8. For each of these pairs of numbers, we can find their product. These are known as **cross products**.

$$\frac{2}{3} = \frac{8}{12}$$

$3 \cdot 8 = 24$ Cross product
$2 \cdot 12 = 24$ Cross product

In proportions, cross products are equal. We can use this idea to solve proportions like $\dfrac{x}{8} = \dfrac{12}{17}$. Find the cross products and set them equal to each other. Then solve the resulting equation.

$$17x = 12 \cdot 8$$
$$17x = 96$$
$$\frac{17x}{17} = \frac{96}{17}$$
$$x \approx 5.6$$

Need more practice?

Solve $\dfrac{4x + 7}{x} = \dfrac{5}{3}$.

Answer: $x = -3$

3. Solve $\dfrac{-3}{x + 1} = \dfrac{5}{x - 1}$.

$x = -\dfrac{1}{4}$

In #3, the resulting equation after setting cross products equal is a linear equation, but equations of higher degrees are possible.

To write and solve a proportion:

1. Form a ratio by comparing two quantities. Include the units in the ratio.
2. Form another ratio by using the unknown variable and one additional fact. Write this ratio with quantities in the same relative location as in the first ratio.
3. Scale one ratio so that either the numerator or denominator matches the other ratio.
4. If scaling is not possible or not preferred, set the cross products equal to each other and solve.
5. State your answer with appropriate units.

EXAMPLE: A mortgage lender requires a borrower's debt-to-income ratio to be no greater than 40%. For example, if a borrower has $40,000 in annual income, she can have no more than $16,000 in debt. Andrea is planning to buy a home and needs a mortgage. If she makes $52,000, what is the maximum amount of debt she can have to qualify for a mortgage?

First, form a ratio of the quantities involved to illustrate what is being compared: $\frac{debt}{income}$.

The ratio of debt to income can be no greater than 40% or $\frac{40}{100}$. Use this ratio or the example that is equivalent to it, $\frac{\$16,000}{\$40,000}$. The relationship with Andrea's maximum debt, d, and her income, $52,000$, must be maintained. Write another ratio like the first, with her maximum debt in the numerator and her income in the denominator, $\dfrac{d}{\$52,000}$.

Equating 40% written as a ratio with Andrea's ratio, the result is $\dfrac{40}{100} = \dfrac{d}{\$52,000}$.

Notice that in each ratio, the debt is written in the numerator and the income is written in the denominator. There is not one single correct arrangement when forming ratios. However, the units in the second ratio must match the units in the first ratio. The first ratio could have been written income/debt as long as the second ratio was also written that way.

Since there is a proportion, set the cross products equal to each other and solve:

$$40(\$52,000) = 100d$$
$$\$2,080,000 = 100d$$
$$\$20,800 = d$$

Andrea can have no more than $20,800 in debt to qualify for the lender's mortgage.

INSTRUCTOR NOTE: Students should focus on comparing units consistently when setting up the proportions.

For any proportional situation, there are multiple ways to make comparisons and set up the equation. As long as the comparisons made in each ratio are consistent, the proportion will be correct.

4. Ben and his roommates order pizza and Chinese takeout for dinner. Ben tips the driver $5 for a $21 pizza. How much should he tip the driver who delivers the Chinese food costing $34 so that he tips each driver the same percent? Round the answer to the nearest dollar. Set up four different proportions to solve the problem. For each proportion, label the quantities being compared.

$8

$$\frac{\$5 \text{ tip}}{\$21 \text{ food cost}} = \frac{\$x \text{ tip}}{\$34 \text{ food cost}}$$

$$\frac{\$21 \text{ food cost}}{\$5 \text{ tip}} = \frac{\$34 \text{ food cost}}{\$x \text{ tip}}$$

$$\frac{\$5 \text{ tip}}{\$x \text{ tip}} = \frac{\$21 \text{ food cost}}{\$34 \text{ food cost}}$$

$$\frac{\$x \text{ tip}}{\$5 \text{ tip}} = \frac{\$34 \text{ food cost}}{\$21 \text{ food cost}}$$

Connect

Apportioning whole-number quantities according to certain ground rules has long been a part of determining seats in the U.S. House of Representatives. The Constitution states that the number of representatives will be determined based on each state's respective number, meaning its population. Each state must get at least one seat. Currently, there can be no more than 435 seats total. According to the 2010 census, the population of the United States was 309,183,463.

5. Use proportionality to find the number of seats in the U.S. House of Representatives for each state in the table. Round the results to the nearest whole number.

State	Population from 2010 Census	Number of Seats in U.S. House
Florida	18,900,773	27
Ohio	11,568,495	16
Texas	25,268,418	36

Data from www.census.gov

Due to issues of rounding, proportions are not used by Congress when determining the number of seats in the House of Representatives for each state. Depending on how a given population is rounded, a state could lose or gain a seat, which sometimes makes a difference in a vote. Certain apportionment methods can favor large or small states through their rounding methods. Thus, a more complicated but also fairer method was developed.

The U.S. Senate uses a much simpler formula: two seats per state. The number of representatives is not proportional to the state population. In a populous state, the ratio of constituents to senators is higher than the ratio in a less populous state.

Reflect

WRAP-UP

What's the point?

Proportions can be used to determine a number that maintains a desired ratio or rate. For any proportional problem, there are multiple ways to write an equation to describe it.

What did you learn?

How to write and solve proportions

2.14 Homework

Skills MyMathLab

First complete the MyMathLab homework online. Then work the two exercises to check your understanding.

1. Solve $\dfrac{-5}{2x-1} = \dfrac{6}{3x-7}$. $\dfrac{41}{27}$

2. A designer is drawing a scale model of a living room. The scale is such that: $\dfrac{1}{4}''$ on the plan represents 1 foot in the living room. If the sofa is 6′8″ long, how long is it on the plan?

$1\dfrac{2}{3}''$

Concepts and Applications

Complete the problems to practice applying the skills and concepts learned in the section.

3. If there are 50 cups of rice for a group of five people who have a combined weight of 880 pounds, determine how many cups each person should get per 100 pounds. Round to the nearest tenth of a cup.

5.7 cups per 100 pounds

4. The state of Wyoming has 5.49 people per square mile and is 97,100 square miles in size. The state of New York has 412.81 people per square mile and is 47,214 square miles in size. *Data from*: wordatlas.com.

a. Use the size of Wyoming and the population density to approximate the population of Wyoming.

533,079 people

b. If the state of Wyoming had the same population density as New York (in people per square mile), how many people would be in Wyoming?
40,083,851 people

5. There are about half a million heart attack deaths in the United States each year, according to womensheart.org.

   a. About how many heart attack deaths are there each hour in the United States?
   About 57 heart attack deaths per hour

   b. About how many heart attack deaths are there each minute in the United States?
   About 1 heart attack death per minute

6. Suppose a town has 75,000 residents and needs to generate $5 million in new tax revenue. If the tax burden is distributed evenly, how much does each resident need to pay?
$66.67

7. If corn yield is typically 200 bushels per acre, what yield would you expect on a 225-acre field? Which of the following proportions are correct ways to represent this situation? There may be more than one right answer.

   a. $\dfrac{200}{1} = \dfrac{x}{225}$ **b.** $\dfrac{200}{x} = \dfrac{1}{225}$ **c.** $\dfrac{200}{225} = \dfrac{1}{x}$

   d. $\dfrac{200}{x} = \dfrac{225}{1}$ **e.** $\dfrac{200}{1} = \dfrac{225}{x}$

   a and b are correct.

8. Looking Forward, Looking Back Simplify.

   a. $3x^4 - 8x + 5x^4 + 2$ $8x^4 - 8x + 2$

   b. $(3x^4)(-8x)(5x^4)(2)$ $-240x^9$

2.15 Three of a Kind: Pythagorean Theorem

Explore

1. For each rectangle shown here, draw a diagonal and estimate its length as accurately as you can. Assume each segment of the grid is one unit. Describe any relationship you see between the lengths of the sides of a rectangle and the length of its diagonal.

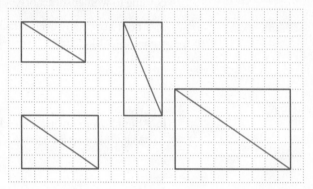

SECTION OBJECTIVES:

- Use the Pythagorean theorem to find the length of a side in a right triangle
- Solve problems using the Pythagorean theorem

Answers will vary. The diagonal is always longer than both sides of a rectangle. It is also shorter than the sum of the two sides.

Discover

In the *Explore*, you could only *estimate* the lengths of the diagonals of the squares, since the diagonals were not horizontal or vertical segments whose lengths could be easily determined using the grid. There are situations in which we need to be able to find a diagonal length more exactly. This section will introduce you to a formula to do this.

As you work through this section, think about these questions:

1 When can you use the Pythagorean theorem?

2 Is it more difficult to solve the Pythagorean theorem for the hypotenuse or for one of the legs of a right triangle?

When a diagonal is drawn in a square, two right triangles are formed. The length of the diagonal can be determined by using a well-known relationship between the lengths of the sides of a right triangle.

Pythagorean Theorem

A triangle that has one 90° angle is known as a right triangle. In a right triangle, the sides that form the right angle are known as the **legs**. The side across from the right angle is called the **hypotenuse** and is the longest side in a right triangle.

Right triangles have a special relationship between the lengths of their sides that is expressed in the **Pythagorean theorem**, which says

$$\text{leg}^2 + \text{leg}^2 = \text{hypotenuse}^2$$

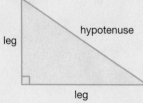

Pythagorean Theorem   (*continued*)

The Pythagorean theorem is an example of a **quadratic equation**, since it involves polynomials of the second degree, in which the variable is squared. The Pythagorean theorem is also commonly written as

$$a^2 + b^2 = c^2$$

where *a* and *b* represent the lengths of the legs and *c* represents the length of the hypotenuse.

FOR EXAMPLE, in a right triangle whose legs have lengths of 3 cm and 4 cm, the hypotenuse is 5 cm since

$$3^2 + 4^2 = 9 + 16$$
$$= 25$$
$$= 5^2$$

Let's see how the Pythagorean theorem can be used to find a distance.

EXAMPLE 1

At John's school, there is a rectangular patch of grass called the *quad* that measures 350 feet by 520 feet. John needs to walk from one corner to the corner diagonally opposite him. He could walk along the sidewalks that line the perimeter of the quad, or he can cut across the quad in a straight line. He knows that the shortest distance between two points is a straight line, but what will that distance be? How much time will he save by cutting across?

INSTRUCTOR NOTE: Consider having a student walk a specified length and then time the walk in seconds. Use their rate to determine the amount of time saved when walking across the quad.

SOLUTION

Let's begin by drawing a picture to represent this situation and finding the length of the diagonal path.

The diagonal path forms a triangle inside the rectangle. We know this is a right triangle because the angle in the rectangle is a right angle. So the Pythagorean theorem can be used to find the length of the diagonal path, *d*.

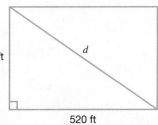

350 ft

520 ft

$$350^2 + 520^2 = d^2$$
$$122{,}500 + 270{,}400 = d^2$$
$$392{,}900 = d^2$$
$$\sqrt{392{,}900} = \sqrt{d^2}$$
$$626.8 \text{ ft} \approx d$$

There are two square roots of 16 and two solutions to the equation $x^2 = 16$, 4 and −4. In general, when you take the square root of a number, both the positive and negative results should be stated unless one does not make sense in the context of the problem.

In comparison, the sidewalk route would involve walking 350 feet + 520 feet = 870 feet. The distance John will save by cutting across the quad is 870 feet − 626.8 feet ≈ 243.2 feet.

To determine how much time John will save by cutting across the quad, we need to know how fast he walks. Let's assume that he walks about 3 feet per second. Since distance is rate times time, we can find the time by dividing the distance by the rate.

$$\text{Time} = \frac{\text{distance}}{\text{rate}}$$
$$= \frac{243.2 \text{ ft}}{3 \text{ ft/sec}}$$
$$\approx 81.1 \text{ sec}$$
$$\approx 1.35 \text{ min}$$

John will save 1.35 minutes by cutting across the quad.

HOW IT WORKS

To find the length of the hypotenuse of a right triangle using the Pythagorean theorem:

1. Substitute the lengths of the legs into the equation $\text{leg}^2 + \text{leg}^2 = \text{hypotenuse}^2$.
2. Simplify the left side of the equation.
3. Take the square root of both sides.

EXAMPLE: If the lengths of the legs of a right triangle are 7 cm and 9 cm, find the length of the hypotenuse.

The solution involves writing and solving a quadratic equation.

$\text{leg}^2 + \text{leg}^2 = \text{hyp}^2$	Substitute the leg lengths into the equation
$7^2 + 9^2 = \text{hyp}^2$	Square the leg lengths
$49 + 81 = \text{hyp}^2$	Add
$130 = \text{hyp}^2$	Square root
$\sqrt{130} = \text{hyp}$	Use your calculator to approximate the square root
$11.4 \text{ cm} \approx \text{hyp}$	

To find the length of a leg of a right triangle using the Pythagorean theorem:

1. Substitute the lengths of the known leg and the hypotenuse into the equation $\text{leg}^2 + \text{leg}^2 = \text{hypotenuse}^2$.
2. Simplify both sides of the equation.
3. Isolate the square of the leg.
4. Take the square root of both sides.

EXAMPLE: If one leg of a right triangle is 4 inches long and the hypotenuse is 6 inches long, find the length of the other leg.

The solution involves writing and solving a quadratic equation. To isolate the length of the leg, undo the operations done to the length of the leg in the reverse order they were applied.

$\text{leg}^2 + \text{leg}^2 = \text{hyp}^2$	Substitute the lengths into the equation
$4^2 + \text{leg}^2 = 6^2$	Square the lengths
$16 + \text{leg}^2 = 36$	Subtract 16 from each side
$\text{leg}^2 = 20$	Square root
$\text{leg} = \sqrt{20}$	Use your calculator to approximate the square root
$\text{leg} \approx 4.47 \text{ in.}$	

INSTRUCTOR NOTE: For additional practice, consider having students use the Pythagorean theorem to check their estimates of the lengths of the diagonals in the *Explore* problem.

2. a. Find the length of the hypotenuse of a right triangle if the legs each measure 10 feet.
$\sqrt{200} \approx 14.1$ feet

b. Find the length of the legs of a right triangle if the legs are the same length and the hypotenuse measures 10 feet. $\sqrt{50} \approx 7.07$ feet

Pythagorean Triple

When three whole numbers satisfy the Pythagorean theorem, they are known as a **Pythagorean triple**.

FOR EXAMPLE, 3, 4, and 5 form a Pythagorean triple. Pythagorean triples are always listed with the numbers in increasing order. The last number is the length of the hypotenuse.

3. a. Use the Pythagorean theorem to complete each of the following Pythagorean triples:

3, 4, __5__

6, 8, __10__

9, 12, __15__

30, 40, __50__

57, 76, __95__

60, 80, __100__

b. What relationship do you see between the first and second Pythagorean triples? Between the first and third?

If you double each number in the first Pythagorean triple, you get the numbers in the second Pythagorean triple. If you triple each number in the first Pythagorean triple, you get the numbers in the third Pythagorean triple.

Notice that if a, b, c is a Pythagorean triple, then so is every multiple of a, b, c that results in whole numbers.

4. a. If 5, 12, 13 is a Pythagorean triple, name two more triples related to it.

Examples will vary. Two possibilities: 10, 24, 26 and 100, 240, 260

b. Complete the following two Pythagorean triples:

8, 15, _____   17

14, 48, _____   50

The Pythagorean theorem says that if a triangle is a right triangle, then the lengths of its sides satisfy the equation $\text{leg}^2 + \text{leg}^2 = \text{hypotenuse}^2$. The converse of this statement is also true. That is, if the lengths of the sides of a triangle satisfy the equation $\text{leg}^2 + \text{leg}^2 = \text{hypotenuse}^2$, then the sides form a right triangle.

EXAMPLE 2

A home builder is trying to square the corners of a property. A string is secured 7 inches out from the corner on one side and 2 feet out on the other side as shown in the diagram.

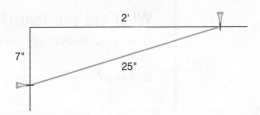

One side is angled until the two marks are exactly 25 inches apart, and then a stake is planted at the property corner and at each mark. Why do these measurements guarantee a square corner? What are some other measurements the builders could use that might be more convenient?

SOLUTION

We can begin by converting the 2 feet to 24 inches so that all three sides of the triangle are measured in the same unit of inches. Then we can substitute the three side lengths into the Pythagorean theorem to see if a true statement results.

$$7^2 + 24^2 \overset{?}{=} 25^2$$
$$49 + 576 \overset{?}{=} 225$$
$$625 = 625$$

Since these values satisfy the Pythagorean theorem, the triangle is a right triangle and the corner is a right angle or "square." Any Pythagorean triple will work to square the corner. 3 feet, 4 feet, 5 feet is a commonly used Pythagorean triple that could work in this application.

Connect

5. Jack has instructions for building a skateboard ramp, and he needs to build a triangular frame on which to nail the ramp board. A 4 foot by 8 foot piece of plywood will serve as the ramp, and Jack wants the plywood oriented so that the ramp is as long as possible. The instructions say to choose the desired height for the ramp and then use the Pythagorean theorem to find the length of the frame that will sit on the ground. He wants the end of the ramp to be 2 feet off the ground.

Draw a picture to get started. How long will the bottom length of the frame be?

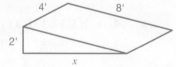

$x^2 + 2^2 = 8^2$
$x^2 = 8^2 - 2^2 = 60$ $x \approx 7.7$ feet or approximately 7 feet, 8.4 inches

Reflect

INSTRUCTOR NOTE: Remind students to read the *Getting Ready* article for the next section.

WRAP-UP

What's the point?

The Pythagorean theorem can be used to find the length of one side of a right triangle when you know the lengths of the other two sides. This theorem has applications in geometry, construction, and other areas of mathematics.

What did you learn?

How to use the Pythagorean theorem to find the length of a side in a right triangle

How to solve problems using the Pythagorean theorem

2.15 Homework

Skills MyMathLab

First complete the MyMathLab homework online. Then work the two exercises to check your understanding.

1. Find the length of the hypotenuse of a right triangle whose legs are each 5 inches long. Round to the nearest tenth of an inch.   7.1 inches

2. Find the length in feet of a leg of a right triangle whose hypotenuse is 10 feet and other leg is 48 inches. Round to the nearest tenth of a foot.   9.2 feet

Concepts and Applications

Complete the problems to practice applying the skills and concepts learned in the section.

3. a. Verify that each set of numbers forms a Pythagorean triple.

8, 15, 17 $8^2 + 15^2 = 289 = 17^2$ 16, 30, 34 $16^2 + 30^2 = 1{,}156 = 34^2$

b. Write one more Pythagorean triple by multiplying each number in the 8, 15, 17 triple by 3.   24, 45, 51

c. Draw a right triangle for each of the Pythagorean triples in parts a and b on the grid. For each triangle, put the right angle at the origin and the shorter leg along the positive x-axis.

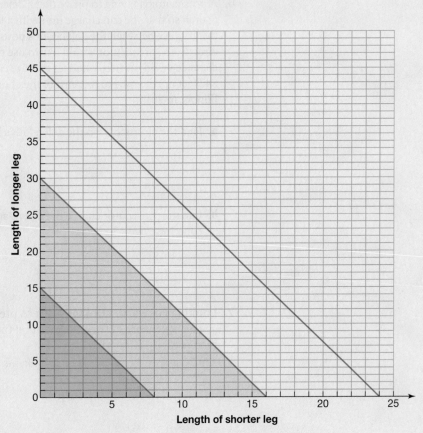

d. What do you notice about the three triangles?   They are similar.

4. The triangle inequality states that for any triangle, the sum of the lengths of the two shorter sides must be greater than the length of the longest side. In other words,

$$a + b > c$$

where a and b are the shorter sides of the triangle and c is the longest side.

a. Will the lengths 3 in., 4 in., and 7 in. make a triangle?
No

b. Will the lengths 3 in., 4 in., and 6 in. make a triangle? Will they make a right triangle?   Yes; no

c. Write three lengths that could be used to form a triangle but would not form a right triangle.
Answers will vary. One possibility: 4 feet, 5 feet, 6 feet

5. Use the Pythagorean theorem to find the length of the diagonal of a square that has an area of 100 square inches. Round to the nearest tenth of an inch.
Each side of the square is 10 inches. The diagonal is about 14.1 inches.

6. A prosecutor is trying to prove that a defendant's house is within 1,000 feet of a church so that she can charge him with a felony rather than a misdemeanor. Since the distance between the defendant's house and the church is difficult to measure directly, the prosecutor would like to use the Pythagorean theorem to find the distance based on two distances that are easily obtained. The defendant's house is 847 feet north and 446 feet east of the church. (This is based on a case in Rockford, Illinois, in 2011.)

a. Find the distance from the defendant's house to the church, to the nearest foot.
About 957 feet

b. Can the prosecutor charge the defendant with a felony within 1,000 feet of a church?   Yes

7. Looking Forward, Looking Back A premature baby was born at 22 weeks and given a 5% chance of survival. Interpret the chance of a survival as a fraction in a sentence.
1 in 20 babies born at 22 weeks will survive.

Getting Ready for Section 2.16

Read the article and answer the questions that follow.

A Statistician's View: What Are Your Chances of Winning the Powerball Lottery?

In November 2012, a Missouri couple and an Arizona man shared the largest Powerball jackpot ever—$587 million. An article about the Missouri couple—Mark and Cindy Hill—appeared in The Huffington Post on February 25 telling a wonderful story about how the couple is using their winnings to benefit their community. Such stories lead us to daydream about what we might do if we won all that money.

What are your chances of winning? A quick look at the Powerball website tells you the probability of winning the jackpot is 1 in 175,223,510. To see where that number comes from, imagine purchasing every number combination. In Powerball, a player first picks five different whole numbers between 1 and 59. One could make a list of all the possibilities, starting with (1, 2, 3, 4, 5), (1, 2, 3, 4, 6), and so on all the way through (55, 56, 57, 58, 59). But it would take a long time to make that list, because it has more than five million entries! Indeed, mathematics tells us the number of ways to choose five distinct numbers from 1 to 59 is 5,006,386.

After choosing the five numbers between 1 and 59, the player then picks another number between 1 and 35 that is called the Powerball. So, we multiply the 5,006,386 by 35 and see that there are 175,223,510 possible Powerball combinations. For simplicity, let's be generous and round off to an even 175,000,000.

Your chance of winning the lottery on a single ticket is one in 175 million. That seems tiny, and it is. In fact, it is so small that it is difficult for us to grasp. Understanding how small this number is provides the key to understanding how likely—or unlikely—it is you will become the next big winner of the Powerball jackpot.

For some reason, we tend to associate unlikely events with a specific physical phenomenon. "I have a better chance of being struck by lightning," we often say. But that does not provide much of a basis for comparison. We realize being struck by lightning is unlikely, but we have no sense of how unlikely, and of course the chance of being struck by lightning is much different for a farmer than a coal miner.

The problem with grasping the smallness of "1 in 175 million" is that we never see 175 million distinct objects. It is easy to grasp 1 in 50, for example, because we can imagine ourselves with 49 other people in a room. We can get our minds around 1 in 75,000 (roughly) by visualizing the crowd of people at the Super Bowl and imagining ourselves being the one person selected from that crowd to win a prize. But one in 175 million cannot be readily visualized.

Here is an example I have used in classrooms all over the country, and it is way more fun than thinking about being struck by lightning! Imagine 175 million freshly minted one-dollar bills are being delivered to my house near Washington, D.C. One of those dollar bills is specially marked as the "lucky dollar bill." You get to pick a dollar bill, and if you happen to pick the lucky dollar bill, you win all the dollar bills.

A straightforward mathematical calculation using the dimensions of a dollar bill reveals it will take two semi-trailers to deliver the 175,000,000 dollar bills to my house. Once these arrive, they will have to be unloaded, of course, so you will have a fair chance to pick the lucky dollar bill. So, we will lay them out end to end. How long will that line of dollar bills go?

If we start from my house, we'll have enough dollar bills to go all the way south to Disney World in Orlando. Then we'll still have enough to go clear across the country to Disneyland! But, even then, we are not out of dollar bills, so we can go north and make it all the way to Portland, Oregon. Still, we have dollar bills, enough to make it all the way east to Portland, Maine. And, fortunately, we'll have enough to make it back to my house near DC, completing the loop.

Do we have any dollars bills left? Yes! We would still have enough dollar bills to go all the way around the loop a second time!

Now imagine that you walk, bike or drive for as long as you want around the double loop, and when you decide to stop, you stoop over and pick up one dollar bill. Your chance of selecting the lucky dollar bill is one in 175 million, the same as your chance of winning the Powerball jackpot!

Your chance of ever winning this big jackpot is impossibly small. It isn't going to happen. But you might say,

(continued)

A Statistician's View: What Are Your Chances of Winning the Powerball Lottery?   *(continued)*

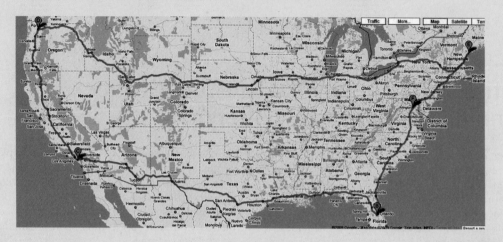

"If the chances of winning are so small, how did Mark and Cindy Hill win?" Or said another way, "If they can do it, why can't I?" I'll explain that in a future post.

It is not my purpose or place to discourage people from buying lottery tickets. I just want everyone to understand their chances as fully and accurately as possible.

Questions

1. The article mentions that there are more than 5 million ways to choose 5 different numbers from the numbers 1 through 59. To get a sense of why there are so many ways to choose the numbers, try it for a much smaller set. List all the ways you can pick sets of 3 numbers from 1 through 5. How many are there?

   1, 2, 3; 1, 2, 4; 1, 2, 5; 1, 3, 4; 1, 3, 5; 1, 4, 5; 2, 3, 4; 2, 3, 5; 2, 4, 5; 3, 4, 5
   10 ways

2. How far would 1,000 dollar bills stretch if you laid them end to end? A bill is about 2.61 inches by 6.14 inches. Convert your answer to a useful unit of length to make sense of it.

   About 512 feet

2.16 What Are the Odds?: Theoretical Probability

Explore

SECTION OBJECTIVES:

- Calculate theoretical probabilities
- Compare theoretical and experimental probabilities

1. A state lottery runs a weekly game in which players must pick five different numbers from the numbers 1 through 45. A player can win by matching at least 2 of the 5 numbers that are drawn for the week. There are lots of ways to match only 2 numbers but fewer ways to match more of the numbers, making those outcomes much less likely. The following table shows the probability of different outcomes for a particular drawing and how much you win in each case.

	Probability	Amount You Win
Match all 5 numbers	$1/1{,}221{,}759 = 0.000000818$	$10,000
Match 4 of the 5 numbers	$1/6{,}109 = 0.000164$	$200
Match 3 of the 5 numbers	$1/157 = 0.00637$	$15
Match 2 of the 5 numbers	$1/12 = 0.0833$	$1

a. Find and interpret the probability of winning at least $200 by playing this lotto game one time.

0.000165, or about 2 out of every 10,000 plays

b. What is the probability that you win something? What is the probability that you win nothing?

About a 9% chance of winning something and a 91% chance of winning nothing

c. If you play this game once a week for a year, and you win something three times, are you doing better or worse than expected?

If you win three times out of 52, then that's about 5.8% of the time. You would expect to win 9% of the time. So you're doing a little worse than expected.

Discover

This section will help you understand how the probabilities in the *Explore* table were found and give you the tools to calculate many other probabilities. As you work through this section, think about these questions:

1 When is a probability theoretical and when is it experimental?

2 How do you calculate a theoretical probability when all the outcomes are equally likely?

The first thing we need to do is clarify the difference between theoretical probability, the focus of this section, and experimental probability, which we discussed in Section 1.5.

Theoretical Probability

An event is an outcome of an experiment that is conducted. The **probability** of an event, *E*, occurring is its chance of happening and is written as P(*E*).

Probabilities are numbers from 0% to 100%, but they can also be stated as simplified fractions or decimal numbers between 0 and 1. A probability cannot be greater than 1 or less than 0. In mathematical notation, this becomes $0 \leq P(E) \leq 1$.

(continued)

Theoretical Probability   (continued)

An event that has a probability of 0% is known as an **impossible** event, while a probability of 100% denotes a **certain** event.

To find the **theoretical probability** of an event E with equally likely outcomes, find this ratio:

$$P(E) = \frac{\text{number of favorable outcomes}}{\text{number of possible outcomes}}$$

When we list all the possible outcomes for an experiment, we are finding the **sample space** for the event. The number of items in the sample space gives the denominator in the ratio above.

FOR EXAMPLE, if we want to find the probability of rolling a 5 when we roll a die, we need to know all the possible outcomes.

Sample space: 1, 2, 3, 4, 5, 6

Now we need to know how many of the outcomes are in the event of interest. Just one outcome is the number 5. So P(5) = $\frac{1}{6}$.

2. Suppose you have a bag with 5 blue marbles, 14 red marbles, and 9 green marbles, and you pick one marble out of the bag.

   a. What is the theoretical probability of picking a blue marble?

   $\frac{5}{28}$

   b. What is the theoretical probability of picking a green marble?

   $\frac{9}{28}$

Writing the sample space, as shown in the next example, can help you find theoretical probabilities.

EXAMPLE

Find the theoretical probability of getting a sum of 5 when two fair dice are rolled.

SOLUTION

First, we list the sample space for rolling two fair dice. It can be helpful to picture the dice as two different colors, one red and one green.

(1, 1)	(1, 2)	(1, 3)	(1, 4)	(1, 5)	(1, 6)
(2, 1)	(2, 2)	(2, 3)	(2, 4)	(2, 5)	(2, 6)
(3, 1)	(3, 2)	(3, 3)	(3, 4)	(3, 5)	(3, 6)
(4, 1)	(4, 2)	(4, 3)	(4, 4)	(4, 5)	(4, 6)
(5, 1)	(5, 2)	(5, 3)	(5, 4)	(5, 5)	(5, 6)
(6, 1)	(6, 2)	(6, 3)	(6, 4)	(6, 5)	(6, 6)

From the organized list of outcomes in the sample space, we can see that there are $6 \cdot 6 = 36$ possible outcomes. It's important to note that each outcome of an individual die has the same probability, which makes these 36 outcomes also equally likely. Of these 36 outcomes, only 4 have a sum of 5. So the theoretical probability of rolling two fair dice and getting a sum of 5 is $\frac{4}{36} = \frac{1}{9}$.

In order to use the approach shown in the last example, we must be sure that the outcomes in the sample space are equally likely. Using the ratio $P(E) = \frac{\text{number of favorable outcomes}}{\text{number of possible outcomes}}$ when the experiment does not have equally likely outcomes can lead to incorrect probabilities.

For example, it is tempting to think about the probability of getting a sum of 5 in the following way: There are 11 different possible sums when you roll two fair dice (2, 3, 4, . . . , 12). Since a sum of 5 is one of the 11 outcomes, the probability of getting a sum of 5 is $\frac{1}{11}$. Clearly this is wrong, since it does not lead to the answer of $\frac{1}{9}$ that we already found. The reason this approach is wrong is that it assumes those 11 possible sums are equally likely when, in fact, they are not. There is only one way to get a sum of 2 but six ways to get a sum of 7, for example.

The probability of an event occurring and the probability of it not occurring sum to 100% or 1. For example, if there is a 70% chance of rain, then there is a 30% chance that it will not rain. If we know that the probability of getting a sum of 5 is $\frac{1}{9}$, then the probability of *not* getting a sum of 5 is $\frac{8}{9}$.

> To find the probability that an event *E* does not happen, use the formula:
>
> $$P(\text{not } E) = 1 - P(E)$$

INSTRUCTOR NOTE: Consider helping students write the sample space for flipping three fair coins.

3. a. How many different outcomes are possible when three fair coins are flipped? Are they equally likely?

8; yes

b. What is the probability of flipping three fair coins and getting all heads? What is the probability of not getting all heads? What is the probability of getting at least one of each heads and tails?

1/8; 7/8; 6/8 = 3/4

c. If you flip three fair coins ten times and get all heads only once, how does this experimental probability compare to the theoretical probability?

1/10; the experimental probability of 10% is smaller than the theoretical probability of 12.5%. You got all heads less often than expected.

The last problem makes the important point that experimental and theoretical probabilities for the same event can be, and often are, different. The experimental probability tells us what percent of the time something actually happened, whereas the theoretical probability tells us what percent of the time we expect the event to happen. Usually, the more times an experiment is conducted, the closer the experimental probability gets to the theoretical probability.

4. Suppose there is a raffle with numbered tickets. One ticket is randomly drawn from a hat. Your ticket is #121. Find the following probabilities of winning. State your answer as a fraction, a decimal, and a percent.

a. If there are 200 tickets, what is your chance of winning?

$$\frac{1}{200} = 0.005 = 0.5\%$$

b. If there are 400 tickets, what is your chance of winning?

$$\frac{1}{400} = 0.0025 = 0.25\%$$

c. If there are 1,000 tickets, what is your chance of winning?

$$\frac{1}{1,000} = 0.001 = 0.1\%$$

d. What happens to your chance of winning as the number of tickets sold increases? Why?

The chance of winning decreases because the numerator is constant and the denominator is increasing.

e. If 150 tickets were sold in a raffle, how many tickets would you need to buy to have a 10% chance of winning?   15

When probabilities are small, it is helpful to use place value to interpret them in a meaningful way. For example, 0.25% is 0.0025, or 25 out of 10,000.

Connect

Suppose a local restaurant is running a dart competition for which contestants will be blindfolded. The owners want to do something fun and are considering three nontraditional options for paper targets. In #5 we show pictures of a traditional dartboard and the other options. Hitting a bull's-eye is defined as hitting the shaded area of the dartboard.

5. For each target, find the chance of hitting a bull's-eye by calculating the ratio of the shaded area to the total area. Assume each dartboard is 18 inches in diameter. Convert each probability to a decimal value, rounded to two places, and then to a percent. For which target did you find the greatest chance of hitting a bull's-eye? Is that what you expected?

a.

6 in.

18 in.

$\dfrac{36}{254.47} \approx 0.14 \approx 14\%$ chance

b.

3 in.

18 in.

$\dfrac{36}{254.47} \approx 0.14 \approx 14\%$ chance

INSTRUCTOR NOTE: Consider showing students how to write the numerator in 5d as a single exact term using π. It reduces the number of calculator keystrokes and illustrates another use of like terms.

c.

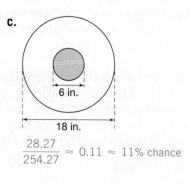

6 in.

18 in.

$\dfrac{28.27}{254.27} \approx 0.11 \approx 11\%$ chance

d.

3 in.

12 in.

14 in.

18 in.

$\dfrac{47.91}{254.47} \approx 0.19 \approx 19\%$ chance

The traditional bull's-eye, *d*

The probabilities found in the *Connect* represent the chances of hitting a bull's-eye for a particular target as well as the percent of the target area that is considered the bull's-eye.

Reflect

WRAP-UP

What's the point?

It is useful to know how likely an event is to happen. Calculating theoretical probabilities can help you determine what to expect when you buy a lottery ticket, apply for a job, or assess your risk of getting a disease.

What did you learn?

How to calculate theoretical probabilities
How to compare theoretical and experimental probabilities

2.16 Homework

Skills MyMathLab

First complete the MyMathLab homework online. Then work the two exercises to check your understanding.

1. Suppose there is a raffle with numbered tickets $1 - 150$. One ticket is randomly drawn from a hat. Find the following probabilities:

   a. Drawing an even number $\dfrac{75}{150} = \dfrac{1}{2} = 0.5 = 50\%$

   b. Drawing a number greater than 100 $\dfrac{50}{150} = \dfrac{1}{3} \approx 0.33 = 33\%$

   c. Drawing an even prime number $\dfrac{1}{150} \approx 0.007 = 0.7\%$

   d. Drawing a negative number $\dfrac{0}{150} = 0 = 0\%$

2. In a lottery game, the chance of winning is given as "5 in 19 will win." Write this as a probability between 0 and 1.   0.26

Concepts and Applications

Complete the problems to practice applying the skills and concepts learned in the section.

3. A sock drawer contains 4 black socks, 10 white socks, and 8 gray socks. Socks are not connected in pairs. What is the probability of pulling out a white sock?

   $\dfrac{5}{11}$

4. Use the spinner to answer the questions that follow. Assume that the spinner wedges are the same size.

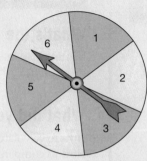

a. What is the probability of the spinner landing on a section with an odd number?

$\frac{1}{2}$

b. Find the probability of the spinner landing on a section with a negative number.

0

5. McDonald's Monopoly game is a popular contest the restaurant chain has offered for several years. The chances of winning a food prize are approximately 1 in 4. Here were the chances in 2011 for collecting the less common game pieces, each of which netted a nonfood prize. All values listed are approximate according to McDonald's.

Mediterranean Avenue ($50 prize, 10,000 prizes)—1 in 61,811

Vermont Avenue ($100 prize, 3,000 prizes)—1 in approx. 206,036

Virginia Avenue ($200 SpaWish® Gift Certificate, 250 prizes)—1 in 2,472,425

Tennessee Avenue (Beaches Resort Vacation, 300 prizes)—1 in 2,060,354

Kentucky Avenue ($50,000, 8 prizes)—1 in 77,263,275

Short Line Railroad (EA Sports Fan Trip, 8 prizes)—1 in 77,263,275

Ventnor Avenue ($100,000, 6 prizes)—1 in 103,017,700

Pennsylvania Avenue (SL MY 12 Nissan Leaf, 2 prizes)—1 in 309,053,100

Boardwalk ($1,000,000, 1 prize)—1 in 618,106,200

Source: playatmcd.com

a. What percent of the time would a player expect to win a food prize? What percent of the time would a player expect to not win a food prize?

25%, 75%

b. Relatively speaking, were the chances of winning a food prize high?

Yes, the chances of winning a food prize were high compared to the chances of winning the other prizes.

c. If a player collected 12 game pieces, how many food prizes should he or she have expected to win? Would the player necessarily have won that many?

In 12 pieces, the player should have won 3 food prizes. While that is what *should* have happened, there was no guarantee that it *would* happen.

d. If a player collected 12 game pieces and won one food prize, what percent of the time did he or she win a food prize? How does that result compare to the answer in part a?

The player won 8.3% of the time, which is a smaller percent of the time than the expected 25%.

6. Suppose there is a weekly raffle with 200 tickets sold each week. You play every week for a year and buy one ticket each time.

    a. If you win only twice, what was your experimental probability of winning the weekly raffle?

    $\frac{2}{52} \approx 0.038$, or approximately 4%

    b. What was your theoretical probability of winning each weekly raffle?

    $\frac{1}{200} = 0.005$, or 0.5%

7. **Looking Forward, Looking Back** Solve $\dfrac{17}{2x + 8} = \dfrac{34}{4x + 16}$.

All real numbers

2.17 Size Up: Volume and Surface Area

Explore

SECTION OBJECTIVE:

- Calculate volume and surface area

1. Imagine you have a sick child and need to give him a dose of 1 teaspoon of antibiotics. You have two choices of measuring devices for medication as shown here. Suppose you are not exact when pouring and go over the 1-teaspoon line by 2 mm in each device. Which measuring device would produce the greater overdose error? Use the volume of each liquid medication to determine the percent overdose.

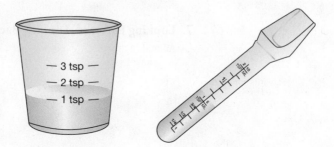

Although the cups are not right circular cylinders, we can approximate the quantities of medication using the following figures. The dimensions shown closely approximate the volumes but are not exactly 1 teaspoon.

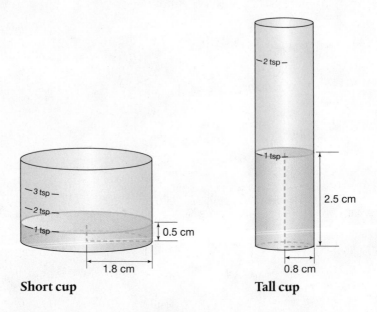

Short cup   **Tall cup**

Use the formula for the volume of a cylinder, $V = \pi r^2 h$, and state the answers in cubic centimeters (cc's).

	Short	**Tall**
1 dose (intended)	5.09 cc	5.03 cc
1 dose (overfilled)	7.13 cc	5.43 cc
Amount over per dose	2.04 cc	0.40 cc
Percent over per dose	40%	8%

The short cup produces the greater error. The child will receive 40% more medication than intended.

Discover

It is common to have situations and problems that necessitate measuring the size of three-dimensional objects. The needed measurement could be the space the object takes up or the area needed to cover the object.

As you work through this section, think about these questions:

1 Do you understand the difference between volume and surface area?

2 Can you find the volume of an object by using a formula or by counting blocks?

3 Can you find the surface area of an object by using a formula or by counting squares?

In this section, you will tie together concepts you have already learned in this cycle with the ideas of volume and surface area. You will also see how changes in the dimensions of a three-dimensional object affect both its volume and surface area. We have seen one example of this in the *Explore*, where one cup had a larger radius and then its value was squared in the volume formula, which contributed a much greater addition in volume than the other cup's overfill amount.

LOOK IT UP

Volume and Surface Area

The **volume** of a three-dimensional object is the amount of space it occupies. Volume is measured in cubic units, such as cubic inches or cubic yards.

The **surface area** of a three-dimensional object is the sum of the areas of its exterior faces. Surface area is measured in square units, such as square meters or square feet.

EXAMPLE: Find the volume and surface area of the rectangular prism shown.

SOLUTION: Notice there are three layers of 9 cubes each for a volume of $3 \cdot 9 = 27$ cubic units. Each face of the prism has an area of 9 square units. Since there are six identical faces in this object, the surface area is $9 \cdot 6 = 54$ square units.

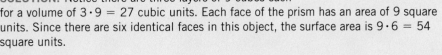

2. Find the volume of each object by counting the cubic units. Assume that there are blocks you cannot see if they are necessary for the structure. Find the surface area of each object by counting square units.

a.

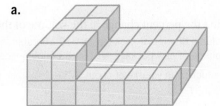

Volume = 32 cubic units
Surface area = 80 square units

b.

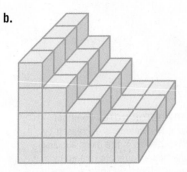

Volume = 44 cubic units
Surface area = 94 square units

In the previous problem, it was possible to count square and cubic units to find the surface areas and volumes. When that is not possible, a formula can be used.

3. Find the volume or surface area as indicated in each problem. Use the formulas provided, and include units in your work.

a. Find the volume of the cone using the formula $V = \frac{1}{3}\pi r^2 h$.

$V = \frac{1}{3}\pi r^2 h = \frac{1}{3}\pi (4 \text{ cm})^2 (6 \text{ cm}) = \frac{1}{3}\pi (16 \text{ cm}^2)(6 \text{ cm})$

$= 32\pi \text{ cm}^3 \approx 100.53 \text{ cm}^3$

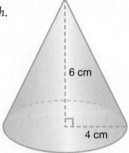

6 cm

4 cm

b. Find the surface area of the sphere using the formula $S = 4\pi r^2$.

$S = 4\pi r^2 = 4\pi (2 \text{ in.})^2 = 4\pi (4 \text{ in.}^2) = 16\pi \text{ in.}^2$

$\approx 50.27 \text{ in.}^2$

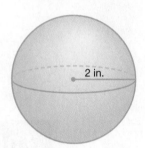

2 in.

c. Find the volume of the pyramid using the formula $V = \frac{1}{3}lwh$.

$V = \frac{1}{3}lwh = \frac{1}{3}(5 \text{ in.})(7 \text{ in.})(12 \text{ in.}) = 140 \text{ in.}^3$

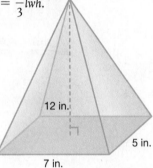

12 in.

5 in.

7 in.

Connect

4. A company intends to sell less of a food in a package and reduce the packaging as well. The company's cereal box currently has these dimensions: 11" tall, $7\frac{5}{8}$" wide, $2\frac{3}{4}$" deep. A vice president asks for a reduction of a half inch on each dimension, with the idea of reducing the size of the package enough to save the company money without being so drastic that the change is noticeable.

a. Find the percent change in the volume of the cereal box.

Volume of original package ≈ 230.66 cubic inches
Volume of new package ≈ 168.33 cubic inches

Reduction in volume of about 27%

INSTRUCTOR NOTE: Point out that if the volume is decreased by 27%, then so is the mass because mass and volume are related by a constant (mass = volume · density).

b. The original cereal box held 12 ounces of a particular type of cereal. Estimate the amount the new box will hold. Find the percent change in the ounces of cereal in the box.

0.73(12) = 8.76 ounces; 27%

c. Find the percent reduction in the amount of packaging needed. Is it the same as the percent reduction in volume?

Surface area of original package ≈ 270.19 in.²
Surface area of new package ≈ 228.94 in.²

Reduction in surface area ≈ 15%. This is not the same as the volume reduction.

Reflect

INSTRUCTOR NOTE: Consider discussing the solution to the focus problem. You might want to show the sample solution, available in MyMathLab, or a correct solution from a student or student group in the class. See the Instructor Guide for more debriefing ideas.

WRAP-UP

What's the point?

There are many contexts in which a three-dimensional object's size, either the space it takes up or its exterior area, needs to be quantified. Small changes in a dimension of a three-dimensional object can have a significant impact on its volume and surface area.

What did you learn?

How to calculate volume and surface area

Focus Problem Check-In
Do you have a complete solution to the focus problem? If your instructor has provided you with a rubric, use it to grade your solution. Are there areas you can improve?

2.17 Homework

Skills MyMathLab

First complete the MyMathLab homework online. Then work the two exercises to check your understanding.

1. Find the volume and surface area of a cube with sides 6 inches long.

   216 cubic inches, 216 square inches

2. Using the cube given in #1, increase the length of each side by 10%. Find the volume and surface area of the new cube.

   About 287.50 cubic inches, 261.36 square inches

Concepts and Applications

Complete the problems to practice applying the skills and concepts learned in the section.

3. A tuna can is a right circular cylinder with a radius of 4.25 centimeters and a height of 3.25 centimeters.

   a. If the volume of a cylinder is given by $V = \pi r^2 h$, find the volume of the can to the nearest cubic centimeter.

      $A \approx 184$ cubic centimeters

b. The label that wraps around the can covers the "lateral" surface area of the can. If you unrolled the label, it would be a rectangle with a length equal to the circumference of the can and a height equal to the height of the can. The lateral surface area is given by $SA = 2\pi rh$. Find the lateral surface area of the can to the nearest square centimeter.

SA ≈ 87 square centimeters

c. If there was a 17% size reduction from 6-oz to 5-oz cans, what percent increase would be needed to go back to 6-oz cans? NOTE: The percent increase from 5 oz to 6 oz will not be the same as the percent decrease from 6 oz to 5 oz.

From 5 oz to 6 oz is 1 oz of increase. Compared against the starting point of 5 gives $\frac{1}{5}$, or a 20% increase.

d. If the can were enlarged to hold 6 ounces, the volume would need to change by the same percent found in part c. If the radius stays the same, how tall would the can need to be?

Height = 3.9 centimeters
This can be found by solving $220.8 = 56.75h$ or by just increasing the height by 20%.

4. In an effort to find the volume of a cylindrical measuring cup, you measure the radius to be about 1.6 cm and the height to be 5.8 cm.

a. Find the volume using the two measurements given and round to the nearest cubic centimeter. Use the formula $V = \pi r^2 h$.

$V \approx 47$ cc

b. If you round the radius and height to the nearest centimeter before calculating the volume, what do you get for the volume rounded to the nearest cubic centimeter?

$V \approx 75$ cc

c. How did rounding before the calculation instead of after the calculation affect your answer?

The volumes are vastly different.

d. By what percent did you overestimate the volume when you rounded before calculating?

About 60%

5. Suppose two different cylinders have the same volume of $V = 36\pi$ cubic centimeters.

a. If one cylinder has a radius of 2 centimeters, what is its height? $h = 9$ cm

b. If one cylinder has a radius of 3 centimeters, what is its height? $h = 4$ cm

c. For a given volume, the smaller the radius, the _____larger_____ the height.

For a given volume, the larger the radius, the _____smaller_____ the height.

6. Suppose two different cylinders have the same volume of $V = 36\pi$ cubic centimeters.

    a. If one cylinder has a height of 9 centimeters, what is its radius? $r = 2$ cm

    b. If one cylinder has a height of 4 centimeters, what is its radius? $r = 3$ cm

    c. For a given volume, the smaller the height, the _____larger_____ the radius.

    For a given volume, the larger the height, the _____smaller_____ the radius.

7. In a right circular cone, there are two heights. The internal height, h, is the distance between the vertex of the cone and the center of the circular base. The external height, s, also called the slant height, is the distance between the vertex and a point on the circumference of the circular base. A right triangle is formed from the height, the slant height, and the radius of the base, r.

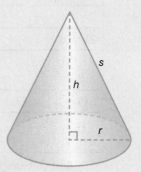

Find the volume of a right circular cone that has a slant height of 13 cm and a radius of 5 cm. Begin by finding the height, h. Round to the nearest tenth of a centimeter.

314.2 cm³

8. Find the volume and surface area of the figure. Assume that there are blocks you cannot see if they are necessary for the structure.

Volume = 22 cubic units

Surface area = 64 square units

9. Looking Forward, Looking Back A tourist paid $19.89 for a museum entrance fee that included a $3 city tax and then a 7.5% sales tax. Find the mistake in the following work to find the price of the original admission ticket.

$$0.075(\$19.89) = \$1.49$$
$$\$19.89 - \$1.49 = \$18.40$$
$$\$18.40 - \$3 = \$15.40$$

The 7.5% sales tax should not be applied to the total but instead to the ticket price +$3.

Cycle 2   Study Sheet

This page is an overview of the cycle, providing a quick snapshot of the material covered to help you as you begin studying for a test. *However, it is not comprehensive.* The *Cycle Wrap-Up* tasks that follow will help you review the cycle in more detail.

EQUATIONS

Solve one-step equations by undoing what's been done to the variable.

Equation	Good First Step to Solve
$2x - 4 = 6$	Add 4 to each side.
$2x = 6 - 3x$	Add $3x$ to each side.
$\dfrac{x - 3}{2} = 14$	Multiply each side by 2.
$3(x - 1) = 2x + 5$	Distribute the 3.
$\dfrac{15}{3 - x} = \dfrac{2}{x}$	Find cross products.

Contradiction

$$2x + 4 = 2x$$
$$\underline{-2x \qquad -2x}$$
$$4 = 0$$

No solution

Identity

$$2x + 4 = 2(x + 2)$$
$$2x + 4 = 2x + 4$$
$$\underline{-2x \qquad -2x}$$
$$4 = 4$$

All real numbers

ORDER OF OPERATIONS

P   Parentheses and grouping symbols

E   Exponents

M   Multiplication
D   Division $\Big\}$ left to right

A   Addition
S   Subtraction $\Big\}$ left to right

$$3 - 15(\sqrt{4} + 2^2)$$
$$3 - 15(2 + 4)$$
$$3 - 15(6)$$
$$3 - 90$$
$$-87$$

APPLICATIONS

Pythagorean theorem

To complete the Pythagorean triple 30, _____, 78:

$$30^2 + \text{leg}^2 = 78^2$$
$$900 + \text{leg}^2 = 6{,}084$$
$$\text{leg}^2 = 5{,}184$$
$$\text{leg} = 72$$

The triple is 30, 72, 78.

Weighted mean

To find the average of 6, 8, 12 if 6 and 8 carry three times the weight as 12:

$$\frac{6 \cdot 3 + 8 \cdot 3 + 12 \cdot 1}{3 + 3 + 1} = \frac{54}{7} \approx 7.71$$

ALGEBRAIC RULES

Properties

Commutative properties $a + b = b + a$
$\qquad\qquad\qquad\qquad\qquad\quad ab = ba$

Associative properties $(a + b) + c = a + (b + c)$
$\qquad\qquad\qquad\qquad\quad (ab)c = a(bc)$

Distributive properties $a(b + c) = ab + ac$
$\qquad\qquad\qquad\qquad\quad a(b - c) = ab - ac$

Exponent Rules

$x^a x^b = x^{a+b}$

$\dfrac{x^a}{x^b} = x^{a-b}$

$x^0 = 1$

$(x^a)^b = x^{ab}$

$(xy)^a = x^a y^a$

$\left(\dfrac{x}{y}\right)^a = \dfrac{x^a}{y^a}$

To simplify $\left(\dfrac{-15x^8y^4}{5xy}\right)^2$:

$$\left(\frac{-15x^8y^4}{5xy}\right)^2 = (-3x^7y^3)^2$$
$$= (-3)^2(x^7)^2(y^3)^2$$
$$= 9x^{14}y^6$$

SELF-ASSESSMENT: REVIEW

For each objective, use the boxes provided to indicate your current level of expertise. If you cannot indicate a high level of expertise for a skill or concept, continue working on it and seek help if necessary.

INSTRUCTOR NOTE: Consider discussing this page before a test. The objectives are the same ones provided at the beginning of the cycle, giving students two points of comparison on their learning. Completing the page can help students see where they need to focus their studying for a test.

SKILL or CONCEPT

Low   High

1. Find and interpret weighted means. *(2.2)*

2. Find the median and mode of a data set. *(2.2)*

3. Apply basic exponent rules. *(2.3)*

4. Use geometric formulas. *(2.3)*

5. Identify and add like terms. *(2.4)*

6. Use the order of operations to simplify expressions. *(2.5)*

7. Apply the commutative and associative properties. *(2.6)*

8. Use and apply the distributive property. *(2.7)*

9. Determine if two expressions are equivalent by using the commutative, associative, and distributive properties. *(2.8)*

10. Recognize when the distributive property can be applied. *(2.9)*

11. Verify a solution to an equation. *(2.10)*

12. Write and solve one-step equations. *(2.11)*

13. Solve two-step and multi-step linear equations. *(2.12)*

14. Write an equation to model a situation. *(2.13)*

15. Solve a problem numerically and algebraically. *(2.13)*

16. Write and solve proportions. *(2.14)*

17. Solve problems using the Pythagorean theorem. *(2.15)*

18. Calculate theoretical probabilities. *(2.16)*

19. Compare theoretical and experimental probabilities. *(2.16)*

20. Calculate volume and surface area. *(2.17)*

WRAP-UP

The measure of success is not whether you have a tough problem to deal with, but whether it is the same problem you had last year.

— John Foster Dulles

INSTRUCTOR NOTE: Discuss the steps included in the *Cycle Wrap-Up* in detail with students so they understand the actions that lead to success on a test.

The *Cycle Wrap-Up* will help you bring together the skills and concepts you have learned in the cycle so that you can apply them successfully on a test. Since studying is an active process, the *Wrap-Up* will provide you with activities to improve your understanding and your ability to show it in a test environment.

STEP 1

GOAL Revisit the cycle content

Action Skim your cycle notes

Action Read the Cycle Study Sheet

Action Complete the Self-Assessment: Review

> As you are studying for the exam, use all five steps of the *Cycle Wrap-Up*. If you did not get the exam grade you wanted during Cycle 1, these steps can help you. They show you how to study math content effectively.

STEP 2

GOAL Practice cycle skills

Action Complete the MyMathLab assignment

Action Address your areas of difficulty

MyMathLab

Complete the MyMathLab assignment. Reference your notes if you are having trouble with a skill. For more practice, you can rework Skills problems from your homework. You can also access your MyMathLab homework problems, even if the assignment is past due, by using the Review mode available through the Gradebook.

STEP 3

GOAL Review and learn cycle vocabulary

Action Use your notes to complete the Vocabulary Check

Action Memorize the answers

Vocabulary Check

Cycle 2 Word Bank

associative property	exponent	median	sample space
base	hypotenuse	mode	solution
certain	identity	operation	surface area
coefficients	impossible	PEMDAS	theoretical probability
commutative property	isolate	proportion	volume
contradiction	legs	Pythagorean theorem	weighted mean
cross products	like terms	Pythagorean triple	
distributive property	linear	ratios	

Choose a word from the word bank to complete each statement.

1. A(n) _____weighted mean_____ is an average in which some values carry more importance than others. To find this average, multiply each data value by its weight and divide by the sum of the weights.

2. The _____median_____ of a set of data is the middle value when the data values are arranged in order.

3. The most common value in a set of data is known as the _____mode_____ of the data.

4. The _____exponent_____ in an expression tells you how many factors of the _____base_____ to multiply.

5. To combine _____like terms_____ in an expression, add the _____coefficients_____ and keep the same name or variable.

6. The order of operations can be remembered with the acronym _____PEMDAS_____.

7. The _____commutative property_____ states that when you add or multiply numbers, the order of the numbers does not matter.

8. The _____associative property_____ states that when you add or multiply three or more numbers, it does not matter how the numbers are grouped.

9. The _____distributive property_____ allows you to rewrite an expression such as $2(x - 5)$ without parentheses.

10. An equation in which the two sides are equal for every possible value of the variable is called a(n) _____identity_____.

11. An equation in which there is no value that will make the two sides equal is called a(n) _____contradiction_____.

12. A(n) _____solution_____ to an equation is a number that can be substituted for the variable in the equation, resulting in a true statement.

13. To solve a one-step equation, undo the _____operation_____ that has been done to the variable.

14. One goal when solving an equation is to _____isolate_____ the variable.

15. A(n) _____linear_____ equation is one that can be written in the form $ax + b = c$.

16. A(n) _____proportion_____ is a statement that two _____ratios_____ are equal.

17. One way to solve proportions is to set the _____cross products_____ equal to each other.

18. The relationship between the lengths of the sides of a right triangle is stated in the _____Pythagorean theorem_____.

19. In a right triangle, the side opposite the right angle is called the _____hypotenuse_____, and the other two sides are called _____legs_____.

20. Three whole numbers that can be used as the lengths of the sides of a right triangle are known as a(n) _____Pythagorean triple_____.

21. If an event has a theoretical probability of 1, then it is called _____certain_____.

22. If an event has a theoretical probability of 0, then it is called _____impossible_____.

23. The _____theoretical probability_____ of an event is found by dividing the number of favorable outcomes by the total number of possible outcomes for the experiment.

24. The _____sample space_____ is a list of all the possible outcomes of an experiment.

25. The _____volume_____ of a three-dimensional object is the amount of space it occupies, while the _____surface area_____ is the sum of the areas of its exterior faces.

STEP 4

GOAL Practice cycle applications

Action Complete the Concepts and Applications Review
Action Address your areas of difficulty

Concepts and Applications Review

Complete the problems to practice applying the skills and concepts learned in the cycle.

1. Suppose you are in your last semester of college and have a 2.9 GPA after 105 credit hours. If you are taking 15 credit hours in your last semester, what GPA do you need in your last semester to graduate with a 3.0 GPA?   3.7

2. Suppose your friend reads the following instructions to you from a book of math tricks:

> Pick a number.
> Subtract 1.
> Multiply by 3.
> Add 6.
> Divide by 3.
> Subtract the original number.

She then correctly guesses that your final number is 1. You repeat this with the same result.

a. Try this trick once with a negative number and once with a fraction. Do you get 1 each time as an answer? Write out the steps of your calculation.

-2

$-2 - 1 = -3$

$-3 \cdot 3 = -9$

$-9 + 6 = -3$

$-3 \div 3 = -1$

$-1 - (-2) = 1$

$\dfrac{1}{2}$

$\dfrac{1}{2} - 1 = -\dfrac{1}{2}$

$-\dfrac{1}{2} \cdot 3 = -\dfrac{3}{2}$

$-\dfrac{3}{2} + 6 = \dfrac{9}{2}$

$\dfrac{9}{2} \div 3 = \dfrac{3}{2}$

$\dfrac{3}{2} - \dfrac{1}{2} = 1$

b. Write the expression that needs to be evaluated if the number you start with is called x.

$\dfrac{3(x - 1) + 6}{3} - x$

c. Simplify the expression in part b.

$$\dfrac{3(x - 1) + 6}{3} - x = \dfrac{3x - 3 + 6}{3} - x = \dfrac{3x + 3}{3} - x = x + 1 - x = 1$$

3. Consider the expression

$$3x^2 - 14x + 3y^3 - 2y + 5x - x^2 + 9x - 1$$

a. Is this expression a polynomial?

Yes

b. Indicate which terms are like terms by drawing a circle around one set of terms that are alike and drawing a square around another set that are alike, for example.

$3x^2$ and $-x^2$ are like terms. $-14x$, $5x$, and $9x$ are like terms.

c. Simplify the expression by combining like terms.

$2x^2 + 3y^3 - 2y - 1$

Work all the applications problems until you can get the correct answer. This may require looking up the topic in the notes and working on more problems of a similar nature.

If something doesn't make sense, work with it until it does. Seek outside help if you are not making progress on your own.

4. Simplify the following expression in each of the described ways:

$$(5 - 14)(2 - 4)$$

a. Simplify within each set of parentheses and then multiply.

$(-9)(-2) = 18$

b. Simplify within the first set of parentheses and then distribute.

$(-9)(2 - 4) = -18 + 36 = 18$

c. Distribute each number in the first set of parentheses to each number in the second set and simplify.

$5(2) + 5(-4) + (-14)(2) + (-14)(-4) = 10 - 20 - 28 + 56 = 18$

5. A preschool classroom contains triangular tables that the teacher wants to arrange in a line for a class party. Assume that a triangular table has room for one chair on each side.

a. How many chairs will there be around the perimeter if 5 tables are arranged this way? Include a picture.

7

b. How many chairs will there be if n tables are arranged this way? Write an unsimplified algebraic expression and then show the work to simplify it.

$n + 2$

c. Explain in words how you counted the chairs.

There is one chair for each table plus two more chairs on the ends.

6. In the following table, complete any missing expressions or translations.

Expression	Translation
$\sqrt{x} + 2$	Take the square root of x and then add 2.
$\sqrt{x + 2}$	Add 2 to x and then take the square root.
$\dfrac{5}{x + 2}$	Divide 5 by the sum of x and 2.
$\dfrac{5}{x} + 2$	Divide 5 by x and then add 2.
$2x^2$	Square x and then multiply the result by 2.
$(2x)^2$	Square the product of 2 and x.
$|x - 1|$	Take the absolute value of the difference of x and 1.
$|x| - 1$	Take the absolute value of x and then subtract 1.

7. a. Verify that the chemical equation is balanced.

$$Al_2(CO_3)_3 + 2H_3PO_4 \rightarrow 2AlPO_4 + 3CO_2 + 3H_2O$$

There are 2 aluminum atoms, 3 carbon atoms, 17 oxygen atoms, 6 hydrogen atoms, and 2 phosphorus atoms on each side of the chemical equation.

b. Verify that $x = -2$ is a solution to $x^3 + 5x^2 + 5x = 2$.

$$(-2)^3 + 5(-2)^2 + 5(-2) \overset{?}{=} 2$$
$$-8 + 20 - 10 \overset{?}{=} 2$$
$$2 \overset{\checkmark}{=} 2$$

8. Solve each equation.

a. $\dfrac{1}{3} + x = \dfrac{2}{5}$ $x = \dfrac{1}{15}$

b. $12 - 3(x - 2) + 4 = 28$ $x = -2$

c. $\dfrac{4x - 1}{3} = \dfrac{x + 6}{2}$ $x = 4$

9. Solve the first equation for the variable a, then use that result to solve the second equation for the variable b, and so on. Continue to solve the equations, in order, using the previous solutions.

a.
$$a + 10 = 30$$
$$a + b - 7 = a - 2$$
$$\frac{b}{-5} = c$$
$$c + d - 4 = -2$$
$$3d - 1 + e - 8 = -8 + 3e$$

$a = 20$, $b = 5$, $c = -1$, $d = 3$, $e = 4$

b.
$$\frac{a}{10} = 20$$
$$a - b = 250$$
$$\frac{a - b}{50} - c = 0$$
$$d(2b - c) = -210$$

$a = 200$, $b = -50$, $c = 5$, $d = 2$

10. Complete the right column with the reason for each step shown in the equation in the left column.

Equation	Reason
$8 + 15x + 7 = 3(x + 2)$	Original equation
$8 + 15x + 7 = 3x + 6$	Distributive property
$15 + 15x = 3x + 6$	Combine like terms
$15x = 3x - 9$	Subtract 15 from both sides
$12x = -9$	Subtract $3x$ from both sides
$x = -\dfrac{3}{4}$	Divide both sides by 12

11. a. Write an equation that can be solved with the following steps:

- Distribute the 2.
- Add 4 to each side.
- Subtract $3x$ from each side.
- Divide both sides by 3.

Answers will vary. One possible answer is $2(3x - 2) = 3x + 2$.

b. Solve the equation you wrote.

Answers will vary. For the sample equation shown, $x = 2$.

12. A real estate agent earns a commission of 3% of the selling price of a house he sells. How expensive of a house does he have to sell to earn $10,000? Round to the nearest whole dollar.

$333,333

13. If an investment account lost 12% of its value and is currently worth $15,000, what was it worth before the loss? Write and solve an appropriate linear equation.

$0.88v = 15,000$
$v \approx \$17,045.45$

14. You are considering buying an MP3 player, and you have $250 to spend on the device and songs. For each part, write and solve an appropriate equation. You may need to round your final answers.

a. If you buy a player for $75, how many songs can you purchase at $1.29 each?

$250 - 75 = 1.29x$
$x = 135$

b. If you buy a player for $125, how many songs can you purchase at $1.29 each?

$250 - 125 = 1.29x$
$x = 96$

c. If you buy a player for $200, how many songs can you purchase at $1.29 each?

$250 - 200 = 1.29x$
$x = 38$

15. The term "aspect ratio" is used to describe the ratio of a television screen's length to its width. Most conventional television sets and some LCD sets have a 4:3 aspect ratio, while widescreen HDTVs have a 16:9 ratio.

The size of a television set is often described by the length of its diagonal. Find the dimensions of a 42" television with each aspect ratio.

a. Draw a rectangle with a 4:3 aspect ratio such that the length is 4 and the width is 3. Find the length of the diagonal of the rectangle.   5

b. Use proportions to find the dimensions of a television that has a 4:3 aspect ratio but a 42" diagonal.
33.6" × 25.2" for the 4:3 aspect ratio

c. Repeat this process but with a 16:9 aspect ratio instead of a 4:3 aspect ratio.
36.6" × 20.6" for the 16:9 aspect ratio

16. Consider the following data from the U.S. Census Bureau:

Illinois	230.8 people per square mile	55,583.58 square miles
California	238.9 people per square mile	155,959.34 square miles
Montana	6.8 people per square mile	145,552.43 square miles

a. Can you conclude from this data, without doing any calculations, that California has a higher population than the other two states listed? If so, how? If not, why not?
Yes; total population is found by multiplying the number of people per square mile by the number of square miles. Since California has higher numbers for both, the product will be larger as well.

b. Find the population of each of these states.
Illinois: 12,828,690 people; California: 37,258,686 people; Montana: 989,757 people

17. David is having a custom shower built during a renovation of his master bathroom. In the corner, he wants to have a triangular seat that comes out 18 inches from the corner on each side of the shower, creating an angled front edge. He is having a tile border put on the front edge and needs to know its length. Find the length and round it to the nearest tenth of an inch.   25.5 inches

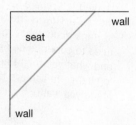

18. a. If a right triangle has one leg of length 16 inches and a hypotenuse of length 34 inches, how long is the other leg?   30 inches

b. If the lengths of the known leg and the hypotenuse are each increased by 20%, does the length of the other leg increase by 20% as well?
Yes; the known leg has a length of 19.2 inches and the hypotenuse has a length of 40.8 inches. Then the other leg has a length of 36 inches, which is 20% greater than 30 inches.

19. A 12-sided die is rolled, and $1 is paid if the die shows a prime number.

a. What is the probability of winning the $1? $\frac{5}{12}$

b. What is the probability of not winning the $1? $\frac{7}{12}$

c. Is this a "fair" game in the sense that you have an equal chance of winning versus losing?   No

d. If you roll the die eighteen times and win the $1 eight of the times, what is the experimental probability of winning the $1? Is it the same as the theoretical probability?

$\frac{4}{9}$; no, but close

20. Draw a circular target that gives a 25% chance of hitting a bull's-eye. Be sure to clearly indicate the dimensions of your target and the dimensions of the bull's-eye. Show all the work to verify that the probability of getting a bull's-eye is 25%.

Answers will vary. One possibility: a circular target with a radius of 20 inches and a circular bull's-eye in the middle with a radius of 10 inches

21. Airlines are requiring passengers to use smaller carry-on bags. They would like passengers to use bags with dimensions no larger than 21.5 inches by 13.5 inches by 7.5 inches instead of the previous dimensions of 22 inches by 14 inches by 9 inches. By how many cubic inches does the volume of carry-on bags change? By what percent does the volume change?

The new bags are 595.125 cubic inches smaller. This is a 21% decrease in volume.

STEP 5

GOAL Simulate a test environment

Action Construct a mock test and take it while being timed

Creating a Mock Test

To pass a math test, you must be able to recall vocabulary, skills, and concepts and apply them to new problems in a fixed amount of time.

Up to this point, you have worked on all the components of the cycle, but they were separated and identified by their topic or type. Although helpful, this is not a test situation. You need to create a test situation to help overcome anxiety and determine where problem areas still exist. Think about a basketball player. She must practice her skills, but until she engages in a practice game, she doesn't really have a feel for a game situation.

The key is to practice problems without assistance under time pressure.

Use the following steps to create a mock test:

- Pick a sample of 5 to 10 problems from your notes and homework and write them on note cards. Choose both skills and applications. On the back of the note card, write the page number where the problem was found.
- Shuffle the cards.
- Set a timer for 20 minutes.
- Work as many problems as you can without outside help.
- Check your work by consulting the page number on the card.

This test will show you where your weaknesses are and what needs more work. It will also give you a feel for your pace. If you completed only a few problems, you are not ready for a test.

Continue to work on the areas that need improvement.

> What did you do that worked to prepare for the first test? Now is the time to think about what you need to continue doing and what you might need to change to achieve the grade you want.

WHEN IS IT
WORTH IT?

zeros
algebraic
equations
graphic
slope-intercept form
quadratic
formula
slope
correlation
distance
models
functions
factoring
exponential
systems of
equations
vertex
comparisons
linear
formulas
representations

SELF-ASSESSMENT: **PREVIEW**

Below is a list of objectives for this cycle. For each objective, use the boxes provided to indicate your current level of expertise.

INSTRUCTOR NOTE: Consider discussing this page when you begin the cycle. The same list is presented at the end of the cycle so that students can note their progress and identify areas that they still need to work on before the test.

SKILL or CONCEPT

Low   High

1. Determine if data has a positive or negative linear correlation.

2. Use the equation of the trendline to make predictions.

3. Find the slope of a line from points, tables, and graphs.

4. Interpret the slope as a rate of change.

5. Use the distance formula to find the distance between two points.

6. Make comparisons using equations, tables, and graphs.

7. Find and interpret the slope and y-intercept from a linear equation.

8. Graph a line using a table and using the slope and y-intercept.

9. Write the equation of a line using a point and the slope or two points.

10. Model with linear functions.

11. Model with exponential functions.

12. Graph exponential functions.

13. Solve nonlinear equations.

14. Solve an equation for a specified variable.

15. Factor an expression using the greatest common factor.

16. Factor quadratic expressions.

17. Use the quadratic formula to solve equations.

18. Solve a 2 × 2 linear system of equations by graphing and substitution.

19. Solve a 2 × 2 linear system of equations by elimination.

20. Find the vertex of a parabola.

CYCLE **THREE**

You learn to speak by speaking, to study by studying, to run by running, to work by working.

—Anatole France

3.1 Deciding to Run
Focus Problem

INSTRUCTOR NOTE: Additional focus problems are available in MyMathLab. Each focus problem is accompanied by a writing template to help students form their solution. A grading rubric is also available. For more information on ways to use the focus problems, see the Instructor Guide at the beginning of the worktext.

If you have ever spent any time in a wilderness area, you have likely wondered about the animals there and your safety. Could you stumble upon a bear on a path in the woods? Could you accidentally come between a moose and her calf? What should you do? It's said that you should never run from a bear, for example, but is that always true? What if the bear is far enough away from you that you think you can get away before it catches you? How far away would you need to be in order to be safe?

Obviously, any serious consideration or calculation has to be done well before you actually encounter a wild animal and must face this decision. We know the top speed at which many animals can run, and we know how long they are able to maintain that top speed. This information is shown in the table for a few animals.

Animal	Top Speed (miles per hour)	Distance at Top Speed
Polar bear	20	1 mile
Black bear	25	2 miles
Lion	30	48 meters
Moose	35	400 meters
Rhinoceros	35	100 meters

Let's assume that the average person can run 10 miles per hour (at least for short distances while being chased). Let's also assume that a person and an animal each achieve top speed immediately. Determine how much of a head start you would need in order to escape from each animal listed in the table if it was chasing you at full speed. From which animal would you need the greatest head start? Consider that each

animal can maintain its top speed for only a certain distance. You can use that distance to determine how long the animal can run and, therefore, how much time it has to catch you. How far can you run in that time? When you calculate the head start that you would need, you are determining the minimum safe distance to be away from that animal. This is useful information to know before heading into that animal's natural habitat.

Choose one of the animals listed in the table and write an equation to model its distance run vs. time. Write a separate equation to model your distance from the animal vs. time. Graph both of these equations on the same coordinate grid. Explain how the graphs illuminate the situation. What does the slope represent? Where do the graphs intersect, and how do you interpret that point? What do the y-intercepts represent?

Finally, choose another animal that you are interested in and find the minimum safe distance from it.

When we're considering running from a wild animal or heading into its natural habitat, we have to consider . . .

WHEN IS IT WORTH IT?

This focus problem can be solved in many different but related ways. You might begin by trying to understand the issues numerically by "running the numbers." At some point, though, you might discover that representing the information in equations might be more efficient. And, if you're trying to explain what you've discovered about the issues to someone else, a graph might be the most useful way to do it. It's important that you can use multiple representations not only to solve a problem but also to explain your solution to an audience. This is true for the focus problem as well as for many other problems that you'll solve during this cycle.

The question "When is it worth it?" will be the focus of the third cycle of the book. Sometimes the focus will be on

determining when it is worth it to use algebraic methods. Other times, the question will relate to particular contexts, such as buying a hybrid vs. gas-powered car or buying an e-reader vs. books. By the end of the cycle, you'll have many tools that you can use to answer this type of question.

You will see periodic sticky notes throughout the cycle to encourage you to keep working on your solution to this problem as you learn new methods to apply.

INSTRUCTOR NOTE: If you are using groups and would like more detailed information on using them effectively, please consult the Instructor Guide at the front of the worktext. It contains techniques and resources for group work.

3.2 What's Trending: Correlation

SECTION OBJECTIVES:

- Determine if data has a positive or negative linear correlation
- Graph the equation of the trendline
- Use the equation of the trendline to make predictions

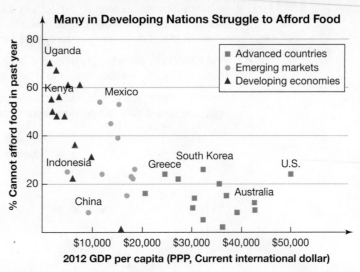

Data from www.pewresearch.org

1. Use the graph to answer the questions that follow.

   a. Write a sentence to summarize the trend shown in the graph.

   Countries with higher GDPs have lower percents of citizens who cannot afford food.

   b. Does the United States follow the trend shown in the graph, or is the United States an exception? Explain your response.

   The United States appears to be an exception. Countries with GDPs as high as that of the United States usually have a much lower percent of citizens who cannot afford food.

   c. Estimate the percent of citizens who cannot afford food in the past year in a country that has a GDP of $48,000.

   5–10%

Discover

We have used scatterplots to illustrate relationships between variables. In this section, we will look at scatterplots that appear to have a linear pattern and then use that pattern to make predictions.

As you work through this section, think about these questions:

❶ Can you determine if variables are positively or negatively correlated?

❷ For which *x*-values should the equation of the trendline be used to make predictions?

❸ How do you find and interpret a residual?

Let's look at scatterplots in which the points look like they fall close to a line.

Linear Correlation

When a scatterplot has a linear pattern, we say there is a **linear correlation** between the two variables. This means that the points generally fall close to a line. The points may be tightly clustered or even lie perfectly in a line. The points may also be more loosely arranged. The best line that can be drawn through the data points is known as the **trendline** or **best-fit line**. It does not necessarily pass through all, or even any, of the data points.

FOR EXAMPLE, since the graph on the left has a trendline that is going up as you look from left to right, the data is said to have a **positive correlation**.

Since the graph on the right has a trendline that is going down as you look from left to right, the data is said to have a **negative correlation**.

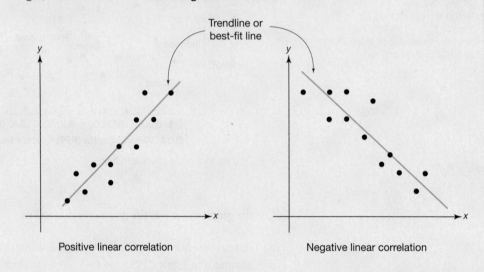

A trendline always goes through the point (mean of the x-values, mean of the y-values).

2. For each pair of variables, determine if you expect a positive or negative correlation.

   a. The number of customers a company has and the total revenue for the company
   Positive

   b. The number of hours of overtime worked and number of hours spent with family
   Negative

   c. The cost of a car and the number of people willing to pay for it
   Negative

   d. The number of hours a student spends at an ACT prep course and her ACT score
   Positive

When two variables are linearly correlated, the relationship between them can be described by the trendline. We can write a linear function to represent this trendline and use it to make predictions for x-values that fall within the scope of the x-values in the original data.

EXAMPLE

In theory, an adult's armspan and height should be the same, but in reality that is not always the case. The following table lists the armspans and heights for a set of adults. Make a scatterplot of the data to graph the height vs. armspan. Graph the trendline for the data on the scatterplot, and use its equation to predict the height of a person who has an armspan of 67 inches.

Armspan (in.)	64	61	63	72	73	65	62	62	65	60	63	70	71	72	71	75	65	69	74	60
Height (in.)	65	62	66	74	75	64	62	61	64	60	65	70	72	71	70	74	65	70	71	63

Tech TIP

A spreadsheet program or graphing calculator can be used to graph a scatterplot and find the equation of the trendline. See the appendix for instructions on using Excel. In a graphing calculator, enter the x-values into the first list in the list editor and the y-values in the second list. Then choose STAT, CALC, 4:LinReg($ax + b$) to get the values of a and b to write the equation of the line.

SOLUTION

To begin, notice that the independent variable is armspan, which will be graphed on the x-axis. The dependent variable is height, which is graphed on the y-axis. An appropriate scale needs to be chosen for each axis. Since we are choosing to begin each axis at a value greater than zero, a jagged start is included. Once the axes are established and labeled, the points are plotted to form the scatterplot.

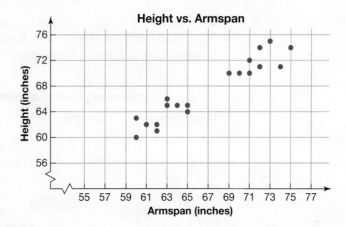

Using Excel or a graphing calculator, we can find the equation of the trendline, $y = 0.9x + 7.3$, where x is the armspan and y is the height. To make it easier to remember the variables, we can choose meaningful variables, such as A for armspan and H for height, and rewrite the equation as $H = 0.9A + 7.3$. To graph the line, we choose at least two values of A between 60 and 74 to be substituted into the equation. Pair the values in ordered pairs and plot them. Draw a line through them.

A	$0.9A + 7.3 = H$	(A, H)
62	$0.9(62) + 7.3 = 63.1$	$(62, 63.1)$
65	$0.9(65) + 7.3 = 65.8$	$(65, 65.8)$
70	$0.9(70) + 7.3 = 70.3$	$(70, 70.3)$

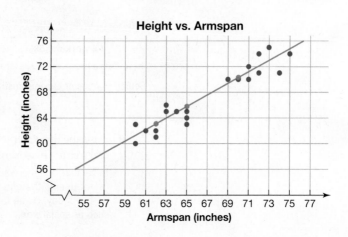

We can use the graph to estimate the height of a person whose armspan is 67 inches, but the equation provides a more accurate estimate. To use the equation, substitute the value of 67 inches for A in the equation.

$$H = 0.9A + 7.3$$
$$= 0.9(67) + 7.3$$
$$= 60.3 + 7.3$$
$$= 67.6 \text{ in.}$$

The trendline equation predicts that a person who has an armspan of 67 inches is 67.6 inches tall. In general, this trendline equation predicts a height for an adult that is close to his or her armspan.

When you use a trendline to make predictions, it is important to use only x-values that are within the range of the x-values in the data. If you use x-values outside that range, the pattern shown may not continue to apply and the resulting predictions may not make sense in the given context. If we consider an infant who has an armspan of 24 inches, the equation in Example 1 predicts a height, or length, of 28.9 inches, which is unlikely. The infant's length is likely to be closer to 25 inches, not 29 inches. Although this difference may appear to be small, it is significant considering the infant's size. The model is not accurate for armspans in this range.

3. The double-bar graph from the Bureau of Labor Statistics shows the unemployment rate and median weekly earnings based on the type of degree earned. Answer the questions that follow to investigate the relationship between the two variables.

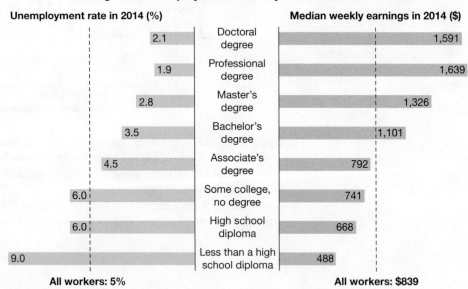

Earnings and Unemployment Rates by Educational Attainment

Unemployment rate in 2014 (%)		Median weekly earnings in 2014 ($)
2.1	Doctoral degree	1,591
1.9	Professional degree	1,639
2.8	Master's degree	1,326
3.5	Bachelor's degree	1,101
4.5	Associate's degree	792
6.0	Some college, no degree	741
6.0	High school diploma	668
9.0	Less than a high school diploma	488

All workers: 5%   All workers: $839

Note: Data are for persons age 25 and over. Earnings are for full-time wage and salary workers.

Source: Bureau of Labor Statistics, www.bls.gov

a. Describe the general trend you see in the graph. Are there exceptions to the general trend? Give examples if possible.

In general, the more education you have, the more money you make and the lower your risk of unemployment. Yes, there are exceptions. This graph does not account for the field of study. Some fields require a college degree, but do not lead to high earnings, such as social work.

b. Notice that under the last bar on each side of the graph is an average. These numbers are not the simple means of the values listed on the bars for each category (unemployment and median weekly earnings). How do you think they were calculated?

Each average listed is a weighted average whose weights are the number of people in each category.

c. To analyze the data's relationship further, convert the earnings from weekly to hourly earnings, assuming 40 hours of work per week. Record the unemployment rates and the hourly earnings in the following table.

Educational Attainment	Unemployment Rate (%)	Hourly Earnings ($)
Doctoral degree	2.1	39.78
Professional degree	1.9	40.98
Master's degree	2.8	33.15
Bachelor's degree	3.5	27.53
Associate's degree	4.5	19.80
Some college, no degree	6.0	18.53
High school diploma	6.0	16.70
Less than a high school diploma	9.0	12.20

d. Make a scatterplot for the median hourly earnings vs. the unemployment rates; that is, graph the unemployment rates on the *x*-axis and the hourly earnings on the *y*-axis. This choice is arbitrary, since we could pick either variable to be *x* or *y*.

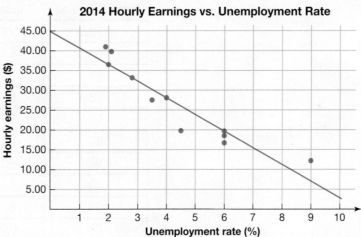

e. What does your scatterplot suggest about the relationship between the earnings and unemployment rates? What kind of correlation is suggested?

The scatterplot shows that the unemployment rate tends to be lower if the earnings are higher; the correlation is negative.

f. Graph the equation of the trendline

$$H = -4.2U + 44.9$$

where *U* represents the unemployment rate and *H* represents the hourly earnings, on the scatterplot. To do so, complete the table to find three points on the trendline. Plot them and draw a line through them.

U	H
2	36.5
4	28.1
6	19.7

g. Predict the hourly earnings for a job level that has an unemployment rate of 5% ($U = 5$).

$23.90 per hour

Connect

When we use a trendline, there is often a difference between the value it predicts and the true value. That difference is called the residual.

LOOK IT UP

Residual

A **residual** is the difference between an actual, observed value and the corresponding value predicted by the equation of the trendline.

$$\text{Residual} = (\text{observed value}) - (\text{predicted value})$$

FOR EXAMPLE, if your actual height is 71 inches and the equation of the trendline predicts your height to be 72 inches, the residual is −1 inches. Since you are shorter than predicted, the residual is negative.

4. A professor is wondering if his final exam really provides any additional information about what his students have learned during the semester. To help him decide, he looks at his students' grades from the previous semester, both before the final exam and on the final exam.

Overall Grade before Final Exam	Final Exam Grade	Predicted Final Exam Score	Residual
72	73	68.8	4.2
68	54	64.8	−10.8
81	76	78.0	−2
88	86	85.1	0.9
78	75	74.9	0.1
80	81	76.9	4.1
92	90	89.1	0.9
70	52	66.8	−14.8
66	82	62.7	19.3
75	70	71.9	−1.9

After graphing the paired data (overall grade, final exam grade), he finds the following trendline equation to model the relationship in the data:

$$\text{Final exam grade} = 1.0138(\text{overall grade}) - 4.1629$$

a. Use the equation of the trendline to find each student's predicted final exam score rounded to the nearest tenth. Fill in the third column of the table with these values.

b. Find the residuals by subtracting the predicted final exam scores from the actual final exam scores. Fill in the fourth column of the table with these values.

c. What does a positive residual tell you about that student's performance on the final exam?

The student performed better than expected.

d. What does a negative residual tell you about that student's performance on the final exam?

The student performed worse than expected.

e. How well do you think the model, given by the equation of the trendline, predicts the final exam score? Does the professor still need to give a final exam? Why or why not?

Some of the final exam scores are much higher or lower than the equation predicts. A final exam is probably needed.

Reflect

INSTRUCTOR NOTE: Section 4.5 on standard deviation can be done after this section without loss of continuity.

WRAP-UP

What's the point?

Graphing paired data allows us to draw conclusions about the relationship between the variables. The trendline describes the linear relationship that might exist and provides a way to predict for other values of the independent variable.

What did you learn?

How to determine if data has a positive or negative linear correlation
How to graph the equation of the trendline
How to use the equation of the trendline to make predictions

3.2 Homework

Skills MyMathLab

First complete the MyMathLab homework online. Then work the two exercises to check your understanding.

1. For each pair of variables, determine if you expect a positive or negative correlation.

  a. Salary and years of education

  Positive

  b. Unemployment rates and years of education

  Negative

2. Does the following scatterplot indicate a linear trend in the data? If so, is the correlation positive or negative?

Yes; positive

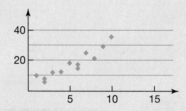

Concepts and Applications

Complete the problems to practice applying the skills and concepts learned in the section.

3. In this section, we saw a relationship between armspans and height. There are other linear relationships between body measurements. One such relationship is between the circumference of a wrist and the circumference of a neck. Wrap your hand around your wrist to get a sense of the circumference. The distance around your wrist is approximately the same as the circumference of the circle created with your thumb and middle finger. Now take your hands and wrap them around your neck. For many people, the circumference of their neck is approximately twice the circumference of their wrist.

The following table lists wrist and neck circumferences for a group of college students.

Wrist Circumference (in.)	5.5	6	6	6.5	7	7	7.5	8	8.5	9
Neck Circumference (in.)	10	13	12.5	12	13.5	14	14	13	18	21

a. Use this data to create a scatterplot. Graph neck circumference versus wrist circumference.

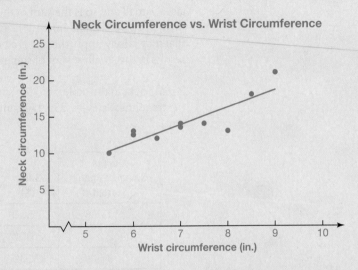

b. Graph the trendline $NC = 2.4WC - 2.8$ on the scatterplot. NC represents neck circumference and WC represents wrist circumference. The trendline equation illustrates that neck circumference is approximately twice wrist circumference.

c. Use the equation of the trendline to estimate the neck circumference in inches for someone whose wrist circumference is 6.25 inches.

12.2 inches

d. Find the residual for the predicted neck circumference from part c if the actual neck circumference is 12.5 inches. Interpret the value.

0.3 inch; the person's neck circumference is 0.3 inches larger than the circumference predicted by the equation.

4. A linear model describes the relationship between the weight in pounds and the age in months for Great Dane puppies: Weight = 16.25(age in months) − 9.5. The model was created by using weights for puppies aged 2 to 6 months.

a. Use the model to estimate the weight of a puppy at 2 weeks (half a month), 1 year, and 5 years.

−1.375 pounds, 185.5 pounds, 965.5 pounds

b. Does the model make sense for ages outside of 2 to 6 months? Why?

No; a puppy cannot have a negative weight, and the growth rate of a puppy does not continue for very long. Two weeks, 1 year, and 5 years are outside of the scope of ages used to create the model.

5. A way to measure how well the trendline fits the data is by looking at the sum of the squared residuals. When a trendline is close the points on the scatterplot, the residuals are small and so is the sum of the squared residuals. Squaring the residuals makes them positive, which removes the canceling effect of positive and negative values that may falsely imply the line is a better fit than it is. The trendline is the line that results in the smallest sum of squared residuals.

For the data in the table, find the sum of the squared residuals using the equation of the trendline: Sales $= 255.95(\text{months}) + 1285.7$. Round any decimals to the nearest tenth.

Months Since Company Started	Sales ($)	Predicted Value	Residual	(Residual)^2
1	1,500	1,541.7	−41.7	1,734.7
2	2,000	1,797.6	202.4	40,965.8
3	2,100	2,053.6	46.4	2,157.6
4	2,400	2,309.5	90.5	8,190.3
5	1,800	2,565.5	−765.5	585,913.7
6	3,000	2,821.4	178.6	31,898.0
7	3,500	3,077.4	422.7	178,633.0
8	3,200	3,333.3	−133.3	17,768.9

Sum of squared residuals = 867,261.9

6. A student keeps track of how long it takes her to write her signature. She collects the data in one sitting so that she can see if there is a change in her writing rate over time. She wants to know if it takes her a longer time due to fatigue or a shorter time due to repetition. Her data are given in the table.

Number of Signatures	Time (seconds)
1	4.42
2	8.66
3	13.51
4	16.26
5	23.6
6	28.11
7	30.17
8	32.56
9	33.68
10	35.87

When using Excel to find the equation of the trendline, she has an option for the type of model. She chooses a linear model but notices that it doesn't seem to model the pattern closely. Next, she chooses polynomial models of different degrees to see how closely they fit the points. Each graph and its equation follow. Use each equation to predict how long it will take to write 7 signatures. Then find the residual for each prediction.

a.

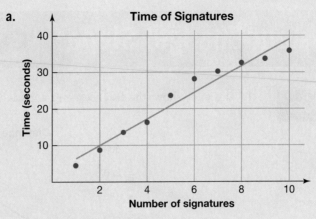

$$y = 3.6344x + 2.6947$$

Prediction = 28.1355 seconds, residual = 2.0345 seconds

b.

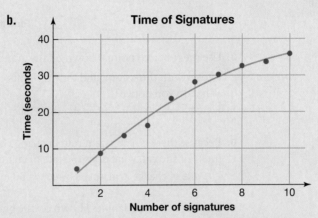

$$y = -0.2492x^2 + 6.3753x - 2.787$$

Prediction = 29.6293 seconds, residual = 0.5407 second

c.

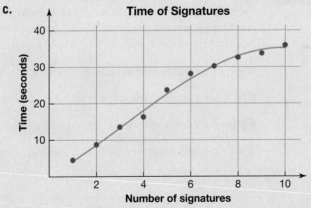

$$y = -0.0352x^3 + 0.3324x^2 + 3.6928x + 0.2373$$

Prediction = 30.3009 seconds, residual = −0.1309 second

7. **Looking Forward, Looking Back** Simplify.

$$\frac{-9 + 7}{3 - (-5)} \quad -\frac{1}{4}$$

3.3 Constant Change: Slope

Explore

SECTION OBJECTIVES:

- Find the slope of a line from points, tables, and graphs
- Interpret the slope as a rate of change

INSTRUCTOR NOTE: Consider mentioning to students that the triangles drawn are congruent when they are drawn between points with the same difference in *x*-values, but the triangles are only similar otherwise.

1. The graph shows the distance a car travels as a function of the time it travels.

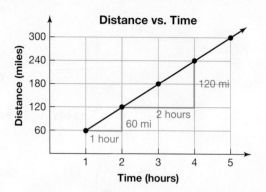

a. Use the first two points shown on the graph to find this ratio: $\frac{\text{increase in distance}}{\text{increase in time}}$.
Repeat with the second and fourth points on the graph. Simplify the ratios. What do you notice?

For both pairs of points, the ratio is 60 miles/1 hour.

b. Because there is a constant rate of change of distance to time, the points fall in a line. For the car, what does the rate of change represent?

The rate of change is its speed, which is 60 mph.

c. Explain why the triangles shown on the graph are similar.

Their sides are proportional.

Discover

The rate of change in the distance and time data in the *Explore* is known as the slope of the line through the points. The slope measures the steepness of the line but also gives the speed for the car in this context. In this section, we will explore slope in several contexts and from many perspectives: physical, numerical, and graphical. You will learn how to calculate the slope as well as how to interpret it in context.

INSTRUCTOR NOTE: The slope formula will be developed later in this section. Some students might remember the formula immediately, but the goal is to develop a conceptual foundation and connect it to ideas already developed.

As you work through this section, think about these questions:

1 Can you interpret the slope in physical contexts and as a rate of change?

2 Does it matter which point we use as the "first" point in the slope formula?

3 What is the difference between a zero slope and an undefined slope?

Slope

Slope is a numeric measure of the steepness of a line. It can be found by computing the ratio of the rise of the line to its run. Slope is usually written with the symbol *m*. Slope may be given as a fraction, decimal, or percent.

$$m = \frac{\text{rise}}{\text{run}}$$

FOR EXAMPLE, to say a road has a 5% grade means there are 5 units of vertical change for every 100 units of horizontal change. This is the same as 1 unit of rise for every 20 units of run. The units can be any measure of length such as feet, miles, or meters. The rise can be up or down, depending on whether the road is ascending or descending.

Slope also describes the rate of change of a linear relationship. For a line, the rate of change between the *y*-values and the *x*-values is constant.

FOR EXAMPLE, if a company gains 2 clients per month, the rate of change between the number of clients and months is $\frac{2 \text{ clients}}{1 \text{ month}}$.

The letter *m* is often used to represent slope, although the origin of that abbreviation is unknown. It is possibly related to *modulus of slope* or *monter*, the French word for "to climb."

Let's look at how slope can be used to measure the steepness of a line on a coordinate system, as we did in the *Explore*.

EXAMPLE

Plot the points $(-1, -1)$, $(0, 1)$, and $(1, 3)$. Find the slope of the line that passes through the points.

SOLUTION

We begin by plotting the points on a graph and drawing a line through them. The slope of the line is the ratio of rise to run between the points. Visually, rise is vertical change, and run is horizontal change. If we move to the right or up, we use positive numbers to describe the movement. If we move to the left or down, we use negative numbers. This is consistent with the approach we use when plotting points.

We can travel from $(-1, -1)$ to $(0, 1)$ to find the slope of the line. On the graph, a path with arrows is shown along with the amount and direction of the movement. Since we go right one unit and up two units, the slope is $\frac{2}{1} = 2$. Notice that when we read from left to right, the line looks like it is going up and its slope is positive.

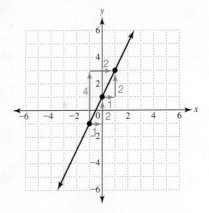

Another way to find the slope is to travel from $(0, 1)$ to $(1, 3)$. Since we go right one unit and up two units, the slope is $\frac{2}{1} = 2$. A third path can be used, starting from $(-1, -1)$ and moving to $(1, 3)$. Since we go right two units and up four units, the slope is $\frac{4}{2} = 2$.

Note that the slope of the line is the same regardless of which points on the line are used to obtain it. Also, we can travel above or below the line and get the same result for the slope. Further, we can travel up the line or down the line and the slope will be the same. If we had traveled down the line from the third point to the first point, the slope would have been $\frac{-4}{-2} = 2$, the same value we found earlier. This means that we can travel from one point to another point, starting with any point we like. We'll still get the same value for the slope as long as we pay attention to the rise and the run and in which direction we're moving so that the signs are correct.

2. Use the graph to find the slope of the line passing through $(-3, -2)$ and $(1, -5)$. Begin by plotting the points and drawing a line through them.

$-\dfrac{3}{4}$

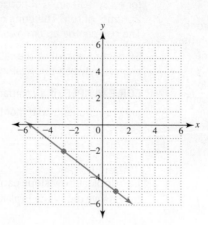

Although a graph is a helpful visual aid, we might not always want to draw the graph and plot the points to find a slope. Let's look at the coordinates of the points to find how we can use them to get the rise and run without a graph. Remember that on the Cartesian coordinate system, rise or vertical change involves y, and run or horizontal change involves x. So another way of calculating slope is

$$\text{Slope} = m = \frac{\text{rise}}{\text{run}} = \frac{\text{change in } y\text{-values}}{\text{change in } x\text{-values}} = \frac{\Delta y}{\Delta x}$$

The symbol Δ is the Greek letter delta. In mathematics, it stands for "change in."

Notice in the graph of the line passing through $(-3, 4)$ and $(2, -2)$ that the rise is -6 and the run is 5 when we move down the line. Since the rise is the vertical distance we move, that value can be found by subtracting the y-values: $-2 - 4 = -6$. Similarly, the run, which is the horizontal distance we move, can be found by subtracting the x-values. But we must be careful with the order of the subtraction. With the y-values, we started with -2 from $(2, -2)$. With the x-values, we need to start with the x-coordinate of the same point $(2, -2)$. So $2 - (-3) = 2 + 3 = 5$.

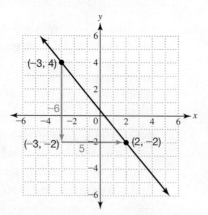

A negative sign on a fraction can be in **one** of three places: in the numerator, in the denominator, or in front of the fraction. The third option, with the negative sign in front of the fraction, such as $-\frac{6}{5}$, is the preferred form.

So $m = \dfrac{-2 - 4}{2 - (-3)} = \dfrac{-6}{5} = -\dfrac{6}{5}$.

To find the slope, we could have traveled from $(2, -2)$ up to $(-3, 4)$, giving a slope of $\frac{6}{-5}$. The fraction $\frac{-6}{5}$ is the same as $\frac{6}{-5}$, which is also the same as $-\frac{6}{5}$. They are all equivalent forms of the same number. Notice that when we read from left to right, the line looks like it is going down and its slope is negative.

Generalizing the process of using coordinates to find the rise and run, we have the graph and points shown. To specify the values of x and y needed in the slope formula, the two points are often indicated with the notation (x_1, y_1) and (x_2, y_2), as shown here.

The slope formula can then be written as follows:

$$m = \frac{y_2 - y_1}{x_2 - x_1}$$

We can use the formula with the points $(-3, 4)$ and $(2, -2)$. If we consider $(-3, 4)$ to be the first point, then $x_1 = -3$ and $y_1 = 4$. If we consider $(2, -2)$ to be the second point, then $x_2 = 2$ and $y_2 = -2$. We get the following calculation and result.

$(-3, 4)$ and $(2, -2)$

$$m = \frac{y_2 - y_1}{x_2 - x_1} = \frac{-2 - 4}{2 - (-3)} = \frac{-6}{5} = -\frac{6}{5}$$

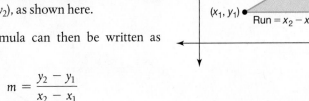

HOW IT WORKS

To find the slope of a line using two points on the line and the formula $m = \frac{y_2 - y_1}{x_2 - x_1}$:

1. Choose which of the points is the first point and which is the second.
2. Subtract the y-values. The result becomes the rise or numerator.
3. Subtract the x-values, using the same order for the subtraction. This result becomes the run or denominator.
4. Simplify the fraction.

EXAMPLE: Find the slope of the line passing through $(-5, -7)$ and $(1, 2)$.

To find the slope, the point $(-5, -7)$ can be treated as (x_1, y_1) and the point $(1, 2)$ as (x_2, y_2). The slope is then calculated as follows:

$$m = \frac{2 - (-7)}{1 - (-5)} = \frac{2 + 7}{1 + 5} = \frac{9}{6} = \frac{3}{2} \quad \text{Follow the correct order of operations to simplify.}$$

Alternatively, the order of the points can be reversed and the slope can be found by using

$$m = \frac{-7 - 2}{-5 - 1} = \frac{-9}{-6} = \frac{3}{2}$$

As long as there is consistency with the order of the subtraction in the numerator and denominator, the result will be the same.

Need More Practice?

Find the slope of the line passing through each pair of points.

1. $(-3, 10)$ and $(7, -6)$
2. $(5, -8)$ and $(-8, 5)$

Answers:

1. $-\dfrac{8}{5}$ 2. -1

Subtract in an order that makes the calculations easy. If you need to find the slope through $(6, 8)$ and $(1, 6)$, it is easier to subtract the second point from the first, $m = \frac{8 - 6}{6 - 1} = \frac{2}{5}$, than to subtract the first from the second, $m = \frac{6 - 8}{1 - 6} = \frac{-2}{-5} = \frac{2}{5}$.

3. For each pair of points, find the slope of the line that passes through them by using the slope formula. Draw a sketch of the points and the line connecting them to help you make sense of the slope.

a. $(1, 1)$ and $(6, 4)$

$$m = \frac{3}{5}$$

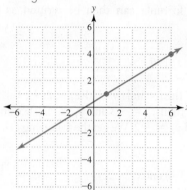

b. $(-3, 5)$ and $(3, -5)$

$$m = \frac{-10}{6} = -\frac{5}{3}$$

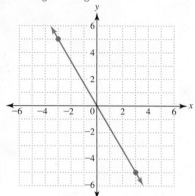

c. $(2, 3)$ and $(5, 3)$

$$m = \frac{0}{3} = 0$$

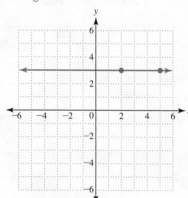

d. $(2, 3)$ and $(2, 5)$

$$m = \frac{2}{0} = \text{undefined}$$

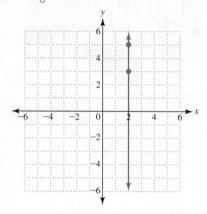

Lines that are increasing from left to right have positive slopes, and lines that are decreasing have negative slopes. Vertical lines do not have a slope. Instead, we state that the slope is undefined. Horizontal lines do have a slope, but its value is zero.

> In Section 3.2, you learned about correlation. Trendlines that model positive correlation have a positive slope. Likewise, trendlines that model negative correlation have a negative slope.

Slope	Line
Positive	Increasing
Negative	Decreasing
Zero	Horizontal
Undefined	Vertical

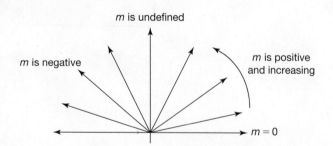

Keep these properties in mind when finding a slope. If you get a negative slope but the graph shows that the line through the points is increasing, that indicates a mistake in your calculation or graph.

In the definition of slope, the rise is in the numerator. Let's look at why that is the case.

Roofs are described in terms of their steepness or pitch. A $\frac{5}{12}$ pitch means 5 inches of vertical rise for every 12 inches or foot of horizontal run.

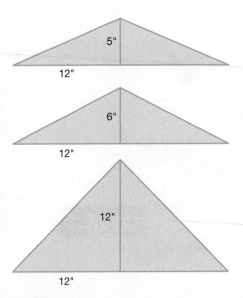

Notice in the figures that the roof with the $\frac{5}{12}$ pitch is the flattest of the three roofs and the roof with the $\frac{12}{12}$ pitch is the steepest. The roof with the $\frac{6}{12}$ pitch is between the other two in steepness. Numerically, $\frac{5}{12} < \frac{6}{12} < \frac{12}{12}$. The $\frac{12}{12}$ roof is the steepest because it has the greatest rise of the three for the same run of 12 inches. As the roof gets steeper, the value of the slope increases. Intuitively, it makes sense for the greatest slope to be associated with the steepest roof. For this to happen, the rise must be in the numerator.

In applications, we want to know how the dependent variable is changing relative to the independent variable, not the other way around. So we want the change in the y-values to be in the numerator of the slope fraction and the change in the x-values to be in the denominator.

4. A recommendation for wheelchair ramps is that the rise-to-run ratio be no greater than 1:12. This ensures that the ramps are navigable by people in wheelchairs. If a business needs to install a concrete ramp to the entrance of their building, which is 12 inches off the ground, how long should the ramp be and how far out horizontally on the ground will it protrude from the building? Give answers in feet and inches, such as 5 feet, 4 inches. What percent grade is the ramp?

   The ramp should be approximately 12 feet, 0.5 inch long and protrude 12 feet out from the building; 8.3%.

So far, we have found the slopes of lines on a graph as well as the slopes of physical objects like ramps, roads, and roofs. Let's look at a context that involves slope but not as obviously. In the next problem, you will explore slope as the rate of change.

5. A hair stylist is analyzing her profits, which were uneven when she began her business. In the third year, she had an annual profit of $45,000. Then her business began to stabilize, and each year after that her profit increased by the same amount each year. Currently, she is in her eighth year and has $61,000 in profit. At what rate is her annual profit increasing?

Her profit is increasing by $3,200 each year.

Connect

In Cycle 1, you learned how to look at the change in the y-values of a set of points and determine if the points show a relationship that is linear, exponential, or neither. In each set of points, the x-values were consecutive integers like those shown in the table.

x	y
1	−5
2	6
3	17
4	28

$+1$ between x values and $+11$ between y values for each step.

To check for linearity, we looked for a constant amount added to or subtracted from the y-values. In the previous table, the y-values increase by 11 each time the x-values increase by 1, which indicates a linear relationship. Now, with an understanding of slope, you can assess linearity even if the points do not have consecutive integer x-values. Since lines have a constant rate of change between the points on them, a constant slope is enough to determine if the points fall in a line.

Notice in the table that follows that the x-values are increasing by 3. This is the change in the x's, or the run between points on the graph. The y-values are decreasing by 5. This is the change in the y's, or the rise between points on the graph. The slope is just the ratio of those changes.

> Consider using a table to find the rate of change instead of graphing the points or using the slope formula.

x	y
−12	12
−9	7
−6	2
−3	−3

$+3$ between x values and -5 between y values for each step.

So the slope is $m = \dfrac{\Delta y}{\Delta x} = \dfrac{-5}{3} = -\dfrac{5}{3}$. Since the rate of change is constant, the points fall in a line. So we know the pattern of the points from the rates of change of the x's and y's, and we know that the line will appear to be going down when we read from left to right, since the slope is negative.

6. For each set of points, determine if the data is linear without drawing a graph. If the data is linear, give the rate of change.

a.

x	y
11.26	−11
27.06	−7.6
42.86	−4.2
58.66	−1

No

b.

x	y
−140	0
−123	5
−106	10
−89	15

Yes; $m = \dfrac{5}{17}$

Reflect

WRAP-UP

What's the point?

Slope describes the constant rate of change between two variables in a linear relationship, such as a steady increase in the number of inches a tree grows each year. It is also a measure of the steepness of a line and the incline of physical objects like roads and ramps.

What did you learn?

How to find the slope of a line from points, tables, and graphs
How to interpret the slope as a rate of change

3.3 Homework

Skills MyMathLab

First complete the MyMathLab homework online. Then work the two exercises to check your understanding.

1. a. Find the slope of the line that passes through (−8, 11) and (2, −3). $-\dfrac{7}{5}$

b. Find the slope of the line shown in the graph. $-\dfrac{3}{5}$

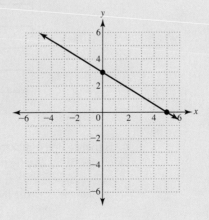

2. Find the rate of change in the following table: -5

x	y
−2	15
−1	10
0	5
1	0

Concepts and Applications

Complete the problems to practice applying the skills and concepts learned in the section.

3. In Section 2.8, "Seat Yourself," we modeled the number of chairs that could be set around tables placed adjacent to one another. Two examples of that configuration are shown here.

Number of tables: 2

Number of tables: 3

a. Complete the following sentence to interpret the rate of change in the number of chairs: Every time one table is added, _the number of chairs increases by 2_ _____ .

b. Complete the following table:

Number of Tables Arranged in a Line	Number of Chairs
1	4
2	6
3	8
4	10
5	12
6	14
10	22
11	24

c. Is there a linear relationship between the number of tables and the number of chairs? Why or why not?

Yes; the relationship is linear. Every time the number of tables increases by 1, the number of chairs increases by 2.

d. Plot this data as points on a coordinate system.

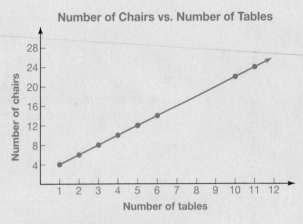

Number of Chairs vs. Number of Tables

e. Use the graph to find the slope of the line.

2

f. Pick two points from the table and find the slope by using the slope formula.

$m = 2$

g. Interpret the slope in words.

For every 1 table added, 2 chairs are added.

4. Building codes typically require 7 inches in rise for every 11 inches in tread for standard stairs. Although stairs can have up to 8 inches in rise and 12 inches in tread, each change makes the stairs less comfortable to navigate. In a new home being built, the builder needs to construct stairs to code for a lower deck. The dimensions of the full staircase will be 70 inches vertically and 110 inches horizontally.

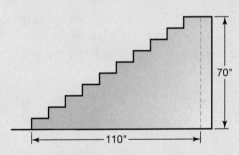

a. Find the slope of the overall staircase. $\dfrac{7}{11}$

b. Find the slope of each step. $\dfrac{7}{11}$

c. If a banister runs diagonally the full length of the staircase, find its length in feet and inches.   Approximately 10 feet, 10 inches

5. Guidelines for wheelchair ramps used inside and outside of buildings specify a preferred slope of 1:20, but 1:16 and 1:12 are also allowable. Which ramp slope is steepest, and which is flattest? Explain your choice.

1:12 is steepest because it has the shortest run of the three ratios for the same rise of 1.

As fractions, $\dfrac{1}{12}$ is the largest, then $\dfrac{1}{16}$, then $\dfrac{1}{20}$.

6. Imagine a road on which you lose 250 feet in elevation for every mile driven.

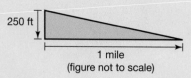

250 ft

1 mile
(figure not to scale)

a. Draw pictures that show a loss of 500 feet in elevation over 2 miles of horizontal distance and 1,000 feet over 4 miles of horizontal distance.

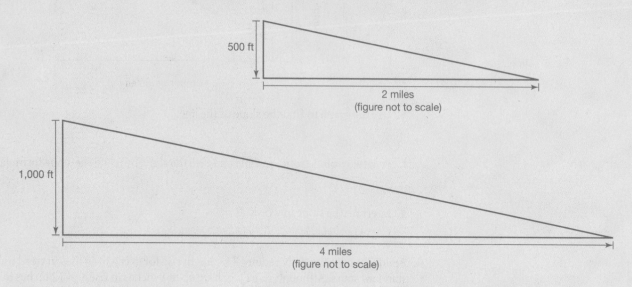

500 ft

2 miles
(figure not to scale)

1,000 ft

4 miles
(figure not to scale)

b. Find the slope of the road, using each of the three triangles. Simplify each slope. What do you notice?

1. Slope $= \dfrac{-250 \text{ ft}}{1 \text{ mile}} = -250$ ft/mile   2. Slope $= \dfrac{-500 \text{ ft}}{2 \text{ miles}} = -250$ ft/mile

3. Slope $= \dfrac{-1,000 \text{ ft}}{4 \text{ mile}} = -250$ ft/mile

We get the same slope no matter which triangle is used.

c. A 6% grade is the maximum steepness that a truck is able to navigate under company safety guidelines. Can the truck drive on this road?

Yes; the slope is 1,000 feet/4 miles or 1,000 feet/21,120 feet ≈ 4.7%. The road is within the grade guidelines.

7. Students were asked to calculate the slope of the line that connects the points $(-2, 4)$ and $(2, 10)$. Explain the mistake in each student's work.

a. $m = \dfrac{2 - -2}{10 - 4} = \dfrac{4}{6} = \dfrac{2}{3}$

The student put the change in x on the top and the change in y on the bottom.

b. $m = \dfrac{10 - 4}{-2 - 2} = \dfrac{6}{-4} = -\dfrac{2}{3}$

The student subtracted in a different order in the numerator than in the denominator.

c. $m = \dfrac{10 - 4}{2 - -2} = \dfrac{6}{4}$

The student did not simplify the final fraction to $\dfrac{3}{2}$.

8. a. Find the slope of the line shown in the graph by traveling from $(4, -3)$ up to $(-2, 5)$.

$m = \dfrac{4}{-3} = -\dfrac{4}{3}$

b. Find the slope of the line shown by traveling from $(-2, 5)$ down to $(4, -3)$.

$m = \dfrac{-4}{3} = -\dfrac{4}{3}$

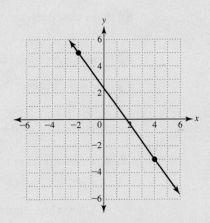

9. Jana's son had a growth spurt that started at age 12 and continued until he was 15 years old. He grew from 5'3" to 6'2". What was his average growth per year?

3.67 inches

10. Looking Forward, Looking Back For each triangle in homework problem #6, find the length of the hypotenuse. Round your answer to the nearest tenth.

First triangle: 5,285.9 feet; second triangle: 10,571.8 feet; third triangle: 21,143.7 feet

3.4   Shortest Path: Distance Formula

Explore

SECTION OBJECTIVE:

• Use the distance formula to find the distance between two points

1. Assume the desks in your classroom are all 5 feet apart, measured from the centers. You have one friend who sits 2 desks ahead and 2 desks to the right of you. You have another friend who sits 3 desks behind and 1 desk to the left of you. Determine how far apart these two friends' desks are in feet. Use the grid provided to draw a sketch of the desk locations.

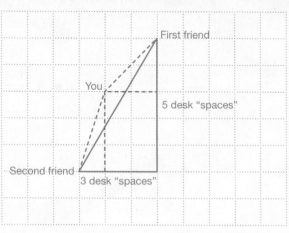

Approximately 5.83 desk spaces, or 29.15 feet apart

Discover

Some distances, like the one in the *Explore*, can be found by a straightforward application of the Pythagorean theorem. However, there are situations in which you want to find a distance but don't necessarily want to draw and label a right triangle. In that case it's more convenient to use a formula to calculate a distance. This section will focus on developing such a distance formula.

As you work through this section, try to answer the following questions:

1 When should you use the Pythagorean theorem and when should you use the distance formula?

2 In what order do you perform the operations to simplify the distance formula?

3 How is the distance formula related to the slope formula?

Let's begin by finding horizontal and vertical distances using coordinates.

2. a. Plot the points $(4, 5)$ and $(-3, 5)$.

b. Draw a line segment connecting them.

c. Find the distance between the points. The distance in units is the length of the line segment that connects the points.

7 units

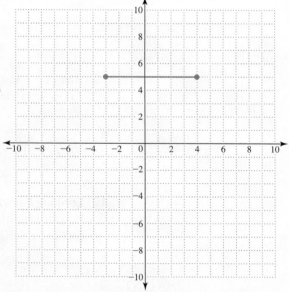

You probably found the distance between the points in #2 by simply counting squares on the grid. Notice, however, that you could have found the distance by subtracting the *x*-coordinates of the points.

3. a. Plot the points (6, −3) and (6, 8).

b. Draw a line segment connecting them.

c. Find the distance between the points. Give the distance in units.
11 units

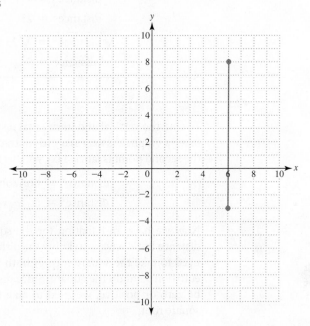

Notice in #3 that you could have found the distance by subtracting the *y*-coordinates of the points. If the points are aligned vertically or horizontally, we can simply count squares or subtract coordinates to find the distance. Next we will consider how to find the distance between two points that are not aligned vertically or horizontally.

EXAMPLE

Find the distance between the points (1, 5) and (4, 9).

SOLUTION

We begin by plotting the points and connecting them with a line segment. Since this line segment is on a slant and not vertical or horizontal, we can't find its length by simply counting squares. We can make this line segment the hypotenuse of a right triangle by adding vertical and horizontal segments to form a right triangle as shown. We can then find the distance by using the Pythagorean theorem. It's important to notice how we find the lengths of the legs of this right triangle.

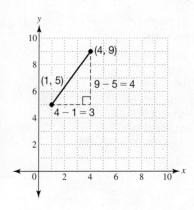

To find the length of the vertical leg, we subtract the y-coordinates of the points: $9 - 5 = 4$. To find the length of the horizontal leg, we subtract the x-coordinates of the points: $4 - 1 = 3$. We can then find the distance between the points using the Pythagorean theorem:

$$\text{hypotenuse}^2 = \text{leg}^2 + \text{leg}^2$$
$$\text{distance}^2 = \text{horizontal distance}^2 + \text{vertical distance}^2$$
$$\text{distance}^2 = 3^2 + 4^2$$
$$\text{distance}^2 = 9 + 16$$
$$\text{distance}^2 = 25$$
$$\sqrt{\text{distance}^2} = \sqrt{25}$$
$$\text{distance} = 5$$

To find the vertical and horizontal distances involved in the distance calculation without using a graph, it is necessary to subtract coordinates. If the first point is labeled (x_1, y_1) and the second point is labeled (x_2, y_2), this distance formula becomes

$$\text{distance}^2 = \text{horizontal distance}^2 + \text{vertical distance}^2$$
$$\text{distance}^2 = (x_2 - x_1)^2 + (y_2 - y_1)^2$$

Now we have an equation for the square of the distance but we want to know the distance. So we need to take the square root of both sides of the equation to undo the square. Since distance must be zero or positive, we need to take the positive square root.

$$\text{distance} = \sqrt{(x_2 - x_1)^2 + (y_2 - y_1)^2}$$

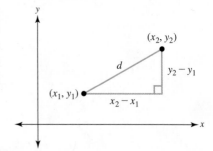

HOW IT WORKS

To find the distance, d, between two points (x_1, y_1) and (x_2, y_2), use the distance formula:

$$d = \sqrt{(x_2 - x_1)^2 + (y_2 - y_1)^2}$$

EXAMPLE: Find the distance between $(-1, 4)$ and $(11, -1)$.

To find the distance, substitute the coordinates from the ordered pairs into the distance formula and simplify.

$$d = \sqrt{(11 - (-1))^2 + (-1 - 4)^2} \qquad \text{Substitute the coordinates into the formula. Subtract.}$$

$$d = \sqrt{(12)^2 + (-5)^2} \qquad \text{Square.}$$

$$d = \sqrt{144 + 25} \qquad \text{Add.}$$

$$d = \sqrt{169} \qquad \text{Square root.}$$

$$d = 13$$

So the distance between the points is 13 units.

Be careful with negative numbers!

Squaring a real number never results in a negative number. If you have a negative number under the square root sign after you've squared both numbers, you have made a mistake.

Remember to add before finding the square root.

4. Use the distance formula to find the distance between $(2, 10)$ and $(-5, 4)$.

$d \approx 9.22$ units

5. Find the length of the diagonal of a rectangle if the rectangle has a length of 10 inches and a width of 5 inches.

$d \approx 11.18$ inches

Connect

We have now seen both slope and distance from several perspectives.

In general, given the points (x_1, y_1) and (x_2, y_2) on a graph, we can draw a right triangle and label the sides, as shown in the graph.

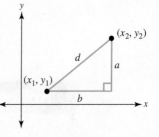

Doing so allows us to write expressions for the distance, d, between the points and the slope, m, of the line that passes through the points. There are multiple expressions that can be written for distance and slope.

6. a. Write an expression for the distance and an expression for the slope using a and b from the graph.

$$d = \sqrt{a^2 + b^2}$$

$$m = \frac{a}{b}$$

b. Write an expression for the distance and an expression for the slope using the coordinates shown in the graph.

$$d = \sqrt{(x_2 - x_1)^2 + (y_2 - y_1)^2}$$

$$m = \frac{y_2 - y_1}{x_2 - x_1}$$

Since the distance formula can look intimidating, write the shell of the formula to get started:

$$d = \sqrt{(\ \ )^2 + (\ \ )^2}$$

Then fill in the *x*'s and *y*'s.

Or draw a graph with the points and use the Pythagorean theorem directly.

Remember: The distance formula is just a version of the Pythagorean theorem.

7. Find the distance between these points as well as the slope of the line connecting them: $(-5, 3), (7, -1)$. Label the points, draw the line, and label the distance on the graph.

$$m = -\frac{1}{3}; d \approx 12.6 \text{ units}$$

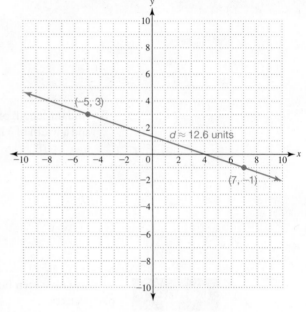

Reflect

WRAP-UP

What's the point?

The Pythagorean theorem and the distance formula are two forms of the same relationship. It's not always practical to find a distance by first drawing a right triangle. Knowing the distance formula provides a fast, accurate way to find distance when you do not want to or are not able to find it physically by measurement.

What did you learn?

How to use the distance formula to find the distance between two points

3.4 Homework

Skills MyMathLab

First complete the MyMathLab homework online. Then work the two exercises to check your understanding.

1. Find the distance between these points and round to the nearest tenth: (2, 6) and (3, 18).

12.0 units

2. If you drive 5 miles east and then 4 miles south, how far are you, to the nearest tenth of a mile, from your starting point?

   6.4 miles

Concepts and Applications

Complete the problems to practice applying the skills and concepts learned in the section.

3. Students are trying to find the distance between the two points $(2, 18)$ and $(-1, 16)$. Find the mistake in each student's work:

   Student 1:

   $$\begin{aligned} \text{Distance} &= (-1 - 2)^2 + (16 - 18)^2 \\ &= (-3)^2 + (-2)^2 \\ &= 9 + 4 \\ &= 13 \end{aligned}$$

   This student forgot the square root.

   Student 2:

   $$\begin{aligned} \text{Distance} &= \sqrt{(2 - 1)^2 + (18 - 16)^2} \\ &= \sqrt{(1)^2 + (2)^2} \\ &= \sqrt{1 + 4} \\ &= \sqrt{5} \end{aligned}$$

   This student subtracted 1 instead of -1 for the x-coordinates.

   Student 3:

   $$\begin{aligned} \text{Distance} &= \sqrt{(2 - -1)^2 + (18 - 16)^2} \\ &= \sqrt{(3)^2 + (2)^2} \\ &= 3 + 2 \\ &= 5 \end{aligned}$$

   This student did not do the operations in the correct order. He needed to square and add before taking the square root.

   Show how to correctly find the distance between the two given points.

   $$\begin{aligned} \text{Distance} &= \sqrt{(-1 - 2)^2 + (16 - 18)^2} \\ &= \sqrt{(-3)^2 + (-2)^2} \\ &= \sqrt{9 + 4} \\ &= \sqrt{13} \approx 3.6 \end{aligned}$$

4. Find the distance between these points in two ways: $(-2, 5)$ and $(4, 13)$.

   a. Use the point $(-2, 5)$ as (x_1, y_1) and the point $(4, 13)$ as (x_2, y_2) in the distance formula. $d = 10$

   b. Use the point $(4, 13)$ as (x_1, y_1) and the point $(-2, 5)$ as (x_2, y_2) in the distance formula. $d = 10$

c. If your answers are the same, explain why that happened. If your answers are not the same, look for a mistake in part a or b.

When you square the differences in the distance formula, you get the same positive number regardless of the order of the subtraction.

5. Find the length and slope of the segment that connects the points in the graph.

Length $\approx$ 12.04; slope $= \dfrac{9}{8}$

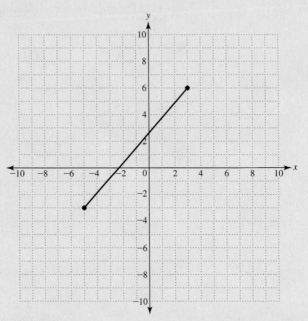

6. Find four points that are 10 units away from the point (1, 6). Consider the fact that 3, 4, 5 is a Pythagorean triple.

Answers will vary. Possible answers include (7, 14), (9, 12), (−5, −2), and (−7, 0).

7. Another student explains to you that the order of the subtraction doesn't really matter in either the slope or the distance formula. Explain whether his statement is correct.

The order of the subtraction does not matter in the distance formula; the same answer is obtained regardless of the order. The order of the subtraction does not matter in the slope formula, but the same order must be used in both the numerator and the denominator.

8. Looking Forward, Looking Back Complete the Pythagorean triple: 21, 72, _____.

75

3.5 More or Less: Linear Relationships

SECTION OBJECTIVE:

- Make comparisons using equations, tables, and graphs

1. Ava has multiple coupon codes to use for discounts at an online clothing store. She is debating between using a $15-off coupon or a 20%-off coupon but is not sure which will give her the greatest discount. For what range of prices is each coupon best?

If she buys less than $75, the $15-off coupon gives a greater discount. If she buys more than $75, the 20%-off coupon gives a greater discount. If she buys exactly $75, the discount is the same with either coupon.

Discover

The problem in the *Explore* is an example of a comparison problem, in which we need to judge when one option is better than another. We have already seen problems like this when we compared pay raises in Section 1.12 and in this cycle's focus problem. When you approach comparison problems, you may use a guess-and-check strategy, but that approach can be tedious and time consuming. In this section you will learn three ways to solve comparison problems using skills you already have.

As you work through this section, think about these questions:

1 What do the numeric and graphic methods provide that the algebraic method does not?

2 Which method—numeric, algebraic, or graphic—makes the most sense to you?

Let's look at a comparison problem from each of following perspectives: numeric, algebraic, and graphic.

INSTRUCTOR NOTE: The primary example of the section is lengthy since it illustrates three ways to solve the problem. It allows you to focus on analyzing the tables, equations, and graphs instead of constructing and analyzing them. Students have multiple problems to work after the example to practice applying the ideas in it.

EXAMPLE

With gas prices high, consumers often shop around for the lowest price. Some gas stations offer per-gallon discounts with the purchase of a car wash. At your local gas station, these options are available:

Option 1: $3.65 per gallon with a $5.50 car wash
Option 2: $3.85 per gallon

Use numeric, algebraic, and graphic methods to analyze the options and determine if and when it is worth it to get the car wash. Assume the driver does not care about the car wash but is interested only in paying less for gas.

SOLUTION

We will use three different approaches to solve the problem. Each approach offers something unique.

Numeric approach: To use a numeric approach, we pick several gas volumes and find the cost with each option. This approach is sometimes referred to as "running the numbers." We use a table to display the results. A table shows the costs in an easy-to-read way, but it may take some refinement of our choices of gallons to find exactly when one option becomes a better choice than the other. The last line of the table shows the generalized version of the cost calculation under each option.

INSTRUCTOR NOTE: Consider loading Excel and using it to create the data shown. To increase student interaction, consider asking students to find the cost of 15 gallons under each option. Ask students to use slope to determine if the options are linear. Also, can they see approximately when each option is preferable?

Option 1
$3.65 per gallon of gas with a $5.50 car wash

Gallons	Calculation	Cost
0	3.65(0) + 5.50	$5.50
1	3.65(1) + 5.50	$9.15
2	3.65(2) + 5.50	$12.80
3	3.65(3) + 5.50	$16.45
10	3.65(10) + 5.50	$42.00
20	3.65(20) + 5.50	$78.50
25	3.65(25) + 5.50	$96.75
30	3.65(30) + 5.50	$115.00
35	3.65(35) + 5.50	$133.25
g	$3.65g + 5.50$	

Option 2
$3.85 per gallon of gas

Gallons	Calculation	Cost
0	3.85(0)	$0.00
1	3.85(1)	$3.85
2	3.85(2)	$7.70
3	3.85(3)	$11.55
10	3.85(10)	$38.50
20	3.85(20)	$77.00
25	3.85(25)	$96.25
30	3.85(30)	$115.50
35	3.85(35)	$134.75
g	$3.85g$	

To make the process of completing the table faster, consider using Excel. You can find each cost by beginning with the starting value, either $5.50 or $0, and adding the cost of an additional gallon using the *Fill Down* command. This approach finds a cost, the dependent variable in this scenario, using a previous cost. Another option is to find each cost in terms of the number of gallons and then use the *Fill Down* command. This approach finds a cost in terms of the number of gallons, the independent variable in this scenario. Specific instructions for these and other spreadsheet commands are provided in the appendix.

Before we determine when each option is preferable, let's look more closely at the relationships shown in the table. Understanding them will help us as we work with the problem.

The last line of the table shows an expression for the cost, which is a generalized form of the calculations. But we can get to the expression for the cost without completing a table if we notice how the cost is calculated. Every time a gallon of gas is added, the price with Option 1 goes up by the cost of the gallon of gas, $3.65. Similarly, every time a gallon of gas is added, the price with Option 2 goes up by the cost of the gallon of gas, $3.85. In each case, there is a constant rate of change, which means each cost option is a linear function of the number of gallons of gas. Each cost calculation follows the format of a linear function: *Starting value* + *rate of change* · *independent variable*. Before we plot any points, we know the graph will be a line.

Based on the values in the tables, Option 2 appears to be less expensive for fewer gallons of gas, and Option 1 seems to be less expensive for more gallons of gas. But where is the exact cutoff between "fewer" and "more" gallons? Notice that the difference in the costs lessens as we near 25 gallons. Between 25 and 30 gallons, we switch which option costs less. In that range of gallons will be a number of gallons at which the costs are the same. This is known as the **break-even point**.

<table>
<tr><td colspan="2" align="center">Option 1</td><td colspan="2" align="center">Option 2</td></tr>
<tr><td>*g*</td><td>*C*</td><td>*g*</td><td>*C*</td></tr>
<tr><td>25</td><td>$96.75</td><td>25</td><td>$96.25</td></tr>
<tr><td>27.5</td><td>$105.88</td><td>27.5</td><td>$105.88</td></tr>
<tr><td>30</td><td>$115.00</td><td>30</td><td>$115.50</td></tr>
</table>

To determine numerically the number of gallons for which both options have the same cost, consider that you save $0.20 for each gallon of gas with Option 1 but you have to pay an extra $5.50 for the car wash. So we need to know how many gallons to purchase to make up for the cost of the car wash. Dividing $5.50 by $0.20, we get 27.5 gallons.

Therefore, if you are buying more than 27.5 gallons of gas, you will pay less for the gas if you also get the car wash.

The numeric approach provides a way to see approximately when each option is better, but finding the exact break-even point may require additional calculations. To quickly find the break-even point, the algebraic method might be preferred.

Algebraic approach: Each of the generalized cost expressions in the table can be made into a cost function or model by setting the expression equal to a cost variable.

Cost with Option 1: $C = 3.65g + 5.50$ Cost with Option 2: $C = 3.85g$

To use algebra to find the break-even point, we need an equation stating that the cost is the same for the two options. Substituting the expressions that compute the costs, we get the following equation. When we solve it, the result is 27.5 gallons, which is the same value as when we approached the problem numerically.

$$\text{Cost with Option 1} = \text{Cost with Option 2}$$
$$3.65g + 5.50 = 3.85g$$
$$5.50 = 0.20g$$
$$27.5 = g$$

We can verify that the number of gallons is correct by finding the cost of this number of gallons under each cost option.

Cost with Option 1:   Cost with Option 2:

$\quad C = 3.65g + 5.50$ $\quad C = 3.85g$

$\quad = 3.65(27.5) + 5.50$ $\quad = 3.85(27.5)$

$\quad = 105.88$ $\quad = 105.88$

Each option gives a cost of $105.88 for 27.5 gallons of gas.

The algebraic approach provides the point at which the costs are the same, but it does not show when each option is better. The table shows that the first option is better for values of *g* greater than 27.5 gallons and the second option is better for values of *g* less than 27.5 gallons. Another approach, which uses graphs, also allows us to see when one option is better than another.

Graphic approach: A graph can also be used to solve this problem with the gas-pricing options. To draw the graph, we need to graph each line on the same coordinate system so that we can compare them. We can plot points, which are already available in the table. Another option is to use Excel to create a graph. Earlier, we noticed that each cost option is a linear function of the number of gallons of gas. Since the functions are linear, lines can be drawn through each set of points. We get the following graph. Since the cost of gas depends on the number of gallons purchased, the number of gallons is the independent variable and is graphed on the horizontal axis. The cost is the dependent variable and is graphed on the vertical axis. Ordered pairs are in the form (g, C).

INSTRUCTOR NOTE: Consider loading Excel and have it graph the data. Ask students if they can see when each option is preferable.

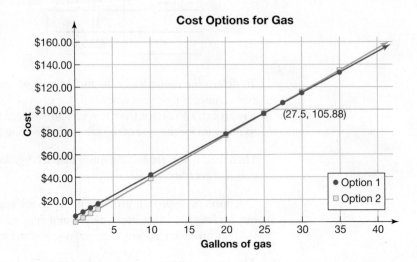

We know from the algebraic approach that the break-even point is 27.5 gallons at a cost of $105.88. We have seen that this point, (27.5, 105.88), appears in both tables. It is also on both lines, making it the point at which the lines intersect.

Until $g = 27.5$, the points for Option 2 are vertically below those for Option 1, which means that the costs are lower with Option 2 until 27.5 gallons. However, the differences between the costs of the two options get smaller near 27.5 gallons. This graph is difficult to read and use to discern the better option because the lines are so close to each other. This is another reason why multiple approaches to problems like this are valuable.

Summarizing, if you need fewer than 27.5 gallons, don't get the car wash. If you need more than 27.5 gallons, the car wash is worth it for the lower cost per gallon of gas.

Let's use the three methods to solve some other comparison problems.

2. Dena and her sister Emily own a small home-based business creating gift boxes with special wrapping details on each item in a box. Emily can pack a box in 6 minutes. Dena works faster, at 4 minutes per box, but she has to join their packaging session 20 minutes after Emily begins.

   a. How much time will Emily have worked when Dena has completed as many boxes as Emily has? How many boxes will each have completed at that point? Use the graph shown to answer the questions.

   b. Write time functions to describe each line.

   c. Write an equation to find when Dena and Emily have packed the same number of boxes, and solve it to verify your answer from the graph.

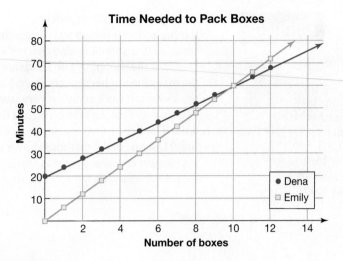

a. 60 minutes; 10 boxes

b. Emily: $T = 6n$, where n is the number of boxes and T is the amount of time in minutes she has worked

   Dena: $T = 20 + 4n$

c. $6n = 20 + 4n$; after 1 hour, each will have packed 10 boxes.

INSTRUCTOR NOTE: Consider asking students if they can determine why the second store's prices are always lower than the first store's prices in #3.

3. Nate's wife recently had triplets. Knowing that diapers are going to be a part of their lives for a few years, he looks into options for buying them inexpensively. A jumbo box of a brand he likes costs $30 in a town with a 6% sales tax. At another store in a town nearby, he can buy the same jumbo box for $26 with a 7.5% sales tax. If he buys online, he can get a jumbo box for $25 with no sales tax, but there is a flat $8.50 shipping cost if he buys one or several boxes. Nate uses Excel to look at the numbers. Which option is best and for how many boxes of diapers?

	A	B	C	D
	Number of boxes	First store ($30/box + 6% sales tax)	Second store ($26/box + 7.5% sales tax)	Online store ($25/box + $8.50 flat rate shipping)
2	1	$31.80	$27.95	$33.50
3	2	$63.60	$55.90	$58.50
4	3	$95.40	$83.85	$83.50
5	4	$127.20	$111.80	$108.50
6	5	$159.00	$139.75	$133.50
7	6	$190.80	$167.70	$158.50
8	7	$222.60	$195.65	$183.50
9	8	$254.40	$223.60	$208.50
10	9	$286.20	$251.55	$233.50
11	10	$318.00	$279.50	$258.50
12	11	$349.80	$307.45	$283.50
13	12	$381.60	$335.40	$308.50
14	13	$413.40	$363.35	$333.50
15	14	$445.20	$391.30	$358.50
16	15	$477.00	$419.25	$383.50
17	16	$508.80	$447.20	$408.50
18	17	$540.60	$475.15	$433.50
19	18	$572.40	$503.10	$458.50
20	19	$604.20	$531.05	$483.50
21	20	$636.00	$559.00	$508.50
22	21	$667.80	$586.95	$533.50
23	22	$699.60	$614.90	$558.50

If he needs only one or two boxes, it is cheaper to go to the second store. If he buys three or more boxes at one time, the online store is the cheapest option.

Connect

Do you prefer algebraic or numeric approaches? Does seeing a graph help you visualize information? Knowing how you prefer to solve problems can help you be successful in this course.

4. Lauren spends $140 on groceries per weekly trip on average. She is considering joining a shopping club with prices that are 20% lower than the prices at her grocery store, but the shopping club charges a $45 annual fee to join the club. Lauren has looked at the shopping club's inventory and finds that she can't buy everything she normally buys at her grocery store. She expects she will still need to spend $20 of the $140 each week at her grocery store if she joins the club and buys the rest of her items there. Will she save enough at the shopping club to make up the cost of the annual fee before the year is over, or should she buy all her items at her grocery store only? Use one of the methods shown in this section to answer the question.

Yes, she will save enough at the shopping club, but it will take two visits before she recoups the cost of the annual fee.

Reflect

WRAP-UP

What's the point?

We can solve problems in which a comparison is being made with equations, tables, and graphs. Each method allows particular insights. Think about which method is the clearest and most helpful to you. If the numeric approach does not make as much sense, the algebraic or graphic method might be better for you.

What did you learn?

How to make comparisons using equations, tables, and graphs

Focus Problem Check-In
Use what you have learned so far in this cycle to work toward a solution on the focus problem. If you are working in a group, work with your group members to create a list of any remaining tasks that need to be completed. Determine who will do each task, what can be done now, and what will have to wait until you have learned more.

3.5 Homework

First complete the MyMathLab homework online. Then work the two exercises to check your understanding.

1. Complete the table by continuing the pattern. Then use the values in the table to create a graph.

x	y
0	−7
1	−4
2	−1
3	2
4	5
5	8

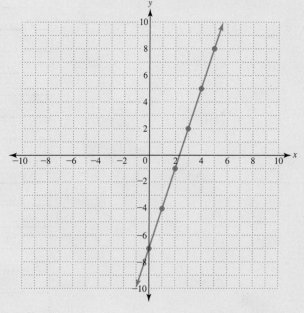

2. A store charges $12.50 for 5 boxes of cookies. A second store charges $8.00 for 4 boxes of cookies, but you have to buy a gallon of milk for $3.00 to get that deal. Write the cost model equations for both stores, and find the number of boxes of cookies that makes the costs equal.

   6 boxes

Concepts and Applications

Complete the problems to practice applying the skills and concepts learned in the section.

3. Amazon offers a Prime membership for $99 per year, which qualifies a buyer for a reduced shipping rate. With the Prime membership, one-day shipping costs as little as $2.99 per shipment, but for many items the cost is $3.99 per shipment. Without the membership, it is $10.99 per shipment. How many shipments would you need to order in one year to make it worth paying the $99 fee? Assume the higher possible shipment cost with the membership.

a. Complete the following table. How many shipments would you need to have each year to make up for the cost of the Prime membership?   15 shipments

Number of 1-Day Shipments per Year	Cost with Prime Membership ($)	Cost without Prime Membership ($)
0	99.00	0.00
1	102.99	10.99
2	106.98	21.98
3	110.97	32.97
4	114.96	43.96
5	118.95	54.95
6	122.94	65.94
7	126.93	76.93
8	130.92	87.92
9	134.91	98.91
10	138.90	109.90
11	142.89	120.89
12	146.88	131.88
13	150.87	142.87
14	154.86	153.86
15	158.85	164.85
16	162.84	175.84
17	166.83	186.83
18	170.82	197.82

b. Write a function to model the cost for x 1-day shipments per year under each plan.

Cost with Prime membership: $C = 99 + 3.99x$

Cost without Prime membership: $C = 10.99x$

c. Solve an equation to determine how many shipments you would need to have each year to make up for the cost of the Prime membership.

15 shipments

d. For each function, create a graph by plotting points from your table. Connect the points to form a line for each equation.

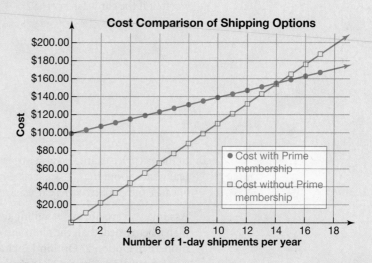

e. Use the graph to determine how many shipments you would need to have each year to make up for the cost of the Prime membership.

15 shipments

f. Compare your answers to the question: When is it worth it to buy a Prime membership on Amazon? Do your answers from the table, equation, and graph all agree?

The answers should all be 15 shipments. It is worth it to buy a Prime membership if you have 15 or more shipments in a year.

g. The Prime membership includes free 2-day shipping, which is usually $7.99 per shipment. If you were using only the 2-day shipping, how many shipments would you need to order to justify the Prime membership? Use algebra or a numeric calculation.

$99 = 7.99x$

$x = \dfrac{99}{7.99} \approx 12$ shipments

4. Nolan is considering buying stamps online. Stamps cost 49 cents each and are sold in books of 20 stamps. Each order has a $1.30 handling charge added to the cost of the stamps. The stamps cost the same at Nolan's local post office, but he will have to drive there, which creates a cost. Assume that gas costs $3.22 per gallon and the distance to the post office from his office is 15 miles. If Nolan's truck gets 16 miles per gallon, is it cheaper for him to buy the stamps online or make the round trip to buy them? Assume he buys the same number of books of stamps either way. What if he drives a car that gets 21 miles per gallon?

Whether he drives his truck or a car, it's cheaper to buy the stamps online.

5. Imagine that the price per gallon of gas with a $7 car wash is $3.19 and the price without the car wash is $3.39. When is it worth it to buy the car wash? When is it worth it if the car wash costs $2?

35 or more gallons; 10 or more gallons

6. In the gas price example earlier in the section, we assumed that the driver did not really care about the car wash. The two options are not truly equivalent, however, because the option with a car wash gives the driver something more than just gas.

Assume you are normally willing to pay $3 for a car wash. When is the car wash plan a better deal than the alternative? In other words, when is Option 1 only $3 more than Option 2? Option 1 includes a $5.50 car wash with gas that costs $3.65 per gallon. Option 2 has no car wash and gas that costs $3.85 per gallon.

First answer this question using only the table provided in the numeric solution. Estimate your answer. Then write an equation to answer the question.

From the table, at 10 gallons the car wash option costs $3.50 more than the alternative. At 15 gallons, it's $2.50 more. Splitting the difference, we find that 12.5 gallons seems to be a good approximation.

To find the exact value, we want to know when the difference between the options is $3, that is, $3.65g + 5.50 - 3.85g = 3$. Solving this equation yields $g = 12.5$ gallons.

7. A local pizza place sends you online coupons. You can't combine the coupons, and you are trying to choose from two deals: $6.99 per two-topping medium pizza or 20% off the regular price of $11 for a medium two-topping pizza. Which deal is better and when?

$6.99 per pizza is always a better deal.

8. Looking Forward, Looking Back Complete these sentences using the two cost options for gas from earlier in the section.

Cost with Option 1: $C = 3.65g + 5.50$ Cost with Option 2: $C = 3.85g$

a. As the number of gallons of gas increases by 1, the cost for Option 1 increases by ___$3.65___.

b. As the number of gallons of gas increases by 1, the cost for Option 2 increases by ___$3.85___.

c. What is the slope of the line for Option 1? Write it as a fraction with units. $\dfrac{\$3.65}{1\ \text{gallon}}$

d. What is the slope of the line for Option 2? Write it as a fraction with units. $\dfrac{\$3.85}{1\ \text{gallon}}$

e. Where does the slope appear in the cost formulas?

It is the coefficient of the independent variable.

3.6 | Get in Line: Slope-Intercept Form

SECTION OBJECTIVES:
- Find and interpret the slope and *y*-intercept from a linear equation
- Graph a line using a table and using the slope and *y*-intercept

Explore

1. On a comprehensive final exam, a professor asked each student to state the number of hours spent studying for the exam. She then used the data on student exam scores and study times to create the following model that can predict an exam score based on the number of hours studied:

$$T = 4h + 62$$

a. Use this model to complete the following table:

Study Hours	Exam Score
0	62
1	66
2	70
3	74
4	78
5	82

b. Since the exam scores increase by ___4___ every time the study hours increase by ___1___, we can say that there is a slope of ___4/1 = 4___ and a ___linear___ relationship between the number of hours studied and the exam score.

Since 0 study hours leads to an exam score of ___62___, we can say that the starting or initial value is ___62___.

c. Use the model or table of ordered pairs to graph the equation $T = 4h + 62$.

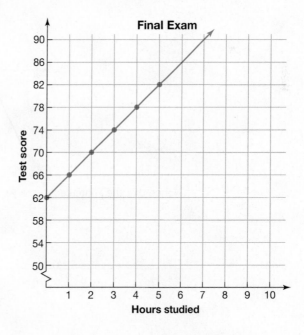

d. Where do the slope and starting value appear on the graph?

The slope is the ratio of the rise to the run when you move between two points on the line. The starting value is where the graph intersects the *y*-axis.

e. Look again at the equation $T = 4h + 62$. Where do the slope and starting value appear in the equation?

The slope is the coefficient of the independent variable, h. The starting value is the constant term.

Discover

In the *Explore*, you considered a linear relationship from three perspectives: graphically, numerically, and algebraically. This section will focus on helping you identify and describe a linear relationship when you are given an equation or function.

As you work through this section, consider these questions:

1 Can you identify slope from a table, graph, or equation?

2 Can you quickly write a linear equation if you know the slope and y-intercept?

An important detail that you can find from a linear equation is where the graph of the equation will intersect the axes.

LOOK IT UP

Intercept

The point at which a graph crosses an axis is called an **intercept**. If the graph crosses the x-axis, the point of intersection is called an x-intercept. If the graph crosses the y-axis, the point of intersection is called a y-intercept.

FOR EXAMPLE, this graph has an x-intercept of $(-2, 0)$ and a y-intercept of $(0, 4)$.

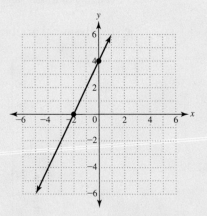

Notice that the x-intercept has a y-coordinate of 0 and the y-intercept has an x-coordinate of 0.

x-intercept: (x-value, 0)

y-intercept: (0, y-value)

2. Complete the table for this equation and then identify the slope and the intercepts of the graph of the equation: $2x + 4y = 12$.

x	y
0	3
1	$\frac{5}{2}$
2	2
3	$\frac{3}{2}$
4	1
5	$\frac{1}{2}$
6	0

Slope $= -\frac{1}{2}$; x-intercept $= (6, 0)$; y-intercept $= (0, 3)$

The equation $2x + 4y = 12$ can be written as $y = -\frac{1}{2}x + 3$, which is a very useful form because it clearly shows the slope and y-intercept of the graph. This form also makes it easier to substitute a value for x and calculate the corresponding y-value. The equation $y = -\frac{1}{2}x + 3$ is in a form referred to as the slope-intercept form. Later in the cycle, you'll learn how to rewrite a linear equation in this form.

Slope-Intercept Form

Linear relationships can be written in the form

$$y = mx + b$$

which is called **slope-intercept form**. When an equation is in slope-intercept form, the coefficient of x is the slope. The constant b is the y-coordinate of the point where the line intersects the y-axis. The ordered pair (0, b) is the y-intercept.

FOR EXAMPLE, if a savings account starts with $200 and is increasing in value by $50 each month, the equation $y = 50x + 200$ gives the value, y, in the account after x months. In this equation, $m = 50$ and $b = 200$. So the corresponding line has a slope of 50 and a y-intercept of (0, 200).

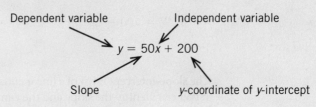

The slope-intercept form is usually written as $y = mx + b$, but it might make more sense to you as $y = b + mx$, since we tend to think about the starting value first and then apply the rate of change.

The slope-intercept form is essentially the same as the equation for a linear function that was given in Section 1.15: *Dependent variable = starting value + rate of change · independent variable.* We now know that the starting value is the *y*-coordinate of the *y*-intercept and the rate of change is the slope of the graph.

Need More Practice?

Identify the slope and *y*-intercept of each line.

1. $y = 2 - 6x$

2. $y = -9x$

Answers:

1. $m = -6$; *y*-intercept: (0, 2)

2. $m = -9$; *y*-intercept: (0, 0)

EXAMPLE 1

For each linear equation, identify the slope and *y*-intercept.

a. $y = -3x - 2$

b. $y = \dfrac{x}{2} - 8$

c. $y = 4 - x$

SOLUTION

a. Notice that this equation is already in $y = mx + b$ form. So the coefficient of *x*, -3, is the slope. And the constant term, -2, tells us the *y*-intercept is $(0, -2)$.

  Be sure to recognize that the slope is only the coefficient of *x*, not the expression $-3x$.

b. The equation $y = \dfrac{x}{2} - 8$ can be rewritten as $y = \dfrac{1}{2}x - 8$ to make it easier to identify the slope.

  We can see that the slope is $\dfrac{1}{2}$ and the *y*-intercept is $(0, -8)$.

c. This equation can be rewritten as $y = -x + 4 = -1 \cdot x + 4$.

  We can see that the slope is -1 and the *y*-intercept is $(0, 4)$.

3. For each linear equation, identify the slope and *y*-intercept. Remember that the *y*-intercept is an ordered pair, not just a number.

  a. $y = -6.5x + 2.5$ $m = -6.5$; *y*-intercept: (0, 2.5)

  b. $y = -\dfrac{x}{3}$ $m = -\dfrac{1}{3}$; *y*-intercept: (0, 0)

  c. $y = 4$ $m = 0$; *y*-intercept: (0, 4)

  d. $W = 100 - 2t$ $m = -2$; *y*-intercept: (0, 100)

The slope-intercept form of a linear equation is also very useful for graphing the equation. If you can identify the slope and the *y*-intercept of the equation, these two pieces of information are all you need to quickly draw a graph.

EXAMPLE 2

Graph the linear equation $y = -5x - 6$ by first plotting points and then using the slope and y-intercept to draw the graph.

SOLUTION

To graph the equation by plotting points, we need to choose x-values to substitute into the equation. We can choose any x-values we like, but small integer values are the easiest to graph.

x	y	(x, y)
-1	$y = -5 \cdot -1 - 6 = 5 - 6 = -1$	$(-1, -1)$
0	$y = -5 \cdot 0 - 6 = 0 - 6 = -6$	$(0, 6)$
1	$y = -5 \cdot 1 - 6 = -5 - 6 = -11$	$(1, -11)$

We can now plot the three ordered pairs shown in the table and connect them with a line. Technically, only two points are needed to plot a line. Plotting a third point is a useful check though. If the three points do not lie on a line, you know that at least one of them is wrong and you can look for a mistake.

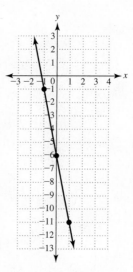

Now let's graph the equation again but using the slope and y-intercept instead of plotting random points. Notice that the equation is in slope-intercept form and that $m = -5$ and the y-intercept is $(0, -6)$.

We can begin by plotting the y-intercept, $(0, -6)$. Then we interpret the slope as the ratio of rise to run to move down 5 and right 1 from $(0, -6)$ to another point on the line.

$$m = -5 = \frac{-5}{1} = \frac{\text{rise}}{\text{run}}$$

You can continue to move down 5 and right 1 to find more points on the line, or you can move up 5 and left 1 instead by interpreting the slope fraction differently.

$$m = -5 = \frac{5}{-1} = \frac{\text{rise}}{\text{run}}$$

Notice that this graph is the same as the one we obtained by plotting the points from the table.

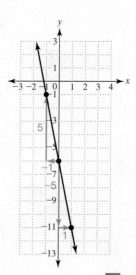

Tech TIP

Linear functions can be graphed on a graphing calculator. First, enter the function on the Y = screen, then choose an appropriate window and select Graph.

4. Use the slope and y-intercept to graph this linear equation: $y = -\dfrac{3}{4}x + 5$.

$m = -\dfrac{3}{4}$; y-intercept: (0, 5)

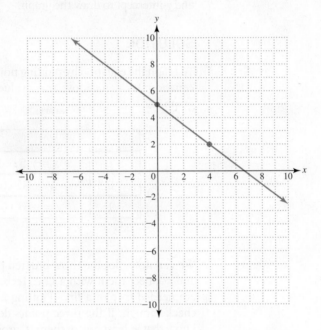

When working in a context we need to be able to interpret the slope and y-intercept, not just find them. Keep in mind that the slope tells how the dependent variable is changing relative to the independent variable.

EXAMPLE 3

Kudzu vines are invasive and can exhibit incredible growth in favorable conditions. Suppose a vine's length is given as $L = 525 + 9t$, where L is in inches and t is the number of days since July 1. Identify and interpret the slope and y-intercept of this linear function.

SOLUTION

The slope is the coefficient of the independent variable, t.

$$m = 9 = \frac{\text{change in } L}{\text{change in } t} = \frac{9 \text{ inches}}{1 \text{ day}}$$

The slope tells us that the vine is growing at a rate of 9 inches per day. The y-intercept is the ordered pair (0, 525), which tells us the vine was 525 inches long on July 1.

5. For each problem, use meaningful variables to write a linear function to model the situation. Then interpret the slope and *y*-intercept in words. Identify which variable would be on the horizontal axis and which would be on the vertical axis if you graph the function.

a. Model the cost if you buy an iPod for $200 and download songs for $1.29 each.

$C = 1.29s + 200$ gives the cost, *C*, of the iPod and *s* songs. The slope gives the price per song of $1.29. The *y*-intercept is (0, 200), which gives the cost of the iPod with 0 songs downloaded. The number of songs, *s*, would be on the horizontal axis. The cost, *C*, would be on the vertical axis.

b. Model the weekly salary a salesman earns if his salary is a combination of his base salary and commissions on his sales. His base weekly salary is $500 and he makes $50 for each sale.

$S = 500 + 50n$, where *n* is the number of sales he makes and *S* is his weekly salary. The slope gives the commission per sale of $50. The *y*-intercept is (0, 500), which gives his base salary if he made 0 sales. The number of sales, *n*, would be on the horizontal axis. *S* would be on the vertical axis.

Keep in mind that slope-intercept form does not have to use the letters *y* and *x*. If you see an equation in this form, it is in slope-intercept form:

$$\begin{matrix} \text{Dependent} \\ \text{variable} \end{matrix} = \text{A number} \cdot \begin{matrix} \text{Independent} \\ \text{variable} \end{matrix} + \text{A number}$$

$$y \quad = \quad mx \quad + \quad b$$

Connect

In Cycle 2, we solved linear equations of the form $ax + b = c$, where *a* is not zero. In this section we have learned how to graph linear equations or functions of the form $y = mx + b$. It's important to understand how solving an equation relates to graphing a function.

6. Suppose $y = 2x + 4$.

a. Graph the equation using the slope and *y*-intercept.

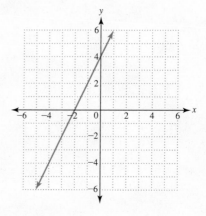

b. Pick an ordered pair that appears to be on the graph of the line in part a and confirm that the ordered pair is a solution to the equation.

Answers will vary.

c. Pick another ordered pair that appears to be on the graph of the line in part a and confirm that the ordered pair is a solution to the equation.

Answers will vary.

d. How are the points on the line related to the equation $y = 2x + 4$?

The coordinates of the points on the line are the solutions of the equation.

e. Determine if the point $(-100, -196)$ is on the graph of the line $y = 2x + 4$. Explain your answer.

The point is on the graph, since the ordered pair is a solution to the equation.

Reflect

WRAP-UP

What's the point?

It's very useful to be able to identify the slope and y-intercept in a linear function. This information can help you graph the function and interpret the rate of change and the starting value in a model.

What did you learn?

How to find and interpret the slope and y-intercept from a linear equation
How to graph a line using a table and using the slope and y-intercept

3.6 Homework

Skills MyMathLab

First complete the MyMathLab homework online. Then work the two exercises to check your understanding.

1. State the slope and y-intercept for $y = 3 - 2x$.

$m = -2$; y-intercept: (0, 3)

2. Use the slope and y-intercept from homework problem #1 to graph the equation.

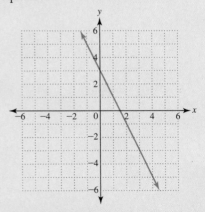

Concepts and Applications

Complete the problems to practice applying the skills and concepts learned in the section.

3. For each situation, write a linear function using meaningful variables. Find and interpret the slope and *y*-intercept.

   a. For each dollar a driver spends on gas, she can drive 15 miles. Assume she bought her car with 20,000 miles on it already. Write a linear function to model how many miles are on her car based on how much she spends on gas.

   $m = 15d + 20{,}000$, where m = miles on car and d = dollars spent on gas

   The slope is 15 miles/dollar, which means that she can drive 15 miles for each dollar she spends on gas.

   The *y*-intercept is (0,20000). This means that the car has 20,000 miles on it before she spends any money on gas.

   b. A man has a starting weight of 205 pounds and is losing 10 pounds each month. Model his weight with a linear function.

   $W = 205 - 10m$, where m = number of months and W = man's weight

   The slope is -10, which means that he is losing 10 pounds per month.

   The *y*-intercept is (0, 205). This means that before he starts losing weight, when no months have passed, his weight is 205 pounds.

4. A company is offering newly hired employees three salary structures. One option computes raises by adding $1,000 and then a 3% increase. Another option computes raises by adding a 3% increase and then $1,000. A final option computes raises by adding a 5% increase.

   a. Write a function for computing the new salary under each plan, where NS = new salary and S = old salary.

Add $1,000; increase by 3%	$NS_1 = 1.03(S + 1{,}000)$
Increase by 3%; add $1,000	$NS_2 = 1.03S + 1{,}000$
Increase by 5%	$NS_3 = 1.05S$

   b. In each function, what is the independent variable? S, the old salary in dollars

   What are the dependent variables? NS_1, NS_2, NS_3

   c. Using the functions you created, complete the following table. All amounts are in dollars.

Old Salary, S	$NS_1 = 1.03(S + 1{,}000)$	$NS_2 = 1.03S + 1{,}000$	$NS_3 = 1.05S$
10,000	11,330	11,300	10,500
20,000	21,630	21,600	21,000
30,000	31,930	31,900	31,500
40,000	42,230	42,200	42,000
50,000	52,530	52,500	52,500

d. Simplify the first function and list the three functions here. What do you notice about their slopes and *y*-intercepts?

$NS_1 = 1.03S + 1,030$

$NS_2 = 1.03S + 1,000$

$NS_3 = 1.05S$

The slopes are the same for Options 1 and 2. The first function's value of *b* is exactly $30 more than the value of *b* in the second function. The third function has a different slope and *y*-intercept than either of the first two.

e. Of Options 1 and 2, which will always be better and why?

When the functions are simplified, we can see that the rate of change is the same, but Option 1 will always be $30 more than Option 2.

f. Since the three lines are very close to each other, a specific graph is not helpful in this case. Instead, use only the *y*-intercepts and slopes to make a rough sketch of all three lines on the same graph. Use the fact that lines that have the same slope are parallel.

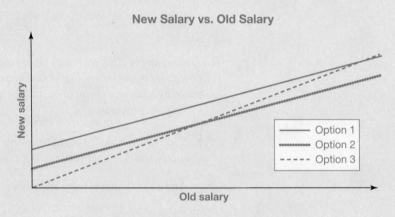

New Salary vs. Old Salary

g. Solve an equation to determine for which original salary amount the first two options produce equal new salaries.

$1.03(S + 1,000) = 1.03S + 1,000$; no solution

There are no salaries for which the first two plans produce equal new salaries.

h. Solve an equation to determine for which original salary amount the second and third options produce equal new salaries.

$1.03S + 1,000 = 1.05S$; $S = \$50,000$

For an original salary of $50,000, the second and third options produce the same new salary.

5. a. Do all lines have a *y*-intercept? Why or why not?

No; vertical lines do not have a *y*-intercept (unless the vertical line is the *y*-axis).

b. What kinds of lines are not functions?

Vertical

6. Decide if each equation is linear. If necessary, make a table to help you decide. If the equation is linear, identify the slope and the *y*-intercept.

a. $y = x - 4$ Linear; $m = 1$; *y*-intercept: $(0, -4)$

b. $y = 4 - x$ Linear; $m = -1$; *y*-intercept: $(0, 4)$

c. $y = \dfrac{1}{x} - 4$ Not linear

d. $y = \dfrac{1}{x - 4}$ Not linear

e. $y = \dfrac{1}{4}x$ Linear; $m = \dfrac{1}{4}$; *y*-intercept: $(0, 0)$

f. $y = \dfrac{x + 1}{4}$ Linear; $m = \dfrac{1}{4}$; *y*-intercept: $\left(0, \dfrac{1}{4}\right)$

7. Looking Forward, Looking Back Complete the pattern.

x	*y*
0	−4
1	−1
2	2
x	−4 + 3*x*

3.7 Chain, Chain, Chain: Writing Linear Equations

Explore

Alkanes are a type of molecule that has only hydrogen (H) and carbon (C) atoms joined with single bonds. In the structural formula, these single bonds are shown by a line segment connecting a C with an H. The first straight-chain alkanes are shown in the following table, starting with methane, the smallest alkane.

SECTION OBJECTIVES:
- Write the equation of a line using a point and the slope
- Write the equation of a line using two points
- Create a linear model in an applied problem

1. Draw the structures for the rest of the alkanes in the table. Write the formula for each alkane by first counting the number of carbon atoms and the number of hydrogen atoms in the structural formula. Then write these numbers as subscripts on the element symbols. Finally, complete the Ordered Pair column by representing the number of C atoms and the number of H atoms as an ordered pair, as shown. The first two alkanes have been done as an example.

INSTRUCTOR NOTE: Some students may notice that this problem is similar to the tables and chairs problem in Section 2.8.

Molecule Name	Structural Formula	Chemical Formula	Ordered Pair (number of C, number of H)
Methane	H | H—C—H | H	CH_4	(1, 4)
Ethane	H H | | H—C—C—H | | H H	C_2H_6	(2, 6)
Propane	H H H | | | H—C—C—C—H | | | H H H	C_3H_8	(3, 8)
Butane	H H H H | | | | H—C—C—C—C—H | | | | H H H H	C_4H_{10}	(4, 10)
Pentane	H H H H H | | | | | H—C—C—C—C—C—H | | | | | H H H H H	C_5H_{12}	(5, 12)
Hexane	H H H H H H | | | | | | H—C—C—C—C—C—C—H | | | | | | H H H H H H	C_6H_{14}	(6, 14)
Heptane	H H H H H H H | | | | | | | H—C—C—C—C—C—C—C—H | | | | | | | H H H H H H H	C_7H_{16}	(7, 16)

2. a. Complete the following chemistry formula for the *n*th alkane by generalizing the pattern shown in the Chemical Formula column.

$$C_nH_{2n+2}$$

b. Write a linear function to find the number of hydrogen atoms for a given number of carbon atoms by using the ordered pairs in the Ordered Pair column.

H = 2C + 2

c. Interpret the slope and *y*-intercept of the function.

The slope tells us that there are two additional atoms of H for each additional atom of C in the molecule. The *y*-intercept of 2 has no physical meaning in this situation, since it's impossible to have an alkane with 0 carbon atoms.

d. If a straight-chain alkane has 10 carbon atoms, how many hydrogen atoms does it have? If a straight-chain alkane has 10 hydrogen atoms, how many carbon atoms does it have? Is it possible to have a straight-chain alkane with an odd number of H atoms? Why or why not?

22; 4

No; the model says that the number of H atoms is always twice the number of C atoms (an even number) plus 2 (maintaining an even number). The result for the number of H atoms is always an even number.

In the *Explore*, you wrote a linear function that gives the number of hydrogen atoms for a given number of carbon atoms. There are a variety of ways you might have approached that problem. This section will focus on making the process of writing a linear equation faster and more streamlined.

As you work through this section, think about these questions:

1 What is the first step in finding the equation of a line through two points?

2 How do you write an equation to represent a vertical or horizontal line?

3 When do you need to solve an equation to find the *y*-intercept and when can you find it numerically?

We will begin by considering ways to find the equation of the line that passes through the points listed in the table. We need to find the slope and *y*-intercept to write the equation in $y = mx + b$ form.

x	y
2	−15
4	−12
6	−9
8	−6

+2 ⟶ +3 (between rows)

To find the slope, *m*, we need to find how the *y*-values change relative to the *x*-values. In this list of ordered pairs, the *x*-values are increasing by 2, and the *y*-values are increasing by 3. So $m = \dfrac{3}{2}$. This constant rate of change confirms that the equation will be linear.

Working backward to find the *y*-intercept for a table of values is like working backward to find the 0th term in an arithmetic sequence.

To find the value of *b*, we need to know the *y*-coordinate of the *y*-intercept. The *y*-intercept is the point that has an *x*-coordinate of 0. If we can use the changes in the *x*- and *y*-values going down the table, we can also use them going up the table to find the ordered pair $(0, -18)$. So the value of *b* is -18.

The equation of the line that passes through these points is $y = \dfrac{3}{2}x - 18$.

x	y
0	−18
2	−15
4	−12
6	−9
8	−6

The method we used with the values in the table necessitates an understanding of how to apply the slope backward and forward. It is sometimes the fastest way to find the *y*-intercept. However, if the ordered pairs are not integers or if the change indicated by the slope makes it difficult to get to an *x*-value of 0, this process can be more difficult and you might want an alternative.

We can instead choose any two points from the table, find the slope of the line that passes through the points, and then solve for *b* in the equation $y = mx + b$. If we use the ordered pairs $(2, -15)$ and $(4, -12)$, we get the following slope:

$$m = \frac{-12 - (-15)}{4 - 2} = \frac{-12 + 15}{2} = \frac{3}{2}$$

We can then substitute one ordered pair and the slope into the equation $y = mx + b$. If we use $(2, -15)$ and $m = \dfrac{3}{2}$, we get

$$y = mx + b$$
$$-15 = \frac{3}{2}(2) + b$$
$$-15 = 3 + b$$
$$b = -18$$

So we arrive at the same slope-intercept equation: $y = \dfrac{3}{2}x - 18$.

HOW IT WORKS

To write the equation of a line in slope-intercept form:

1. Write the general slope-intercept form of a line: $y = mx + b$.
2. Find the slope of the line.
3. Substitute the slope for *m* and an ordered pair on the line for *x* and *y*. Solve the resulting equation for *b*.
4. Write the slope-intercept form with the values of *m* and *b* included.

EXAMPLE: Find an equation of the line that passes through $(-2, 6)$ and $(3, 13)$.

Since the slope is not given, it needs to be found.

$$m = \frac{\Delta y}{\Delta x} = \frac{13 - 6}{3 - (-2)} = \frac{7}{5}$$

Since the *y*-intercept is not given, substitute one of the ordered pairs into the equation to find the value of *b*. $(3, 13)$ might be easier to work with than $(-2, 6)$, since its coordinates are positive.

Substitute $m = \dfrac{7}{5}$ and the point $(x, y) = (3, 13)$ into $y = mx + b$ and solve for b:

$$13 = \frac{7}{5} \cdot 3 + b$$

$$13 = \frac{21}{5} + b$$

$$13 - \frac{21}{5} = b$$

$$\frac{65}{5} - \frac{21}{5} = b$$

$$\frac{44}{5} = b$$

The linear equation is $y = \dfrac{7}{5}x + \dfrac{44}{5}$.

If you need to write an equation in slope-intercept form, the goal is to find m and b. Sometimes the slope will simply be provided. If it's not, then you need to find it using a table, a graph, or two ordered pairs. If you are not given the y-intercept, you can solve for b as shown in the *How It Works* or you can use a table or graph to find it. If you are solving an applied problem, it's best to use meaningful variables instead of x and y.

3. Write an equation for each line.

    a. Line passes through $(0, -8)$ with $m = 4$

       $y = 4x - 8$

    b. Line passes through $(1, -8)$ with $m = 4$

       $y = 4x - 12$

    c. Line passes through $(-8, 6)$ and $(-4, 4)$

       $y = -\dfrac{1}{2}x + 2$

d. Line passes through the points with the following ordered pairs:

x	y
0	5
1	2
2	−1
3	−4

$y = -3x + 5$

EXAMPLE

a. Write an equation of the horizontal line that passes through $(-3, 5)$.

b. Write an equation of the vertical line that passes through $(-3, 5)$.

SOLUTION

a. The horizontal line that passes through $(-3, 5)$ is shown on the graph.

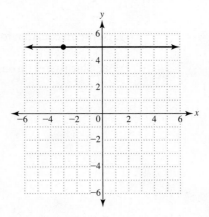

Notice that every point on the line has the same y-coordinate of 5. If we make a table of ordered pairs for this graph with a few points from the line, each x-value is paired with a y-value of 5.

x	y
−1	5
0	5
1	5
2	5

Since every point on the line has a y-coordinate of 5, the equation of the line is simply $y = 5$. We could also find this equation by calculating the slope and observing the y-intercept.

$$m = \frac{5 - 5}{2 - 1} = \frac{0}{1} = 0$$

The y-intercept is $(0, 5)$, which means that b is 5. So the slope-intercept form of the equation for this line is $y = 0x + 5$ or just $y = 5$. Every horizontal line has an equation of the form $y = a$ number. Every horizontal line also represents a function, one in which every input is paired with the same output.

b. The vertical line that passes through $(-3, 5)$ is shown on the graph.

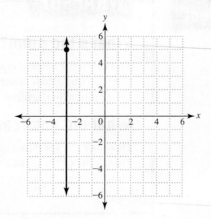

Notice that every point on the line has the same x-coordinate of -3. If we make a table of ordered pairs for this graph with a few points from the line, each y-value is paired with an x-value of -3.

x	y
−3	−1
−3	0
−3	1
−3	2

Since every point on the line has an x-coordinate of -3, the equation of the line is simply $x = -3$. This equation cannot be written in slope-intercept form. The slope of this, and every, vertical line is undefined.

$$m = \frac{2 - 1}{-3 - (-3)} = \frac{1}{-3 + 3} = \frac{1}{0}$$

Since the slope is undefined, there is nothing to substitute into $y = mx + b$. Every vertical line has an equation of the form $x = a$ number. Vertical lines never represent functions, since one input (x-value) is paired with infinitely many outputs (y-values).

> Horizontal lines have equations of the form $y = a$, and vertical lines have equations of the form $x = b$, where a and b are constants.

4. Write the equation of a horizontal line and the equation of a vertical line through the point $(5, 8)$.

$y = 8$; $x = 5$

Connect

5. Two friends are running through a state park where a 10-mile path is labeled with mile markers. They are experienced runners and maintain a consistent pace. John passes mile marker 2 at 1:18 p.m. and mile marker 5 at 1:45 p.m. Mike started later than John and is trying to catch up. He passes mile marker 2 at 1:30 p.m. and mile marker 5 at 1:54 p.m. Write a linear function to describe each man's location, D, on the path (which mile marker) as a function of the time since 1:00 p.m., t.

John: $D = \dfrac{1}{9}t$ Mike: $D = \dfrac{1}{8}t - 1.75$

INSTRUCTOR NOTE: Remind students to read the *Getting Ready* article for the next section.

Reflect

WRAP-UP

What's the point?

You can write a linear equation if you know the slope and one point on the line or if you know two points on the line. You should be able to quickly write a linear model in the form $y = mx + b$.

What did you learn?

How to write the equation of a line using a point and the slope
How to write the equation of a line using two points
How to create a linear model in an applied problem

3.7 Homework

Skills MyMathLab

First complete the MyMathLab homework online. Then work the two exercises to check your understanding.

1. Write an equation of the line that passes through the point $(0, 5)$ with a slope of $\frac{2}{3}$.

$$y = \frac{2}{3}x + 5$$

2. Write the equation of the line that passes through $(6, -3)$ and $(-5, 2)$.

$$y = -\frac{5}{11}x - \frac{3}{11}$$

Concepts and Applications

Complete the problems to practice applying the skills and concepts learned in the section.

3. Assuming the data in the following table follows a linear pattern, complete the table and write the equation of the line that passes through the points.

x	−1	0	1	2	3	4
y	−35	−26	−17	−8	1	10

$$y = 9x - 26$$

4. For the values in each table, write the equation of the line that passes through the points.

a.

x	y
−3	21
1	−7
2	−14
4	−28

$y = -7x$

b.

x	y
−4	−26
−2	−16
4	14
8	34

$y = 5x - 6$

5. Two of your friends are attending a college that you are considering. One friend is taking 10 credit hours and paying $3,800 in tuition and fees for the semester. Another friend is taking 20 credit hours and paying $6,850 in tuition and fees for the semester. Write a function to model the total cost for the semester, *T*, of taking *h* credit hours. Use it to determine how much you will pay if you take 18 credit hours.

$T = 305h + 750$; $6,240

6. Carbon chains also exist in a cyclic form with double bonds, as shown in the following table. These are known as alkenes.

a. For each structure shown in the table, write the formula indicating the number of carbon and hydrogen atoms. Then complete the last column in the table by writing an ordered pair to indicate the number of carbon atoms and hydrogen atoms in each structure.

Name	Structural Formula	Chemical Formula	(C, H)
Cyclopropene	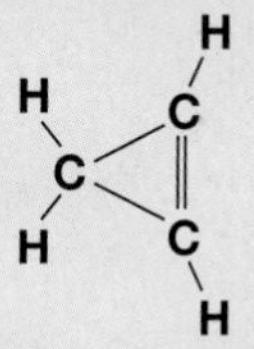	C_3H_4	(3, 4)
Cyclobutene	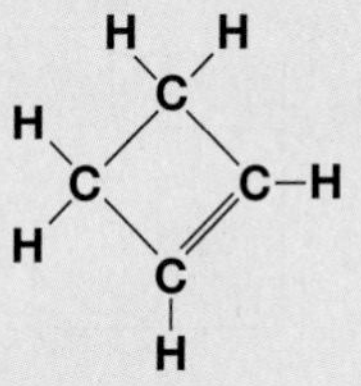	C_4H_6	(4, 6)
Cyclopentene	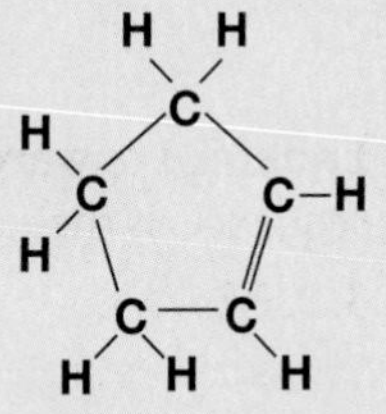	C_5H_8	(5, 8)
Cyclohexene	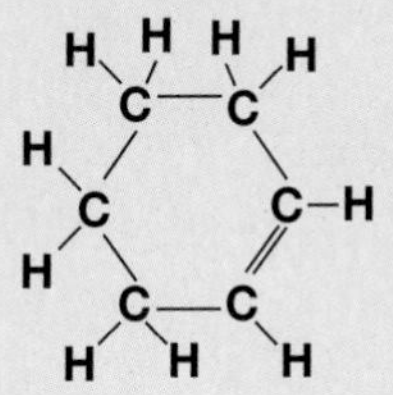	C_6H_{10}	(6, 10)

b. Write a linear function that gives the number of hydrogen atoms for a given number of carbon atoms.

$H = 2C - 2$

c. If a cyclic alkene has 12 carbon atoms, how many hydrogen atoms does it have?

22

d. If a cyclic alkene has 12 hydrogen atoms, how many carbon atoms does it have?

7

e. Is it possible for a cyclic alkene to have an odd number of hydrogen atoms?

No

f. How many carbon atoms are necessary for the linear model to make sense?

There must be at least 3 carbon atoms to form a cyclic arrangement.

g. On the following grid, create a graph that shows the ordered pairs from the table and the function.

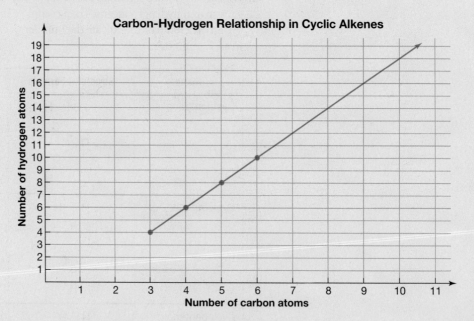

Carbon-Hydrogen Relationship in Cyclic Alkenes

7. Looking Forward, Looking Back Write an expression for the nth term of the sequence. Is the sequence arithmetic, geometric, or neither?

3, 9, 27, ...

3^n; geometric

Getting Ready for Section 3.8

Read the article and answer the questions that follow.

Recognizing and Coping with Exponential Change In Marketing

Who could have predicted, 10 years ago, that self-driving cars would be on the road?

What if I told you Facebook, Google and cell phones would be 100 percent obsolete by 2020? You'd probably ask me to check in with the nearest psychiatrist for some help.

In truth, it's hard for me to imagine that such a prediction would be right, but that's precisely the point noted futurist Ray Kurzweil makes in *The Singularity Is Near*: We are notoriously bad at predicting the speed of change.

Before debating the impending demise of Facebook and Google, let me start by explaining Kurzweil's theory, which he calls the "the law of accelerating returns."

Here's how he described the theory in a blog post more than 10 years ago:

An analysis of the history of technology shows that technological change is exponential, contrary to the common-sense "intuitive linear" view. So we won't experience 100 years of progress in the 21st century—it will be more like 20,000 years of progress (at today's rate). The "returns," such as chip speed and cost-effectiveness, also increase exponentially. There's even exponential growth in the rate of exponential growth. Within a few decades, machine intelligence will surpass human intelligence, leading to The Singularity—technological change so rapid and profound it represents a rupture in the fabric of human history.

It sounds crazy, doesn't it? Twenty thousand years of progress in 100 years? But think about the technical changes that have happened in the last century alone:

- It took humans tens of thousands of years to invent the wheel, but in the 20th century, we progressed from the automobile to the propeller plane to the jet plane to the rocket and now to the reusable rocket (SpaceX, Space Shuttle). In the next ten years, we may be traveling via rockets and self-driving vehicles.

- Information was spread via print (books and newspapers) until the radio, then the TV, then the internet. Ninety percent of the world's data has been created in the last two years.

- Computations were once performed by abacus; the 20th century brought us calculators, and eventually supercomputers. An Apple iPhone 4 has 2.7 times the processing power of an IBM mainframe in 1985. From 1956 to 2015, we've seen a one-trillion-fold increase in computer performance.

Exponential Change In Online Marketing

Given that online marketing is wholly dependent on technology (the internet, cell phones, big data, attribution and so on), it seems reasonable to expect exponential change in our industry, as well.

Indeed, if you look at just the last 15 years of online marketing, change does seem to be accelerating. Let's look at a few trends, starting with online marketing.

Way back in 1997, online marketing spend accounted for about 0.67 percent of all advertising in the US. In 2015, that percentage increased to 31.6 percent, and digital marketing is expected to overtake TV as the number one advertising medium in 2017.

Next, let's chart the rise of mobile marketing as an important part of online marketing.

In 2006, mobile marketing accounted for about three percent of all online marketing. Most studies project mobile marketing to exceed and eventually dominate online marketing spend in the next year.

In other words, it took mobile less than ten years to go from afterthought to primary channel.

How about social media marketing? In 2007, Facebook managed to drive $153 million in revenue (almost entirely from advertising). That number shot up to $12.4 billion in 2014 and $7.6 billion in the first six months of 2015 alone.

To put that in perspective, in 2014 Facebook drove 2.3 percent of all advertising spend in 2014 and 8.9 percent of all digital advertising.

So you could conclude from these statistics that Facebook, Google, the internet in general, and mobile in particular, will continue to dominate both media consumption and advertising spend, right?

Not necessarily. This is where you have to understand the nuance of Kurzweil's point.

If you assume that the recent growth of online marketing (as we know it today) will continue into the future, you are making a fundamental error, according to Kurzweil. You are taking a "linear" view of the future, rather than an exponential one. As Kurzweil writes:

Even for those of us who have been around long enough to experience how the pace increases over time, our unexamined intuition nonetheless provides the impression that progress changes at the rate that we have experienced recently. From the mathematician's perspective, a primary reason for this is that

(continued)

Recognizing and Coping with Exponential Change In Marketing   *(continued)*

an exponential curve approximates a straight line when viewed for a brief duration. So even though the rate of progress in the very recent past (e.g., this past year) is far greater than it was ten years ago (let alone a hundred or a thousand years ago), our memories are nonetheless dominated by our very recent experience. It is typical, therefore, that even sophisticated commentators, when considering the future, extrapolate the current pace of change over the next 10 years or 100 years to determine their expectations. This is why I call this way of looking at the future the "intuitive linear" view.

What does this mean for online marketing? Well, consider some of the trends we discussed above:

- Will mobile marketing be the dominant force in advertising in the next few decades, or will a new technology replace mobile (usage and advertising) at a rate even faster than online marketing's growth against traditional and mobile marketing's growth against desktop?

  Is it in fact more likely that mobile advertising will cease to even exist in a few decades (perhaps replaced by the Internet of Things but more likely by something we cannot even currently contemplate)?

- Will Facebook continue to grow exponentially, or will other social media sites take its place? While it is true that Facebook's revenue growth has been exponential since 2007, it's also seen exponential competition.

  Snapchat, for example, which didn't even launch until 2011, has grown to more than 70 million monthly active users (MAUs) in less than three years. WhatsApp (acquired by Facebook) had only a few million users in 2010 and now has more than 700 million MAUs.

The social channels that will dominate in the near future have likely not been invented (assuming that social will be relevant at all).

The challenge for online marketers is to figure out the balance between what works today and what will work tomorrow. Know when to invest in new trends, when to fully pivot, and when to ignore the noise. And recognize that change occurs more quickly than we think!

Questions

1. Consider 2007 to be the starting year for Facebook in which it had $153 million in revenue. In 2014, it had $12.4 billion in revenue and $7.6 billion in revenue in the first six months of 2015. Assume, conservatively, that Facebook had the same performance in the second half of 2015 as it did in the first half. Create a table of values using *x* to represent the years since Facebook began and *y* to represent the revenue in millions. Is the data linear? Explain. Graph the data to confirm your conclusion.

x	y
0	153
7	1,240
8	1,520

No; the slope of the line through (0, 153) and (7, 1,240) is $155.29/year, while the slope of the line through (7, 1,240) and (8, 1,520) is $280/year.

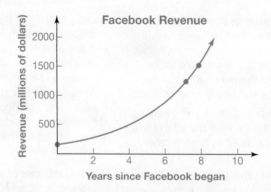

2. The article mentioned that chip speed increases exponentially. A computer executive stated in 1975 that chip processing speeds should double every 2 years. Write an exponential function to model chip speeds using a starting year of 1975 assuming a processor speed of 2 megahertz in 1975.

$S = 2 \cdot 2^t$, where t = years since 1975 and S equals speed of a chip processor in megahertz

3.8 Going Viral: Exponential Functions

SECTION OBJECTIVES:

- Write the equation of an exponential function using a starting value and rate of change
- Model with exponential functions
- Graph exponential functions

Remember?

Exponential functions have the form

$y = $ *starting value* $\cdot$
$(multiplier)^x$

1. Match the functions and the graphs.

a. $y = \left(\dfrac{1}{4}\right)^x$ **b.** $y = 10 \cdot 3^x$ **c.** $y = 7 \cdot 2^x$ **d.** $y = 6 \cdot \left(\dfrac{1}{3}\right)^x$

   ii   iii   i   iv

i.

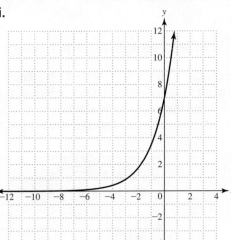

ii.

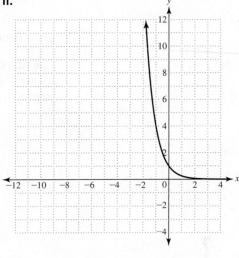

iii.

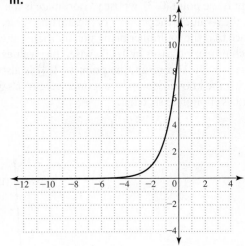

iv.

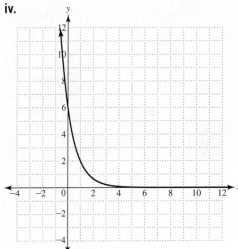

2. a. Where does the starting value in the exponential function appear on the graph?

The starting value is the *y*-coordinate of the *y*-intercept of the graph.

b. Describe the value of the multiplier when the *y*-values on the graph are increasing. Describe the value of the multiplier when the *y*-values on the graph are decreasing.

Increasing exponential functions have positive multiplier greater than 1. Decreasing exponential functions have positive multipliers less than 1.

Discover

Exponential change appears in many contexts, including social networking and online marketing. A website, story, or link can "go viral" and reach millions of people in a very short period of time. We explored exponential change and functions in Section 1.15, and now you will learn more about exponential functions and how their equations relate to their graphs.

As you work through this section, think about these questions:

1 Can you tell from the equation of an exponential function if the graph will be increasing or decreasing?

2 Can you use the starting value and rate of change of an exponential function to sketch its graph?

Let's continue the work started in the *Explore* by looking closer at exponential graphs.

Consider the exponential function $y = 3 \cdot 4^x$. We know that it's exponential because it follows the format of exponential functions, which is $y = $ *starting value* $\cdot$ *(multiplier)*^{x}. The starting value is 3 and the multiplier is 4, which means that the y-values are quadrupling. We can use this information, or the equation, to create a table of ordered pairs.

x	$3 \cdot 4^x$	y
0	$3 \cdot 4^0 = 3 \cdot 1$	3
1	$3 \cdot 4^1 = 3 \cdot 4$	12
2	$3 \cdot 4^2 = 3 \cdot 16$	48
3	$3 \cdot 4^3 = 3 \cdot 64$	192

The y-intercept is the point (0, 3), whose y-coordinate is the starting value in the equation. When $x = 1$, the y-value is the product of the starting value, 3, and the multiplier, 4, which is 12. We notice that the multiplier being greater than 1 means the y-values increase by a factor of 4 each time x increases by 1. Likewise, the y-values decrease by a factor of 4 each time x decreases by 1. As a result, the point $\left(-1, \frac{3}{4}\right)$ is also on the curve. Plotting a few points and using the knowledge of the exponential curve's general shape, as we have seen in Section 1.15, is enough to sketch it.

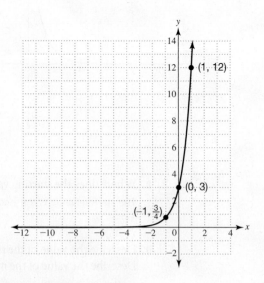

Like the equation of a linear function, there are two important numbers in the equation of an exponential function that can be used to draw its graph quickly: a starting value and a rate of change.

Exponential Functions and Graphs

An **exponential function** can be represented by an equation that has the following form, where the variable is in the exponent:

$$y = a \cdot b^x$$

The value of *a* is the starting value and *b* is the multiplier. The variable *a* must be a nonzero number. The variable *b* must be a positive number that is not 1. If $b > 1$, the graph will be increasing and show exponential growth. If $0 < b < 1$, the graph will be decreasing and show exponential decay. The exponent *x* can be any real number. The points $(0, a)$ and $(1, ab)$ will always be on the graph. An exponential function of the form $y = a \cdot b^x$ will approach the *x*-axis, but will not touch or cross it.

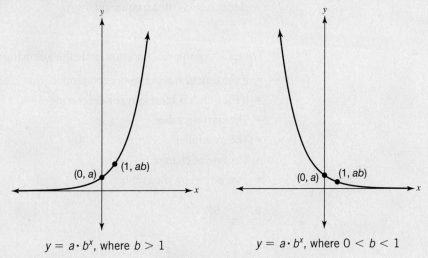

$y = a \cdot b^x$, where $b > 1$ $y = a \cdot b^x$, where $0 < b < 1$

FOR EXAMPLE, the following graphs represent exponential functions.

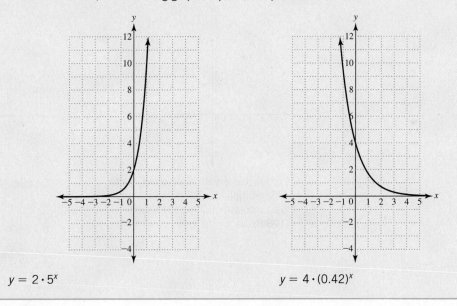

$y = 2 \cdot 5^x$ $y = 4 \cdot (0.42)^x$

The value of *b* in an exponential function is not the *y*-coordinate of the *y*-intercept, like the value of *b* in $y = mx + b$.

If $b = 1$ in the exponential equation, the equation becomes $y = a \cdot 1 = a$, which is a constant function. The graph would then be a horizontal line, not an exponential curve.

Remember that we can interpret the multiplier to understand the rate of change in an exponential function. Multipliers that are whole numbers like 2 or 3 indicate that the y-values are doubling or tripling. They can also be interpreted as a percent. For example, a multiplier of 2 is the same as multiplying by 200% or increasing by 100%. If the multiplier is a decimal, the rate of change is often interpreted as a percent.

3. a. Suppose y is an exponential function of x with an initial value of 7 and a 10% growth rate. Write an equation for y.

$y = 7(1.1)^x$

b. Suppose y is an exponential function of x with an initial value of 7 and a 10% decay rate. Write an equation for y.

$y = 7(0.9)^x$

4. For each exponential function, state the following and then graph the function.

- If the function represents exponential growth or decay
- If the graph is increasing or decreasing
- The starting value
- The multiplier
- The rate of change
- Two points on the graph

a. $y = 8(1.1)^x$

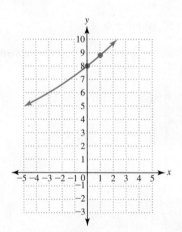

Exponential growth, increasing, starts at 8, multiplier is 1.1, rate of change is an increase of 10%, (0, 8) and (1, 8.8) are on the graph

b. $y = 5\left(\dfrac{1}{2}\right)^x$

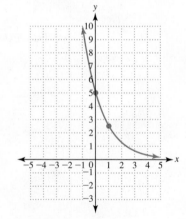

Exponential decay, decreasing, starts at 5, multiplier is $\dfrac{1}{2}$, rate of change is a decrease of 50%, (0, 5) and $\left(1, \dfrac{5}{2}\right)$ are on the graph

Let's use our knowledge of exponential functions to model a real situation.

5. Candy Crush Saga is a popular app in which colored candies are matched. The app is free to download but requires purchases at a certain point to advance in the game. It also offers in-app purchases to progress faster in the game. One year after its debut, it had been downloaded more than 500 million times.

Data from company.king.com

a. In a recent month, the app brought in $43 million in revenue after Apple and Google took 30% of its revenue. How much revenue did the app earn before paying Apple and Google?

$61.4 million

Data from seekingalpha.com

b. If the app has been downloaded 500 million times already and gains 8 million new downloads per month, write a linear model of the form $y = mx + b$ to represent the number of downloads. Let D be the number of downloads in millions and m be the number of months.

$D = 500 + 8m$

INSTRUCTOR NOTE: Students will need to use trial and error or a spreadsheet or graphing calculator to estimate the time needed to reach 1 billion downloads with the exponential model in 5d.

c. If the app has been downloaded 500 million times already and the number of downloads grows 2% each month, write an exponential model of the form $y = a \cdot b^x$ to represent the number of downloads, using the same variables as part b.

$D = 500(1.02)^m$

d. Use each model to predict how long it will take to reach 1 billion downloads at the rate stated.

With the linear model, it will take 62.5 months or 5 years, 2.5 months. With the exponential model, it will take approximately 35 months or 2 years 11 months.

Connect

When you need to write the equation of a line for which you are given two ordered pairs, neither of which is the y-intercept, you can use the slope to work backward or forward to the y-intercept. Once you know the slope and y-intercept, you have enough information to write an equation for the line.

Similarly, when you are given a set of ordered pairs that show exponential change, you can use the rate of change to work backward to the starting value if it's not known. If you have consecutive y-values, you can divide them to find the multiplier. Otherwise, you can divide any two y-values to find the multiplier that gets you from the first y-value to the second y-value. Use the difference in the x-values and the corresponding root to determine the multiplier for one step.

For example, suppose the ordered pairs in the following table show exponential change. To find the multiplier, begin by dividing the given y-values: $\frac{48}{6} = 8$. To move three steps, you would apply the multiplier three times, which is the same as cubing the multiplier. To work backward to find the multiplier, you need to take the cube root: $\sqrt[3]{8} = 2$. The multiplier is 2.

INSTRUCTOR NOTE: Consider having students fill in the table with the rest of the ordered pairs.

x	y
0	
1	6
2	
3	
4	48

3 steps · 8

You can use the multiplier to work backward to the starting value by dividing: $\frac{6}{2} = 3$. This value completes the ordered pair (0, 3). You can use the multiplier to work forward and complete the table with the ordered pairs (2, 12) and (3, 24). The exponential function that represents the relationship shown in the table is $y = 3 \cdot 2^x$.

6. Complete each table of ordered pairs assuming exponential change. Write a function to represent the relationship shown in the table.

a.

x	y
0	6
1	18
2	54
3	162
4	486
5	1,458
6	4,374

$y = 6 \cdot 3^x$

b.

x	y
0	16
1	4
2	1
3	$\frac{1}{4}$
4	$\frac{1}{16}$

$y = 16\left(\frac{1}{4}\right)^x$

Reflect

INSTRUCTOR NOTE: Section 4.7 on direct variation and Section 4.8 on inverse variation can be done after this section without loss of continuity.

WRAP-UP

What's the point?

Exponential functions exhibit a type of change in which a quantity increases or decreases by a constant multiplier over time. Like linear functions, the starting value and rate of change can be seen in both the graph and the equation.

What did you learn?

How to write the equation of an exponential function using a starting value and rate of change
How to model with exponential functions
How to graph exponential functions

3.8 Homework

Skills MyMathLab

First complete the MyMathLab homework online. Then work the two exercises to check your understanding.

1. Assume an exponential function has a starting value of 12 and a growth rate of 8%. Write an equation to model the situation.

$y = 12(1.08)^x$

2. Assume an exponential function has a starting value of 12 and a decay rate of 8%. Write an equation to model the situation.

$y = 12(0.92)^x$

Concepts and Applications

Complete the problems to practice applying the skills and concepts learned in the section.

3. Assume that approximately 3 million bacteria can fit on a pinhead and that the number of bacteria doubles every 6 hours.

a. Complete the following table to show the population of bacteria over time:

Number of 6-Hour Periods, t	Time (hours)	Population in Millions, P
0	0	3
1	6	6
2	12	12
3	18	24
4	24	48
5	30	96
6	36	192
7	42	384
8	48	768

b. Graph the data in the table as ordered pairs of the form (t, P) and connect the points with a smooth curve.

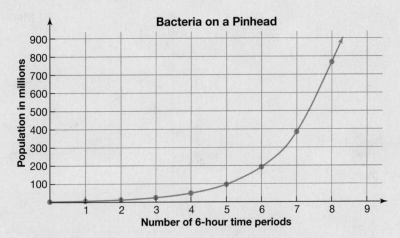

c. Use the table to write a model for the number of bacteria (in millions) after t 6-hour periods.

$P = 3 \cdot 2^t$

d. Use the model to find the number of bacteria after one week. (HINT: First figure out how many 6-hour periods this is.)

$t = 28$; $P = 805,306,368$ million bacteria

e. Use your model to estimate how many days it takes the population to reach 1 trillion. You can use Excel to speed up the process.

5 days

4. Suppose a certain stock is losing 15% of its value every week.

   a. If the stock started at $45 per share, complete the following chart to show its value each week:

Number of Weeks, w	Value of Stock, V ($)
0	45.00
1	38.25
2	32.51
3	27.64
4	23.49
5	19.97
6	16.97
7	14.43
8	12.26
9	10.42
10	8.86

   b. Graph the data in the table as ordered pairs of the form (w, V) and connect the points with a smooth curve.

   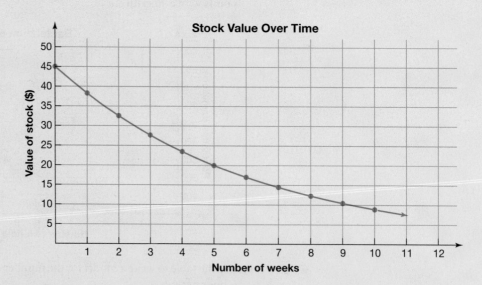

   c. Use the table to write a formula for the value of the stock after w weeks.

   $V = 45(0.85)^w$

   d. Use the model to find the value of the stock after 20 weeks if this trend continues.

   $1.74

   e. When is the stock worth 1 dollar?

   24 weeks

5. Complete each table assuming the type of change indicated. Then use the table to write the function.

a. Linear change

x	y
−3	12
−2	7.5
−1	3
0	−1.5
1	−6
2	−10.5

$$y = -\frac{9}{2}x - \frac{3}{2}$$

b. Exponential change

x	y
−3	12
−2	6
−1	3
0	1.5
1	0.75
2	0.375

$$y = 1.5(0.5)^x$$

6. Looking Forward, Looking Back For each equation, determine if it is linear. If it is, state the slope and y-intercept.

a. $y = 3x^2$

No

b. $y = 3x + 2$

Yes; $m = 3$, y-intercept: $(0, 2)$

c. $y = 3^x$

No

Mid-Cycle Recap

Complete the problems to practice applying the skills and concepts learned in the first half of the cycle.

INSTRUCTOR NOTE: The Mid-Cycle Recap provides students a chance to see how they are accomplishing the goals of the cycle so far. A quiz is available in MyMathLab as an option for assessing students' progress on the objectives at this point. See the Instructor Guide for more assessment ideas.

Skills

1. Write the equation of the function with the given information.

a. Linear function whose slope is $-\frac{5}{2}$ and y-intercept is (0, 8)

$y = -\dfrac{5}{2}x + 8$

b. Exponential function with a starting value of 11 and y-values that increase at a rate of 3%

$y = 11(1.03)^x$

Concepts and Applications

2. If a stretch of mountain road covers a horizontal distance of 0.5 mile and has a vertical drop of 200 feet, find the percent grade of the road and the length of the actual stretch of road.

Grade ≈ 7.6; length of road ≈ 2,647.6 feet

3. The temperature inside a car on a hot day can be much higher than the outside air temperature. The temperatures listed in the table, in degrees Fahrenheit, show a relationship between the outside air temperature and the temperature inside a car that has four windows slightly open. As the temperature increases outside, it also increases inside the car. Does the inside temperature increase by a constant amount, or a constant percent, or neither?

Outside Temperature	Inside Car Temperature
84°	98°
90°	108°
95°	113°
101°	114°
110°	123°
115°	132°

Neither; the inside temperature do not increase by either a constant amount or percent.

4. A technician earns a salary of $37,200 plus $40 for each hour of overtime worked. Write a function to model her salary using meaningful variables.

$S = \$37{,}200 + \$40h$, where S is her salary and h is the number of hours of overtime she works.

3.9 Untangling the Knot: Solving Nonlinear Equations

Explore

1. Solve each equation by undoing the operations that have been done to the variable.

a. $t^2 = 400$
$t = \pm 20$

b. $4t^2 = 400$
$t = \pm 10$

SECTION OBJECTIVE:

• Solve nonlinear equations

c. $\sqrt{y} = 9$
$y = 81$

d. $\sqrt{3y} = 9$
$y = 27$

e. $\dfrac{1}{x} = 8$
$x = \dfrac{1}{8}$

f. $\dfrac{1}{2x} = 8$
$x = \dfrac{1}{16}$

Discover

In Cycle 2 you learned how to solve *linear* equations by undoing the operations that had been done to the variable. That technique also works on many equations that are not linear. This section will give you the opportunity to solve other types of equations by expanding on the strategies you learned in the previous cycle.

As you learn the material in this section, consider the following questions:

1 How do you decide what to do first when you solve an equation?

2 Is a nonlinear equation necessarily more difficult to solve than a linear equation?

We will start with an example that illustrates how to solve a quadratic equation that you encountered earlier in the text.

Quadratic Equation

A **quadratic equation** can be written in the form $ax^2 + bx + c = 0$, where $a \neq 0$. A quadratic equation always contains a term that has the variable squared and no terms with exponents greater than 2.

FOR EXAMPLE, $3x^2 = -2x - 5$ is a quadratic equation, since it can be written as $3x^2 + 2x + 5 = 0$.

EXAMPLE 1

In Cycle 2, you learned that a Pythagorean triple is a set of three whole numbers that satisfy the Pythagorean theorem. Complete this Pythagorean triple: 8, _____, 17.

SOLUTION

We will begin by choosing a variable, n, to represent the unknown number. Then we can write an equation using the Pythagorean theorem:

$$8^2 + n^2 = 17^2$$

Notice that this is a quadratic equation because the only variable is squared. We can still solve it with our usual strategy of undoing operations to isolate the variable.

$$64 + n^2 = 289$$

Notice that two operations are represented on the left side of the equation. We need to undo these operations in the reverse order to solve the equation for n. First, we undo the addition with subtraction. Then, we undo the squaring with a square root since they are inverse operations.

Operations done to n:	Steps to solve for n:
1. Square	1. Subtract 64
2. Add 64	2. Square root

$$64 + n^2 = 289$$
$$\underline{-64 \qquad\quad -64}$$
$$n^2 = 225$$
$$\sqrt{n^2} = \sqrt{225}$$
$$n = 15$$

The Pythagorean triple is 8, 15, 17.

Our work in the preceding example involved taking the square root of both sides of an equation. When you do this, you have to allow for both the positive and negative square roots. In the last example, we wanted only the positive square root, since the unknown number was part of a Pythagorean Triple and represented the length of a side of a right triangle. Unless there is a reason to omit one of the solutions, you will want to state both the positive and negative square roots as solutions. For example, note the solution to the following equation:

$$x^2 = 64$$
$$\sqrt{x^2} = \pm\sqrt{64}$$
$$x = \pm 8$$

Whenever you take an even root of a positive number to solve an equation, you have to state both the positive and the negative solutions. When you solve an equation by taking an odd root, the solution always has the same sign as the number under the radical. For example,

$$x^3 = 64 \qquad\qquad\qquad x^3 = -64$$
$$\sqrt[3]{x^3} = \sqrt[3]{64} \qquad\qquad \sqrt[3]{x^3} = \sqrt[3]{-64}$$
$$x = 4 \qquad\qquad\qquad\quad x = -4$$

2. Solve each equation.

a. Square root equation:

$$\sqrt{x} + 8 = 64 \quad \text{4,088}$$

b. Quadratic equation:

$$x^2 + 8 = 129 \quad \pm 11$$

It is important that your work has sufficient detail and is neat enough for others to follow. Pay close attention to the amount and quality of the work you show in your homework and on the tests.

c. Cubic equation: $\dfrac{x^3 - 1}{2} = 62 \quad$ 5

In the previous problem, you solved three types of nonlinear equations. In the next example, you will learn how to solve another nonlinear equation with a variable in the denominator.

EXAMPLE 2

The following equation gives the time, T, needed to travel 300 miles based on the speed, R:

$$T = \frac{300}{R}$$

This is an example of a **rational equation** with a variable in the denominator. What speed is required for a 300-mile drive to take 4 hours?

SOLUTION

We can begin by substituting 4 hours for T.

$$4 = \frac{300}{R}$$

Notice that the variable that we want to isolate is in the denominator of a fraction. One way to proceed is to multiply both sides of the equation by R to get the variable out of the denominator.

$$R \cdot 4 = \frac{300}{R} \cdot R \quad \text{Multiply both sides by } R.$$

$$4R = 300 \quad \text{Divide both sides by 4 to isolate } R.$$

$$\frac{4R}{4} = \frac{300}{4}$$

$$R = 75$$

The required speed is 75 mph.

Another way to begin the solution is to take the reciprocal of each side of the equation. This leads to the same solution.

$$4 = \frac{300}{R}$$

$$\frac{4}{1} = \frac{300}{R} \quad \text{Take the reciprocal of both sides.}$$

$$\frac{1}{4} = \frac{R}{300} \quad \text{Multiply both sides by 300 to isolate } R.$$

$$300 \cdot \frac{1}{4} = \frac{R}{300} \cdot 300$$

$$75 = R$$

A third possible way to solve the equation is to set the cross products equal after the left side has been written as a fraction.

3. Solve the rational equation.

$$\frac{64}{x + 8} = 1 \quad 56$$

Need more practice?

Solve each equation.

1. $\sqrt{x} + 2 = 12$

2. $3x^2 = 48$

3. $5x^3 - 8 = -48$

4. $\dfrac{3x}{x + 1} = 8$

Answers:

1. 100   2. ± 4

3. -2 4. $-\dfrac{8}{5}$

Although we have now solved many types of equations, we cannot extend our current equation-solving methods to all equations. For example, we cannot use the algebraic methods developed so far to solve this equation: $x^2 + 7x + 12 = 0$. You will, however, learn techniques in the following sections of this cycle for solving quadratic equations like this.

Other, more advanced techniques are required to solve some equations. For now, though, many can be solved numerically or graphically if our algebraic equation-solving process is not sufficient.

Connect

4. Consider this quadratic equation:

$$x^2 + 2x - 4 = 0$$

a. Try to solve the quadratic equation by isolating x. What do you notice?

Answers will vary. The variable appears in two terms and cannot be isolated with our current techniques.

b. Use the values in the table to estimate the solutions of the quadratic equation.

Answers will vary; $x \approx -3.2, x \approx 1.2$

x	$x^2 + 2x - 4$	x	$x^2 + 2x - 4$
−4	4	−0.8	−4.96
−3.8	2.84	−0.6	−4.84
−3.6	1.76	−0.4	−4.64
−3.4	0.76	−0.2	−4.36
−3.2	−0.16	0	−4
−3	−1	0.2	−3.56
−2.8	−1.76	0.4	−3.04
−2.6	−2.44	0.6	−2.44
−2.4	−3.04	0.8	−1.76
−2.2	−3.56	1	−1
−2	−4	1.2	−0.16
−1.8	−4.36	1.4	0.76
−1.6	−4.64	1.6	1.76
−1.4	−4.84	1.8	2.84
−1.2	−4.96	2	4
−1	−5	2.2	5.24

INSTRUCTOR NOTE: If students get stuck on part c, show them how the polynomial expression can be called y. The appropriate columns in the table can then be labeled y. That may illuminate the location on the graph of each solution to the equation.

More work on zeros of functions will be done in later sections of Cycle 3.

c. Use the graph to estimate the solutions of the quadratic equation.

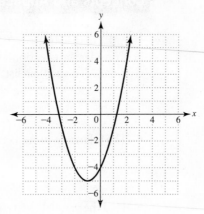

Answers will vary; $x \approx -3.3, x \approx 1.2$

d. How do your answers to parts b and c compare?

Answers will vary.

Reflect

INSTRUCTOR NOTE: Remind students to read the *Getting Ready* article for the next section.

WRAP-UP

What's the point?

You can solve many different types of equations by undoing operations to isolate the variable. This technique works not only on linear equations of the form $ax + b = c$, but also on some nonlinear equations. However, this technique of undoing operations does have its limitations, and there are many types of equations that can't be solved this way. Some additional equation-solving techniques will be developed later in this cycle.

What did you learn?

How to solve nonlinear equations

3.9 Homework

Skills MyMathLab

First complete the MyMathLab homework online. Then work the two exercises to check your understanding.

1. Solve $\sqrt{x} - 5 = 1$. To help you know where to begin solving, list the operations done to x.

6

2. Find the length of a leg of a right triangle if the length of the hypotenuse is 14 meters and the length of the other leg is 6 meters.

Approximately 12.6 meters

Concepts and Applications

Complete the problems to practice applying the skills and concepts learned in the section.

3. Consider the equation $3x^2 - 5 = 16$. List the operations performed on the variable x, in order, and then list, in order, the operations necessary to solve the equation.

Operations on x: square, multiply by 3, subtract 5

Operations to solve: add 5, divide by 3, square root

4. The following formula is used in physics for parallel resistance:

$$\frac{1}{R} = \frac{1}{R_1} + \frac{1}{R_2}$$

Find the parallel resistance, R, when R_1 and R_2 are each 6 ohms.

3 ohms

5. Complete each equation with a number so that the equation has the stated solution(s).

a. $x^2 + 7 = x^2 +$ _____ ; no solution

Answers will vary; any number other than 7 is correct.

b. $x^2 + 7 = x^2 +$ _____ ; all real numbers

7

c. $x^2 + 7 =$ _____ ; $x = \pm 3$

2

6. Solve each equation.

a. $\frac{1}{2}(2x - 4) = 12 + 3x$

−7

b. $4(x + 2) - 6x = 8 - 2x$

All real numbers

c. $\frac{4x + 8}{2} = 4 - 2x$

0

7. **Looking Forward, Looking Back** Is the equation linear? If it is, state the slope and y-intercept.

a. $y = -5x^2 - 8x + 17$

No

b. $y = -\frac{5}{8}x + 17$

Yes; $m = -\frac{5}{8}$, $(0, 17)$

c. $y = \frac{3 - x}{2}$

Yes; $m = -\frac{1}{2}$, $\left(0, \frac{3}{2}\right)$

d. $y = \sqrt{-5x - 8} + 17$

No

Getting Ready for Section 3.10

Read the article below and answer the questions that follow.

Estimating Appliance and Home Electronic Energy Use

Determining how much electricity your appliances and home electronics use can help you understand how much money you are spending to use them. Use the information below to estimate how much electricity an appliance is using and how much the electricity costs so you can decide whether to invest in a more energy-efficient appliance.

There are several ways to estimate how much electricity your appliances and home electronics use:

• Reviewing the Energy Guide label. The label provides an estimate of the average energy consumption and cost to operate the specific model of the appliance you are using. Note that not all appliances or home electronics are required to have an Energy Guide.

• Using an electricity usage monitor to get readings of how much electricity an appliance is using

• Calculating annual energy consumption and costs using the formulas provided below

• Installing a whole house energy monitoring system.

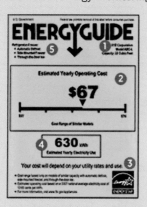

Electricity Usage Monitors

Electricity usage monitors are easy to use and can measure the electricity usage of any device that runs on 120 volts. (But it can't be used with large appliances that use 220 volts, such as electric clothes dryers, central air conditioners, or water heaters.) You can buy electricity usage monitors at most hardware stores for around $25-$50. Before using a monitor, read the user manual.

To find out how many watts of electricity a device is using, just plug the monitor into the electrical outlet the device uses, and then plug the device into the monitor. It will display how many watts the device uses. If you want to know how many kilowatt-hours (kWh) of electricity the

devices uses in an hour, or a day, or longer, just leave everything set up and read the display later.

Monitors are especially useful for finding the amount of kWh used over any period of time for devices that don't run constantly, like refrigerators. Some monitors will let you enter the amount your utility charges per kilowatt-hour and provide an estimate how much it cost to run the device since it was plugged into the monitor.

Many appliances continue to draw a small amount of stand-by power when they are switched "off." These "phantom loads" occur in most appliances that use electricity, such as televisions, stereos, computers, and kitchen appliances. Most phantom loads will increase the appliance's energy consumption a few watt-hours, and you can use a monitor to estimate those too. These loads can be avoided by unplugging the appliance or using a power strip and using the switch on the power strip to cut all power to the appliance.

Calculating Annual Electricity Consumption and Costs

Follow these steps for finding the annual energy consumption of a product, as well as the cost to operate it.

1. **Estimate the number of hours per day an appliance runs.** There are two ways to do this:

— *Rough estimate*
If you know about how much you use an appliance every day, you can roughly estimate the number of hours it runs. For example, if you know you normally watch about 4 hours of television every day, you can use that number. If you know you run your whole house fan 4 hours every night before shutting it off, you can use that number. To estimate the number of hours that a refrigerator actually operates at its maximum wattage, divide the total time the refrigerator is plugged in by three. Refrigerators, although turned "on" all the time, actually cycle on and off as needed to maintain interior temperatures.

— *Keep a log*
It may be practical for you to keep a usage log for some appliances. For example, you could record the cooking time each time you use your microwave, work on your computer, watch your television, or leave a light on in a room or outdoors.

(*continued*)

Estimating Appliance and Home Electronic Energy Use

(continued)

2. **Find the wattage of the product.** There are three ways to find the wattage an appliance uses:

— *Stamped on the appliance*
The wattage of most appliances is usually stamped on the bottom or back of the appliance, or on its nameplate. The wattage listed is the maximum power drawn by the appliance. Many appliances have a range of settings, so the actual amount of power an appliance may consume depends on the setting being used. For example, a radio set at high volume uses more power than one set at low volume. A fan set at a higher speed uses more power than one set at a lower speed.

— *Multiply the appliance ampere usage by the appliance voltage usage*
If the wattage is not listed on the appliance, you can still estimate it by finding the electrical current draw (in amperes) and multiplying that by the voltage used by the appliance. Most appliances in the United States use 120 volts. Larger appliances, such as clothes dryers and electric cooktops, use 240 volts. The amperes might be stamped on the unit in place of the wattage, or listed in the owner's manual or specification sheet.

— *Use online sources to find typical wattages or the wattage of specific products you are considering purchasing.*

3. **Find the daily energy consumption using the following formula:**
*(Wattage × Hours Used Per Day) ÷ 1000
= Daily Kilowatt-hour (kWh) consumption*

4. **Find the annual energy consumption using the following formula:**
*Daily kWh consumption × number of days used per year
= annual energy consumption*

5. **Find the annual cost to run the appliance using the following formula:**
*Annual energy consumption × utility rate per kWh
= annual cost to run appliance*

Source: energy.gov

Questions

1. Power, *P*, in watts is found by multiplying the current, *I*, in amps times the voltage, *V*, in volts:

$$P = I \cdot V$$

   a. If a heater uses 900 watts and has a voltage of 110 volts, what current does it draw?

   Approximately 8 amps

   b. If a clothes dryer uses 4,400 watts and has a voltage of 220 volts, what current does it draw?

   20 amps

2. Use your work in #1 to write a formula for current as a function of power and voltage.

   $I = \dfrac{P}{V}$

3.10 Hot and Cold: Rewriting Formulas

Explore

SECTION OBJECTIVE:

- Solve an equation for a specified variable

1. The table lists some temperatures in degrees Celsius and the corresponding temperatures in degrees Fahrenheit.

Celsius	Fahrenheit
0	32
5	41
10	50
15	59
20	68

a. Write a linear function to give the Fahrenheit temperature for a given Celsius temperature.

$F = \dfrac{9}{5}C + 32°$

b. Use the function from part a to calculate the Fahrenheit temperature that corresponds to a Celsius temperature of 95°.

203°

c. Use the function from part a to calculate the Celsius temperature that corresponds to a Fahrenheit temperature of 95°.

35°

d. If you want to convert several more Fahrenheit temperatures to Celsius temperatures, what operations should you perform on each Fahrenheit temperature? Look at your work in part c. Consider the fact that dividing by a fraction is the same a multiplying by its reciprocal.

Subtract 32°; multiply by $\dfrac{5}{9}$.

Use this list of operations to write a function that gives the Celsius temperature for a given Fahrenheit temperature.

$C = \dfrac{5}{9}(F - 32°)$

Confirm that your formula works by checking your answer to part c.

Discover

Sometimes we need a formula in a different form than the one presented to us. For example, if we want to know the area of a circle as a function of its radius, then the formula $A = \pi r^2$ is in the right form. If we instead want to know the radius of a circle as a function of its area, then we'd rather have the formula in the form $r = \sqrt{A/\pi}$. The two formulas describe the same relationship between radius and area, but the form desired depends on what question you are trying to answer. This section will focus on how to rearrange a formula for a specified variable.

As you work through this section, consider the following questions:

1 When would you need to solve a formula for a different variable?

2 How does the process of solving a formula for a specified variable compare to solving an equation for x?

We solve a formula for a specified variable the same way we solve any equation for a variable, that is, by undoing the operations that have been done to that variable.

INSTRUCTOR NOTE: You might need to show students how to simplify the radical involved in the solution to #2.

2. You have used the Pythagorean theorem, $a^2 + b^2 = c^2$, to find the lengths of the sides of a right triangle. If a right triangle has two legs that are the same length, then it is isosceles and this formula becomes

$$a^2 + b^2 = c^2$$
$$a^2 + a^2 = c^2$$
$$2a^2 = c^2$$

Solve this formula for c to get a formula for the hypotenuse of an isosceles right triangle. Then use this new formula to find the length of the hypotenuse of an isosceles right triangle if the legs are each 5 inches long.
$c = \sqrt{2a^2} = \sqrt{2}a$; 7.07 inches

EXAMPLE 1

The formula $P = 2L + 2W$ gives the perimeter, P, of a rectangle based on its length, L, and width, W. Solve this formula for L.

SOLUTION

We want to isolate the L in the formula. To begin, notice that we can remove a $2W$ term from the side with the $2L$. Then we divide both sides by 2 to isolate L. We are thinking of L as the only variable and any other variables are being treated as constants.

$$P = 2L + 2W$$

$$\frac{-2W \qquad\qquad -2W}{P - 2W = 2L} \qquad \text{Subtract } 2W \text{ from each side.}$$

$$\frac{P - 2W}{2} = \frac{2L}{2} \qquad \text{Divide each side by 2.}$$

$$\frac{P - 2W}{2} = L$$

The formula can be left in this form, or each term in the numerator can be divided by 2 to obtain the alternative form:

$$L = \frac{P - 2W}{2} = \frac{P}{2} - \frac{2W}{2} = \frac{P}{2} - W$$

HOW IT WORKS

To solve a formula for a specified variable:

1. Determine the variable that will be treated as the unknown.
2. Treat all other variables as constants.
3. Solve the equation for the unknown.

EXAMPLE: The formula $I = Prt$ is used for computing simple interest, I, from the principal, P, interest rate, r, and time, t. Solve the formula for P.

Since the equation will be solved for P, it becomes the unknown, and all the other variables are treated as though they are constants.

Since P is multiplied by rt, undo that operation by dividing each side by rt.

$$I = Prt$$

$$\frac{I}{rt} = \frac{P\cancel{rt}}{\cancel{rt}}$$

$$\frac{I}{rt} = P$$

The result is a new formula for P in terms of I, r, and t.

Consider writing a list of the operations done to the variable. Then write the corresponding list of steps needed to solve. Another option is to write a version of the equation with numbers substituted for the other variables, such as $6 = 5 + 4t$ for the equation in part b.

3. For each formula, solve for the variable specified.

a. $D = RT$ for T

$$T = \frac{D}{R}$$

b. $A = P + Prt$ for r

$$r = \frac{A - P}{Pt}$$

c. $S = 4\pi r^2$ for r

$$r = \sqrt{\frac{S}{4\pi}}$$

Solving a linear equation for y can help you identify important information about its graph, as shown in the next example.

EXAMPLE 2

Find the slope and y-intercept of the graph of the equation $4x - 3y = 12$.

SOLUTION

It will be easier to identify the slope and y-intercept if the equation is in slope-intercept or $y = mx + b$ form. To get this form, we need to solve for y.

$$
\begin{array}{ll}
4x - 3y = 12 & \text{Isolate } -3y \text{ by subtracting } 4x \text{ from each side} \\
\underline{-4x \qquad\quad -4x} & \text{of the equation.} \\
-3y = -4x + 12 \\
\dfrac{-3y}{-3} = \dfrac{-4x}{-3} + \dfrac{12}{-3} & \text{Isolate } y \text{ by dividing all terms by } -3. \\
y = \dfrac{4}{3}x - 4 & \text{Simplify.}
\end{array}
$$

From the equation we can see that the slope is $m = \dfrac{4}{3}$ and the y-intercept is $(0, -4)$.

---

4. Solve the equation $6x + 3y = 15$ for y. Identify the slope and y-intercept, and then use them to graph the equation.

   $y = -2x + 5$; $m = -2$; (0, 5)

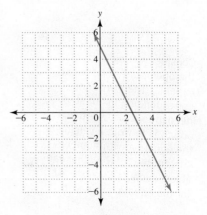

Connect

5. You have learned how to use the slope-intercept form, $y = mx + b$, to write an equation of a line. Let's consider another method for finding the equation of a line. We will begin by writing a version of the slope formula you saw in Section 3.3, but with one of the specific points (x_2, y_2) replaced with a generic point (x, y).

$$
m = \frac{y - y_1}{x - x_1}
$$

   a. Solve this formula for $y - y_1$. The resulting formula is called the **point-slope form** of the equation of a line.

   $y - y_1 = m(x - x_1)$

b. Substitute a specific point (2, 5) for (x_1, y_1) and use a value of $m = 3$. Then solve the equation for y to get the slope-intercept form of the equation.

$y = 3x - 1$

c. Use the point and slope given in part b and the slope-intercept form, $y = mx + b$, to find the equation of the line again. You should get the same result as you did in part b. Which method did you prefer?

$y = 3x - 1$; answers will vary.

Reflect

WRAP-UP

What's the point?

Sometimes a formula is easier to use if you first solve it for the variable of interest. Solving a formula for a specific variable is similar to solving an equation that has only one variable. In both cases, you undo the operations that have been done to the variable of interest.

What did you learn?

How to solve an equation for a specified variable

3.10 Homework

Skills MyMathLab

First complete the MyMathLab homework online. Then work the two exercises to check your understanding.

1. Rewrite this equation in slope-intercept form: $3x + 4y = 12$.

$y = -\dfrac{3}{4}x + 3$

2. Profit is found by subtracting costs from revenue that is, $P = R - C$. Solve this formula for C.

$C = R - P$

Concepts and Applications

Complete the problems to practice applying the skills and concepts learned in the section.

3. The diagram compares Celsius and Fahrenheit temperatures.

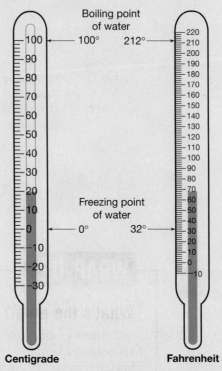

a. Notice that the label on the left thermometer is Centigrade, another name for Celsius. Why do you think the Celsius scale is also known by this name?

There are 100 degrees between the freezing and boiling points of water.

b. Normal body temperature is considered to be 98.6° Fahrenheit. Find this temperature in Celsius degrees by estimating from the diagram and by using a formula to find the exact answer.

37° Celsius

c. Which is larger: a degree in Celsius or a degree in Fahrenheit? Why?

A Celsius degree is larger. From the diagram, we can see that we need to increase by only 100 degrees in Celsius to go from the freezing point to the boiling point of water, whereas we have to increase 180 degrees in Fahrenheit to do the same. Since it takes fewer Celsius degrees, each Celsius degree must be larger than each Fahrenheit degree.

4. In statistics, the formula $E = t\dfrac{s}{\sqrt{n}}$ can be used to calculate a margin of error. Solve this formula for n, the sample size.

$n = \dfrac{t^2 s^2}{E^2}$

5. In statistics, the formula $E = z\sqrt{\dfrac{pq}{n}}$ can be used to calculate the margin of error for a political poll. Solve this formula for n, the sample size.

$n = \dfrac{pqz^2}{E^2}$

6. The formula $Z = \dfrac{\text{data value} - \text{mean}}{\text{standard deviation}}$ will be used in Cycle 4 to calculate a z-score. Solve this formula for "data value."

Data value = mean + standard deviation $\cdot Z$

7. The following formula gives the area, A, of a trapezoid based on its height, h, and the length of its two bases, b and B:

$$A = \frac{1}{2}h(b + B)$$

a. Solve the formula for the height, h.

$h = \dfrac{2A}{b + B}$

b. Solve the formula for the length of the base b.

$b = \dfrac{2A - hB}{h}$ or $b = \dfrac{2A}{h} - B$

c. Solve the formula for the length of the base B.

$B = \dfrac{2A - hB}{h}$ or $B = \dfrac{2A}{h} - b$

8. Use the point-slope form of the equation of a line to find the equation of the line that has a slope $m = -5$ through the point $(-1, 7)$. Solve the equation for y.

$y = -5x + 2$

9. **Looking Forward, Looking Back** Becky lost 10% of her body weight and now weighs 153 pounds. Find how much she weighed before she lost weight. Check your answer.

170 pounds

3.11 A Common Goal: Greatest Common Factors

Explore

SECTION OBJECTIVE:

• Factor an expression using the greatest common factor

INSTRUCTOR NOTE: Consider reminding students how to make a factor tree to write a number in prime-factored form.

1. Write the numerator and denominator of each fraction in prime-factored form, such as $3 \cdot 3 \cdot 5 \cdot 7$. Then state the greatest common factor of the numerator and denominator. Divide off the greatest common factor to simplify each fraction.

a. $\dfrac{4{,}620}{8{,}160} = \dfrac{2 \cdot 2 \cdot 3 \cdot 5 \cdot 7 \cdot 11}{2 \cdot 2 \cdot 2 \cdot 2 \cdot 2 \cdot 3 \cdot 5 \cdot 17} = \dfrac{77}{136}$ GCF = 60

b. $\dfrac{x^5 y^7}{x^2 y^{10}} = \dfrac{x \cdot x \cdot x \cdot x \cdot x \cdot y \cdot y \cdot y \cdot y \cdot y \cdot y \cdot y}{x \cdot x \cdot y \cdot y \cdot y \cdot y \cdot y \cdot y \cdot y \cdot y \cdot y \cdot y} = \dfrac{x^3}{y^3}$ GCF = $x^2 y^7$

Discover

The **greatest common factor**, or GCF, is used to simplify fractions, but it also has other uses. One such use is to rewrite formulas in order to make them easier to use or to solve for a variable that appears multiple times in a formula.

As you work through this section, think about these questions:

1 Do you know the difference between the factored and nonfactored forms of an expression?

2 Can you find the GCF of the terms of a variable expression without writing the prime factors?

Let's begin by looking at factored and nonfactored forms of expressions.

The formula to convert Fahrenheit to Celsius is $C = \dfrac{5}{9}(F - 32°)$. The right side of the formula is one term that is a product with factors $\frac{5}{9}$ and $(F - 32°)$. Quantities that are written as products are said to be in **factored form**. So the right side of the formula is in factored form.

We can rewrite the right side of the formula by using the distributive property. The new right side, $\dfrac{5}{9}F - \dfrac{160°}{9}$, is an expression with two terms and is not in factored form.

$$C = \underbrace{\dfrac{5}{9}(F - 32°)}_{\text{Factored form}} = \underbrace{\dfrac{5}{9}F - \dfrac{160°}{9}}_{\text{Nonfactored form}}$$

Notice that evaluating the factored form of the formula for $212°$ is simpler because the fraction occurs only once.

Evaluate with factored form:	**Evaluate with nonfactored form:**
$C = \dfrac{5}{9}(F - 32°)$	$C = \dfrac{5}{9}F - \dfrac{160°}{9}$
$= \dfrac{5}{9}(212° - 32°)$	$= \dfrac{5}{9}(212°) - \dfrac{160°}{9}$
$= \dfrac{5}{9}(180°)$	$= \dfrac{1060°}{9} - \dfrac{160°}{9}$
$= \dfrac{5}{\cancel{9}} \cdot \dfrac{\overset{20°}{\cancel{180°}}}{1}$	$= \dfrac{900°}{9}$
$= 100°$	$= 100°$

We used the distributive property to change the factored form of the expression into the nonfactored form. To reverse that process and write an expression in factored form, we factor the expression.

LOOK IT UP

Factoring

Factoring is the process of writing a sum or difference as a product.

FOR EXAMPLE, the expression $3x + 6$ is not factored, since it is the sum of two terms. However, $3(x + 2)$ is an equivalent expression but in factored form, since it is the product of the quantities 3 and $(x + 2)$.

There are multiple factoring methods; each depends on the traits of the expression at hand. This section will focus on factoring out the greatest common factor (GCF), which is the reverse process of applying the distributive property.

Distributing

$$3(x + 2) = 3x + 6$$

Factoring out the GCF

Factoring out a common factor involves taking a common factor from a set of terms and stating that factor only once, instead of in each term. In an algebraic expression, the common factor is written in front of parentheses. The result is useful not only in mathematics, as seen in the next problem, but also in real life. When two people share a last name, like Will Smith and Ella Smith, it is easier and faster to say Will and Ella Smith.

Factoring out the GCF involves finding *all* the factors that are common to the terms and then rewriting the expression with the product of those factors in front of parentheses. Inside the parentheses are the terms necessary to create the original expression if the distributive property is used.

2. Factor the following formula for the area of a trapezoid in three different ways:

$$A = \frac{1}{2}hb_1 + \frac{1}{2}hb_2.$$

Which of the factored forms involved factoring out the GCF?

$A = \frac{1}{2}h(b_1 + b_2)$ or $A = \frac{1}{2}(hb_1 + hb_2)$ or $A = h\left(\frac{1}{2}b_1 + \frac{1}{2}b_2\right)$

The form $A = \frac{1}{2}h(b_1 + b_2)$ has the GCF factored out.

We need to further understand how to identify the GCF of an expression and how to use it to factor the expression.

EXAMPLE

Identify and factor out the GCF: $3x^2 - 6x^3y$.

SOLUTION

To begin factoring out the greatest common factor, we first need to identify the factors that are common to each term. For each term in the expression, we list the prime factors. Since there are exponents on the variables, we list the simplest factors for each variable factor. This will allow us to see common factors, if there are any.

$$3x^2 - 6x^3y$$

Simplest factors of each term: $3, x, x;$ $2, 3, x, x, x, y$

Notice that the factors that appear in both lists, in other words, the factors the terms have in common, are 3, x, and x. The product of these factors is known as the greatest common factor of the expression, $3x^2$.

Next, we write the GCF in front of parentheses. We fill in the parentheses with two terms (the number of terms we started with) so that if the expression is simplified using the distributive property and multiplication, it will look like the original expression.

$$3x^2(\underline{\quad} \ \underline{\quad})$$

To complete the parentheses, we determine what $3x^2$ needs to be multiplied by to get the first term of the original expression, $3x^2$. The answer is 1, which becomes the first term inside the parentheses. Then we determine what $3x^2$ needs to be multiplied by to get the second term of the original expression, $-6x^3y$. We need to multiply by $-2xy$, which becomes the second term inside the parentheses. The final factored form of the expression is:

$$3x^2(1 - 2xy)$$

We can check our work by distributing.

$$3x^2(1 - 2xy) = 3x^2 - 6x^3y$$

We end up with the original expression: $3x^2 - 6x^3y$.

We can generalize which factors contribute to the GCF so that we can factor out the GCF more quickly. Notice in the previous example that the coefficient of the GCF is the GCF of the coefficients in the individual terms. The variable factors in the GCF are the variables that are common to all the terms, raised to the lowest exponent that appears on each variable.

3. Identify the GCF of the terms in each expression.

   a. $16y^4 + 100x^5y^2$ $4y^2$

   b. $51yz + 34yzw + 68wz$ $17z$

   c. $6x + 4x^2 + 8x^3 + 16x^4 + 32x^5$ $2x$

HOW IT WORKS

To factor out the greatest common factor (GCF):

1. Find the greatest common factor of the terms. This is the product of the GCF of the coefficients in the individual terms and the variables common to each term, raised to the lowest exponent that appears on each variable.
2. Write the GCF in front of a set of parentheses.
3. Fill in the parentheses by determining the terms that are necessary to create the original expression if the distributive property is used.

NOTES:

1. The number of terms in the original expression is the same as the number of terms inside the parentheses.
2. If the terms do not have any common factors other than 1, the expression cannot be factored using the GCF.

HOW IT WORKS

EXAMPLE: Factor out the GCF in the expression $9a^4b + 12a^5b^2c + 15a^8b^5$.

To begin, find the GCF of the coefficients of the terms. It is 3, since 3 is the largest common factor of 9, 12, and 15.

Notice that a and b are the variables common to all terms. The lowest exponent on a among all the terms is 4. The lowest exponent on b among all the terms is 1.

The GCF of the expression is $3a^4b$.

Write the GCF in front of a set of parentheses, and then fill in the parentheses with the three terms that will result in the original expression if the distributive property is used.

So, the factored form is $3a^4b(3 + 4abc + 5a^4b^4)$.

The terms inside the parentheses should have a greatest common factor of 1 if you've factored out the GCF correctly.

4. Factor out the GCF in each expression.

a. $30x^2 + 75xy$ $\quad 15x(2x + 5y)$

b. $14xwz - 28wz^3 + 7wz$ $\quad 7wz(2x - 4z^2 + 1)$

c. $24 - 8x + 12x^3 + 16x^5$ $\quad 4(6 - 2x + 3x^3 + 4x^5)$

The GCF can be used to rewrite a side of a formula so that a variable appears only once, which allows it to be isolated.

5. Simple interest, I, is calculated with the formula $I = Prt$, where P is the invested amount, r is the interest rate, and t is a period of time. The final value, A, of an investment with simple interest is given by $A = P + I$. Substituting for I, we get a new formula for the final value of an investment: $A = P + Prt$.

a. Solve the formula $A = P + Prt$ for P, the invested amount. You need to first factor the right side of the formula so that P appears only once.

$A = P(1 + rt); P = \dfrac{A}{1 + rt}$

b. How much should you invest to have $50,000 after 10 years at 4% interest? Use r in decimal form.

$35,714.29

Connect

Another method of factoring is known as **factoring by grouping**. It is used when there are four or more terms that do not have a factor common to all the terms but do have common factors when considered in groups. For example, the first two terms in the following expression have a common factor of x, and the last two terms have a common factor of y:

$$\underbrace{ax + bx}_{\substack{x \text{ is a} \\ \text{common factor}}} + \underbrace{ay + by}_{\substack{y \text{ is a} \\ \text{common factor}}}$$

Factoring out those GCFs from their respective pairs of terms, we get the following:

$$ax + bx + ay + by = x(a + b) + y(a + b)$$

The terms have a common factor of $(a + b)$, which is necessary to factor further. Factoring that out, we get the following product:

$$x(a + b) + y(a + b) = (a + b)(x + y)$$

6. Use the method of factoring by grouping to factor each expression.

a. $x^2 + 5x + 9x + 45$ $(x + 5)(x + 9)$

b. $xy + 11x + 10y + 110$ $(y + 11)(x + 10)$

Reflect

WRAP-UP

What's the point?

Factoring out the greatest common factor and factoring by grouping are two important factoring techniques. These methods can be used to rewrite expressions as products instead of sums or differences. They can also be used to rewrite expressions and formulas in more useful forms. In the next section of the cycle, you will learn additional methods of factoring.

What did you learn?

How to factor an expression using the greatest common factor

3.11 Homework

Skills MyMathLab

First complete the MyMathLab homework online. Then work the two exercises to check your understanding.

1. State the GCF of the terms in the expression $24x^3y^6 - 144x^2y^2$. Then factor out the GCF.
   $24x^2y^2$; $24x^2y^2(xy^4 - 6)$

2. Factor the right side of each formula.

   a. $S = 2\pi r^2 + 2\pi rh$ $S = 2\pi r(r + h)$

   b. $S = \pi r^2 + \pi rl$ $S = \pi r(r + l)$

Concepts and Applications

Complete the problems to practice applying the skills and concepts learned in the section.

3. The expression $-2x^3 + 6x^3y - 12x^2$ is factored in two different ways as shown below.

Original expression:	$-2x^3 + 6x^3y - 12x^2$
Expression with the GCF factored out:	$2x^2(-x + 3xy - 6)$
Expression with the opposite of the GCF factored out:	$-2x^2(x - 3xy + 6)$

   The third form is often used in mathematics. It is not better than the others, just more common, since the negative sign appears with the GCF instead of with the first term inside the parentheses.

   a. How do the terms in the parentheses change when the opposite of the GCF is factored out?
      Each term has the opposite sign of the term in the original expression.

   b. Evaluate the original expression and the two factored forms when $x = -1$ and $y = 2$ without using a calculator.

      Original: -22

      First factored form: -22

      Second factored form: -22

   c. Which of the three forms was easiest to use in part b? Why?
      Answers will vary.

4. Although temperatures are commonly measured in Fahrenheit or Celsius degrees, they can also be indicated on the Kelvin scale. Temperatures on the Kelvin scale are about 273 higher than Celsius temperatures. For example, 0°C is 273 K.

a. Write a formula for Kelvin temperatures in terms of Celsius temperatures.

$K = C + 273$

b. Convert 104° Fahrenheit to Kelvin by first converting to Celsius using $C = \frac{5}{9}(F - 32)$ and then converting to Kelvin using the formula in part a.

104° Fahrenheit = 40° Celsius = 313 Kelvin

c. Repeat the conversion for 86° Fahrenheit.

86° Fahrenheit = 30° Celsius = 303 Kelvin

d. The conversion process we used is cumbersome, especially if we need to do it multiple times. Since we want to convert from Fahrenheit to Kelvin, let's use the formulas we have to create a new formula. Replace the C in the Kelvin equation from part a with $\frac{5}{9}(F - 32)$.

$K = \frac{5}{9}(F - 32) + 273$

e. Distribute and simplify the right-hand side of the formula in part d. State the answer using fractions, not decimals.

$K = \frac{5}{9}F + \frac{2{,}297}{9}$

f. The fraction at the end of the formula makes this form difficult to use. Typically the formula is rewritten using two steps: (1) Factor out $\frac{5}{9}$, and then (2) rewrite the fraction remaining inside the parentheses as a decimal.

$K = \frac{5}{9}(F + 459.4)$

g. Check that this formula works by converting 104° Fahrenheit to Kelvin. You should get the same answer as you did in part b.

313 Kelvin

5. In Cycle 2, you learned how to use a box to multiply polynomial expressions. To multiply the expression $2x(5x + 2x^4 - 1)$ with a box, the work is as follows:

$$\begin{array}{c|c|c|c} & 5x & 2x^4 & -1 \\ \hline 2x & 10x^2 & 4x^5 & -2x \end{array}$$

So $2x(5x + 2x^4 - 1) = 10x^2 + 4x^5 - 2x$.

A box can also be used to factor expressions. The expression being factored is written inside the box, and the GCF of the terms is written as one of the dimensions of the box.

a. Use the box method to factor out the GCF of $15x^4 + 15x + 15xz$. Write the GCF on the left side of the box and the remaining terms that would be inside the parentheses on the top of the box. State the factored form of the expression.

$$\begin{array}{c|c|c|c} & x^3 & 1 & z \\ \hline 15x & 15x^4 & 15x & 15xz \end{array}$$

$15x^4 + 15x + 15xz = 15x(x^3 + 1 + z)$

b. Fill in the missing terms outside or inside the box. Then write the product indicated and the result.

$$\begin{array}{c|c|c|c} & 6x^3 & -x^2 & 5 \\ \hline 10x & 60x^4 & -10x^3 & 50x \end{array}$$

$10x(6x^3 - x^2 + 5) = 60x^4 - 10x^3 + 50x$

6. Looking Forward, Looking Back Find the products.

a. $(7x - 8)(7x + 1)$ $49x^2 - 49x - 8$

b. $(7x - 8)(7x + 8)$ $49x^2 - 64$

c. $(7x - 8)(7x - 8)$ $49x^2 - 112x + 64$

3.12 Thinking Outside the Box: Factoring Quadratic Expressions

Explore

SECTION OBJECTIVE:
• Factor quadratic expressions

Recall the box method for multiplying binomials. Each binomial is written as a dimension on the outside of a 2×2 box. The corresponding products are found, written inside the box, and then added to get the result.

The following box shows how to multiply $(x + 4)(x + 5)$:

	x	5
x	x^2	$5x$
4	$4x$	20

Area of box: $\underbrace{(x + 4)(x + 5)}_{\text{Factored form}} = x^2 + 4x + 5x + 20 = \underbrace{x^2 + 9x + 20}_{\text{Nonfactored form}}$

We used the box method before to multiply polynomials, but now we will use it for the reverse process of factoring polynomials.

1. Use the box method to factor $x^2 + 10x + 21$. Write the area of the box as a sum and as a product.

	x	3
x	x^2	$3x$
7	$7x$	21

$(x + 7)(x + 3) = x^2 + 10x + 21$

Discover

INSTRUCTOR NOTE: For students who will be taking more algebra, speed with factoring is necessary. However, most non-STEM students will not take much more algebra. To support these students and the philosophy of a pathways course, this section will focus on understanding factoring and making connections. If you would like to include more factoring practice, additional factoring exercises are included in the MyMathLab course that accompanies the text.

In Section 3.11 you learned how to factor an expression using the greatest common factor. This section will explore additional techniques for transforming an expression that is written with multiple terms into an expression with multiple factors. We will focus particularly on factoring quadratic expressions of the form $ax^2 + bx + c$, where a, b, and c are real numbers and a is not zero. Many times these expressions will be trinomials, but if they are missing the x term, as in $x^2 - 16$, or the constant term, as in $x^2 + 8x$, they are binomials.

As you work through this section, think about these questions:

❶ Can you factor a quadratic expression using the box method as well as the guess-and-check method?

❷ Do you know how to determine if a trinomial is prime?

❸ Can you combine factoring techniques to factor an expression completely?

Remember?

FOIL is an acronym for the four products involved when you multiply two binomials: First, Outer, Inner, and Last.

Let's begin by looking at quadratic expressions like the ones in the *Explore* to find relationships that can be used to factor the expressions.

Consider these products. Each can be found using the box method or FOIL.

$$(x + 2)(x + 3) = x^2 + 3x + 2x + 6 = x^2 + 5x + 6$$

$$(x - 4)(x - 6) = x^2 - 6x - 4x + 24 = x^2 - 10x + 24$$

$$(x - 5)(x + 1) = x^2 + x - 5x - 5 = x^2 - 4x - 5$$

Notice that when we multiply binomials of the form $(x + $ a number$)(x + $ a number$)$, where the numbers can be positive or negative, we get a quadratic expression of this form $x^2 + ($sum of numbers$)x + ($product of numbers$)$. So we can factor an expression of the form $x^2 + ($sum of numbers$)x + ($product of numbers$)$ with factors $(x + $ a number$)(x + $ a number$)$. Let's practice finding pairs of numbers with given sums and products and then use them to factor quadratic expressions.

2. For each row of the table, find a pair of numbers with the given product and sum.

Product	Sum	Pair of Numbers
30	11	5, 6
30	−11	−5, −6
44	−15	−11, −4
−63	−2	−9, 7
−24	5	−3, 8
−24	−5	3, −8

3. Use the pattern

$$x^2 + (\text{sum of numbers})x + (\text{product of numbers}) = (x + \text{a number})(x + \text{a number})$$

to factor each expression. Use FOIL to check the result.

a. $x^2 + 7x + 12$ $(x + 3)(x + 4)$

b. $x^2 - 12x + 20$ $(x - 10)(x - 2)$

c. $x^2 + 2x - 8$ $(x - 2)(x + 4)$

d. $x^2 - 2x - 35$ $(x - 7)(x + 5)$

In each part of #3, you were able to find two numbers that multiplied to the necessary product and added to the necessary sum. There are quadratic expressions for which it is not possible to find numbers that work, however. For example, for $x^2 + 6x + 7$, there are no rational numbers that multiply to be 7 and add to be 6. If that is the case and the expression doesn't have a GCF, the expression is considered to be **prime**.

The previous quadratic expressions had the number 1 as the coefficient of the x^2 term. In those cases, the method you just practiced works well to factor the expressions. However, if the coefficient of the x^2 term is not 1, we will need to adjust our method somewhat. We can strategically guess and check to reverse the FOIL process.

Notice how the product is formed with FOIL.

$$\text{First} = 2x^2 \qquad \text{Last} = -6$$

$$(2x - 1)(x + 6) = 2x^2 + 11x - 6$$

$$\text{Inner} = -1x$$
$$\text{Outer} = 12x$$

To "unFOIL" $2x^2 + 11x - 6$, we need to write two sets of parentheses that will be filled in with two binomials. The first terms in the binomials must multiply to be $2x^2$ but must also work to make the middle term $11x$. So the first terms are $2x$ and x instead of $2x^2$ and 1. The last terms in the binomials must multiply to be -6 but again must work to make the middle term $11x$. We can try pairs of numbers for the last terms of the binomial that multiply to be -6. Each time we try a pair, we must check the inner and outer products to see if their sum is $11x$.

INSTRUCTOR NOTE: Consider having students verify the middle term by finding the inner and outer terms in each option.

$(2x + 3)(x - 2) = 2x^2 - x - 6$	Incorrect
$(2x - 3)(x + 2) = 2x^2 + x - 6$	Incorrect
$(2x + 2)(x - 3) = 2x^2 - 4x - 6$	Incorrect
$(2x - 2)(x + 3) = 2x^2 + 4x - 6$	Incorrect
$(2x + 6)(x - 1) = 2x^2 + 4x - 6$	Incorrect
$(2x - 6)(x + 1) = 2x^2 - 4x - 6$	Incorrect
$(2x + 1)(x - 6) = 2x^2 - 11x - 6$	Incorrect (middle sign is wrong)
$(2x - 1)(x + 6) = 2x^2 + 11x - 6$	Correct

The order of the binomial *factors* does not matter. $(2x - 1)(x + 6)$ is the same as $(x + 6)(2x - 1)$. However, as seen in the several incorrect options, how the terms are paired in the parentheses does matter.

So $2x^2 + 11x - 6$ factors as $(2x - 1)(x + 6)$. The more you do this process, the faster you will get. But it can be time consuming depending on the coefficient of the x^2 term and the constant term.

If the constant term of the quadratic expression is positive, the signs in the binomials will both be positive when the middle term is positive or will both be negative when the middle term is negative. If the constant term is negative, the sign in one binomial factor will be positive and the other will be negative.

Pay attention to the sign patterns to factor more quickly:

$+\ +\ +\ \rightarrow (\ +\ )(\ +\ )$
$+\ -\ +\ \rightarrow (\ -\ )(\ -\ )$
$+\ +\ -\ \rightarrow (\ +\ )(\ -\ )$
$+\ -\ -\ \rightarrow (\ +\ )(\ -\ )$

Constant term of quadratic expression	Signs in binomial factors	Example
Positive	Both positive if the middle term is positive	$x^2 + 5x + 4 = (x + 4)(x + 1)$
	Both negative if the middle term is negative	$x^2 - 5x + 4 = (x - 4)(x - 1)$
Negative	One positive, one negative	$x^2 + 3x - 4 = (x + 4)(x - 1)$ $x^2 - 3x - 4 = (x - 4)(x + 1)$

EXAMPLE

Factor $3x^2 - 16x + 5$.

SOLUTION

We begin by checking for a GCF to see if the expression can be written in a simpler way. The only GCF for this expression is 1, so we will try the unFOILing approach. Continue by writing two sets of parentheses $(\ )(\ )$. The first terms must multiply to be $3x^2$ but contribute to make a middle term of $-16x$. $3x$ and x are the only choice if we are using integers.

$$(3x\quad)(x\quad)$$

Next, we need numbers that multiply to be 5 but make $-16x$ when the inner and outer products are added. The numbers are -5 and -1, since the product is positive but the eventual sum is negative. We need to determine the location for each number. Trying -5 in the first factor and -1 in the second factor, we get the following:

$$(3x - 5)(x - 1) = 3x^2 - 3x - 5x + 5 = 3x^2 - 8x + 5 \quad \text{Incorrect}$$

Since the middle term is not correct, we swap the -5 and -1.

$$(3x - 1)(x - 5) = 3x^2 - 15x - x + 5 = 3x^2 - 16x + 5 \quad \text{Correct}$$

The x terms add to $-16x$. Thus, the original expression has been factored as $(3x - 1)(x - 5)$.

4. Factor each quadratic expression.

    a. $4x^2 + 5x + 1 = (4x + 1)(x + 1)$

    b. $2x^2 + x - 6 = (2x - 3)(x + 2)$

The guess-and-check approach can be fast with some quadratic expressions and slow with others. Another approach to factor a quadratic expression uses the box method you saw in the *Explore*. The box method has a few more steps, but it is more systematic and requires less guess and check. When you encounter a quadratic expression to factor, choose the method that makes sense to you.

To use the box method to factor a quadratic expression, we need to understand how the box is used to multiply two first-degree binomials so that we can undo that process and find the outside of the box.

Consider the product $(3x + 2)(x + 5)$. The box method has been used to *multiply* these binomials. Notice that multiplying two first-degree binomials results in a second-degree polynomial.

Multiplying: Start outside the box and work in

Factoring: Start inside the box and work out

Factored form $(3x + 2)(x + 5)$

	$3x$	2
x	$3x^2$	$2x$
5	$15x$	10

Nonfactored form $3x^2 + 17x + 10$

So $(3x + 2)(x + 5) = 3x^2 + 17x + 10$.

Now let's reverse this process and *factor* the expression $3x^2 + 17x + 10$ using a box. To do that, notice the following relationships between the nonfactored form of the quadratic expression, $3x^2 + 17x + 10$, and the terms *inside* the box:

- The box dimensions are two rows by two columns.
- The squared term of the expression, $3x^2$, and the constant term, 10, are in the box diagonally opposite each other.
- The x term, $17x$, is not in the box. Instead there are two terms, $15x$ and $2x$, that sum to $17x$.

Next, we need to understand how to find $15x$ and $2x$, since there are many pairs of terms that sum to $17x$. If we look more closely at the terms inside the box, we see the following:

- The constants diagonally opposite each other have the same product.

That is, $3 \cdot 10 = 30$ and $2 \cdot 15 = 30$. So we need to find two numbers that have the same product, 30, but have the sum of 17 when added. In this case, those numbers are 2 and 15. We can fill in the box with $2x$ and $15x$.

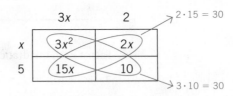

The last part of factoring involves finding the dimensions on the outside of the box. Notice the following:

- Each term on the outside of the box is the GCF of its row or column.

For example, x is the GCF of the first row terms $3x^2$ and $2x$. We can use this idea of the GCF to complete the outside of the box. The result is a product of those dimensions: $(3x + 2)(x + 5)$.

5. Use the box method to factor $5x^2 + 23x + 12$.

	x	4
$5x$	$5x^2$	$20x$
3	$3x$	12

$(5x + 3)(x + 4)$

HOW IT WORKS

To factor a quadratic expression using the box method:

1. Draw a 2 × 2 box.
2. Place the squared term and the constant term inside the box diagonally opposite each other.
3. Find the product of the coefficient of the squared term and the constant term.
4. Find two numbers that have the product from Step 3 and add to the coefficient of the x term in the original expression. These numbers are the coefficients of the x terms inside the box.
5. Write the x terms from Step 4 in the remaining boxes inside the box.
6. Factor out the GCF of each row and column and write them outside the box.
7. State the dimensions of the box as a product.

EXAMPLE: Use the box method to factor $2x^2 - 19x + 24$.

Begin by drawing a box with two rows and two columns. Place the squared term, $2x^2$, and the constant term, 24, inside the box diagonally opposite each other.

$2x^2$	
	24

Find the product of the constant term and the coefficient of the x^2 term in the box: $2 \cdot 24 = 48$. The other terms are x terms whose coefficients multiply to 48 and add to -19. The numbers that satisfy those conditions are -3 and -16.

Place $-3x$ and $-16x$ in either of the empty boxes.

$2x^2$	$-3x$
$-16x$	24

Next, write the GCF of each row and each column outside the box. Be careful with the signs so that the dimensions written result in the products inside the box. For example, either 3 or -3 can be factored out of $-3x$ and 24. If x is written for the GCF of the first row, then -3 will need to be above the second column. This is the approach used in the box that follows. 3 could be factored out first and placed above the second column. In that case, then $-x$ will be the GCF of the first row. The final answer's form will look different, $(-2x + 3)(-x + 8)$, but it is equivalent to $(2x - 3)(x - 8)$.

	$2x$	-3
x	$2x^2$	$-3x$
-8	$-16x$	24

Last, write the factored form: $(2x - 3)(x - 8)$. You can check the result by multiplying the factors to see if you get the original nonfactored expression, $2x^2 - 19x + 24$.

6. Use the box method to factor each expression.

a. $14x^2 + 11x - 15$

	$2x$	3
$7x$	$14x^2$	$21x$
-5	$-10x$	-15

$(7x - 5)(2x + 3)$

b. $4x^2 + 20x + 25$

	$2x$	5
$2x$	$4x^2$	$10x$
5	$10x$	25

$(2x + 5)(2x + 5)$ or $(2x + 5)^2$

c. $x^2 + 0x - 81$

	x	-9
x	x^2	$-9x$
9	$9x$	-81

$(x - 9)(x + 9)$

In #6b, the polynomial factored into a binomial squared. When that happens, the original nonfactored polynomial is called a **perfect square trinomial**. In #6c, the original non-factored polynomial was a **difference of two squares**, a type of quadratic expression that will be explored more in the homework.

In Section 3.11 you learned how to factor out a greatest common factor if an expression has one. In this section, you learned how to factor a quadratic expression. It's possible to have polynomials where both of these factoring methods can be used.

7. Factor the expression completely: $5x^2 - 10x - 40$. That is, factor out a GCF if one exists and then factor the remaining quadratic expression using any method from this section.

$5(x - 4)(x + 2)$

Connect

As we saw in Section 3.9, some quadratic equations can be solved by isolating the variable. For those that cannot, factoring may be a way to solve the equation. To use factoring to solve a quadratic equation, we need to understand products of 0.

In each of the following products, at least one factor is 0 and the resulting product is 0:

$$0 \cdot 4 = 0$$
$$-7 \cdot 0 = 0$$
$$0 \cdot 0 = 0$$

The **zero-product property** states that if a product of two numbers is 0, then at least one of the numbers *must* be 0. This idea can be combined with factoring to solve certain quadratic equations. If a quadratic equation has one side equal to 0 and the other side is factorable, we can factor the side with the quadratic expression. Since the factors multiply to be 0, at least one of the factors must be 0. Each factor can be set equal to 0, and the new equations can be solved.

For example, the quadratic equation $x^2 - 11x + 18 = 0$ can be solved as follows:

$$x^2 - 11x + 18 = 0 \qquad \text{Make one side of the equation 0.}$$
$$(x - 9)(x - 2) = 0 \qquad \text{Factor the other side of the equation.}$$
$$x - 9 = 0 \ \text{ or } \ x - 2 = 0 \qquad \text{Set each factor equal to 0.}$$
$$x = 9 \ \text{ or } \qquad x = 2 \qquad \text{Solve each equation.}$$

8. Solve $x^2 + 12x + 35 = 0$.

$-5, -7$

Reflect

WRAP-UP

What's the point?

Factoring quadratic expressions is the reverse process of multiplying them. You learned in this section how to use a guess-and-check strategy as well as a box method. Factoring can be used to solve certain kinds of quadratic equations. However, in practice most quadratic expressions do not factor due to the types of numbers involved in realistic situations. Additional techniques will be needed in those cases. You will explore one of those techniques in the next section.

What did you learn?

How to factor quadratic expressions

3.12 Homework

Skills MyMathLab

First complete the MyMathLab homework online. Then work the two exercises to check your understanding.

1. Find two numbers that multiply to -48 and sum to -2.
−8, 6

2. Factor $9x^2 - 12x + 4$.
$(3x - 2)(3x - 2)$ or $(3x - 2)^2$

Concepts and Applications

Complete the problems to practice applying the skills and concepts learned in the section.

3. Factor each expression.

a. $x^2 - 15x$ $x(x - 15)$

b. $x^2 - 36$ $(x - 6)(x + 6)$

c. $10x - 100$ $10(x - 10)$

4. Students are trying to factor quadratic expressions using the guess-and-check method. Find the mistake in each student's work. Then give the correct answer.

a. $x^2 - x - 72 = (x - 8)(x + 9)$
The signs on the last terms should be switched so that the inner and outer products add to −1; $(x + 8)(x - 9)$.

b. $3x^2 - 19x + 28 = (3x - 4)(x - 7)$
The last terms in the binomials should be switched; $(3x - 7)(x - 4)$.

c. $6x^2 + 25x + 4 = (2x + 1)(3x + 4)$

The first terms should be changed to $6x$ and $1x$; $(6x + 1)(x + 4)$.

5. Quadratic expressions of the form $a^2 - b^2$ are known as the difference of two squares. They factor as follows:

$$a^2 - b^2 = (a - b)(a + b)$$

For example, $x^2 - 25$ can be rewritten as $x^2 - 5^2$. Doing so allows us to see that $a = x$ and $b = 5$. Using the previous formula, $x^2 - 25$ factors as $(x - 5)(x + 5)$. The expression $x^2 - 25$ could also be factored using the box method, but recognizing its form as a difference of two squares allows for a faster approach. If an expression is a sum of two squares and the expression does not have a GCF, it is prime.

Factor each expression that is a difference of two squares using the formula shown. If one of the resulting binomials can be factored further, do so.

a. $x^2 - 100 = (x - 10)(x + 10)$

b. $16x^2 - 49 = (4x - 7)(4x + 7)$

c. $x^4 - 1 = (x - 1)(x + 1)(x^2 + 1)$

6. Expressions of the form of $a^3 - b^3$ are known as the difference of two cubes. They factor as follows:

$$a^3 - b^3 = (a - b)(a^2 + ab + b^2)$$

Similarly, expressions that are the sum of two cubes factor as follows:

$$a^3 + b^3 = (a + b)(a^2 - ab + b^2)$$

In the sum and difference of cubes formulas is a quadratic expression that will not factor further.

For example, $x^3 - 8$ can be factored as $x^2 + 2x + 4$ using the formula as follows:

$$x^3 - 8 = x^3 - 2^3 \qquad a = x, b = 2$$
$$= (x - 2)(x^2 + 2x + 4)$$

Factor each expression that is a sum or difference of two cubes using the formula shown.

a. $x^3 + 64 = (x + 4)(x^2 - 4x + 16)$

b. $y^3 - 125 = (y - 5)(y^2 + 5y + 25)$

7. Each of the following equations contains an expression with a GCF. Factor it out, and then factor the remaining quadratic expression. If the GCF is a constant, divide both sides by the constant before using the zero-product property. Otherwise, set all of the factors equal to zero and solve.

a. $2x^2 + 34x + 60 = 0$

$-15, -2$

b. $x^3 - 8x^2 + 12x = 0$

$0, 6, 2$

8. Solve this perfect square trinomial equation by each stated method: $25x^2 - 10x + 1 = 0$.

a. Factor the left side. Use the zero-product property.

$\dfrac{1}{5}, \dfrac{1}{5}$

b. Factor the left side. Write the factors as a binomial squared so that the equation can be solved by taking the square root of both sides.

$\dfrac{1}{5}$

9. For each box, fill in the missing terms outside and inside the box. Then write the product indicated and the result. State at least two possibilities.

a.

$$2$$

x^2	

Answers will vary; possible answers include:

$(x + 2)(x + 1) = x^2 + 3x + 2$

$(x + 2)(x - 4) = x^2 - 2x - 8$

$(x + 2)(x + 10) = x^2 + 12x + 20$

b.

$$x$$

x^2	
	24

Answers will vary; possible answers include:

$(x + 4)(x + 6) = x^2 + 10x + 24$

$(x - 4)(x - 6) = x^2 - 10x + 24$

$(x + 3)(x + 8) = x^2 + 11x + 24$

$(x - 2)(x - 12) = x^2 - 14x + 24$

10. Looking Forward, Looking Back Simplify.

$$\frac{-7 + \sqrt{7^2 - 4(1)(10)}}{2}$$

-2

3.13 A Formula for Success: The Quadratic Formula

Explore

SECTION OBJECTIVE:

• Use the quadratic formula to solve equations

1. a. Solve the following equation by isolating the squared variable and then taking square roots:

$$3x^2 - 5 = 967$$

± 18

b. Solve the following equation by factoring:

$$x^2 - 2x - 3 = 0$$

3, −1

c. Can the following equation be solved using the techniques from parts a and b? Explain why or why not.

$$x^2 - 3x - 11 = 0$$

No; answers will vary. The equation cannot be easily solved with the square root method because there is an *x* term, and it does not factor with rational numbers.

Discover

In Section 3.12 you learned how to solve quadratic equations by factoring. Factoring is a fast technique and is the best method to use when it works. Unfortunately, most quadratic expressions do not factor, which limits the usefulness of the technique. In this section, you will learn and practice a method that always works for solving quadratic equations. This method, which uses the quadratic formula, usually takes longer than factoring and may seem more involved, but it has the distinct advantage of working on every quadratic equation.

As you learn the material in this section, consider the following questions:

1 When do you *need* to use the quadratic formula?

2 What is the correct order of operations for simplifying a solution with the quadratic formula?

We will begin by stating the quadratic formula and looking at an example of how to use it. We will not discuss the derivation of this formula, although you might in a future course.

You must memorize the quadratic formula!

Quadratic Formula

If a quadratic equation is in the form $ax^2 + bx + c = 0$, where $a \neq 0$, then its solutions are given by the **quadratic formula**:

$$x = \frac{-b \pm \sqrt{b^2 - 4ac}}{2a}$$

FOR EXAMPLE, to solve the equation $x^2 - 3x + 1 = 0$, we notice that $a = 1$, $b = -3$, and $c = 1$. When these values are substituted into the quadratic formula, we get

$$x = \frac{-(-3) \pm \sqrt{(-3)^2 - 4(1)(1)}}{2(1)} = \frac{3 \pm \sqrt{9 - 4}}{2} = \frac{3 \pm \sqrt{5}}{2}$$

So there are two solutions to the equation:

$$\frac{3 + \sqrt{5}}{2} \approx 2.618 \quad \text{and} \quad \frac{3 - \sqrt{5}}{2} \approx 0.382$$

2. Use the quadratic formula to solve the following equation. Then solve the equation by factoring to confirm your solutions.

$$2x^2 + 11x - 6 = 0$$

Quadratic formula: Factoring:

$$x = \frac{1}{2}, -6$$

> The quadratic formula can always be used to solve a quadratic equation. Sometimes the quadratic formula is easier than factoring.

When you are simplifying the solutions from the quadratic formula, you must proceed with the correct order of operations.

i. Simplify the expression under the square root. Square b, multiply $4ac$, and then subtract.

ii. Take the square root if it results in a whole number. Otherwise, leave the square root notation.

iii. Perform the addition or subtraction in the numerator if possible.

iv. Divide by the denominator if it's possible to divide both terms in the numerator.

If an equation is not already in the correct form to begin the quadratic formula, then you must first get all the terms on one side so that 0 is on the other side, and the terms must be in the correct order.

INSTRUCTOR NOTE: In this book, we will focus on providing approximate solutions to quadratic equations if the solutions are irrational. If time permits, you might consider discussing how to simplify the radicals produced by the quadratic formula.

3. Use the quadratic formula to solve the equation. Round the solutions to three decimal places if necessary.

$$2x^2 - 2 = -5x$$

$$x = \frac{-5 + \sqrt{41}}{4} \approx 0.351 \text{ and } x = \frac{-5 - \sqrt{41}}{4} \approx -2.851$$

To solve a quadratic equation using the quadratic formula:

1. Write the equation in the form $ax^2 + bx + c = 0$.
2. Identify the values of a, b, and c.
3. Substitute the values of a, b, and c into the quadratic formula,

$$x = \frac{-b \pm \sqrt{b^2 - 4ac}}{2a}.$$

4. Simplify.

EXAMPLE: Solve $4x^2 - 10x - 1 = 0$ using the quadratic formula.

First notice that $a = 4$, $b = -10$ and $c = -1$. Find the solutions by substituting these values into the quadratic formula. Simplify.

(continued)

HOW IT WORKS

(continued)

$$x = \frac{-(-10) \pm \sqrt{(-10)^2 - 4(4)(-1)}}{2(4)}$$

$$= \frac{10 \pm \sqrt{100 + 16}}{8}$$

$$= \frac{10 \pm \sqrt{116}}{8}$$

$\frac{10 + \sqrt{116}}{8} \approx 2.596$ and $\frac{10 - \sqrt{116}}{8} \approx -0.096$

If the solution involves the square root of a negative number, then the equation has no real solutions.

4. Use the quadratic formula to solve the equation. Round the solutions to three decimal places if necessary.

$$3x^2 + 6x + 7 = 0$$

No real solutions

Connect

The solutions to the equation $2x^2 - 3x - 2 = 0$ are the same values that make the function $y = 2x^2 - 3x - 2$ equal to 0.

Zero of a function

The **zero of a function** is an input value that makes the function output equal to 0.

FOR EXAMPLE, the function $y = 2x + 6$ has a zero of $x = -3$, since that x-value produces a y-value of 0.

$$y = 2(-3) + 6$$
$$= -6 + 6$$
$$= 0$$

Tech TIP

To find the zeros of a function using a graphing calculator, enter the equation in Y_1. Press Graph, and adjust the window if needed so that the x-intercepts can be seen. Press 2nd Trace to access the Calc menu and then choose 2: Zero. Use the arrow keys to move the cursor to the left of a zero and press Enter. Then move the cursor to the right of the zero and press Enter. Press Enter a third time.

5. A quadratic function is one that can be written in the form $y = ax^2 + bx + c$. Find the zeros of the quadratic function $y = 15x^2 - 26x + 8$.

$x = \frac{4}{3}, \frac{2}{5}$

Note that you find *zeros* of a quadratic *function*, but *solutions* of a quadratic *equation*.

Reflect

WRAP-UP

What's the point?

Some quadratic equations cannot be solved by factoring with rational numbers. The quadratic formula might seem more complicated than factoring, but it can be used to solve any quadratic equation. In order to use the quadratic formula, you must memorize it and be able to simplify the solutions correctly.

What did you learn?

How to use the quadratic formula to solve equations

INSTRUCTOR NOTE: Consider providing the focus problem grading rubric and writing template to students at this point to help them form their written solution. Both are available in MyMathLab.

Focus Problem Check-In
Use what you have learned to finish solving the focus problem. Write a rough draft to explain your solution, showing and explaining all mathematical work. Edit your draft into a final write-up by making any needed corrections and proofreading for grammar, spelling, and mathematical errors. Have someone verify the mathematical work and proofread the final draft for mistakes.

3.13 Homework

Skills MyMathLab

First complete the MyMathLab homework online. Then work the two exercises to check your understanding.

1. Write the quadratic equation in the correct form to use the quadratic formula. Write it so that the coefficient of the squared term is positive. Then identify a, b, and c.

$$7 - 6x = 3x^2$$

$3x^2 + 6x - 7 = 0$
$a = 3,\ b = 6,\ c = -7$

2. Solve the quadratic equation by using the quadratic formula and then again by factoring. Do the solutions agree?

$$25x^2 - 15x + 2 = 0$$

$x = \dfrac{1}{5}$, $x = \dfrac{2}{5}$; yes

Concepts and Applications

Complete the problems to practice applying the skills and concepts learned in the section.

3. Consider the quadratic equation $2x^2 + 14x + 2 = 0$.

   a. Solve the equation by using the quadratic formula.

       $\dfrac{-7 + \sqrt{45}}{2} \approx -0.146$, $\dfrac{-7 - \sqrt{45}}{2} \approx -6.854$

   b. Factor out the GCF of the left side of the equation and then divide both sides by that GCF. Solve the resulting equation. You should get the same solutions as you did in part a.

       $\dfrac{-7 + \sqrt{45}}{2} \approx -0.146$, $\dfrac{-7 - \sqrt{45}}{2} \approx -6.854$

4. a. Write a quadratic equation that can be solved by factoring and does not require the use of the quadratic formula.

      Answers will vary; one possible answer: $x^2 + 7x + 12 = 0$

   b. Write a quadratic equation that *cannot* be solved by factoring and then solve it using the quadratic formula.

      Answers will vary; one possible answer: $x^2 + 12x + 7 = 0$

5. Identify the mistake in the following solution using the quadratic formula to solve the equation $x^2 - 6x - 10 = 0$:

$$\frac{-(-6) \pm \sqrt{(-6)^2 - 4(1)(-10)}}{2(1)} = \frac{6 \pm \sqrt{-36 + 40}}{2}$$

$$= \frac{6 \pm \sqrt{4}}{2}$$

$$= \frac{6 \pm 2}{2}$$

$$\frac{6 + 2}{2} = \frac{8}{2} = 4 \text{ and } \frac{6 - 2}{2} = \frac{4}{2} = 2$$

The student did not square −6 correctly. The radical should be $\sqrt{36 + 40}$.

6. Recall that a Pythagorean triple is a set of three whole numbers that satisfy the equation $a^2 + b^2 = c^2$. One Pythagorean triple with consecutive numbers is 3, 4, 5. Are there any other consecutive Pythagorean triples?

Solve the equation $x^2 + (x + 1)^2 = (x + 2)^2$ to find all consecutive Pythagorean triples. You first need to square the binomials and collect like terms.

3, 4, 5 is the only consecutive Pythagorean triple. The other solution to the equation is negative and can't be part of a Pythagorean triple.

7. Looking forward, looking back Solve each equation.

a. $-3(x + 1) + 7x + 8 = 5x - x + 6 - 2$

No solution

b. $-3(x + 1) + 7x + 8 = 5x - x + 6 - 1$

All real numbers

3.14 Systematic Thinking: Graphing and Substitution

Explore

SECTION OBJECTIVES:
- Solve a 2 × 2 linear system of equations by graphing
- Solve a 2 × 2 linear system of equations by substitution

INSTRUCTOR NOTE: Consider completing the first row with students.

1. Suppose you are considering attending a local college. College A is your top choice, but you want to compare your cost per semester at college A to your costs at each of three other colleges you are considering.

Compare college A to the other colleges as follows: Begin by modeling each college's tuition, C, as a function of the number of credit hours, h. Then use the graph of the functions and an equation to determine the number of credit hours for which the costs are the same and what that cost is.

Verbal	Graphic	Algebraic
College A has $400 in fees, and tuition is $100 per credit hour. College B has no fees, but tuition is $200 per credit hour. **Models:** $C_A = 400 + 100h$ $C_B = 200h$	(graph showing lines B and A, Tuition vs Number of credit hours)	$C_A = C_B$ $400 + 100h = 200h$ $h = 4$ credit hours, $800
College A has $400 in fees, and tuition is $100 per credit hour. College C has $200 in fees, and tuition is also $100 per credit hour. **Models:** $C_A = 400 + 100h$ $C_C = 200 + 100h$	(graph showing lines A and C, Tuition vs Number of credit hours)	$C_A = C_C$ $400 + 100h = 200 + 100h$ The costs will never be the same.
College A has $400 in fees, and tuition is $100 per credit hour. College D has $600 in fees, and tuition is $150 per credit hour. A scholarship will cover $\frac{1}{3}$ of the total costs. **Models:** $C_A = 400 + 100h$ $C_D = \frac{2}{3}(600 + 150h) = 400 + 100h$	(graph showing line A, D, Tuition vs Number of credit hours)	$C_A = C_D$ $400 + 100h = 400 + 100h$ The costs will always be the same, regardless of the number of hours.

2. Use the results from #1 to draw a conclusion about the costs for each pair of colleges compared.

Colleges A and B: If you know you'll be taking more than 4 credit hours a semester, then college A is more affordable than college B; if fewer than 4, then college B is more affordable.

Colleges A and C: College A will always be more expensive than college C.

Colleges A and D: College D will always have the same cost as college A, regardless of the number of credit hours.

When two or more equations must be solved at the same time, the equations form a system. We have seen some situations in this book that involved systems of equations, such as the comparisons made between two gas price options in Section 3.5. Let's learn about the vocabulary and solving techniques for systems of equations.

As you learn the material in this section, consider the following questions:

1 What do parallel lines on a graph indicate about a system of equations?

2 What do infinitely many solutions to a system imply about the graph?

3 Can you explain an advantage and a disadvantage of the graphing and substitution methods?

INSTRUCTOR NOTE: Consider having students identify each of the three systems in the *Explore* as either consistent or inconsistent.

System of Equations

A **system of equations** is a set of two or more equations that need to be solved simultaneously. If there are two equations with two unknowns, the system is sometimes referred to as a 2×2 system, or a "two-by-two system." The first number is the number of equations and the second number is the number of variables or unknowns. So a system that has two equations and three unknowns is referred to as a 2×3 system.

When a system of equations has a solution, it is called **consistent**. When it does not have a solution, it is called **inconsistent**.

When the equations in a system produce different graphs, they are called **independent**. When the equations are equivalent and produce the same graph, they are called **dependent**.

A **solution** of a 2×2 system of equations is an ordered pair that satisfies each of the equations in the system. The point lies on both graphs.

FOR EXAMPLE, the following equations form a system of equations when considered together.

$$\begin{cases} y = 2x + 2 \\ y = 4x \end{cases}$$

The point $(1, 4)$ is the solution to the system, since it satisfies both equations. The point $(1, 4)$ also lies on both lines in the graph.

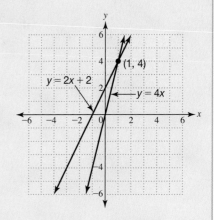

We will primarily focus on systems that are composed of two linear equations. A system of linear equations can have one solution, which occurs at the point of intersection of the lines, or no solutions, or infinitely many solutions.

3. Use the graph for each system to describe it.

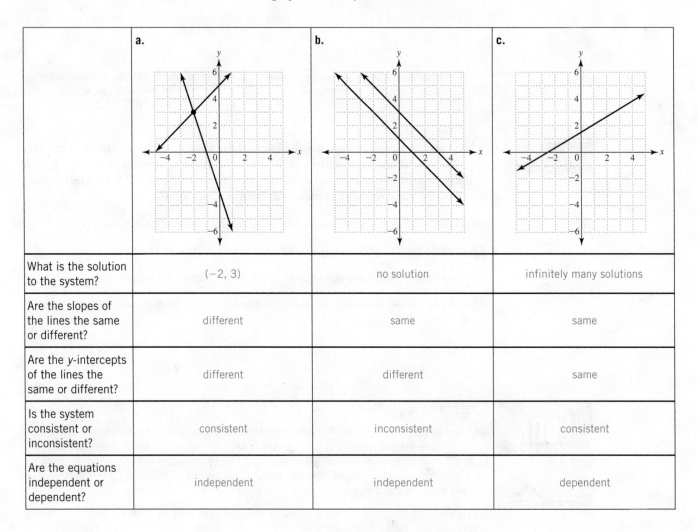

	a.	b.	c.
What is the solution to the system?	(−2, 3)	no solution	infinitely many solutions
Are the slopes of the lines the same or different?	different	same	same
Are the *y*-intercepts of the lines the same or different?	different	different	same
Is the system consistent or inconsistent?	consistent	inconsistent	consistent
Are the equations independent or dependent?	independent	independent	dependent

4. Verify that $\left(9, \dfrac{2}{3}\right)$ is a solution to the following system of equations: $\begin{aligned} 5x + 6y &= 49 \\ 3y - x &= -7 \end{aligned}$

$$5x + 6y = 49 \qquad\qquad 3y - x = -7$$

$$5(9) + 6\left(\frac{2}{3}\right) \overset{?}{=} 49 \qquad\qquad 3\left(\frac{2}{3}\right) - 9 \overset{?}{=} -7$$

$$45 + 4 \overset{\checkmark}{=} 49 \qquad\qquad 2 - 9 \overset{\checkmark}{=} -7$$

We were able to easily solve the systems in the *Explore* algebraically because both equations were solved for the same variable. Since both equations were equal to the same variable, the cost, the expressions for the costs were easily set equal to each other in a new equation. However, solving systems algebraically may not always be this easy. Additional techniques are needed.

One way to solve a 2 × 2 system of equations is to graph the equations and look for their point(s) of intersection. Another option is to solve for one variable in one of the systems and then substitute that value into the other equation. This method is known as **substitution**. It takes two equations and two unknowns and condenses them into one new equation with one unknown that can be solved by methods you have learned. Once one variable is found, its value is substituted into one of the original equations to find the value of the other variable. Both of these methods are illustrated in the next example.

EXAMPLE

Solve the system: $2x = 4y - 8$
$5x - 6y = 0$

SOLUTION

This is a 2 × 2 system of linear equations. It can be solved by graphing or algebraically with the substitution method.

Graphing Approach

To solve the system with graphs, we need to graph both lines and then look for any points of intersection. It is easier to graph lines when their equations are in slope-intercept form, so we solve each equation for y.

We can use the slope and y-intercept to graph each line. We need to find points that are on both lines. Any point that is on both lines is a solution to the system. In this system, the lines intersect at $(6, 5)$. So the solution to the system is $(6, 5)$.

$$2x = 4y - 8 \quad \rightarrow \quad y = \frac{1}{2}x + 2$$

$$5x - 6y = 0 \quad \rightarrow \quad y = \frac{5}{6}x$$

Tech TIP

A graphing calculator can be used to find the point(s) of intersection of graphs. Write each equation so that it is solved for y, and enter the first equation in Y_1 and the second equation in Y_2. Press Graph, and adjust the window if needed so that the point of intersection can be seen. Press 2nd Trace to access the Calc menu and then choose 5: Intersect. A cursor will blink on the first line. Press Enter. It will then blink on the second line. Press Enter. You can enter a guess for the intersection point or just press Enter a third time.

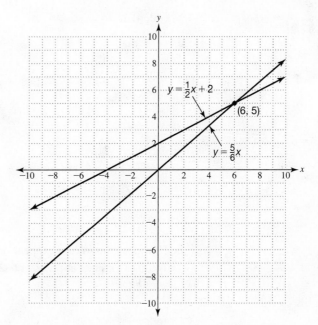

Substitution Approach

Step 1: Solve for a variable.

One equation needs to be written in terms of one of the variables, such as $y =$ or $x =$. In this system, it requires the least amount of work to solve for x in the first equation.

$$\frac{2x}{2} = \frac{4y}{2} - \frac{8}{2}$$
$$x = 2y - 4$$

Step 2: Substitute and solve.

Since x is equivalent to $2y - 4$, we replace x in the other equation with this expression and solve.

$$5x - 6y = 0$$
$$5(2y - 4) - 6y = 0$$
$$y = 5$$

Step 3: Find the remaining variable.

Find the value of the remaining variable by substituting $y = 5$ into one of the equations. The equation we first worked with is easier to use, since it is a formula for x.

$$x = 2y - 4$$
$$x = 2(5) - 4$$
$$x = 6$$

Step 4: Write the solution as an ordered pair.

The solution is (6, 5). This point is on both lines and is a solution to both equations.

For either method, we conclude by checking the solution. We substitute the ordered pair into *both* original equations and check that both equations are true.

$$2x = 4y - 8 \qquad 5x - 6y = 0$$
$$2(6) \stackrel{?}{=} 4(5) - 8 \qquad 5(6) - 6(5) \stackrel{?}{=} 0$$
$$12 \stackrel{\checkmark}{=} 20 - 8 \qquad 30 - 30 = 0$$

When you use the substitution method, it is easiest to solve for a variable that has a coefficient of 1. Also, look ahead at the steps you'll need to solve and which equation and variable will result in the fewest number of steps or the fewest number of fractions. More fractions will mean more work in the next step when you substitute.

5. a. Solve the system using substitution: $\begin{aligned} 2x + y &= 0 \\ 4x + 7y &= 20 \end{aligned}$

(−2, 4)

b. How does the solution appear on the graph?

The lines intersect at the point (−2, 4).

HOW IT WORKS

To solve a 2 × 2 system of equations by graphing:

1. Graph both equations on the same coordinate system.
2. Determine the points that lie on both graphs. Any point of intersection represents a solution.
3. Verify the solution satisfies both original equations.

To solve a 2 × 2 system of equations by substitution:

1. Solve for one variable in one equation.
2. Substitute the expression for the variable into the other equation. Solve.
3. Substitute the first value into one of the equations to find the remaining value. The equation found in the first step will be the easiest to use.
4. Write an ordered pair using the two values. This point is the solution to the system.
5. Verify the solution satisfies both original equations.

EXAMPLE: Solve the system: $3x + 2y = -7$
$$x - y = 1$$

To solve the system by graphing, begin by writing each equation in slope-intercept form. Then use the slope and y-intercept to graph the lines.

In slope-intercept form, the equations are $y = -\dfrac{3}{2}x - \dfrac{7}{2}$ and $y = x - 1$. The lines intersect at $(-1, -2)$, which is the solution to the system.

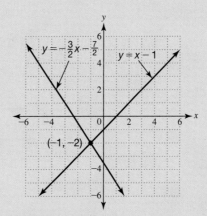

To solve the system by substitution, begin by solving the second equation for x.

$$3x + 2y = -7$$
$$x - y = 1 \quad \rightarrow \quad x = y + 1$$

Next, substitute the expression $y + 1$ for x into the first equation and solve. Then find the value for the remaining variable.

$3x + 2y = -7$	$x = y + 1$
$3(y + 1) + 2y = -7$	$x = -2 + 1$
$3y + 3 + 2y = -7$	$x = -1$
$5y = -10$	
$y = -2$	

Write the solution as an ordered pair: $(-1, -2)$. Then check the solution in both original equations.

$3x + 2y = -7$	$x - y = 1$
$3(-1) + 2(-2) \overset{?}{=} -7$	$-1 - (-2) \overset{?}{=} 1$
$-3 + -4 \overset{\checkmark}{=} -7$	$-1 + 2 \overset{\checkmark}{=} 1$

When you check your solution, it's a good idea to use the original equations because you might have made a mistake with the later equations at some point.

6. Solve each system by substitution. Explain what your result says about the system of equations and how the graph would appear.

a. The following system contains two equations that are equivalent: $y = 2x + 4$
$$3y = 6x + 12$$

The substitution method results in the equation $0 = 0$, which means that the system has infinitely many solutions and the lines coincide in the graph.

b. The following system of equations is inconsistent: $y = 2x + 4$
$$y = 2x + 3$$

The substitution method results in the equation $4 = 3$, which means that the system has no solutions and the lines are parallel in the graph.

7. a. Solve the system graphically: $-3x + 4y = 12$
$$6x - 8y = 0$$

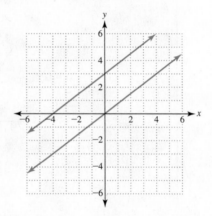

No solution

b. Solve the system using substitution: $-3x - 4y = 12$
$$6x + 8y = 0$$

No solution

If you solve a 2×2 system of linear equations algebraically and the result is a true statement, then there are infinitely many solutions. The graph is only one line, and every point on the line is a solution. If you solve a 2×2 system of linear equations algebraically and the result is a false statement, then there are no solutions. The graph contains parallel lines with no intersection points.

Connect

When applied problems have multiple unknowns, one way to approach the problem is to write a system of equations with at least as many equations as unknowns. We can then use the methods developed in this section to solve the system.

8. Emily is working with a nutritionist to design a more healthful meal plan. She is advised to aim for 1,200 calories a day, with twice as many calories coming from carbohydrates as from protein and with 30–40 grams of fat. She decides to eat 40 grams of fat per day. How many grams of protein should she eat each day? She knows that her total calories come from calories in carbohydrates (4 calories/gram), protein (4 calories/gram), and fat (9 calories/gram).

70 grams of protein

When you solve a system of equations, you might need to state your answer as an ordered pair. However, in applied situations, you need to read the question carefully to determine which value in the ordered pair answers the question. Include units in your answers whenever possible.

Reflect

WRAP-UP

What's the point?

Sometimes a problem involves a system of two or more equations that must be solved simultaneously. Graphing a system of equations allows you to see the situation and understand the solution to a system, but it can be difficult to locate the exact solution on the graph. It can also be time-consuming. Solving a system of equations algebraically allows you to get the exact solution and is faster, but it may not be as clear what is going on to produce that solution or what the solution means. Using the methods together helps you find the solution and make sense of it.

What did you learn?

How to solve a 2 × 2 linear system of equations by graphing
How to solve a 2 × 2 linear system of equations by substitution

3.14 Homework

Skills | MyMathLab

First complete the MyMathLab homework online. Then work the two exercises to check your understanding.

1. Verify that $(3, -10)$ is a solution to the following system:

$$4x - 19y = 202$$
$$\frac{4}{3}x + \frac{1}{2}y = -1$$

$$
\begin{array}{ll}
4x - 19y = 202 & \frac{4}{3}x + \frac{1}{2}y = -1 \\[2mm]
4(3) - 19(-10) \overset{?}{=} 202 & \frac{4}{3}(3) + \frac{1}{2}(-10) \overset{?}{=} -1 \\[2mm]
12 + 190 \overset{?}{=} 202 & 4 + (-5) \overset{?}{=} -1 \\[2mm]
202 \overset{\checkmark}{=} 202 & -1 \overset{\checkmark}{=} -1
\end{array}
$$

2. Solve the system by each method stated: $-2(x + 7) + 8y = -20$
$$x - 4y = \quad 3$$

   a. Solve by graphing.

   Infinitely many solutions

   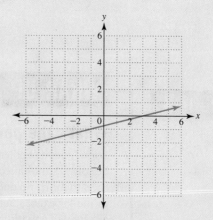

   b. Solve by substitution.

   Infinitely many solutions

Concepts and Applications

Complete the problems to practice applying the skills and concepts learned in the section.

3. Suppose Lauren is working two jobs to help pay for college. One job pays more per hour than the other. Since her boss at the higher-paying job wouldn't give her as many hours as she wanted, she picked up the second job to generate more income. With her course load, 20 hours a week of work is all she can manage, but she wants to make as much money as possible in those 20 hours. Her goal is to get twice as many hours a week at the higher-paying job than at the lower-paying one. How many hours should she work at each job?

a. Write a system of equations and solve using substitution. Let H = number of hours worked per week at the higher-paying job and L = number of hours worked at the lower-paying job.

$H + L = 20$

$H = 2L$

$L = \dfrac{20}{3} \approx 6.7 \text{ hours} \quad H = 2L \approx 2(6.7) = 13.4 \text{ hours}$

b. State the complete solution to the problem in words.

Lauren should work 6.7 hours at the lower-paying job and 13.4 hours at the higher-paying job.

c. Graph these equations. Let L be on the horizontal axis and H be on the vertical axis.

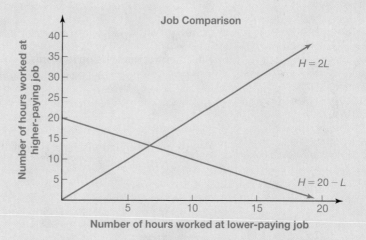

d. Which method, graphic or algebraic, is more precise in this case?

The lines intersect once at (6.7, 13.4), which cannot be seen exactly on the graph. Thus, the algebraic method is more precise.

4. Suppose that next semester Lauren plans to move into an apartment when she transfers to a different college. The cost of rent, along with books, fees, and spending money, makes a job necessary again. She can get a $10/hour job, but the drive is significant, which limits how many hours she wants to work at that job. Taking a second job at $8/hour with a short drive lowers her gas costs. Her rent is $1,250 a semester, which is 15 weeks long. Books cost $500 a semester, and she would like to have $800 for the semester just in case. How many hours should she work at each job to cover her costs if she wants to work only 20 hours a week?

It will be easiest to work with a timeframe of a week. Let $H = $ hours worked per week at the $10/hour job and $L = $ hours worked per week at the $8/hour job.

a. Write an equation for the total number of hours worked.

$H + L = 20$

b. Lauren's costs are $2,550 per semester, which equals $170 a week. To cover those costs, she must have a weekly income equivalent to that amount. Write an equation for Lauren's costs.

$10H + 8L = 170$

c. Solve the system. Give the solution as an ordered pair (L, H).

$(15, 5)$

d. Interpret the solution in words.

Each week, she should work 15 hours at the $8/hour job and 5 hours at the $10/hour job.

e. Graph the equations in parts a and b. Let L be on the horizontal axis and H be on the vertical axis.

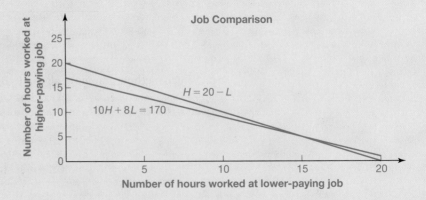

f. Which method, graphic or algebraic, do you prefer in this case?

Answers will vary. The graph is somewhat hard to read because the slopes are similar.

5. Parallel lines have the same slope and different *y*-intercepts, and perpendicular lines have slopes that are opposite reciprocals. For example, if two different lines each have slopes of 7, then the lines are parallel. If two lines have slopes of 7 and $-\frac{1}{7}$, the lines are perpendicular.

For each system, determine if the lines are parallel, perpendicular, or neither. Instead of graphing, write the equations in slope-intercept form and compare the slopes and the *y*-intercepts. What does this tell you about the solution(s) to the system?

a. $y = \dfrac{1}{3}x$

$3x + y = 8$

Perpendicular; one solution

b. $10x + 2y = 4$

$-5x + \ y = 8$

Neither; one solution

c. $y = x$

$3y - 3x = 12$

Parallel; no solution

6. Write an equation to form a system with the equation $y = 2x - 8$ so that the system satisfies the stated condition.

a. The system has the solution $(3, -2)$.

Answers will vary; possible answer: $y = 3x - 11$

b. The system has no solution.

Answers will vary; possible answer: $y = 2x - 9$

c. The system has infinitely many solutions.

Answers will vary; possible answer: $2y = 4x - 16$

7. Looking Forward, Looking Back Find the mistake(s) in a student's work in which she is asked to solve the following quadratic equation: $4x^2 - 8x + 20 = 0$

$$x = \frac{-(-8) \pm \sqrt{(-8)^2 - 4(4)(20)}}{2(4)}$$

$$= \frac{8 \pm \sqrt{64 + 320}}{8}$$

$$= \frac{8 \pm \sqrt{384}}{8}$$

$$= 1 \pm \sqrt{384}$$

$$= 1 \pm 19.6$$

$$= 20.6, -18.6$$

She added instead of subtracting the numbers under the square root symbol. She divided the first term in the numerator by 8 but not the second to get the solutions.

3.15 Opposites Attract: Elimination

Explore

1. a. Solve the system of equations using substitution: $-3x + 7y = -51$
$$3x + 4y = -15$$

$(3, -6)$

SECTION OBJECTIVE:

- Solve a 2 × 2 linear system of equations by elimination

b. Why is it time-consuming to solve the system using substitution?

None of the coefficients is 1 or -1, so multiple steps are needed to solve for x or y in either equation. The resulting equation also has fractions, which take more time to deal with when they are substituted into the other equation.

Discover

In Section 3.14 you learned how to solve systems of equations using their graphs as well as algebraically using substitution. There are systems, such as in the one in the *Explore*, for which substitution can be tedious and lengthy. Another method, known as the **elimination** or addition method, works well for certain systems of equations. This section will explore when the elimination method is most appropriate and how to use it to solve systems of equations.

As you work through this section, think about these questions:

1 Can you tell from the appearance of a system which method will be fastest to solve it?

2 Do you know how to solve a system with all three methods: graphing, substitution, and elimination?

The elimination method involves creating opposite terms that will be eliminated when the equations are added. Let's look at an example to see how the method is used.

EXAMPLE 1

Solve the system using elimination: $-3x + 7y = -51$
$$3x + 4y = -15$$

SOLUTION

In this system from the *Explore*, the coefficients of the *x* terms are opposites and the like terms and equal signs are aligned. A system of equations in this format is ideally suited for the elimination method. Because the left and right sides of the second equation are equal, we can add the left side of that equation to the left side of the first equation and the right side to the right side of the first equation. As long as we add the same quantity to both sides of an equation, the equality remains true. Since the coefficients of the x terms are opposites, the x terms add to zero and are therefore *eliminated*.

$$-3x + 7y = -51$$
$$\underline{3x + 4y = -15} \quad \text{Add the equations vertically.}$$
$$11y = -66 \quad \text{Solve the resulting equation.}$$
$$y = -6$$

Once we have the value of one of the variables, we can substitute it into either original equation to find the value of the other variable. This is similar to how the substitution method continues after we find the value of the first variable. Choose the equation that will require the least amount of work once the value is substituted.

$$3x + 4y = -15$$
$$3x + 4(-6) = -15 \quad \text{Replace } y \text{ with } -6 \text{ in the second equation.}$$
$$3x - 24 = -15 \quad \text{Solve.}$$
$$3x = 9$$
$$x = 3$$

The solution is $(3, -6)$. As with substitution or graphing, to check the solution, you must check that it works in both original equations.

In the previous example, solving the system with elimination was very fast not only because the coefficients of the x terms were opposites but also because of the form of the equations. Both linear equations were in **standard form**: $Ax + By = C$ where A, B, and C are integers. Equations in this form have the x and y terms on one side of the equal sign and a constant term on the other side. However, as long as like terms and the equal signs in the equations are aligned, elimination can be easily used. For example, elimination is a reasonable method to solve this system even though it is not in standard form.

$$7x = 5y - 15$$
$$-7x = 2y + 2$$

Standard form can make solving by graphing or substitution more difficult because we need to solve for one of the variables. We saw this when trying to use substitution to solve the system in the *Explore*. When the equations are in standard form and the coefficients of either the x or y terms are opposites, as in the following problem, elimination is probably the fastest method.

2. Solve the system of equations using elimination: $8x - 12y = 114$
$$17x + 12y = 261$$

$$\left(15, \frac{1}{2}\right)$$

If the coefficients of the *x* terms or the *y* terms are not opposites, we can multiply one or both equations by a nonzero constant to create opposites.

EXAMPLE 2

Solve the system using elimination: $2x + 4y = -46$
$-8x + 3y = 51$

SOLUTION

Since both equations are in standard form and solving for one of the variables for substitution would take multiple steps, elimination is an ideal method to use to solve this system. Notice that the *x* terms have opposite signs but not opposite coefficients. However, if we multiply the first equation by 4, the *x* term becomes 8*x*, the opposite of the *x* term in the second equation.

$$4(2x + 4y) = 4(-46) \qquad \rightarrow \qquad 8x + 16y = -184$$
$$-8x + 3y = 51 \qquad \rightarrow \qquad \underline{-8x + 3y = 51}$$
$$19y = -133$$
$$y = -7$$

Substitute -7 for *y* in the first equation and solve.

$$2x + 4y = -46$$
$$2x + 4(-7) = -46$$
$$2x - 28 = -46$$
$$2x = -18$$
$$x = -9$$

The solution to the system is $(-9, -7)$.

Another way to solve the system of equations is to eliminate the *y* terms. To do so, we would find the least common multiple of the coefficients 3 and 4, which is 12. Next, we would multiply the first equation by 3 and the second equation by -4 to get 12*y* and $-12y$. We would then add the resulting equations and finish the solution.

When you are multiplying an equation by a constant, be careful to multiply *all* the terms of the equation, not just the term you are eliminating.

INSTRUCTOR NOTE: For additional practice, consider having students solve the system again but by eliminating the *y* terms as described in the last paragraph of the Example 2.

3. Solve each system of equations using elimination. State how each solution would appear on the graph of the system.

a. $20x - 12y = 160$
$17x - 2y = 95$
(5, −5); the lines would intersect at (5, −5).

b. $-30x + 5y = -35$
$-24x + 4y = -20$
No solution; the lines would be parallel.

Notice that if you eliminate the *x* terms, you will find *y* first and then substitute to find *x*. Similarly, if you eliminate the *y* terms, you will find *x* first and then substitute to find *y*.

HOW IT WORKS

To solve a 2 × 2 system of equations by elimination:

1. Write each equation in standard form: $Ax + By = C$.
2. If necessary, multiply one or both equations by a nonzero constant so that the coefficients of the *x* terms or the *y* terms are opposites.
3. Add the equations.
4. Solve the resulting equation.
5. Substitute the value you found for the variable into one of the original equations to find the value of the other variable.
6. Form an ordered pair using the two values. This point is the solution to the system.
7. Check the solution by substituting it into both original equations. Verify that each equation is true.

EXAMPLE: Solve the system: $3x + 2y = -7$
$5x - y = -3$

This system is well suited to be solved by elimination, since each equation is already in standard form and it's easy to eliminate one of the variables. First, make the coefficients of the *y* terms opposites by multiplying the second equation by 2.

$$3x + 2y = -7 \quad\quad \rightarrow \quad 3x + 2y = -7$$
$$2(5x - y) = 2(-3) \quad \rightarrow \quad 10x - 2y = -6$$

Then add the equations and solve the resulting equation.

$$3x + 2y = -7$$
$$\underline{10x - 2y = -6}$$
$$13x = -13$$
$$x = -1$$

Substitute $x = -1$ into either original equation and solve to find the remaining variable.

$$3x + 2y = -7$$
$$3(-1) + 2y = -7$$
$$-3 + 2y = -7$$
$$2y = -4$$
$$y = -2$$

The solution is $(-1, -2)$.

Finally, check the solution in both original equations.

$$3x + 2y = -7$$
$$3(-1) + 2(-2) \stackrel{?}{=} -7$$
$$-3 + (-4) \stackrel{\checkmark}{=} -7$$

$$5x - y = -3$$
$$5(-1) - (-2) \stackrel{?}{=} -3$$
$$-5 + 2 \stackrel{\checkmark}{=} -3$$

Connect

Now that you've learned three methods to solve a system of equations, let's use each method on the same system.

4. Solve the system of equations by the three indicated methods:

$$3x = 2y - 8$$
$$x - y = -2$$

a. Graphing

$(-4, -2)$

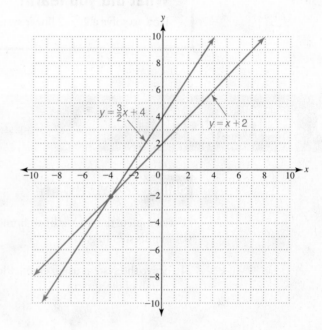

b. Substitution

$(-4, -2)$

c. Elimination

$(-4, -2)$

Reflect

WRAP-UP

What's the point?

When a system has equations in standard form, it can be faster to use the elimination method rather than graphing or substitution. The elimination method requires creating opposite coefficients of the x terms or the y terms so that a variable can be eliminated when the equations are added.

What did you learn?

How to solve a 2×2 linear system of equations by elimination

3.15 Homework

Skills MyMathLab

First complete the MyMathLab homework online. Then work the two exercises to check your understanding.

1. Solve the system using elimination. State how the solution would appear on the graph of the system.

$$3x + 2y = -4$$
$$x + y = -3$$

(2, −5); the lines would intersect at (2, −5).

2. Solve the system using elimination. State how the solution would appear on the graph of the system.

$$3x + 2y = -4$$
$$6x + 4y = -8$$

Infinitely many solutions; the graph would appear to have only one line.

Concepts and Applications

Complete the problems to practice applying the skills and concepts learned in the section.

3. For each system, determine which method of solving (graphing, substitution, or elimination) you prefer to use. Explain your choice.

   a. $\quad\quad y = -8x$

   $4x - 2y = 0$

   Answers will vary. Since one equation is solved for a variable, substitution is a good choice.

   b. $10x + \dfrac{5}{2}y = 15$

   $-5x - y = -7$

   Answers will vary. Since the equations are in standard form and a variable can easily be eliminated, elimination might be preferred.

   c. $y = 3x - 6$

   $y = 3x - 11$

   Answers will vary. Since both equations are solved for a variable, substitution is a good choice.

4. Each system has a first step completed to start the solving process. State the next step for each system.

   a. $7x + y = -7 \quad\quad \rightarrow \quad y = -7x - 7$

   $5x - 3y = 18$

   Substitute $-7x - 7$ for y in the second equation.

   b. $9x + 2y = 3 \quad\quad\quad\quad 9x + 2y = 3$

   $7x + 3y = 0 \quad\quad \rightarrow \quad 2(7x + 3y) = 2(0)$

   Multiply the first equation by -3.

5. Solve the system using the requested approach: $\quad 15x + 6y = 165$

$$5x + 7y = 55$$

   a. Eliminate the x terms.

   $(11, 0)$

b. Eliminate the y terms.

(11, 0)

6. A student is solving the system $\begin{array}{l} 4x + 6y = -6 \\ 5x + 5y = -\dfrac{5}{2} \end{array}$.

Find the mistakes in the student's work. Then solve the system correctly and state the solution.

$5(4x + 6y) = -6 \quad \rightarrow \quad 20x + 30y = -6 \qquad\qquad 4x + 6y = -6 \qquad\qquad \left(\dfrac{17}{20}, -\dfrac{13}{30}\right)$

$-6(5x + 5y) = -\dfrac{5}{2} \quad \rightarrow \quad \dfrac{-30x - 30y = -\dfrac{5}{2}}{} \qquad 4\left(\dfrac{17}{20}\right) + 6y = -6$

$-10x = -\dfrac{17}{2} \qquad\qquad \dfrac{17}{5} + 6y = -6$

$x = \dfrac{17}{20} \qquad\qquad \dfrac{17}{5} + 6y + \dfrac{17}{5} = -6 + \dfrac{17}{5}$

$6y = -\dfrac{13}{5}$

$y = -\dfrac{13}{30}$

The student didn't multiply either equation's right side by the constant when creating opposite terms. Also, the student added $\dfrac{17}{5}$ to both sides when solving for y instead of subtracting it. The correct solution is $\left(\dfrac{3}{2}, -2\right)$.

7. Looking forward, looking back Find the zeros of the quadratic function:

$$y = 2x^2 - 11x + 5$$

$\dfrac{1}{2}, 5$

3.16 The Turning Point: Quadratic Functions

Explore

SECTION OBJECTIVES:
* Identify a quadratic pattern in data
* Find the vertex of a parabola

1. a. Use the table of data for a quadratic function to identify the zeros and the x-intercepts and y-intercepts of the graph.

x	−3	−2	−1	0	1	2	3	4	5
y	7	0	−5	−8	−9	−8	−5	0	7

Zeros: −2, 4; x-intercepts: (−2, 0) and (4,0); y-intercepts: (0, −8)

Remember?

The zeros of a function are values, but intercepts are ordered pairs.

b. What do you notice about the smallest function value shown in the table and its location?

The smallest function value occurs at $x = 1$, halfway between the two zeros.

Discover

Earlier in the cycle you learned how to solve a quadratic equation and find the zeros for a quadratic function. In this section you will add to your knowledge of quadratic functions by learning the shape of their graphs and how to find key details of those graphs.

As you progress through this section, consider these questions:

1 Can you recognize a quadratic function from a table of data or a graph?

2 How do you find the zeros of a quadratic function vs. the vertex?

We will begin by discussing the graph of a quadratic function and one of its important features, the vertex.

Parabola

A quadratic function, $y = ax^2 + bx + c$ where $a \neq 0$, produces a vertical U-shaped graph called a **parabola**. The turning point on a parabola is called the **vertex**.

FOR EXAMPLE, the quadratic function $y = x^2 - 2x - 3$ generates the following parabola when graphed. The vertex of the parabola is the point (1, 2).

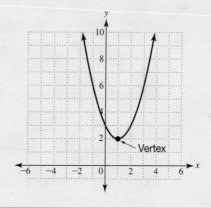

We will focus on identifying the vertex of a parabola and how its position relates to the zeros of the function. In this section, quadratic functions will be written in **standard form**, $y = ax^2 + bx + c$.

EXAMPLE

Use the table of data for a quadratic function to graph the function. Identify the x-intercepts and the vertex.

x	−1	0	1	2	3	4	5
y	−15	−8	−3	0	1	0	−3

SOLUTION

By plotting the ordered pairs listed in the table and connecting them with a smooth curve, we can draw the parabola shown here.

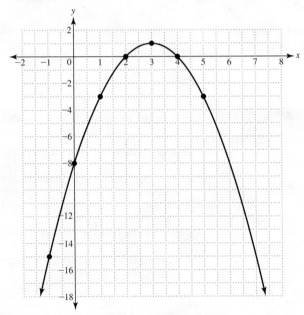

Notice in the table that the y-value is 0 when $x = 2$ and when $x = 4$. So there are x-intercepts at $(2, 0)$ and $(4, 0)$. These points can also be seen in the graph of the function. The vertex can be identified as $(3, 1)$ in the graph. Notice that the x-coordinate of the vertex is halfway between the two zeros at $x = 2$ and $x = 4$. In fact, the x-coordinate of the vertex can always be found by averaging the zeros of a quadratic function.

The two zeros of a quadratic function are given by the quadratic formula as:

$$x = \frac{-b + \sqrt{b^2 - 4ac}}{2a} \text{ and } x = \frac{-b - \sqrt{b^2 - 4ac}}{2a}.$$

If these two values are averaged to find the x-coordinate of the vertex, we get

$$\frac{\dfrac{-b + \sqrt{b^2 - 4ac}}{2a} + \dfrac{-b - \sqrt{b^2 - 4ac}}{2a}}{2} = \frac{-\dfrac{b}{2a} - \dfrac{b}{2a}}{2} = \frac{-\dfrac{b}{a}}{2} = -\frac{b}{2a}.$$

So, the x-coordinate of the vertex of a parabola can always be found with the formula $x = -\dfrac{b}{2a}$.

Remember?

A zero is an x-value that makes the y-value 0.

2. Use the quadratic formula to find the zeros of the function, and then average the zeros to find the x-coordinate of the vertex of the parabola. Evaluate the function at the x-coordinate of the vertex to find the y-coordinate of the vertex. Use the zeros and the vertex to graph the parabola.

$$y = x^2 + 8x + 11$$

Zeros: -1.76, -6.24;
vertex: $(-4, -5)$

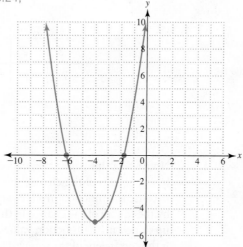

HOW IT WORKS

To find the vertex of a quadratic function:

1. Write the function in the form $y = ax^2 + bx + c$.
2. Identify a, b, and c.
3. Find the x-coordinate of the vertex by using $x = -\dfrac{b}{2a}$.
4. Find the y-coordinate of the vertex by substituting the value of the x-coordinate into the quadratic function.

EXAMPLE: Find the vertex of the parabola for $y = 3x^2 - 12x$.

Start by finding x: $x = -\dfrac{b}{2a} = \dfrac{-(-12)}{2(3)} = \dfrac{12}{6} = 2$

Substitute 2 for x in the equation:

$$\begin{aligned} y &= 3(2)^2 - 12(2) \\ &= 3(4) - 24 \\ &= 12 - 24 \\ &= -12 \end{aligned}$$

So the vertex of the parabola is at $(2, -12)$, as seen in the graph.

The x-coordinate of the vertex of a parabola can also be found by averaging the zeros of the quadratic function. Notice that in this case the zeros are 0 and 4 and the x-coordinate of the vertex is

$$x = \frac{0 + 4}{2} = \frac{4}{2} = 2.$$

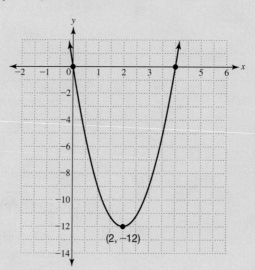

Tech TIP

You can use a graphing calculator to find the vertex of a parabola. Graph the quadratic function. Then select CALC, then Minimum or Maximum to find the vertex. The calculator might ask you to select a point to the left of the vertex, another to the right of the vertex, and then a third point as a guess.

Let's consider again the data shown in the *Explore*. How can you tell that this data represents a quadratic function, as opposed to a linear or exponential function? Notice that as the x-value increases by 1 from the vertex of $(1, -9)$, the y-value increases by 1. As the x-value increases by 2, the y-value increases by 4. As the x-value increases by 3, the y-value increases by 9. It appears that the y-value always increases by the square of the increase in the x-value.

INSTRUCTOR NOTE: Consider having students find the pattern on the left side of the vertex in the table and the graph.

x	−3	−2	−1	0	1	2	3	4	5
y	7	0	−5	−8	−9	−8	−5	0	7

Notice that this pattern repeats on the left side of the vertex as well. The y-value increases by the square of the change in the x-value from the vertex. This should seem reasonable because the function involves squaring x. The function values could increase or decrease by the square of the change in the x-coordinate, based on whether the parabola is opening up or down. The function values could also increase by a multiple of the squared number if the coefficient of the squared term in the function is not 1.

This quadratic pattern can also be seen on this graph of the function.

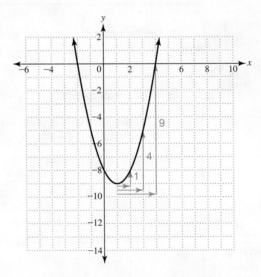

From the table and graph, we can see that the function has zeros at $x = -2$ and $x = 4$. We can use these zeros to write the function in factored form. Each zero corresponds to a factor.

$$y = (x + 2)(x - 4) = x^2 - 2x - 8$$

Notice that this function produces the table of values above.

3. Suppose a parabola has x-intercepts at $(5, 0)$ and $(11, 0)$.

   a. What two factors would be present in the quadratic function?
   $(x - 5), (x - 11)$

b. Multiply these factors to write a quadratic function in $y = ax^2 + bx + c$ form.

$y = (x - 5)(x - 11) = x^2 - 16x + 55$

c. Write two more functions with the given x-intercepts. Leave results in factored form.

Answers will vary. Possible answers include $y = 2(x - 5)(x - 11)$ and $y = -3(x - 5)(x - 11)$.

Connect

Earlier in the book, you learned how to identify linear and exponential patterns in a table.

4. Identify the pattern in each table of data as linear, exponential, quadratic, or none of these.

a.

x	−3	−2	−1	0	1	2	3
y	3	7	11	15	19	23	27

Linear

b.

x	−3	−2	−1	0	1	2	3
y	3	8	11	12	11	8	3

Quadratic

INSTRUCTOR NOTE: Section 4.11 on vertex form of parabolas can be done after this section without loss of continuity. Additional work on quadratic patterns is included.

c.

x	−3	−2	−1	0	1	2	3
y	3	4	6	9	13	18	24

None of these

INSTRUCTOR NOTE: Consider discussing the solution to the focus problem. You might want to show the sample solution, available in MyMathLab, or a correct solution from a student or student group in the class. See the Instructor Guide for more debriefing ideas.

d.

x	−3	−2	−1	0	1	2	3
y	3	6	12	24	48	96	192

Exponential

Reflect

Focus Problem Check-In
Do you have a complete solution to the focus problem? If your instructor has provided you with a rubric, use it to grade your solution. Are there areas you can improve?

WRAP-UP

What's the point?

Quadratic functions have a pattern that can be recognized from either the data or the graph, just like linear and exponential functions. You should be able to identify the zeros and vertex of a quadratic function.

What did you learn?

How to identify a quadratic pattern in data
How to find the vertex of a parabola

3.16 Homework

First complete the MyMathLab homework online. Then work the two exercises to check your understanding.

1. Find the vertex of the parabola $y = 4x^2 - 24x + 2$.

(3, −34)

2. Find the *x*-coordinate of the vertex of a parabola if the *x*-intercepts are $(3, 0)$ and $(-9, 0)$.

$x = -3$

Concepts and Applications

Complete the problems to practice applying the skills and concepts learned in the section.

3. Identify the vertex and the zeros for the quadratic function represented in each table.

a.

x	−9	−8	−7	−6	−5	−4	−3
y	−5	0	3	4	3	0	−5

Zeros: −8, −4; vertex: (−6, 4)

b.

x	−8	−7	−6	−5	−4	−3
y	0	5	8	9	8	5

Zeros: −8, −2; vertex: (−5, 9)

4. **a.** Complete the table using the quadratic pattern if the point (3, 6) is the vertex of the parabola.

x	1	2	3	4	5	6	7
y	10	7	6	7	10	15	22

b. Graph the parabola.

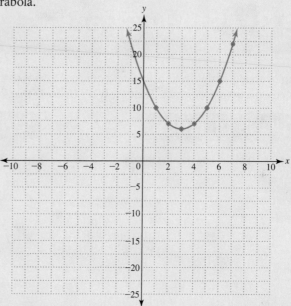

5. Identify the pattern in each table of data as linear, exponential, quadratic, or none of these.

a.

x	−1	0	1	2	3
y	$\frac{5}{3}$	5	15	45	135

Exponential

b.

x	−3	−2	−1	0	1	2
y	14	15	14	11	6	−1

Quadratic

c.

x	−3	−2	−1	0	1	2	3
y	22	17	12	7	2	−3	−8

Linear

d.

x	−3	−2	−1	0	1	2	3
y	5	5	5	5	5	5	5

Linear

6. Looking forward, looking back Find the student's error in the following solution:

$$\begin{cases} 2x - 6y = 20 \\ x + 3y = 9 \end{cases} \rightarrow x = -3y + 9$$

$$x + 3y = 9$$
$$(-3y + 9) + 3y = 9$$
$$9 = 9$$

Infinitely many solutions

The student substituted the expression for y back into the same equation from which it came.

Cycle 3   Study Sheet

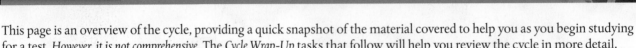

This page is an overview of the cycle, providing a quick snapshot of the material covered to help you as you begin studying for a test. *However, it is not comprehensive.* The *Cycle Wrap-Up* tasks that follow will help you review the cycle in more detail.

EQUATIONS OF LINES

To find the equation of the line passing through $(-3, 5)$ and $(1, 13)$:

1. Find the slope.

$$m = \frac{13 - 5}{1 - (-3)} = \frac{8}{4} = 2$$

2. Solve for b.

$$y = mx + b$$
$$13 = 2(1) + b$$
$$11 = b$$

3. Write the equation.

$$y = 2x + 11$$

To graph the line using the slope and y-intercept:

1. Plot the y-intercept $(0, 11)$.

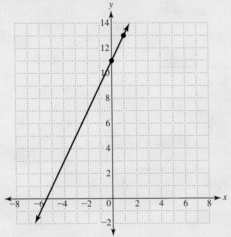

2. Use the slope to move from the y-intercept to another point on the line. Since $m = 2 = \frac{2}{1} = \frac{\text{rise}}{\text{run}}$, move up 2 and right 1 from the y-intercept.

QUADRATIC EQUATIONS & FUNCTIONS

To find the zeros of $y = x^2 - 3x - 28$:

Factoring

$$x^2 - 3x - 28 = 0$$
$$(x - 7)(x + 4) = 0$$
$$x - 7 = 0 \text{ or } x + 4 = 0$$
$$x = 7 \text{ or } \qquad x = -4$$

Quadratic Formula

$$x^2 - 3x - 28 = 0$$
$$a = 1, b = -3, c = -28$$
$$x = \frac{-(-3) \pm \sqrt{(-3)^2 - 4(1)(-28)}}{2(1)}$$
$$x = \frac{3 \pm \sqrt{121}}{2} = \frac{3 \pm 11}{2} = \frac{14}{2} \text{ or } \frac{-8}{2}$$
$$x = 7 \quad \text{or } -4$$

The zeros of the function are 7 and -4.

The x-intercepts of the function are $(7, 0)$ and $(-4, 0)$.

To find the vertex of the parabola:

$$x = -\frac{b}{2a} = -\frac{-3}{2(1)} = \frac{3}{2} \text{ or average the zeros: } \frac{7 + (-4)}{2} = \frac{3}{2}$$
$$y = \left(\frac{3}{2}\right)^2 - 3\left(\frac{3}{2}\right) - 28 = -\frac{121}{4}$$

The vertex is $\left(\frac{3}{2}, -\frac{121}{4}\right)$.

SOLVING EQUATIONS

To solve $\sqrt{5x + 16} = 6$:

$\sqrt{5x + 16} = 6$	Undo the operations done to x
$(\sqrt{5x + 16})^2 = 6^2$	Square both sides
$5x + 16 = 36$	Subtract 16 from both sides
$5x = 20$	Divide both sides by 5
$x = 4$	

To solve $S = \pi r^2 + \pi r l$ for l:

$S = \pi r^2 + \pi r l$	Subtract πr^2 from both sides
$S - \pi r^2 = \pi r l$	Divide both sides by πr
$\dfrac{S - \pi r^2}{\pi r} = l$	

SYSTEMS OF EQUATIONS

System	Possible approach to solve and first step
$y = x - 1$ $y = 2x - 4$	Graphing: Use m and b to graph the lines.
$7x - y = 7$ $2x = 2y + 2$	Substitution: Solve for x in the second equation.
$3x - 2y = 4$ $x + 10y = 44$	Elimination: Multiply the first equation by 5 to create opposite y terms.

A linear system with no solution has parallel lines.
A linear system with infinitely many solutions has one line.

SELF-ASSESSMENT: **REVIEW**

For each objective, use the boxes provided to indicate your current level of expertise. If you cannot indicate a high level of expertise for a skill or concept, continue working on it and seek help if necessary.

INSTRUCTOR NOTE: Consider discussing this page before a test. The objectives are the same ones that are provided at the beginning of the cycle, giving students two points of comparison on their learning. Completing the page can help students see where they need to focus their studying for a test.

SKILL or CONCEPT

Low → High

1. Determine if data has a positive or negative linear correlation. *(3.2)*

2. Use the equation of the trendline to make predictions. *(3.2)*

3. Find the slope of a line from points, tables, and graphs. *(3.3)*

4. Interpret the slope as a rate of change. *(3.3)*

5. Use the distance formula to find the distance between two points. *(3.4)*

6. Make comparisons using equations, tables, and graphs. *(3.5)*

7. Find and interpret the slope and *y*-intercept from a linear equation. *(3.6)*

8. Graph a line using a table and using the slope and *y*-intercept. *(3.6)*

9. Write the equation of a line using a point and the slope or two points. *(3.7)*

10. Model with linear functions. *(3.7)*

11. Model with exponential functions. *(3.8)*

12. Graph exponential functions. *(3.8)*

13. Solve nonlinear equations. *(3.9)*

14. Solve an equation for a specified variable. *(3.10)*

15. Factor an expression using the greatest common factor. *(3.11)*

16. Factor quadratic expressions. *(3.12)*

17. Use the quadratic formula to solve equations. *(3.13)*

18. Solve a 2 × 2 linear system of equations by graphing and substitution. *(3.14)*

19. Solve a 2 × 2 linear system of equations by elimination. *(3.15)*

20. Find the vertex of a parabola. *(3.16)*

WRAP-UP

Our intuition about the future is linear. But the reality of information technology is exponential, and that makes a profound difference. If I take 30 steps linearly, I get to 30. If I take 30 steps exponentially, I get to a billion.

—Ray Kurzweil

INSTRUCTOR NOTE: Discuss the steps included in the *Cycle Wrap-Up* in detail with students so they understand the actions that lead to success on a test.

The *Cycle Wrap-Up* will help you bring together the skills and concepts you have learned in the cycle so that you can apply them successfully on a test. Since studying is an active process, the *Wrap-Up* will provide you with activities to improve your understanding and your ability to show it in a test environment.

STEP 1

GOAL Revisit the cycle content

Action Skim your cycle notes

Action Read the Cycle Study Sheet

Action Complete the Self-Assessment: Review

STEP 2

GOAL Practice cycle skills

Action Complete the MyMathLab assignment

Action Address your areas of difficulty

MyMathLab

Complete the MyMathLab assignment. Reference your notes if you are having trouble with a skill. For more practice, you can rework Skills problems from your homework. You can also access your MyMathLab homework problems, even if the assignment is past due, by using the Review mode available through the Gradebook.

STEP 3

GOAL Review and learn cycle vocabulary

Action Use your notes to complete the Vocabulary Check

Action Memorize the answers

Vocabulary Check

Cycle 3 Word Bank

consistent	factors	quadratic formula	vertex
distance formula	GCF	slope	*y*-intercept
elimination	inconsistent	slope-intercept	zero
exponential	linear correlation	standard	zero-product property
factoring	parabola	system of equations	

Choose a word from the word bank to complete each statement.

1. When a scatterplot has a linear pattern, we say there is a(n)
   _____ linear correlation _____ .

2. The ratio of rise to run of a line is called its _____ slope _____ .

3. The _____ distance formula _____ can be used instead of measuring to find the distance between two points.

4. An equation is in _____ slope-intercept _____ form if it has the form
   $y = mx + b$.

5. In the equation $y = 5x - 6$, the number 5 represents the slope of the graph of the equation, and the point $(0, -6)$ is the _____ y-intercept _____ .

6. A(n) _____ exponential _____ function is an equation of the form $y = a \cdot b^x$.

7. The process of writing a sum or difference of terms as a product is called
   _____ factoring _____ .

8. The _____ GCF _____ is the greatest of all the
   _____ factors _____ that a list of terms has in common.

9. The _____ zero-product property _____ states that if a product of two numbers is zero, then at least one of the numbers must be zero.

10. The _____ zero _____ of a function is an input value that makes the function output equal to zero.

11. The _____ quadratic formula _____ is a formula that can be used to solve any quadratic equation.

12. A(n) _____ system of equations _____ is a set of two or more equations that need to be solved simultaneously.

13. When a system of equations has a solution, it is called _____ consistent _____ .

14. When a system of equations has no solution, it is called _____ inconsistent _____ .

15. An equation in $Ax + By = C$ form is said to be in _____ standard _____ form.

16. The _____ elimination _____ method of solving systems involves finding opposite terms and adding them.

17. The turning point on a quadratic graph is called the _____ vertex _____ .

18. A quadratic function has a graph called a(n) _____ parabola _____ .

STEP 4

GOAL Practice cycle applications

Action Complete the Concepts and Applications Review

Action Address your areas of difficulty

Concepts and Applications Review

Complete the problems to practice applying the skills and concepts learned in the cycle.

1. a. Explain the order of the operations involved when you apply the slope formula.

First, subtract the *y*-values.

Second, subtract the *x*-values in the same order.

Third, divide the difference in the *y*-values by the difference in the *x*-values.

b. Explain the order of operations involved when you use the distance formula.

First, subtract the *x*-coordinates and the *y*-coordinates.

Second, square each of those differences.

Third, add the squared differences.

Fourth, take the square root.

2. What is the steepest line possible, and what is its slope? Justify your response.

Since we can measure steep slopes with positive numbers, there is no steepest line. For any positive value, we can always find a larger positive value. A vertical line may seem to be the steepest line, but by definition, it does not have a slope.

3. Give the coordinates of two ordered pairs so that the distance between them is 5 units, and the slope of the line connecting them is $-\frac{3}{4}$. Begin by drawing the situation on a coordinate plane.

Answers will vary; one possible answer: (1, 6) and (5, 3)

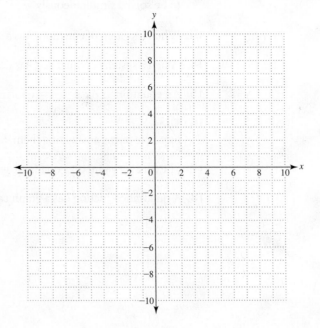

4. Suppose a college professor tracks her students' absences and investigates how the number of absences is related to a student's final grade in the course. After creating a scatterplot, the professor finds the following equation to model the trend: $G = 90 - 2A$, where G = grade and A = number of absences.

a. What is the slope of the line represented by this equation?

−2

b. Write a sentence to interpret the slope in context.

For every absence, the student's grade will drop 2 points.

c. What is the y-intercept of the line represented by this equation?

(0, 90)

d. Write a sentence to interpret the y-intercept in context.

If a student does not miss any classes at all, his grade will be 90.

e. If a student misses 5 classes, what score do you expect him to receive?

80

f. If a student wants to keep his grade at or above 65, how many classes can he afford to miss?

12 classes

5. Many professionals in the health care industry claim that a healthy rate of weight loss is about 2 pounds per week. Suppose a man currently weighs 260 pounds and would like to get down to 200 pounds in a healthy manner.

a. Assume the man will be successful in meeting his weight loss goal. Record his healthy weight loss rate in the following table, where m = number of months and w = his weight. Assume there are 4 weeks in a month.

m	w
0	260
1	252
2	244
3	236
4	228
5	220

b. Write a linear function to model this man's weight.

$w = 260 - 8m$

c. Graph the linear equation using either points or the slope and *y*-intercept.

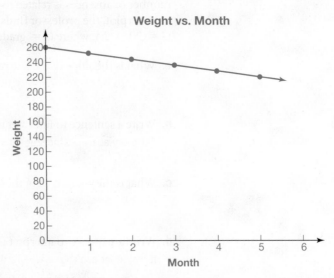

Weight vs. Month

d. When will the man reach his target weight of 200 pounds? Explain your reasoning, show your calculation, or solve an appropriate equation.

7.5 months

6. Suppose a worker earns $660 in a week in which he works 4 hours of overtime. He earns $900 in a week in which he works 20 hours of overtime.

a. Write a linear equation to model the worker's weekly salary based on the number of overtime hours he works that week.

$S = 600 + 15h$

b. What is the slope of the linear equation, and what does it mean?

Slope = 15; he earns $15 an hour for overtime work.

c. What is the *y*-intercept of the linear equation, and what does it mean?

y-intercept = (0,600); he earns a base salary of $600 per week.

d. Use your linear equation to determine how many overtime hours the worker would need to work in order to earn $700 in one week.

Approximately 6.7 hours

e. Graph the linear equation.

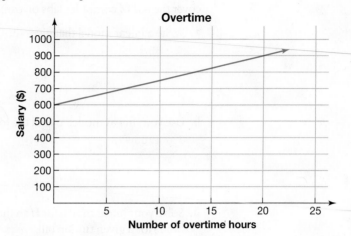

7. Identify the type of change, linear or exponential as well as growth or decay, illustrated in each table.

a.

x	y
0	1
1	1.1
2	1.21
3	1.331

Exponential growth

b.

x	y
0	1
1	0.9
2	0.8
3	0.7

Linear decay

c.

x	y
0	1
1	0.9
2	0.81
3	0.729

Exponential decay

d.

x	y
0	1
1	1.1
2	1.2
3	1.3

Linear growth

8. Solve each equation. Consider the steps done to the variable to determine what operations should be used and the order in which they should be used to solve each equation.

a. $x - 4 = 8$ 12

b. $\sqrt{x - 4} = 8$ 68

c. $\dfrac{1}{x - 4} = 8$ $\dfrac{33}{8}$

d. $x^2 - 4 = 8$ about ± 3.46

9. Suppose tuition at a community college is $85 per credit hour, plus a $100 lab fee to cover the use of computer labs on campus.

    a. Write a linear model that gives the tuition total (*T*) for *H* credit hours.

       $T = 85H + 100$

    b. Use your linear model to find the total tuition for 9 credit hours.

       $T = \$865$

    c. Solve your linear model for *H* so that you can determine the number of credit hours for a given tuition bill.

       $H = \dfrac{T - 100}{85}$

    d. Use this model to determine when the total tuition will be $1,460.

       $H = 16$ credit hours

10. The sum of the interior angles in a polygon can be found by dividing a polygon into nonoverlapping triangles. Each triangle's interior angles have a sum of 180°. The formula for the sum, *S*, of the interior angles in a polygon with *n* sides is given by $S = (n - 2) \cdot 180°$.

    a. For the hexagon shown, use the formula to find the sum of its interior angles. Notice that it has six sides and four nonoverlapping interior triangles within it.

    720°

    b. The interior angles in a regular polygon have the same measure. To find the measure of one interior angle in a regular polygon, find the sum of the interior angles and divide it by the number of angles, *n*. Generalize this approach to state a formula for one interior angle, *A*, in a regular polygon with *n* sides.

       $A = \dfrac{(n - 2) \cdot 180}{n}$

c. Solve the formula in part b for n.

$$n = \frac{360°}{180° - A}$$

d. Use the formula in part c to find the number of sides for a regular polygon whose interior angles each measure 144°.

10

11. Write each of the following physics formulas in factored form by factoring out the GCF.

a. $I = mv - mu$ Impulse $= m(v - u)$

b. $s = ut + \frac{1}{2}at^2$ $s = t\left(u + \frac{1}{2}at\right)$

12. One way to quickly square any two-digit number ending in 5 is to take the tens digit and multiply it by the number that is one greater than it and then follow that with 25. For example, to find 65^2, we multiply 6 (the tens digit) by 7 (one greater than it). The product is $6 \cdot 7 = 42$. The answer to 65^2 is 4,225, the product you got (42) followed by 25.

The shortcut takes the tens digit multiplied by the number that is one greater than it and then multiplies that product by 100 to move the product to the correct place value. 25 is added to the result.

Verify that this shortcut works in general using the steps provided. Suppose T is a whole number from 0 to 9 and represents the tens digit. Let $10T + 5$ represent a two-digit number ending in 5.

a. Square $10T + 5$, the original two-digit number ending in 5.

$100T^2 + 100T + 25$

b. Factor out the GCF from the terms that contain T.

$100T(T + 1) + 25$

Notice that your result from part b describes the shortcut for multiplying the tens digit, T, by the number that is one greater than it and then by 100. Multiplying by 100 puts the product in the thousands and hundreds places. Adding 25 to this result creates a four-digit number ending in 25.

13. Solve $16x^2 = 256$ using each stated approach.

a. Solve by listing the operations done to x and undoing each one.

$x = 4, -4$

b. Solve by factoring.

$x = 4, -4$

c. Solve by using the quadratic formula.

$x = 4, -4$

14. a. Write a 2×2 system of equations that has no solution.

Answers will vary; one possibility: $2x + 3y = 5$
$2x + 3y = 6$

b. Write a 2×2 system of equations that has infinitely many solutions.

Answers will vary; one possibility: $2x + 3y = 4$
$4x + 6y = 8$

15. For each function, state if it is linear, exponential, quadratic, or none of these.

    a. $y = 3 - 2x$ Linear

    b. $y = 3x^2$ Quadratic

    c. $y = 3 \cdot 2^x$ Exponential

    d. $y = 3x$ Linear

    e. $y = 3 \cdot 2^{x-1}$ Exponential

    f. $y = \dfrac{3}{2x}$ None of these

    g. $y = \dfrac{3}{2}x$ Linear

16. Suppose the quadratic function with values in the table has a zero at 3 and its vertex is $(7, -16)$.

x	1	2	3	4	5	6	7
y			0				-16

    a. Find the other zero.

    11

    b. Find a quadratic function in $y = ax^2 + bx + c$ form with these zeros and the other value shown in the table.

    A possible function: $y = (x - 3)(x - 11) = x^2 - 14x + 33$

STEP 5

GOAL Simulate a test environment

Action Construct a mock test and take it while being timed

Creating a Mock Test

To pass a math test, you must be able to recall vocabulary, skills, and concepts and apply them to new problems in a fixed amount of time.

Up to this point, you have worked on all the components of the cycle, but they were separated and identified by their topic or type. Although helpful, this is not a test situation. You need to create a test situation to help overcome anxiety and determine where problem areas still exist. Think about a basketball player. She must practice her skills, but until she engages in a practice game, she doesn't really have a feel for a game situation.

The key is to practice problems without assistance under time pressure.

Use the following steps to create a mock test:

- Pick a sample of 5 to 10 problems from your notes and homework and write them on note cards. Choose both skills and applications. On the back of the note card, write the page number where the problem was found.
- Shuffle the cards.
- Set a timer for 20 minutes.
- Work as many problems as you can without outside help.
- Check your work by consulting the page number on the card.

This test will show you where your weaknesses are and what needs more work. It will also give you a feel for your pace. If you completed only a few problems, you are not ready for a test.

Continue to work on the areas that need improvement.

Look over your previous tests and the types of mistakes you made. How can you avoid making them on this test? By analyzing your mistakes, you might be able to avoid making them again on this test.

List two specific things that you learned from looking over your previous tests:

1.

2.

WHAT ELSE CAN WE DO?

dimensional analysis

quadratic models

order of magnitude

domain

functions

range

z-scores

trigonometry

standard deviation

variation

vertical line test

scientific notation

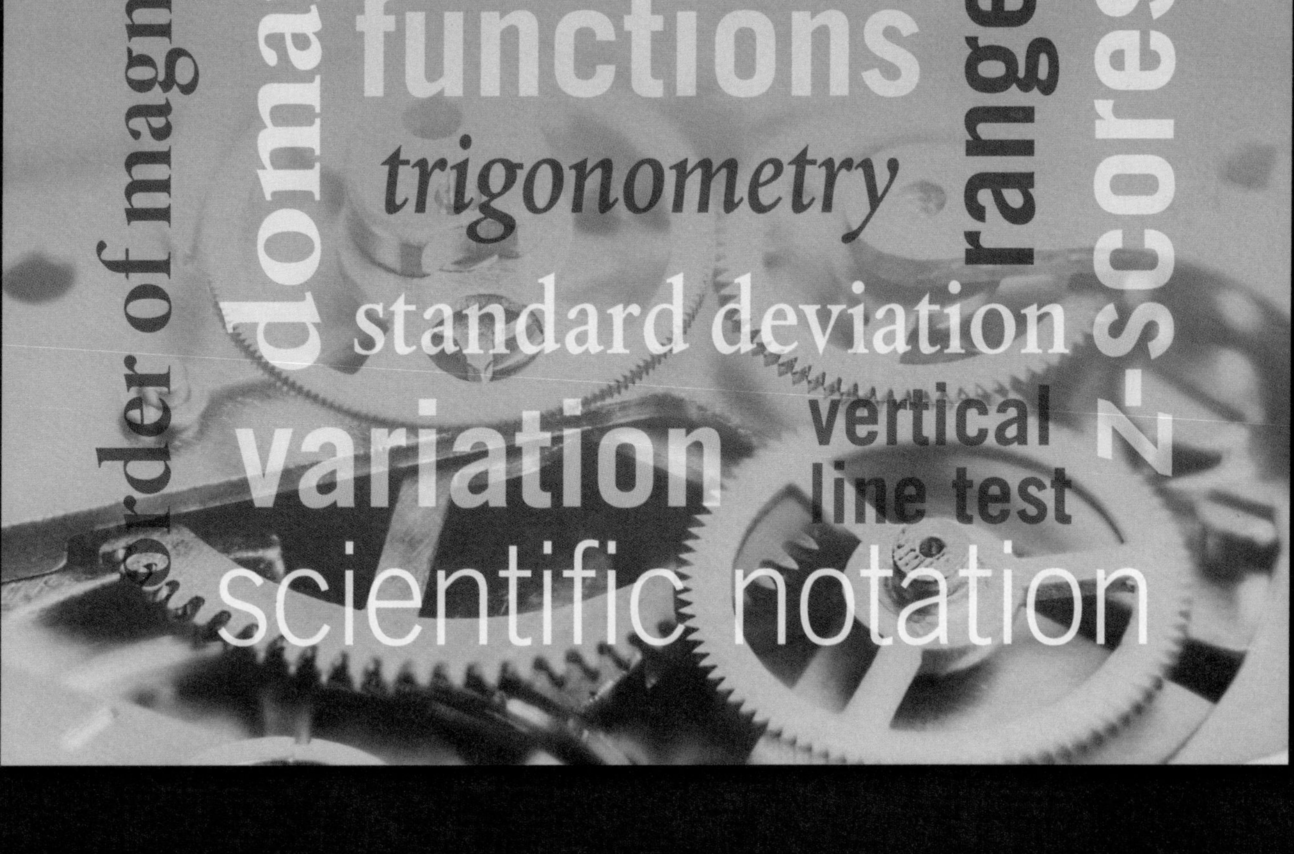

SELF-ASSESSMENT: PREVIEW

Below is a list of objectives for this cycle. For each objective, use the boxes provided to indicate your current level of expertise.

INSTRUCTOR NOTE: Consider discussing this page when you begin the cycle. The same list is presented at the end of the cycle so that students can note their progress and identify areas that they still need to work on before the test.

SKILL or CONCEPT

Low → High

1. Convert units using dimensional analysis.

2. Convert numbers between scientific and standard notation.

3. Convert numbers into and out of engineering notation.

4. Use exponent rules to simplify expressions that have negative exponents.

5. Find the standard deviation of a data set.

6. Interpret the standard deviation of a data set.

7. Interpret logarithmic scales.

8. Identify direct variation from a graph, table, or equation.

9. Solve direct variation problems.

10. Identify inverse variation from a table.

11. Solve inverse variation problems.

12. Use function notation.

13. Find a function input or output given the other.

14. Apply the vertical line test.

15. Find the domain and range from a graph.

16. Identify the vertex of a quadratic function in vertex form.

17. Graph a quadratic function in vertex form.

18. Write the vertex form of a quadratic function given the vertex and a point.

19. Write the six trigonometric ratios for an acute angle.

20. Use trigonometric functions to find the measures of sides and angles of a right triangle.

Statistical thinking will one day be as necessary a qualification for efficient citizenship as the ability to read and write.

— H. G. Wells

4.1 **Measuring Temperature Variability**
Focus Problem

When people consider global warming, they tend to think primarily about the most publicized effect of higher temperatures. It is generally well accepted that mean temperatures are on the rise. However, the effect of global warming on temperature variability is also of concern. Will rising temperatures make temperatures more or less variable? This is still being studied, but evidence suggests that temperature variability has remained steady or even decreased slightly in spite of the common perception that the temperature variability will increase as temperatures rise.

Although the research needed to explore climate change is complicated, you can get a sense of what is involved by collecting and analyzing some data for your city. One way to approach this research is to collect a sample of temperatures and calculate the standard deviation. You can find historical weather data on Weather Underground or other websites. Specifically, choose a month from the last year and record the daily high temperatures for your city. Then record the daily high temperatures for that same month 10 years ago, 20 years ago, 30 years ago, and so on until you have at least six sets of data. For each data set, calculate the mean and standard deviation of the temperatures. How has the mean temperature changed over the decades? How has the standard deviation changed? What do these two measures of daily high temperatures tell you when you consider them together? Write a paragraph to address these questions.

Other than measuring the variation in the daily high temperatures, how else could you collect data to determine if the variability in temperatures is changing? Could you look at the variation in the daily low temperatures over the years? Or could you look at the variation in the daily temperature range over the years? What other data might lend some insight to the study of temperature change?

Other than the models you've learned so far in this book, how else could we measure or model change?

WHAT ELSE CAN
WE DO?

If linear and exponential models are not appropriate for a particular situation, what other types of algebraic models might you use to try to understand data? In this cycle, you will be introduced to other types of functions that can be used as models, in addition to learning more about models that you already know.

You will see periodic sticky notes throughout the cycle to encourage you to keep working on your solution to this problem as you learn more.

INSTRUCTOR NOTE: Additional focus problems are available in MyMathLab. Each focus problem is accompanied by a writing template to help students form their solution. A grading rubric is also available. For more information on ways to use the focus problems, see the Instructor Guide at the beginning of the worktext.

INSTRUCTOR NOTE: If you are using groups and would like more detailed information on using them effectively, please consult the Instructor Guide at the front of the worktext. It contains techniques and resources for group work.

4.2 A Matter of Change: Dimensional Analysis

Explore

SECTION OBJECTIVE:
- Convert units using dimensional analysis

1. The current world-record holder in the 60-meter sprint can run that distance in only 6.39 seconds.

   a. Convert 60 meters to centimeters, then inches, then feet, and then miles. How many miles are in 60 meters?
   About 0.0373 mile

   b. Convert 6.39 seconds to minutes and then hours. How many hours are in 6.39 seconds?
   About 0.001775 hour

   c. How fast, in miles per hour, is this runner sprinting?
   About 21 mph

Discover

When you have to deal with many different units, as in the *Explore*, it can be difficult to determine when to multiply and when to divide. Conversions can become unwieldy, and it can also be a challenge to display your work clearly for someone else. For complicated conversions, using a process called dimensional analysis can make the conversion more organized and clear.

As you work through this section, consider these questions:

1 When is it worth using dimensional analysis?

2 How do you convert squared or cubed units?

We will begin with a definition and example of the process of dimensional analysis.

Dimensional Analysis

Dimensional analysis is an organized process for converting units which involves multiplying by conversion factors. A conversion factor is a fraction equal to 1 and is created by writing a conversion fact as a fraction.

EXAMPLE: Convert 3 hours to minutes using dimensional analysis.

SOLUTION: We begin by finding a conversion fact that relates hours to minutes and writing it as a conversion factor.

$$60 \text{ minutes} = 1 \text{ hour}$$

$$\frac{60 \text{ minutes}}{1 \text{ hour}} \quad \text{or} \quad \frac{1 \text{ hour}}{60 \text{ minutes}}$$

Since we want to eliminate hours and hours is in the numerator, we write the conversion factor with hours in the denominator. Then we multiply 3 hours by the conversion factor and divide off the unit of hours.

$$3 \ \cancel{hr} \cdot \frac{60 \text{ min}}{1 \ \cancel{hr}} = 180 \text{ min}$$

With dimensional analysis, we can use one or more conversion factors to accomplish the conversion.

INSTRUCTOR NOTE: Consider having your students determine other paths for the conversions done in this section.

EXAMPLE 1

Convert 5 miles to inches.

SOLUTION

It can be helpful to plan the path of the conversion before you start any calculations. We cannot convert directly from miles to inches without knowing how many inches are in a mile. So, in this case, we will convert miles $\rightarrow$ feet $\rightarrow$ inches.

$$5 \text{ mi} \cdot \frac{5{,}280 \text{ ft}}{1 \text{ mi}} \cdot \frac{12 \text{ in.}}{1 \text{ ft}} = 316{,}800 \text{ in.}$$

Finally, we should consider whether the answer makes sense. Since we converted from a relatively large unit of miles to a much smaller unit of inches, the large number of inches makes sense. It takes a lot of the smaller inches to cover the larger miles.

Dimensional analysis can be used to convert one or both units involved in a rate, as seen in the next problem.

2. The acceleration due to gravity is 32 ft/sec². Convert this measurement to m/sec².
   9.8 m/sec²

If both units in a rate need to be converted, you can change them one at a time in the same dimensional analysis calculation. As conversions get more complicated, dimensional analysis makes more sense than just trying to multiply or divide as needed.

EXAMPLE 2

Convert 65 mph to centimeters per second.

SOLUTION

Since this problem involves several conversions, it an excellent place to use dimensional analysis.

Notice that we are converting a rate, not a number. The rate has two units that must be changed (miles and hours), so two conversions are involved. It doesn't matter which one we do first; both conversions can be accomplished in one calculation.

Path for miles to centimeters: miles $\rightarrow$ feet $\rightarrow$ inches $\rightarrow$ centimeters

Path for hours to seconds: hours $\rightarrow$ minutes $\rightarrow$ seconds

We convert the miles and then the hours.

$$\frac{65 \text{ mi}}{1 \text{ hr}} \cdot \frac{5{,}280 \text{ ft}}{1 \text{ mi}} \cdot \frac{12 \text{ in.}}{1 \text{ ft}} \cdot \frac{2.54 \text{ cm}}{1 \text{ in.}} \cdot \frac{1 \text{ hr}}{60 \text{ min}} \cdot \frac{1 \text{ min}}{60 \text{ sec}} \approx 2{,}905.76 \text{ cm/sec}$$

Thus, 65 miles per hour is the same as 2,905.76 centimeters per second.

Tech TIP

On a calculator, multiply all the numerators, and then divide by the denominators, using a division symbol before each one. Press = or ENTER. This can save time and improve accuracy, since you don't have to switch repeatedly between multiplication and division.

As you saw in the last example, we can string together several conversion factors because the calculation is still equivalent to multiplying by 1. Each fraction is equal to 1 because the quantities in the numerator and denominator are equal. Multiplying a number by 1 does not change its value. However, it does allow us to change the units being used.

INSTRUCTOR NOTE: Consider having students use dimensional analysis to convert the rate in the *Explore* in one step instead of the three-step process.

3. The record-holding time is 43.18 seconds for the 400-meter dash. Convert this runner's speed to miles per hour using dimensional analysis.

20.72 miles per hour

Sometimes the units we need to convert are squared or cubed. Dimensional analysis can be particularly helpful in these situations. Consider the following illustration of a square yard:

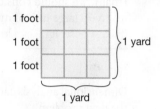

The square has an area of

$$1 \text{ yard} \cdot 1 \text{ yard} = 1 \text{ yard}^2$$

But the area can also be calculated in square feet.

$$3 \text{ feet} \cdot 3 \text{ feet} = 9 \text{ feet}^2$$

So 1 square yard is equivalent to 9 square feet. This makes sense when you look at the figure and observe that there are 9 smaller square feet inside the 1 square yard.

When you need a conversion factor for squared units, just write the conversion factor for the units and square it.

$$\left(\frac{1 \text{ yard}}{3 \text{ feet}}\right)^2 = \frac{1^2 \text{ yard}^2}{3^2 \text{ feet}^2} = \frac{1 \text{ yard}^2}{9 \text{ feet}^2}$$

Similarly, a conversion factor for cubic units can be found by cubing the conversion factor.

$$\left(\frac{1 \text{ yard}}{3 \text{ feet}}\right)^3 = \frac{1^3 \text{ yard}^3}{3^3 \text{ feet}^3} = \frac{1 \text{ yard}^3}{27 \text{ feet}^3}$$

You can verify that there are 27 cubic feet in 1 cubic yard by observing the cubic yard shown here. It has three layers of 9 cubic feet for a total of 27 cubic feet.

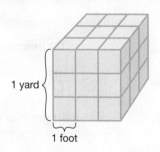

INSTRUCTOR NOTE: To complete #4, students might use the fact that 1 square yard = 9 square feet. Consider showing students how to write the conversion factor with 1 yard and 3 feet, both being squared, as shown in the answer.

4. Jamie wants to carpet her bedroom and knows that the rectangular room has dimensions of 10′ × 11′. If the carpet she needs is sold for $7.50 per square yard, how much will she pay for the carpet? Consider that she needs to buy a whole number of square yards.

$$110 \text{ ft}^2 \cdot \frac{1^2 \text{ yd}^2}{3^2 \text{ ft}^2} \approx 12.22 \text{ yd}^2$$

Jamie will pay 13 yd^2 × $7.50/yd^2 = $97.50.

HOW IT WORKS

To convert units using dimensional analysis:

1. Find a relationship between the units using conversion facts. This may be a direct relationship with only one conversion fact or a path from one unit to the other with more than one conversion fact.
2. Write the calculation, using the conversion factors in such a way that unwanted units will divide off.
3. Divide off units and perform the necessary multiplication and division.
4. Check that your answer is reasonable.

EXAMPLE: Convert 100 yards to meters.

To convert 100 yards to meters, use the conversion facts 1 yard = 3 feet, 1 foot = 12 inches, 1 inch = 2.54 centimeters, and 100 centimeters = 1 meter. Multiply 100 yards by the conversion factors so that the units divide off and only meters remain.

$$100 \text{ yd} \cdot \frac{3 \text{ ft}}{1 \text{ yd}} \cdot \frac{12 \text{ in.}}{1 \text{ ft}} \cdot \frac{2.54 \text{ cm}}{1 \text{ in.}} \cdot \frac{1 \text{ m}}{100 \text{ cm}} = 91.44 \text{ m}$$

So 100 yards is the same as 91.44 meters.

Check that the answer makes sense. Meters are a larger unit of length than yards. So, in converting from 100 yards to meters, fewer meters are needed to cover the same length.

EXAMPLE: Convert 100 cubic yards to cubic meters.

To accomplish this conversion, write the following calculation. Notice that the same conversion factors are used as in the previous example, but each conversion factor is cubed.

$$100 \text{ yards}^3 \cdot \frac{3^3 \text{ ft}^3}{1^3 \text{ yd}^3} \cdot \frac{12^3 \text{ in.}^3}{1^3 \text{ ft}^3} \cdot \frac{2.54^3 \text{ cm}^3}{1^3 \text{ in.}^3} \cdot \frac{1^3 \text{ m}^3}{100^3 \text{ cm}^3} = 76.46 \text{ m}^3$$

If you need a conversion fact for the homework that is not listed here, look it up online.

Commonly used unit conversions:			
1 ft = 12 in.	**1 in. = 2.54 cm**	1 L = 1,000 mL	1 min = 60 sec
1 mi = 5,280 ft	**1 km = 1,000 m**	1 lb = 16 oz	1 hr = 60 min
1 yd = 3 ft	1 m = 100 cm	1 kg = 1,000 g	1 yr = 52 weeks
1 km ≈ 0.62 mi	1 cm = 10 mm	1 g = 1,000 mg	1 week = 7 days
1 mi ≈ 1.61 km		1 ton = 2,000 lb	
		1 gal = 4 qt	
		1 kg ≈ 2.2 lb	

Conversion facts shown in bold provide a conversion between the English and metric systems.

5. Convert 8 cubic meters to cubic inches.

$$8 \ \text{m}^3 \cdot \frac{100^3 \ \text{cm}^3}{1^3 \text{m}^3} \cdot \frac{1^3 \ \text{in.}^3}{2.54^3 \text{cm}^3} \approx 488{,}190 \ \text{in.}^3$$

> When you convert squared or cubed units, square or cube only the numbers and units that are involved in the conversion factor, not the number that you are converting.

6. a. How many yards are in 4 miles?

7,040 yards

b. How many square yards are in 4 square miles?

12,390,400 square yards

Connect

Jake is taking a graduation trip to London for a week in the summer. His parents gave him $750 in spending money. When he arrives, he sees the following currency exchange rate information:

	USD	EUR	GBP	JPY	CAD	AUD
1 USD =	1.000	0.9432	0.6635	122.63	1.3327	1.3870
1 EUR =	1.0602	1.000	0.7034	130	1.4127	1.4704
1 GBP =	1.5071	1.4217	1.000	184.79	2.0082	2.0901
1 JPY =	0.0082	0.0077	0.0054	1.000	0.0109	0.0113
1 CAD =	0.7504	0.7079	0.4980	91.9963	1.000	1.0406
1 AUD =	0.7210	0.6801	0.4784	88.4173	24.6305	1.000

Reading across the first row of the table, he can find what one U.S. dollar is worth in the United States (in dollars), in Europe (in euros), in Great Britain (in pounds), in Japan (in yen), in Canada (in Canadian dollars), and in Australia (in Australian dollars). The remaining rows indicate what one unit of the other currencies is worth in the listed countries.

7. a. Why is there a "1.000" in each row? What is its significance?

It represents what the currency is worth in its own country.

> You can convert units by multiplying or dividing if it's clear to you which you need to do. Otherwise, use dimensional analysis.

b. How much is 1 U.S. dollar worth in British pounds on this day?

0.6635 pound

c. Convert Jake's $750 to pounds. Round to the nearest whole number.

About 498 pounds

d. Analyze the calculation made in part c to determine a quick rule for converting dollars to British pounds. Rules like this help travelers make quick conversions without using paper or possibly even calculators.

Multiply by 0.6635. Essentially, we are taking a little more than two-thirds the number of dollars. A quick mental math approach is to take $\frac{2}{3}$ of a number (divide by 3 and multiply the result by 2).

e. Which is a larger unit of currency: a U.S. dollar or a British pound? How can you tell?

A British pound is larger, since it takes fewer pounds than dollars to make the same amount.

INSTRUCTOR NOTE: Remind students to read the *Getting Ready* article for the next section.

f. If Jake sees an item for sale in London for 200 pounds, what is the price in U.S. dollars? What is a quick way to mentally estimate the dollar equivalent?

$301.42; multiply the price in pounds by one and a half.

Reflect

WRAP-UP

What's the point?

In Cycle 1 you learned how simple conversions can be made by multiplying or dividing. As unit conversions get more complicated, dimensional analysis provides a technique for organizing your work and efficiently conducting the conversion.

What did you learn?

How to convert units using dimensional analysis

4.2 Homework

Skills MyMathLab

First complete the MyMathLab homework online. Then work the two exercises to check your understanding.

1. Convert 20 mph to kilometers per minute.

Approximately 0.54 kilometer per minute

2. Convert 25 cubic yards to cubic inches.

1,166,400 cubic inches

Concepts and Applications

Complete the problems to practice applying the skills and concepts learned in the section.

3. Use the currency table in this section to answer the following problems:

a. Use dimensional analysis to convert $1 from U.S. dollars to euros to pounds to yen.

122.60 yen

b. Use the table to convert $1 directly (in one step) from U.S. dollars to yen.

122.63 yen

c. Are your answers from parts a and b exactly equal? If not, what could account for the differences in the results?

Rounding in the exchange rates

d. Take the reciprocal of 122.63 Japanese yen. What is the result, and what does it mean?

0.0082

$$1 \text{ U.S dollar} = 122.63 \text{ yen, so } 1 = \frac{1 \text{ U.S. dollar}}{122.63 \text{ yen}} \approx \frac{0.0082 \text{ dollar}}{1 \text{ yen}}$$

e. Circle 122.63 and its reciprocal in the table. Describe where they are located relative to each other.

122.63 is in the USD row of the JPY column, and 0.0082 is in the JPY row of the USD column.

f. Find four more pairs of numbers and their reciprocals in the table, and explain the significance of these reciprocal pairs.

0.9432 euro = 1 U.S. dollar, 1 euro = 1.0602 U.S. dollars

0.6635 pound = 1 U.S. dollar, 1 pound = 1.5071 U.S. dollars

1.3327 Canadian dollars = 1 U.S. dollar, 1 Canadian dollar = 0.7504 U.S. dollar

1.3870 Australian dollars = 1 U.S. dollar, 1 Australian dollar = 0.7210 U.S. dollar

A reciprocal pair shows the conversion rates between two currencies.

g. Convert 10 U.S. dollars to euros in three ways: by scaling; by using the conversion factor 1 USD = __0.9432__ euro; and by using the reciprocal of the conversion factor 1 euro = __1.0602__ USD.

9.432

4. A student is trying to convert from miles per day to feet per minute using the following dimensional analysis:

$$\frac{500 \text{ miles}}{1 \text{ day}} \cdot \frac{5,280 \text{ feet}}{1 \text{ mile}} \cdot \frac{1 \text{ day}}{24 \text{ hours}} \cdot \frac{1 \text{ hour}}{60 \text{ min}}$$

To get the result, the student entered $500 \cdot 5,280 \div 24 \cdot 60$ on the calculator. Explain the student's mistake.

The student should have entered $500 \cdot 5,280 \div 24 \div 60$ or $500 \cdot 5,280 \div (24 \cdot 60)$.

5. Looking Forward, Looking Back In the values in the table, is the change linear, quadratic, or exponential?

x	y
0	20
1	5
2	0
3	5
4	20

Quadratic

Getting Ready for Section 4.3

Read the article and answer the questions that follow.

How Many Cells Are In Your Body?

A simple question deserves a simple answer. How many cells are in your body?

Unfortunately, your cells can't fill out census forms, so they can't tell you themselves. And while it's easy enough to look through a microscope and count off certain types of cells, this method isn't practical either. Some types of cells are easy to spot, while others–such as tangled neurons–weave themselves up into obscurity. Even if you could count ten cells each second, it would take you tens of thousands of years to finish counting. Plus, there would be certain logistical problems you'd encounter along the way to counting all the cells in your body–for example, chopping your own body up into tiny patches for microscopic viewing.

For now, the best we can hope for is a study published recently in *Annals of Human Biology*, entitled, with admirable clarity, *"An Estimation of the Number of Cells in the Human Body."*

The authors–a team of scientists from Italy, Greece, and Spain–admit that they're hardly the first people to tackle this question. They looked back over scientific journals and books from the past couple centuries and found many estimates. But those estimates sprawled over a huge range, from 5 billion to 200 million trillion cells. And practically none of scientists who offered those numbers provided an explanation for how they came up with them. Clearly, this is a subject ripe for research.

If scientists can't count all the cells in a human body, how can they estimate it? The mean weight of a cell is 1 nanogram. For an adult man weighing 70 kilograms, simple arithmetic would lead us to conclude that that man has 70 trillion cells. On the other hand, it's also possible to do this calculation based on the volume of cells. The mean volume of a mammal cell is estimated to be 4 billionths of a cubic centimeter. (To get a sense of that size, check out *The Scale of the Universe*.) Based on an adult man's typical volume, you might conclude that the human body contains 15 trillion cells. So if you pick volume or weight, you get drastically different numbers. Making matters worse, our bodies are not packed with cells in a uniform way, like a jar full of jellybeans. Cells come in different sizes, and they grow in different densities. Look at a beaker of blood, for example, and you'll find that the red blood cells are packed tight. If you used their density to estimate the cells in a human body, you'd come to a staggering 724 trillion

cells. Skin cells, on the other hand, are so sparse that they'd give you a paltry estimate of 35 billion cells.

So the author of the new paper set out to estimate the number of cells in the body the hard way, breaking it down by organs and cell types. (They didn't try counting up *all the microbes that also call our body home*, sticking only to human cells.) They've scoured the scientific literature for details on the volume and density of cells in gallbladders, knee joints, intestines, bone marrow, and many other tissues. They then came up with estimates for the total number of each kind of cell. They estimate, for example, that we have 50 billion fat cells and 2 billion heart muscle cells.

Adding up all their numbers, the scientists came up with…drumroll…37.2 trillion cells.

This is not a final number, but it's a very good start. While it's true that people may vary in size–and thus vary in their number of cells–adult humans don't vary by orders of magnitude *except in the movies*. The scientists declare with great confidence that the common estimate of a trillion cells in the human body is wrong. But they see their estimate as an opportunity for a collaboration–perhaps through an online database assembled by many experts on many different body parts–to zero in on a better estimate.

Curiosity is justification enough to ponder how many cells the human body contains, but there can also be scientific benefits to pinning down the number too. Scientists are learning about the human body by building sophisticated computer models of lungs and hearts and other organs. If these models have ten times too many cells as real organs do, their results may veer wildly off the mark.

The number of cells in an organ also has bearing on some medical conditions. The authors of the new study find that a healthy liver has 240 billion cells in it, for example, but some studies on cirrhosis have found the disease organ have as few as 172 billion.

Perhaps most importantly, the very fact that some 34 trillion cells can cooperate for decades, giving rise to a single human body instead of a chaotic war of selfish microbes, is amazing. *The evolution of even a basic level of multicellularity is remarkable enough*. But our ancestors went way beyond a simple sponge-like anatomy, evolving a vast collective made of many different types. To understand that collective on a deep level, we need to know how big it really is.

Source: From *The Loom*, a Blog by Carl Zimmer. Posted Wednesday, 10/23/2013. phenomena.nationalgeographic.com

Questions

1. According to the article, a healthy liver has 240 billion cells. How many *millions* of cells does it have? How many *thousands*?

   240,000 million cells; 240,000,000 thousand cells

2. The article contains several very large and some very small numbers. Write each of the following large or small numbers using a power of 10 instead of a word. For example, 55 thousand is the same as 55(1,000), which is $55(10^3)$. 55 thousandths is the same as $55\left(\frac{1}{1,000}\right)$, or $55\left(\frac{1}{10^3}\right)$.

   a. Number of skin cells: 35 billion

   $35(10^9)$

   b. Number of cells in the human body: 37.2 trillion

   $37.2(10^{12})$

   c. Mean volume of a mammal cell: 4 billionths of a cubic centimeter

   $4\left(\frac{1}{10^9}\right)$

4.3 Little Giants: Scientific and Engineering Notation

Explore

SECTION OBJECTIVES:

- Convert numbers between scientific and standard notation
- Convert numbers into and out of engineering notation

INSTRUCTOR NOTE: Most students will have scientific notation on their calculators and not necessarily know what that is. Consider showing students how to write each answer correctly using appropriate scientific notation, not just the notation displayed on a calculator, which sometimes shows E or EE.

1. According to the U.S. Geological Survey, the Himalaya Mountains rise more than 1 centimeter per year. Find the rate the mountains are rising in miles per hour. Is the speed a small or large rate? What is the sign of the exponent on 10?

   7.09×10^{-10} mph; small; negative

2. A mountain range encompasses glaciers, which store approximately 10,000 cubic kilometers of fresh water. Convert this volume to cubic inches. Is the volume a small or large quantity? What is the sign of the exponent on 10?

   6.10×10^{17} cubic inches; large; positive

Discover

Most calculators will give the answers to the problems in the *Explore* in a form known as scientific notation. Scientific notation may appear as a number like 7×10^{11} or, in a calculator's notation, like 7E10. Scientific notation is a shortened way of writing numbers that are too large or too small for all their digits to show on the calculator. It is also a convenient shorthand for writing large and small numbers even when technology is not used.

As you work through this section, think about these questions:

① Can you tell when a number is in scientific, standard, or engineering notation?

② Do you know how to enter numbers in scientific notation on your calculator?

③ Do you understand the significance of the sign of the exponent on 10 in scientific notation?

Let's look at the form numbers take when written in scientific notation.

LOOK IT UP

Scientific Notation

When a number is written in **scientific notation**, it is in the form

$$A \times 10^N$$

The size of A, the coefficient, must be at least 1 but less than 10. A can be positive or negative. Mathematically, this is written as $1 \leq |A| < 10$. N must be an integer and therefore can also be positive or negative. The multiplication symbol in scientific notation is typically written as an $\times$ instead of a dot ($\cdot$).

FOR EXAMPLE, -5×10^8 is written in scientific notation. The size of the coefficient is 5, which is at least 1 but less than 10, and the exponent on 10 is an integer, 8.

If a number is written using all of its digits and not using exponents, it is considered to be in **standard notation**. The number -5×10^8 is $-500,000,000$ in standard notation.

Remember?

All numbers have a size and a sign. The size, or absolute value, is the distance the number is from 0 on the number line. -5 has a negative sign and a size of 5.

3. Determine if each number is given in scientific notation. If it is not, explain why.

   a. -2.22×10^4 Yes

   b. 550   No; it is in standard notation.

   c. 79×10^{-4} No; the size of the coefficient is not in the correct range.

   d. 0.9×10^{12} No; the size of the coefficient is not in the correct range.

   e. 1.6×10^0 Yes

   f. $1.2 \times 10^{2.3}$ No; the exponent on 10 is not an integer.

   g. 3.79092×10^{-4} Yes

Consider the number 4.1×10^5, which is written in scientific notation. When we write this number without an exponent, we get:

$$4.1 \times 10^5 = 4.1 \cdot 100,000 = 410,000$$

Notice that 410,000 is a large number and the exponent on 10 in the scientific notation is positive. We saw this relationship in #2 in the *Explore*. We will consider numbers with sizes that are greater than or equal to 10 to be large.

When we write the number in both notations with the decimal point visible in each form, we can see that the decimal point moves 5 places when we convert between scientific and standard notation. The exponent on 10 matches the movement of the decimal point when we convert between these two notations. Drawing loops can help us see the movement of the decimal point and where zeros might need to be filled in.

Scientific notation   Standard notation
$$4.1 \times 10^5 = 4.10000 \times 10^5 = 410,000.$$
5 places

Starting decimal point location   Ending decimal point location

Now that we've looked at positive exponents in scientific notation, let's look at the meaning of negative exponents. In #1 in the *Explore*, we saw that a small number in standard notation has a negative exponent on 10 when it is written in scientific notation. We can use this relationship to write 7.6×10^{-2} in standard notation. Since the exponent on 10 matches the movement of the decimal point and the number is small, we move the decimal point 2 places left to create 0.076. If we moved the decimal point to the right, we would end up with 760, which is not a small number. We will consider numbers with sizes that are less than 1 to be small.

Scientific notation   Standard notation
$$7.6 \times 10^{-2} = 07.6 \times 10^{-2} = .076$$
2 places

Ending decimal point location   Starting decimal point location

4. Convert each number in scientific notation to standard notation.

   a. 9.62×10^6 9,620,000

   b. 9.62×10^{-6} 0.00000962

In $3 \times 10^4 = 3(10,000) = 30,000$, the exponent on 10 matches the number of zeros in the number in standard notation. But notice that is not true in this case: $3.8 \times 10^3 = 3.8(1,000) = 3,800$. We cannot conclude that the exponent on 10 always matches the number of zeros. Rather, the size of the exponent always matches the number of places *the decimal point moves.*

HOW IT WORKS

Tech TIP

Explore how to enter the number 5×10^{-3} on your calculator. Use the exponent button to enter the number as it appears, or use your calculator's scientific notation function. Once the number is entered, press $=$ or ENTER. Some calculators will convert this number to standard notation if there is sufficient room in the display.

To convert from scientific to standard notation:

1. Decide if the size of the number is large or small by looking at the exponent on 10.
   - If the exponent is positive, the size is large.
   - If the exponent is negative, the size is small.

2. Move the decimal point the number of positions indicated by the size of the exponent. Choose which direction to move based on whether the size of the number is large or small.

3. Fill in with zeros the empty spaces created by the movement of the decimal point. Keep the sign of the original number.

EXAMPLE: Convert 8.25×10^4 to standard notation.

Notice that 8.25×10^4 is a large number (relative to 10), since the exponent on 10 is positive. Since the exponent is 4, the decimal point must be moved 4 positions. Since the number is large, the decimal point is moved to the right. The two empty spaces should be filled in with zeros.

$$8.25 \times 10^4 \text{ is } 8.2500 \times 10^4, \text{ which is } 82,500.$$

The exponent on the 10 in scientific notation always tells you the number of places and the direction the decimal point moves when you write the number in standard notation. However, it might be simpler to remember the following:

Exponent in scientific notation is negative	⟺	Size of number in standard notation is small (*less than 1*)
Exponent in scientific notation is positive	⟺	Size of number in standard notation is large (*greater than or equal to 10*)
Exponent in scientific notation is zero	⟺	Size of number in standard notation is greater than or equal to 1 but less than 10

In practice, numbers written in scientific notation with an exponent of 0 on 10 are uncommon since scientific notation is used to represent very large and very small numbers. If standard notation is easier to use, the number is simply written in standard notation.

We can use these ideas to convert numbers in standard notation to scientific notation.

EXAMPLE

Convert 7,000,000 to scientific notation.

SOLUTION

To begin, we need to write a coefficient that is at least 1 and less than 10. We move the decimal point until we have 7.0 or 7, which is at least 1 and less than 10. The decimal point moves 6 places, so the exponent on 10 is 6 or −6. Since the number in standard notation, 7,000,000, is large, the exponent is positive.

$$7{,}000{,}000 = 7000000. = 7 \times 10^6$$

Ending decimal point location   Starting decimal point location

In scientific notation, the number is 7×10^6.

---

5. Convert each number to scientific notation.

   a. 5,257 5.257×10^3

   b. 0.5257 5.257×10^{-1}

   c. 0.5257×10^7 5.257×10^6

HOW IT WORKS

To convert from standard to scientific notation:

1. Determine if the size of the number is large or small.
   - If the size of the number is greater than or equal to 10, the size is large and the exponent on 10 will be positive.
   - If the size of the number is less than 1, the size is small and the exponent on 10 will be negative.
2. Move the decimal point until the resulting number, A, is at least 1 and less than 10. Note the number of spaces you move the decimal point. This number of spaces is the size of the exponent on 10 and is denoted N.
3. Write the scientific notation in the form $A \times 10^N$. Keep the sign of the original number.

EXAMPLE: Convert -0.0025 to scientific notation.

Notice that -0.0025 is a small number because its size (0.0025) is less than 1. The exponent on 10 will be negative. Move the decimal point to be between 2 and 5, creating 2.5. The decimal point moved 3 spaces and the size of the original number is less than 1, so the exponent on 10 is -3. The answer is negative, since the original number is negative.

$$-0.0025 \text{ is } -2.5 \times 10^{-3}$$

INSTRUCTOR NOTE: You might need to discuss quadrillions for the last line in the table.

We use the size of the number to determine the movement of the decimal point instead of the idea that negative means moving left and positive means moving right. Those statements are true when you are converting from scientific to standard notation, not the reverse.

6. Complete the table by converting between scientific and standard notation.

Words	Standard Notation	Scientific Notation
Eighty-seven hundred thousandths	0.00087	8.7×10^{-4}
Negative 7	-7	-7×10^0
Eighty-seven thousand, five hundred sixty	87,560	8.756×10^4
5 trillion	5,000,000,000,000	5×10^{12}
1 millionth	0.000001	1×10^{-6}
3 quadrillion	3,000,000,000,000,000	3×10^{15}

If you have a number that is almost in scientific notation, like 12×10^{13}, you can convert it to scientific notation by first changing the number to standard notation. A faster option is to note how many places the decimal point needs to move and then adjust the exponent on 10 accordingly. For example, in 12×10^{13}, the decimal point needs to move one place to the left to change 12 to 1.2. To move a decimal point one place to the left, we divide by 10. We then compensate by multiplying by 10, which changes the exponent to 14. That is:

$$12 \times 10^{13} = \frac{12}{10} \times 10^{13} \cdot 10 = 1.2 \times 10^{14}$$

Engineering notation, like scientific notation, is another system that uses powers of ten to write very large and very small numbers. However, in engineering notation, the exponents on 10 are always multiples of 3 in order to make it easier to read and interpret the number. Our naming convention for numbers changes the base name every three powers of ten. For example, we have three place value positions that are "thousands" and then three that are "millions," followed by three that are "billions." Fifty million in scientific notation is 5×10^7, which is difficult to recognize as millions, since 1 million is 10^6.

In engineering notation, 50 million is written as 50×10^6, which is easier to interpret. Writing a number in engineering notation still provides a compact notation like scientific notation, but it offers the advantage of a better correspondence with the way we name numbers.

There is not a unique form for writing a number in engineering notation. For example, a quarter million can be written as 0.25×10^6 when translated directly from words to engineering notation. But a quarter million is also 250,000 or 250 thousand, which can be written as 250×10^3. This form is also correct in engineering notation, since the exponent on 10 is a multiple of 3.

7. Complete the table by converting between types of numeric notation.

Words	Standard Notation	Scientific Notation	Engineering Notation
10 thousand	10,000	1×10^4	10×10^3
16 trillion	16,000,000,000,000	1.6×10^{13}	16×10^{12}
421 thousandths	0.421	4.21×10^{-1}	421×10^{-3}
650 billion	650,000,000,000	6.50×10^{11}	650×10^9
0.9 million	900,000	9×10^5	0.9×10^6

Connect

Calculations with numbers that are in scientific notation can be done quickly, and sometimes without a calculator, by using exponent rules. For example, to multiply $(3 \times 10^7)(4 \times 10^6)$, treat it as you would if the calculation was $3x^7 \cdot 4x^6 = 12x^{13}$.

$$(3 \times 10^7)(4 \times 10^6)$$
$$= (3 \cdot 4) \cdot 10^7 \cdot 10^6 \qquad \text{Multiply coefficients.}$$
$$= 12 \cdot 10^{13} \text{ or } 12 \times 10^{13} \qquad \text{Add exponents.}$$
$$= 1.2 \times 10^{14} \qquad \text{Write answer in scientific notation.}$$

8. Simplify each expression using exponent rules. Write the answers in scientific notation. Do not use a calculator.

a. $(5 \times 10^4)^2$ 2.5×10^9

b. $\dfrac{18 \times 10^{12}}{3 \times 10^6}$ 6×10^6

Reflect

WRAP-UP

What's the point?

Scientific notation is a way to write very large and very small numbers efficiently. Engineering notation is also a way to write numbers more efficiently but in a form that naturally corresponds to the way numbers are read.

What did you learn?

How to convert numbers between scientific and standard notation
How to convert numbers into and out of engineering notation

4.3 Homework

Skills MyMathLab

First complete the MyMathLab homework online. Then work the two exercises to check your understanding.

1. Convert to standard notation.

   a. 8×10^{-6} 0.000008

   b. 8×10^{6} 8,000,000

2. Convert to scientific notation.

   a. -0.0001 -1×10^{-4}

   b. 52×10^{4} 5.2×10^{5}

Concepts and Applications

Complete the problems to practice applying the skills and concepts learned in the section.

3. Answer true or false for each statement.

   a. If a number in scientific notation has a negative exponent, then the number is negative.
   False

   b. A number that is at least 1 but less than 10 cannot be written in scientific notation.
   False

   c. A negative exponent in the scientific notation form means that the size of the number is less than 1.
   True

   d. In scientific notation, if the exponent on the 10 is 6, then the number is in the millions.
   True

   e. In scientific notation, if the exponent on the 10 is -6, then the number is in the millionths.
   True

4. A state lottery website claims that you have a 1 in 897,106,233 chance of winning their lotto.

   a. Write the probability of winning as a fraction and as a decimal. State the decimal number in both standard and scientific notation.

   $$\frac{1}{897,106,233} \approx 1.115 \times 10^{-9} = 0.000000001115$$

   b. Your friend sees the probability in part a on your calculator and states that your chances are high because the number is greater than 1. Where is the flaw in your friend's conclusion?

   The friend is drawing a conclusion based on the appearance of the number in scientific notation. In standard notation, the probability is 0.00000000115, which is nearly zero. The chances of winning are minuscule.

5. Beth gets the following error message when she tries to email a file as an attachment. By how much does her file exceed the maximum allowed size?

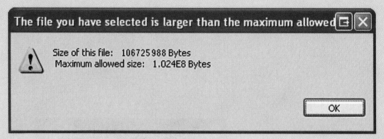

The file you have selected is larger than the maximum allowed

Size of this file: 106 725 988 Bytes
Maximum allowed size: 1.024E8 Bytes

OK

4,325,988 bytes

6. Write each number in standard, scientific, and engineering notation.

Words	Standard Notation	Scientific Notation	Engineering Notation
Half a million	500,000	5×10^5	0.5×10^6
20 trillion	20,000,000,000,000	2×10^{13}	20×10^{12}
3 billion	3,000,000,000	3×10^9	3×10^9
100 thousand	100,000	1×10^5	100×10^3

7. Scientific notation is commonly used in science, technology, engineering, and math, which are known as STEM fields. Assume there are approximately 1.6 million engineers in the United States. If engineers make an average salary of about $80,000 per year, how much income does this group generate collectively? Write the answer in both standard and scientific notation.

1.28×10^{11} per year $=$ $128,000,000,000 per year

8. Simplify each expression using exponent rules. Write each answer in scientific notation. Do not use a calculator.

   a. $(3 \times 10^4)(6 \times 10^6)(2 \times 10^8)$ 3.6×10^{19}

   b. $\dfrac{(1 \times 10^{11})(7 \times 10^5)^2}{7 \times 10^4}$ 7×10^{17}

9. **Looking Forward, Looking Back** Simplify.

   a. $(10x^3)(5x^2)$ $50x^5$

   b. $(10x^3)^2$ $100x^6$

   c. $\dfrac{10x^3}{5x^2}$ $2x$

4.4 A Model Approach: Negative Exponents

Explore

SECTION OBJECTIVE:

• Use exponent rules to simplify expressions that have negative exponents

The diameter of the nucleus of a hydrogen atom is about 1.75 femtometers (1.75×10^{-15} meters). This distance is not only incredibly small, but also incredibly small compared to the overall size of the hydrogen atom when you include the electron. The diameter of the hydrogen atom is about 23,000 times larger than the diameter of the nucleus.

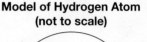

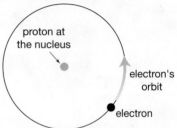

1. Suppose a middle-school class is building a simplified model of a hydrogen atom. They plan to use a Styrofoam ball with a 2-centimeter diameter for the nucleus.

   a. What is the scale of their model? Write a ratio of the model nucleus diameter to the actual nucleus diameter. Do not simplify the ratio.

   Diameter of the model nucleus: 2 cm

   Diameter of an actual nucleus: 1.75×10^{-15} m

   Ratio: $\dfrac{2 \text{ cm}}{1.75 \times 10^{-15} \text{ m}}$

   b. If the diameter of the hydrogen atom is 23,000 times larger than the diameter of the nucleus, find the diameter of the hydrogen atom in the scale model in centimeters.
   46,000 cm

   c. Use dimensional analysis to convert the diameter of the model to feet.
   1,509.2 feet

   d. Is it possible for the class to build this model of the atom to this scale and keep the entire model inside the school?
   No

Discover

You have seen negative exponents in numbers written in scientific notation, but what do they mean exactly? This section will define a negative exponent and show you that negative exponents follow the same basic rules as positive exponents. We will review and practice those exponent rules with both positive and negative exponents.

As you work through this section, think about these questions:

1 Do you understand what a negative exponent means?

2 Does a negative exponent necessarily indicate a negative number?

3 Do you have all the basic exponent rules memorized?

To understand what a negative exponent means, we will begin by simplifying an expression in two different ways.

One way to simplify $\dfrac{x^2}{x^4}$ is to write the factors of x and then divide common factors from the numerator and denominator.

$$\frac{x^2}{x^4} = \frac{\cancel{x} \cdot \cancel{x}}{\cancel{x} \cdot \cancel{x} \cdot x \cdot x} = \frac{1}{x^2}$$

Another way to simplify $\dfrac{x^2}{x^4}$ is to use the quotient rule for exponents.

$$\frac{x^2}{x^4} = x^{2-4} = x^{-2}$$

Since these are both correct ways of simplifying the expression, the final expressions must be equivalent. In other words,

$$x^{-2} = \frac{1}{x^2}$$

It appears that the negative exponent moves the factor of x into the denominator. This holds for any exponent, not just -2.

Another way to see why this equation makes sense is to look at the pattern formed by decreasing exponents. For example, consider the following pattern:

$$3^4 = 81$$
$$3^3 = 27$$
$$3^2 = 9$$
$$3^1 = 3$$
$$3^0 = 1$$

If you notice that each time the exponent decreases by 1, the result is divided by 3, then we can write the next few numbers in this pattern.

$$3^{-1} = \frac{1}{3}$$

$$3^{-2} = \frac{1}{9} = \frac{1}{3^2}$$

Again, it appears that the negative exponent has the result of moving the base to the denominator with a positive exponent. Notice that 3^{-2} is the same as $\dfrac{1}{3^2}$, which is the reciprocal of 3^2. So a negative exponent indicates a reciprocal. We generalize this idea with a definition.

Negative Exponent

A **negative exponent** indicates a reciprocal. So x^{-a} is the reciprocal of x^a. That is,
$x^{-a} = \dfrac{1}{x^a}$.

FOR EXAMPLE, $5^{-1} = \dfrac{1}{5}$.

Tech TIP

The reciprocal button on calculators is usually $\boxed{x^{-1}}$ or $\boxed{\frac{1}{x}}$.

Find the reciprocal of 4 using the appropriate button on your calculator. You should get $\frac{1}{4}$ or 0.25.

2. Rewrite each of the following with only positive exponents. Do not simplify.

a. 2^{-4}

$\dfrac{1}{2^4}$

b. -2^{-4}

$-\dfrac{1}{2^4}$

c. $(-2)^{-4}$

$\dfrac{1}{(-2)^4}$

Now that you have seen the basic definition of a negative exponent, let's consider what it means to have a negative exponent in the denominator of an expression. For comparison, we will rewrite 5^{-2} and $\dfrac{1}{5^{-2}}$ with positive exponents.

We can use the definition of a negative exponent to write $5^{-2} = \dfrac{1}{5^2}$.

To rewrite $\dfrac{1}{5^{-2}}$ with a positive exponent, we use the fact that the fraction bar means division.

$$\dfrac{1}{5^{-2}} = 1 \div 5^{-2}$$

$$= 1 \div \dfrac{1}{5^2} \quad \text{Definition of negative exponent}$$

$$= 1 \cdot \dfrac{5^2}{1} \quad \text{Division is multiplication by the reciprocal}$$

$$= 5^2$$

What can we conclude from this example? If a factor has a negative exponent, we move it from the numerator to the denominator or from the denominator to the numerator and make the exponent positive.

3. Rewrite each of the following with only positive exponents.

a. $x^{-2}y^3$

$\dfrac{y^3}{x^2}$

b. $\dfrac{x^{-4}}{y^{-5}}$

$\dfrac{y^5}{x^4}$

HOW IT WORKS

To simplify expressions with positive or negative exponents:

Assume a and b are integers and x and y are real numbers.

1. $x^a x^b = x^{a+b}$

2. $\dfrac{x^a}{x^b} = x^{a-b}$

3. $x^0 = 1 \ (x \neq 0)$

4. $(x^a)^b = x^{ab}$

5. $(xy)^a = x^a y^a$

6. $\left(\dfrac{x}{y}\right)^a = \dfrac{x^a}{y^a}$

7. $x^{-a} = \dfrac{1}{x^a}$

8. $\dfrac{1}{x^{-a}} = x^a$

EXAMPLE: Simplify $\dfrac{x^{-15}}{(x^2)^3}$.

There are several ways this expression can be simplified. Two methods are shown here.

Method 1		Method 2	
$\dfrac{x^{-15}}{(x^2)^3} = \dfrac{x^{-15}}{x^6}$	Use rule #4.	$\dfrac{x^{-15}}{(x^2)^3} = \dfrac{x^{-15}}{x^6}$	Use rule #4.
$= x^{-15-6}$	Use rule #2.	$= \dfrac{1}{x^{15}x^6}$	Use rule #7.
$= x^{-21}$	Simplify the exponent.		
$= \dfrac{1}{x^{21}}$	Use rule #7 to rewrite with a positive exponent.	$= \dfrac{1}{x^{21}}$	Use rule #1.

> A negative exponent does not mean the expression is negative. Instead, a negative exponent indicates a reciprocal. Also, a negative exponent does not necessarily indicate a number less than 1, as seen in #4d.

4. Simplify each expression. Evaluate the expression completely if possible. If it's not possible, then write the answer with only positive exponents. Do not use a calculator.

a. 3^2 9

b. 3^{-2} $\frac{1}{9}$

c. $\frac{1}{3^2}$ $\frac{1}{9}$

d. $\left(\frac{1}{3}\right)^{-1}$ 3

e. $\frac{1}{3^{-2}}$ 9

f. $3x^{-2}$ $\frac{3}{x^2}$

g. $(3x)^2$ $9x^2$

h. $(3x)^{-2}$ $\frac{1}{9x^2}$

i. $3^{-2}3^5$ 27

j. $(3^{-1})^4$ $\frac{1}{81}$

k. $\frac{3^6}{3^5}$ 3

l. $\frac{3^5}{3^6}$ $\frac{1}{3}$

5. Simplify the expression. Write the final answer with only positive exponents.

$$\frac{(x^2y^3)^{-4}}{xy}$$

$\frac{1}{x^9y^{13}}$

Now that you understand what negative exponents indicate, let's consider again how they are used in scientific notation. Why does a negative exponent on the 10 indicate that you should move the decimal point to the left? Consider this example:

$$6 \times 10^{-3} = \frac{6}{10^3} = \frac{6}{1000} = 0.006$$

The factor of 10 with the negative exponent can be moved to the denominator. So we are really dividing by the power of 10, which produces a smaller number. Specifically, the result is 6 thousandths, or a 6 in the third decimal position. Notice that the exponent on the 10 determines the number of zeros in the power of ten in the denominator, and this determines the decimal position for the 6.

6. Write each of the following numbers as a fraction by using the definition of negative exponents and then write the number in standard notation. How do you read the decimal number?

a. 5.4×10^{-2}

$\dfrac{5.4}{100} = 0.054$; 5.4 hundredths or 54 thousandths

b. 8.91×10^{-4}

$\dfrac{8.91}{10,000} = 0.000891$; 8.91 ten-thousandths or 891 hundred-thousandths

Connect

7. Building a model of the solar system is a popular school project. A good first step is to understand how far each planet is from the sun. Complete the following table by writing the distances in meters in scientific notation and then converting them to kilometers. A scale model will be explored further in the homework. Look for a pattern in the first couple of conversions and then use that pattern to do the remaining conversions.

	Standard Notation	Scientific Notation	
Planet	Average Distance from the Sun (m)	Average Distance from the Sun (m)	Average Distance from the Sun (km)
Mercury	57,910,000,000	5.791×10^{10}	5.791×10^{7}
Venus	108,200,000,000	1.082×10^{11}	1.082×10^{8}
Earth	149,600,000,000	1.496×10^{11}	1.496×10^{8}
Mars	227,940,000,000	2.2794×10^{11}	2.2794×10^{8}
Jupiter	778,330,000,000	7.7833×10^{11}	7.7833×10^{8}
Saturn	1,429,400,000,000	1.4294×10^{12}	1.4294×10^{9}
Uranus	2,870,990,000,000	2.87099×10^{12}	2.87099×10^{9}
Neptune	4,504,300,000,000	4.5043×10^{12}	4.5043×10^{9}

Reflect

WRAP-UP

What's the point?

A negative exponent is another way to indicate a reciprocal. Expressions that have negative exponents can be simplified using the same exponent rules that apply to positive exponents.

What did you learn?

How to use exponent rules to simplify expressions that have negative exponents

4.4 Homework

Skills MyMathLab

First complete the MyMathLab homework online. Then work the two exercises to check your understanding.

1. Simplify $(2^{-3})^2$. $\dfrac{1}{64}$

2. Simplify $\dfrac{(4x)^{-2}}{x^8}$. $\dfrac{1}{16x^{10}}$

Concepts and Applications

Complete the problems to practice applying the skills and concepts learned in the section.

3. Suppose a class is building a model of the solar system. They plan to use a Styrofoam ball with a 10-centimeter diameter for the sun.

a. What is the scale of their model? Write the ratio of the model sun's diameter to the actual sun's diameter. The diameter of the sun $= 1.4$ million kilometers. You do not need to simplify the ratio.

$$\dfrac{10\ \text{cm}}{1.4 \times 10^6\ \text{km}}$$

b. If the model of the entire solar system is to have the same scale, all the distances must follow this ratio.

Find the distance from the sun to each of the planets in the solar system model (in centimeters) by solving the following proportion. Use the actual distances given in the *Connect*. Round each distance to the nearest centimeter. Also convert each distance to feet to make them easier to visualize. Record your answers in the table.

$$\dfrac{10\ \text{cm}}{1.4 \times 10^6\ \text{km}} = \dfrac{\text{model distance in cm}}{\text{actual distance in km}}$$

Planet	Distance from Sun in Model (cm)	Distance from Sun in Model (ft)
Mercury	414	14
Venus	773	25
Earth	1,069	35
Mars	1,628	53
Jupiter	5,560	182
Saturn	10,210	335
Uranus	20,507	673
Neptune	32,174	1,056

    c. Explain why all models that are built of the solar system are not built to scale.

    It is impossible to build a model of the solar system that has a reasonable size for the sun and still has the planets in the same room if you keep to the actual scale of the solar system.

    d. In the model, Neptune is approximately 1,000 feet from the sun. This means the model would have a radius of approximately 1,000 feet. If the model is to fit in a 10-foot-radius space, approximately how big would the model sun need to be?

    0.1 cm or 1 mm in diameter

4. a. How much farther from the sun is Jupiter than Mars? Use the distances in km from the *Connect* to do the calculation.

$7.7833 \times 10^8 - 2.2794 \times 10^8 = 5.5039 \times 10^8$ km

    b. What algebraic concept is used to complete the calculation?

    Like terms or the distributive property

    c. How much farther from the sun is Mars than Earth? Use the distances in km in scientific notation to do the calculation. Write the final answer in scientific notation.

    $2.2794 \times 10^8 - 1.496 \times 10^8 = 0.7834 \times 10^8$ km $= 7.834 \times 10^7$ km

5. Let's explore how Earth compares in size to the sun.

    a. The diameter of the sun is 1.4 million kilometers, and the diameter of Earth is 12,700 kilometers.

    Write each of these diameters in scientific notation in kilometers.

    Sun: 1.4×10^6 km; Earth: 1.27×10^4 km

    b. How many times larger is the sun's diameter than Earth's diameter?

    About 110

c. If we assume that the sun and Earth are both spheres, each volume is given by $V = \frac{4}{3}\pi r^3$. Without a calculator, estimate each volume and divide to find a rough estimate of the number of Earths that would fit in the sun.

$$\frac{\text{Volume of sun}}{\text{Volume of Earth}} \approx \frac{1 \times 10^{18} \text{ cubic km}}{1 \times 10^{12} \text{ cubic km}} = 1 \times 10^6$$

So approximately 1 million Earths would fit in the sun.

d. Use a calculator to find the volume of the sun and the volume of Earth in cubic kilometers.

Volume of the sun: 1.44×10^{18} cubic km

Volume of Earth: 1.07×10^{12} cubic km

e. How many Earths would fit inside the sun exactly?

1.3 million

f. A classmate completes this problem in the following way. Is his work correct? If so, list the reason for each step next to the work. If the work is not correct, explain where he went wrong.

$$\text{Earths in the sun} = \frac{\text{Volume}_{\text{sun}}}{\text{Volume}_{\text{Earth}}} = \frac{\frac{4}{3}\pi (r_{\text{sun}})^3}{\frac{4}{3}\pi (r_{\text{Earth}})^3} \qquad \text{Work is correct.}$$

$$= \frac{(r_{\text{sun}})^3}{(r_{\text{Earth}})^3} \qquad \text{Divide off } \frac{4}{3}\pi.$$

$$= \left(\frac{r_{\text{sun}}}{r_{\text{Earth}}}\right)^3 \qquad \text{Exponent rule 6.}$$

$$= \left(\frac{\dfrac{d_{\text{sun}}}{2}}{\dfrac{d_{\text{Earth}}}{2}}\right)^3 \qquad \text{Radius is half the diameter.}$$

$$= \left(\frac{d_{\text{sun}}}{d_{\text{Earth}}}\right)^3 \qquad \text{Divide off } \frac{1}{2}.$$

$$\approx \left(\frac{1.4 \times 10^6 \text{ km}}{1.27 \times 10^4 \text{ km}}\right)^3 \qquad \text{Substitute values.}$$

$$\approx (1.1 \times 10^2)^3 \qquad \text{Exponent rule \#2}$$

$$= (1.1)^3 (10^6) \qquad \text{Exponent rule \#5}$$

$$\approx 1.33 \times 10^6 \qquad \text{Simplify result.}$$

6. Looking Forward, Looking Back Find the mean: $-5, -4, -3, -2, 0, 2, 3, 4, 5, 6$.

0.6

4.5   Variation on a Theme: Standard Deviation

Explore

SECTION OBJECTIVES:

- Find the standard deviation of a data set
- Interpret the standard deviation of a data set

1. A manager at a fast-food restaurant collects a sample of six order times for several employees who work at the drive-thru window. Which employee has the best order times? Explain your answer.

Employee	Order Times (seconds)					
Ana	124	150	150	156	202	280
Elisa	200	201	202	208	209	210
Jason	135	153	189	241	250	262
Malcolm	139	141	145	146	148	151
Zach	132	135	136	136	200	239

Answers will vary. Ana has the fastest order time but she also has the slowest. She is very inconsistent. Elisa is consistent but has slow order times. Jason's times are very inconsistent and range from fast to slow. Malcolm has fairly fast order times and is consistent. Zach has some of the fastest times and is somewhat consistent but not always.

Discover

INSTRUCTOR NOTE: Because students at this level do not have experience with populations and samples, this section will not use symbols for the mean and the standard deviation in formulas. The formula developed here is for the population standard deviation because it is more natural for students to understand than the sample standard deviation. Students can learn about sample standard deviation later in a statistics course.

The mean describes the center of data, but it does not tell how the data is distributed around the center. As we saw in the *Explore*, data can vary in different ways, but the mean does not describe those differences. Other measures are necessary to understand the spread of data. Specifically, we would like to know how far the data points are from the mean. Since the mean is a balancing point around which the data is clustered, it makes sense to measure the spread relative to that central point. Let's take a closer look at the idea of measuring the variation in a data set with the standard deviation.

As you work through this section, think about these questions:

1 Can you use the standard deviation formula?

2 Do you know what a standard deviation of 0 means?

3 Can you use the standard deviation to judge whether or not a data value is unusual?

The standard deviation of a data set measures the spread of the data about the mean. Let's look at how the formula for standard deviation is developed and see it in action.

Consider the following scenario: At a university, the biology department has a course, Human Heredity, with only four sections per semester. The class sizes for the four sections are 20, 21, 24, and 27 students.

Our goal is to develop a way to measure how spread out the class sizes are from their mean of 23 students.

We need to know how far each data value is from the mean, which is the data value's **deviation from the mean**. To find this, we subtract the mean from each data value. Notice that 20 has a deviation of -3 because 20 is 3 students *below* the mean.

Mean: 23

Data:	20	21	24	27
Deviations from the mean:	-3	-2	1	4

Tech TIP

To square a negative number on your calculator, you need to use parentheses. For example, to square -3, you need to enter $(-3)^2$. It is simpler to use the fact that squaring a nonzero number gives a positive result and instead square 3, which gives the same result as squaring -3.

You might consider averaging the deviations to find a measure of the spread. Since some of the deviations from the mean are positive and some are negative, there is a canceling effect that always gives zero for the average deviation from the mean. This idea will be explored in the homework.

To address the issue of positive and negative deviations adding to zero, we can make all the deviations positive. One way to make a negative number positive is to square it. We will square each deviation from the mean, whether it is positive or negative or zero.

Squared deviations from the mean: 9   4   1   16

Now we will average the squared deviations, which gives us 7.5. We squared the numbers to overcome the issue the negative numbers were presenting, but that operation changed the units in the result. To compensate, we need to undo the squaring by taking the positive square root of the average, 7.5. The result is the standard deviation of the data set.

$$\text{Standard deviation} = \sqrt{7.5} \approx 2.74 \text{ students}$$

Standard Deviation

The **standard deviation** is the square root of the average squared deviations from the mean. The units of the standard deviation are the same as the units of the original data. If two data sets have the same units, then the data set with the larger standard deviation is more spread out from the mean, and the data set with the smaller standard deviation is less spread out from the mean. The smaller the standard deviation, the more consistent the data values are.

FOR EXAMPLE, the data set 4, 7, 8 has a standard deviation of approximately 1.7. The data set 5, 12, 20 has a standard deviation of approximately 6.1. The larger standard deviation of the second data set indicates a greater spread in the data set.

2. The following list summarizes the operations used to find the standard deviation of a data set in the order they are performed. Next to each step, write the mathematical notation. Use the summation symbol Σ as a shortcut when you write a sum. For example, instead of writing "add the data values, x," write Σx.

Step 1: Subtract the mean from each data value, x. $x - \text{mean}$

Step 2: Square the result. $(x - \text{mean})^2$

Step 3: Add the squares. $\Sigma(x - \text{mean})^2$

Step 4: Divide the sum by the number of data values, n. $\dfrac{\Sigma(x - \text{mean})^2}{n}$

Step 5: Take the positive square root of the result. $\sqrt{\dfrac{\Sigma(x - \text{mean})^2}{n}}$

Steps 3 and 4 (addition and division) are the operations necessary to find the average squared deviation.

3. Pastry chefs use accurate scales to measure quantities by weight. The United States is one of the few countries in which people measure volumes instead of weights when they cook. Volume measurements like cups and tablespoons are not as exact and can vary due to humidity, how loosely or densely a cup is packed, and whether it is leveled by hand or a knife.

a. Assume a chef uses a measuring cup to measure 1½ cups of flour three times. The chef weighs the flour to assess the consistency of her measurements and finds these weights: 175 grams, 192 grams, 179 grams. Find the standard deviation of the weights.

7.26 grams

b. If a chef measures flour using a scale and gets exactly 185 grams of flour each time, what is the standard deviation of the weights? Why?

0; there is no variation in the data values.

If you find the standard deviation formula intimidating, setting up a table can make it easier to use the formula and organize your work. But you need to be able to read the formula so that you know how to set up the table, where to begin, and the order in which to proceed.

To use a table, start by drawing a 3-column table with at least as many rows as data values. Write the data values in the first column and find their average. Find the deviations and squared deviations and write them in the second and third columns. Calculate the average of the third column and then take the positive square root of the result. This approach is used in the *How It Works* that follows.

HOW IT WORKS

To find the standard deviation of a data set:

1. Make a table with three column heads: x, $x -$ mean, $(x -$ mean$)^2$. Fill in the data values in the first column.
2. Find the mean of the data set.
3. In the second column, find the deviations from the mean, $x -$ mean.
4. In the third column, find the squares of the deviations from the mean, $(x -$ mean$)^2$.
5. Find the mean of the squared deviations. (Add the numbers in the third column and divide the sum by the number of data values, n.)
6. Take the positive square root of the result.

The formula for the standard deviation is $\sqrt{\dfrac{\Sigma(x - \text{mean})^2}{n}}$.

HOW IT WORKS

EXAMPLE: Find the standard deviation of these ages: 10, 12, 14, 16, 18.

To find the standard deviation, begin by finding the mean age, 14 years. Then complete the table by entering the data values in the first column. Subtract each data value from the mean and enter the results in the second column. Square the results in the second column and enter them in the third column.

	x	$x - $ mean	$(x - $ mean$)^2$
	10	−4	16
	12	−2	4
	14	0	0
	16	2	4
	18	4	16
Average:	14	0	8

The sum of the values in the third column is 40, and their average is 8.

$$\text{Standard deviation} = \sqrt{\frac{\Sigma(x - \text{mean})^2}{n}} = \sqrt{\frac{40}{5}} = \sqrt{8} \approx 2.8$$

The standard deviation is 2.8 years.

Tech TIP

To find the standard deviation of a data set using a graphing calculator, enter the data into the list L_1. Press STAT, then CALC, then 1-Var Stats. Type the list's name, L_1, and press ENTER. The standard deviation developed in this section is usually given by the symbol σx on the calculator.

Calculating the standard deviation by hand, whether you use the formula or a table to organize your work, can be tedious and error-prone for large data sets. In practice, standard deviation is always found using some form of technology.

4. a. At the same university as mentioned earlier, the biology department has another course with four sections per semester, Anatomy and Physiology. The class sizes are 60, 63, 68, and 81. Find the standard deviation for the class sizes.

	x	$x - $ mean	$(x - $ mean$)^2$
	60	−8	64
	63	−5	25
	68	0	0
	81	13	169
Average:	68	0	64.5

8.03 students

b. Complete the table about the two biology courses with the information found in part a about the second course. What do the standard deviations tell you about the class sizes?

Course	Class Sizes				Mean	Standard Deviation
Human Heredity	20	21	24	27	23	2.74
Anatomy and Physiology	60	63	68	81	68	8.03

The Human Heredity class sizes have a smaller standard deviation, since all the class sizes are close to the mean of 23. The Anatomy and Physiology class sizes have more variation and a larger standard deviation. One class size, 81, is much further from the mean than the other class sizes. Because the standard deviation for the Human Heredity class sizes is smaller than for the Anatomy and Physiology sections, the class sizes for Human Heredity are more consistent.

In the previous problem, we compared two data sets and their standard deviations. We can compare standard deviations only when the units are the same, which they were in that case (numbers of students).

INSTRUCTOR NOTE: If you would like to provide more practice for students, consider having them find the standard deviation for the data sets in #5 to verify their answers.

5. Without performing any calculations, match each data set with its standard deviation.

a. 15, 22, 22, 30, 45   ii

b. 15, 22, 22, 30, 31   iv

c. 15, 15, 15, 15, 15   i

d. 21, 40, 55, 58, 59   iii

i. 0

ii. 10.26

iii. 14.51

iv. 5.90

Connect

When one data value is very different from the rest of the data values, it is sometimes considered to be an **outlier**. One way to judge if a value is an outlier is to determine how many standard deviations it is from the mean.

Z-Score

The **z-score** of a data value is the number of standard deviations the value is from the mean.

$$z = \frac{\text{data value} - \text{mean}}{\text{standard deviation}}$$

A z-score has no units and is usually rounded to two decimal places.

FOR EXAMPLE, suppose the average family size in a town is 5 people, with a standard deviation of 2 people. If your family has 3 people, then the z-score for your family size is $z = \frac{3-5}{2} = -1$, which indicates that your family's size is 1 standard deviation below the mean.

We can determine if a value is unusual by considering its z-score. The larger the size of the z-score, the further the value is from the mean and the more unusual it is. Values with z-scores larger than 2 or smaller than −2 are often considered unusual.

6. Assume adult men's heights have a mean of 5′10″ and a standard deviation of 2.8″. The table lists several professional male swimmers' heights. Find the z-score for each height and determine if each height is unusual.

Swimmer	Height	z-Score	Unusual?
Matt Grevers	6′8″	3.57	Yes
Michael Phelps	6′4″	2.14	Yes
Ryan Lochte	6′2″	1.43	No
Nick Thoman	6″1′	1.07	No
Average adult male	5′10″	0	No
Cameron Smith	5′9$\frac{1}{2}$″	−0.18	No
Ryuichi Shibata	5′7$\frac{1}{2}$″	−0.89	No
Tomomi Morita	5′6$\frac{1}{2}$″	−1.25	No

Data from usaswimming.org

Tech TIP

When you calculate a z-score, either press ENTER after the subtraction or put the subtraction in parentheses.

7. Z-scores can be negative, zero, or positive. What does a z-score of 0 mean? What does a positive z-score indicate? What does a negative z-score indicate?

A z-score of 0 means that the data value is equal to the mean. Positive z-scores are associated with values that are above the mean; negative z-scores are associated with values that are below the mean.

8. a. A man says he is really tall at 6′5″. Use his z-score to determine if his claim is correct. Is his height unusual?

Yes; his z-score is 2.5, which is in the unusual range.

b. If a woman is that same height, is her z-score larger or smaller than the man's? Women's heights have a mean of 5′4″ and a standard deviation of 2.4″.

The woman's z-score is even larger than the man's, since a height of 6′5″ is more unusual for a woman than for a man.

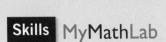

Reflect

WRAP-UP

What's the point?

The mean is a measure of the center of a set of data values; the standard deviation is a measure of the spread of the data values. Standard deviation provides a measure of consistency and also helps determine if a value is unusual compared to the rest of a set of values.

What did you learn?

How to find the standard deviation of a data set
How to interpret the standard deviation of a data set

Focus Problem Check-In
Use what you have learned so far in this cycle to work toward a solution of the focus problem. If you are working in a group, work with your group members to create a list of any remaining tasks that need to be completed. Determine who will do each task, what can be done now, and what has to wait until you have learned more.

4.5 Homework

Skills MyMathLab

First complete the MyMathLab homework online. Then work the two exercises to check your understanding.

1. Find the standard deviation of the data: 10, 50, 50, 100, 200.

x	x − mean	(x − mean)²
10	−72	5,184
50	−32	1,024
50	−32	1,024
100	18	324
200	118	13,924
Average: 82	0	4,296

65.5

2. Find the standard deviation of the data set: 10, 20, 30, 40, 50.

14.14

Concepts and Applications

Complete the problems to practice applying the skills and concepts learned in the section.

3. Here is a list of the operations used when finding the standard deviation of a set of data, $\sqrt{\dfrac{\sum(x - \text{mean})^2}{n}}$.

List the letters to indicate the correct order for the steps.

A. Take the positive square root of the result.

B. Square the deviations from the mean.

C. Divide the sum by the number of data values.

D. Subtract the mean from each data value.

E. Add the squares of the deviations from the mean.

F. Find the mean of the data values.

F, D, B, E, C, A

4. Fill in each blank.

a. Standard deviation is a measure of the spread around the _____mean_____.

b. Standard deviation is always a value that is zero or ___positive___.

c. If the standard deviation is 0, then all the data values are __the same__.

d. The more spread there is in the data, the ___larger___ the standard deviation is.

5. In the section, we saw that if we simply added the deviations from the mean, the sum would be 0. Let's explore why that is always true by using a general case with three numbers: A, B, and C.

a. Write a formula for the mean of A, B, and C. Call the mean D.

$D = \dfrac{A + B + C}{3}$

b. In the table, write an expression for each deviation from the mean.

Data value	Data value − mean
A	$A - D$
B	$B - D$
C	$C - D$
Sum of deviations from the mean	$A - D + B - D + C - D$

c. Simplify the sum of the deviations from the mean by collecting like terms.

$A + B + C - 3D$

d. Replace D in this formula with the expression from part a.

$A + B + C - 3\left(\dfrac{A + B + C}{3}\right)$

e. Simplify. What is the result? Even though only 3 data values were used as an example, this result is the same regardless of the number of data values used.

$A + B + C - 3\left(\dfrac{A + B + C}{3}\right) = A + B + C - (A + B + C) = 0$

6. Another measure of the spread of a data set is called the range. To find the range, subtract the smallest value from the largest value in a data set.

a. Find the range for the data set: 12, 18, 22, 28.

16

b. Although the range is a measure of the spread of a data set, it does not take into account all the values of the data set, since it uses only the largest and smallest values. This is why we need an additional measure like the standard deviation. But the range can be used to estimate the standard deviation in certain cases. Use this formula to estimate the standard deviation for the data set in part a.

$$\text{Standard deviation} \approx \frac{\text{range}}{4}$$

4

c. Find the standard deviation of the data set to see how accurate the estimate in part b is.

5.83

7. Complete the following table, which concerns inequality statements.

	Description of a Set of Real Numbers	List of at Least Five Numbers in the Set	Inequality Describing the Set
a.	Numbers larger than 2	3, 4, 5, 6, 7	$n > 2$
b.	Numbers smaller than 2	−3, −2, −1, 0, 1	$n < 2$
c.	Numbers that are at least 2	2, 3, 4, 5, 6	$n \geq 2$
d.	Numbers that are at most 2	−2, −1, 0, 1, 2	$n \leq 2$
e.	Numbers that are not larger than 2	−2, −1, 0, 1, 2	$n \leq 2$
f.	Numbers that are not smaller than 2	2, 3, 4, 5, 6	$n \geq 2$

8. A compound inequality is a statement that contains more than one inequality, usually joined by "and" or "or."

For example, numbers that are larger than 10 or smaller than 2 can be written as $x > 10$ or $x < 2$.

Numbers that are between 2 and 10 inclusive are written as $2 \leq x \leq 10$. Inclusive means that both the 2 and the 10 are included. To indicate that 2 and 10 are included, we use the "or equal to" version of the inequality symbol.

The inequality $2 \leq x \leq 10$ means $2 \leq x$ AND $x \leq 10$. The inequality $2 \leq x$ is usually turned around as $x \geq 2$ so that it can be read more easily. So $2 \leq x \leq 10$ means numbers that are at least 2 but no larger than 10.

Write each of the following ranges of z-scores using compound inequalities.

a. Unusual values: $z > 2$ or $z < -2$

b. Usual values: $-2 \leq z \leq 2$

9. The following table lists the quiz scores for a class of ten students.

a. Find the mean quiz score and the z-score for each quiz score (rounded to two decimal places). First, find the mean and verify that the standard deviation of the quiz scores is 5.92. Finally, decide if each quiz score is unusual.

Mean = 18.9

Quiz Score	z-Score	Unusual?
15	−0.66	No
20	0.19	No
19	0.02	No
23	0.69	No
25	1.03	No
14	−0.83	No
19	0.02	No
5	−2.35	Yes
24	0.86	No
25	1.03	No

b. What is the largest z-score? What is the smallest z-score? Is either of these scores unusual?

Largest z-score: 1.03; smallest z-score: −2.35(unusually low)

c. How high or low does a quiz score have to be to be considered unusual for this quiz?

Lower than 7.06 or higher than 30.74

d. Write this range as an inequality.

Quiz score < 7.06 or quiz score > 30.74

10. Looking Forward, Looking Back Simplify.

$$\frac{10^5}{10^8} \qquad 10^{-3} = \frac{1}{10^3}$$

4.6 An Order of Magnitude: Understanding Logarithmic Scales

SECTION OBJECTIVE:

- Interpret logarithmic scales

> **Explore**

Sound is measured in decibels (dB) on a scale that is based on powers of ten.

A sound that is ten times more intense than the threshold of hearing is assigned a decibel level of 10 dB.

A sound that is 10^2 or 100 times more intense than the threshold of hearing is assigned a decibel level of 20 dB.

A sound that is 10^3 or 1,000 times more intense than the threshold of hearing is assigned a decibel level of 30 dB.

For every increase of 10 dB, the sound is 10 *times* more intense.

1. The table lists some sounds and the decibel level for each. Answer the questions that follow.

	Decibel Level
Threshold of hearing	0
Breathing	10
Rustling leaves	20
Whisper	30
Refrigerator humming	40
Normal conversation	50
Air conditioning unit	60
Loud traffic	70
Garbage disposal	80
Motorcycle at 25 feet	90
Jet flyover at 1,000 feet; jackhammer	100
Live rock music	110
Average human pain threshold	130
Eardrum rupture	150

a. How many times more intense is live rock music than a jet flyover? Than loud traffic?

Ten times; 10,000 times

b. Explain how to use the decibel level to compare the intensities of two sounds. Use one of the questions from part a as an example.

To compare the intensities of two different decibel levels, subtract the levels and then divide the result by 10. Raise 10 to that power.

For example, $110 - 70 = 40$; $40/10 = 4$; $10^4 = 10,000$

Discover

We know from the *Explore* that the human ear is capable of detecting a huge range of sound intensities. The loudest sound that the human ear can safely detect is actually 1 billion times more intense than the faintest sound it can detect. Since it is challenging to represent such a wide range of numbers on a linear scale, sound intensity is measured on a **logarithmic scale**. This section will focus on helping you understand logarithmic scales and how to use them.

As you work through this section, think about these questions:

1 What does the phrase "order of magnitude" mean?

2 What are some examples of logarithmic scales?

3 Do you understand how to compare numbers on a logarithmic scale?

In order to understand logarithmic scales, it is important to understand what is meant by an order of magnitude.

Order of Magnitude

If a number is rounded to the nearest power of ten, the result is known as an **order of magnitude**. Order of magnitude is often used for convenience when the exact number does not matter and only a rough measure of the size is needed to put it in perspective.

FOR EXAMPLE, the order of magnitude (in meters) for human height is 10^0 because humans are, on average, between 1 and 2 meters tall. The average height is closer to 1 meter than to 10 meters or 0.1 meter.

Logarithmic scales are not linear but are instead based on orders of magnitude. Typically, each step on a logarithmic scale indicates a change by a factor of ten, an order of magnitude. Every two steps on a logarithmic scale indicates a change of 2 orders of magnitude or a change by a factor of 100.

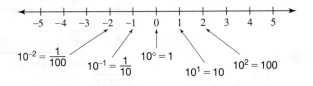

A logarithm is simply another name for an exponent. You will learn more about logarithms if you take additional algebra courses.

2. Consider the following table, which lists some objects and their lengths rounded to the nearest power of ten, in other words, the orders of magnitude for their lengths.

	Order of Magnitude (meters)
Radius of the universe	10^{26}
Distance of sun to nearest star	10^{16}
Radius of the solar system	10^{13}
Distance from Earth to the sun	10^{11}
Radius of the sun	10^{9}
Radius of Earth	10^{7}
Height of an adult person	**10^{0}**
Radius of an amoeba	10^{-4}
Radius of a bacterium	10^{-6}
Radius of an atom	10^{-10}
Radius of a neutron	10^{-15}

a. What is true about the lengths of the objects listed in the table below "Height of an adult person"? How is that evident in the exponent in the order of magnitude?

Those objects are all smaller than the height of an adult person. The exponents in the orders of magnitude are negative.

b. What is true about the lengths of the objects listed above "Height of an adult person"? How is that evident in the exponent in the order of magnitude?

Those objects are all larger than the height of an adult person. The exponents in the orders of magnitude are positive.

Orders of magnitude can be used to compare lengths and distances when exact measurements are unavailable or difficult to determine.

EXAMPLE 1

How does the radius of the sun compare to the radius of Earth? What does that tell us about the sun's volume compared to Earth's volume?

SOLUTION

To see how many times larger the radius of the sun is compared to Earth's radius, we divide their radii.

$$\frac{10^{9}}{10^{7}} = 10^{2} = 100$$

The radius of the sun is 100 times larger than the radius of Earth. We can also say that the sun's radius is two orders of magnitude, two powers of ten, larger than Earth's radius.

If we assume that the sun and Earth are both spheres, we can use the volume formula $V = \frac{4}{3}\pi r^{3}$.

To see how many times larger the sun's volume is, we divide the two volumes.

$$\frac{V_{sun}}{V_{Earth}} = \frac{\frac{4}{3}\pi(10^9)^3}{\frac{4}{3}\pi(10^7)^3} = \frac{\frac{4}{3}\pi(10^{27})}{\frac{4}{3}\pi(10^{21})} = 10^6 = 1,000,000$$

So the sun's volume is about one million times larger than Earth's volume. This means that approximately 1 million Earths would fit inside the sun!

3. The following table lists some objects and their masses rounded to the nearest power of ten, in other words, the orders of magnitude for their masses.

	Order of Magnitude (grams)
Mass of the universe	10^{55}
Mass of the sun	10^{33}
Mass of Earth	10^{27}
Mass of an Egyptian pyramid	10^{13}
Mass of an adult person	10^5
Mass of a grasshopper	**10^0**
Mass of a snowflake	10^{-3}
Mass of an amoeba	10^{-5}
Mass of a bacterium	10^{-12}
Mass of a neutron	10^{-24}
Mass of an electron	10^{-27}

a. What number is given in the table for the order of magnitude of the mass of a grasshopper?

10^0 grams

b. What does this value equal?

1 gram

c. Is this an exact measurement? Explain.

No, but it is accurate to the nearest order of magnitude. The mass of a grasshopper is closer to 1 gram than to 10 grams or 0.1 gram.

In Example 1, we divided to determine that the radius of the sun was 2 orders of magnitude greater than the radius of the Earth. Subtracting the exponents in the orders of magnitude is a faster way to arrive at the same result.

EXAMPLE 2

Remember that an order of magnitude is a power of 10. So 3 orders of magnitude means 3 powers of 10 or 10^3 or 1000.

a. How many orders of magnitude greater is the Egyptian pyramid's mass than the grasshopper's mass? What does this tell us about the mass of the pyramid compared to the mass of the grasshopper?

b. How many orders of magnitude greater is the Egyptian pyramid's mass than the snowflake's mass? How many times greater is the mass of an Egyptian pyramid than the mass of a snowflake?

SOLUTION

a. The order of magnitude for the mass of an Egyptian pyramid is 10^{13} grams, and the order of magnitude for the mass of a grasshopper is 10^0 grams. So the pyramid's mass is $13 - 0 = 13$ orders of magnitude greater than the grasshopper's mass. This means that the pyramid's mass is $10^{13} = 10,000,000,000,000$ times greater.

b. The pyramid's mass is $13 - (-3) = 13 + 3 = 16$ orders of magnitude greater than the snowflake's mass. So the pyramid's mass is $10^{16} = 10,000,000,000,000,000$ times greater.

4. How many orders of magnitude greater is an adult person's mass than an amoeba's mass? How many times greater is the mass of an adult person than the mass of an amoeba?

$5 - (-5) = 10; 10^{10} = 10,000,000,000$

In an Example 1 in the section, we used the radii of the sun and Earth to compare their volumes. In the next problem, you will use their masses to compare their volumes.

5. Use the table with the masses to verify that the sun's volume is 1 million times greater than Earth's volume. (If we assume that the sun and Earth have the same density, then the ratio of their masses is the same as the ratio of their volumes.)

The sun's mass is 6 orders of magnitude greater than Earth's mass: $33 - (27) = 6$. So the sun is $10^6 = 1,000,000$ times larger.

Connect

You might see this definition in a chemistry class: $pH = -\log[H^+]$. It simply means to take the exponent from the hydrogen ion concentration (if it's rounded to the nearest order of magnitude) and then change the sign so that it is positive.

6. The pH scale is a logarithmic scale that indicates how acidic or basic a substance is based on its hydrogen ion concentration. Since hydrogen ion concentrations can be quite small, they are often stated in scientific notation. Since these concentrations vary over a large range of values, it's convenient to convey the information with a logarithmic scale. If the hydrogen ion concentration is rounded to the nearest power of ten (its order of magnitude), then the exponent tells us the pH value. We simply ignore the negative sign on the exponent so that the pH can be stated as a positive value.

For example, a substance with a hydrogen ion concentration of 1×10^{-5} has a pH of 5. A substance with a hydrogen ion concentration of 1×10^{-6} has a pH of 6. If the hydrogen ion concentration **decreases** by a factor of ten, then the pH **increases** by one unit.

Concentration of Hydrogen ions compared to distilled water	pH	Examples of solutions and their respective pH
1/10,000,000	14	Liquid drain cleaner, caustic soda
1/1,000,000	13	Bleaches, oven cleaner
1/100,000	12	Soapy water
1/10,000	11	Household ammonia (11.9)
1/1,000	10	Milk of magnesium (10.5)
1/100	9	Toothpaste (9.9)
1/10	8	Baking soda (8.4), seawater, eggs
1	7	"Pure" water (7)
10	6	Urine (6), milk (6.6)
100	5	Acid rain (5.6), black coffee (5)
1,000	4	Tomato juice (4.1)
10,000	3	Grapefruit & orange juice, soft drink
100,000	2	Lemon juice (2.3), vinegar (2.9)
1,000,000	1	Hydrochloric acid secreted from the stomach lining (1)
10,000,000	0	Battery acid

More basic

More acidic

Based on chemteacher.chemeddl.org

a. How many times greater is the hydrogen ion concentration in lemon juice compared to urine?

Lemon juice is 10^4 or 10,000 times more acidic than urine.

b. Two steps down on the pH scale implies a ___hundred___ times more hydrogen ions.

c. Three steps up on the pH scale implies ___one-thousandth___ the hydrogen ion concentration.

d. Name a substance that is ten times more acidic than milk. Acid rain or black coffee

e. Name a substance that has a hydrogen ion concentration one-thousandth that of toothpaste.

Soapy water

Reflect

WRAP-UP

What's the point?

Many things cannot be easily measured with a linear scale. If the values to be measured vary over a large range, then a logarithmic scale that focuses on the power of ten can be a more useful way to report values.

What did you learn?

How to interpret logarithmic scales

4.6 Homework

Skills MyMathLab

First complete the MyMathLab homework online. Then work the two exercises to check your understanding.

1. a. If a quantity increases by 3 orders of magnitude, by what factor does it increase?

1,000

b. If a quantity decreases by a factor of 1,000,000, by how many orders of magnitude does it decrease?

6

2. What is the decibel level of a sound that is 1,000,000 times more intense than the threshold for human hearing?

60 dB

Concepts and Applications

Complete the problems to practice applying the skills and concepts learned in the section.

3. a. A student argues that the information in the order of magnitude table for masses does not make sense because an adult person is more than 5 grams more massive than a grasshopper. What's wrong with this student's argument?

The exponents differ by 5, but that doesn't mean that the masses differ by 5 grams.

b. What is the correct way to interpret the difference in orders of magnitude for the mass of a grasshopper and the mass of an adult person?

An adult person is $10^5 = 100,000$ times more massive than a grasshopper. We say that the adult person's mass is 5 orders of magnitude greater than the grasshopper's mass.

4. In science, prefix names are used to alter units of length, volume, and mass. The number line shows the orders of magnitude for specific prefixes.

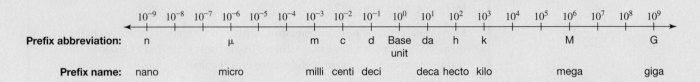

a. Complete all the rows of the second column of the table that follows for grams. Fill in rows in the third and fourth columns when a prefix name exists.

Scientific Notation	Standard Notation	Name of Quantity in Grams	Abbreviation of Quantity in Grams
10^{-9}	0.000000001	nanogram	ng
10^{-8}	0.00000001		
10^{-7}	0.0000001		
10^{-6}	0.000001	microgram	μg
10^{-5}	0.00001		
10^{-4}	0.0001		
10^{-3}	0.001	milligram	mg
10^{-2}	0.01	centigram	cg
10^{-1}	0.1	decigram	dg
10^{0}	1	gram	g
10^{1}	10	decagram	dag
10^{2}	100	hectogram	hg
10^{3}	1,000	kilogram	kg
10^{4}	10,000		
10^{5}	100,000		
10^{6}	1,000,000	megagram	Mg
10^{7}	10,000,000		
10^{8}	100,000,000		
10^{9}	1,000,000,000	gigagram	Gg

b. Use the table to name each quantity.

   i. 5×10^{-6} g   5 micrograms

  ii. 8.23×10^{6} g   8.23 megagrams

 iii. 9.1×10^{-9} g   9.1 nanograms

 iv. 4.9×10^{9} g   4.9 gigagrams

  v. 3.7×10^{3} g   3.7 kilograms

 vi. 52×10^{2} g   52 hectograms or 5.2 kilograms

5. Looking Forward, Looking Back Find the standard deviation of the data set. Round to the nearest hundredth.

$$1.5, 2.8, 6.8, 9.4, 10.2$$

3.47

Mid-Cycle Recap

INSTRUCTOR NOTE: The Mid-Cycle Recap provides students a chance to see how they are accomplishing the goals of the cycle so far. A quiz is available in MyMathLab as an option for assessing students' progress on the objectives at this point. See the Instructor Guide for more assessment ideas.

Complete the problems to practice applying the skills and concepts learned in the first half of the cycle.

Skills

1. Use dimensional analysis to find how many square feet are in a square mile. Use the fact that there are 5,280 feet in 1 mile. Write your answer in both standard notation and scientific notation.

   27,878,400 square feet $= 2.78784 \times 10^7$ square feet

2. Rewrite with only positive exponents: $\dfrac{(2x)^{-2}}{x^{-3}} \cdot \dfrac{x}{4}$

Concepts and Applications

3. Circle T for true or F for false for each statement. If a statement is false, explain why it is false or correct it.

   T   (F)   a. Standard deviation is a measure of the center of data.
   Standard deviation is a measure of the spread of data.

   T   (F)   b. The mean can never equal the median for a data set.
   For some data sets, the mean and median are equal.

   (T)   F   c. Standard deviation can never be negative.

   (T)   F   d. The smaller a z-score, the closer the value is to the mean.

4. Create a data set with ten values. Find the mean, median, and standard deviation. Organize your work with a table and show all the details.
   Answers will vary.

5. A student calculates a distance incorrectly and states the answer as 77.3 meters instead of 773 meters. By how many orders of magnitude is his answer incorrect?
   1 order of magnitude

4.7 Straight to the Point: Direct Variation

SECTION OBJECTIVES:

- Identify direct variation from a graph, table, or equation
- Write models for direct variation problems
- Solve direct variation problems

1. A swimming pool that holds 21,000 gallons of water is empty and is being filled by a garden hose at a constant rate of 300 gallons per hour. The volume of water in the pool is a function of the time the hose has been running.

   a. Complete the table by using the formula *Volume* = *rate* · *time* to find the volume of water in the pool for each time. Do not use a calculator.

Time (hours)	Volume (gallons)
1	300
2	600
3	900
5	1,500
10	3,000
20	6,000
70	21,000

   b. Write a function that models the relationship between volume and time in this scenario. Is the function linear? If so, find the slope and *y*-intercept and interpret each in this context.

   $V = 300h$, where V = volume and h = number of hours; yes

   The slope is $m = 300$ gallons per hour. For each hour, the volume of the pool increases by 300 gallons.

   The *y*-intercept is (0, 0). The pool was empty when the filling process began.

   c. For each line of the table find the quotient volume/time. What do you notice about this quotient? How does this value appear in the equation?

   The quotient is always 300 gal/hr; 300 is the slope in the equation.

   d. Both quantities in the table are increasing. Specifically, whenever the time doubles, the volume _____doubles_____. Whenever the time triples, the volume _____triples_____.

Discover

If two quantities are related, there are different ways they can vary in relation to each other. In the *Explore*, you saw a relationship in which volume and time varied while the rate was held constant at 300 gallons per hour. Because the rate was constant, the volume increased as the time increased and always by the same factor. Also, the quotient volume/time had a constant value. This type of relationship is known as direct variation.

As you work through this section, think about these questions:

1 Can you recognize direct variation from a table, an equation, or a graph?

2 Can you interpret the constant of variation?

3 Do you know how to solve variation problems with equations and using proportionality?

Let's begin by looking at the definition of direct variation.

LOOK IT UP

Direct Variation

When the quotient of two variables is equal to a constant, each variable is said to **vary directly** with the other, or the variables are said to be proportional to each other. The nonzero constant k is called the **constant of proportionality** or the **constant of variation**.

Direct variation can be modeled with a constant quotient or with a product when one variable is isolated.

Variables with a constant quotient: $\dfrac{y}{x} = k$ **Product when one variable is isolated:** $y = kx$

In direct variation, when one variable increases by a factor, the other increases by the same factor. Similarly, if one variable decreases by a factor, the other decreases by the same factor.

FOR EXAMPLE, if you open a savings account and deposit $100 each month, your savings account balance varies directly with the number of months. As an equation, $Balance = \$100 \cdot months$, where $100 per month is the constant of variation, k. Your account balance is always a constant multiple (100) of the number of months.

INSTRUCTOR NOTE: After discussing the definition of direct variation, consider asking students to identify the constant of variation in the *Explore*.

Direct variation is a specific type of linear relationship in which y is a nonzero multiple of x that is, $y = kx$. The slope is the constant of variation. Since $b = 0$, the y-intercept is $(0, 0)$ and the line always passes through the origin. In general, the graph of the relationship between two variables that are directly related has the following appearance:

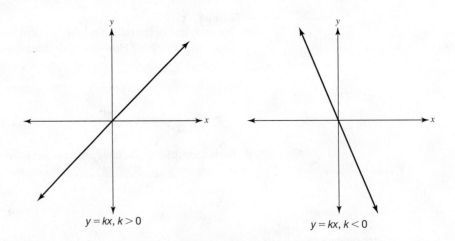

$y = kx, k > 0$ $y = kx, k < 0$

When we need to determine if two quantities vary directly, we can use an equation, a graph, or a table. If we have a table of values, we can use it to check for a constant quotient or we can write the equation and determine if its form shows direct variation.

2. Using the data in the table, write an equation that represents y as a function of x. Does y vary directly with x? If so, state the constant of variation.

x	y
4	14
5	17.5
6	21
7	24.5

$y = 3.5x$, so y varies directly with x; $k = 3.5$

In the previous problem, you wrote the equation to determine if the relationship showed direct variation. Another option is to find the quotient of y and x for each ordered pair. For example, $\frac{14}{4} = 3.5$. In fact, for the values in the table, every quotient is 3.5, which confirms that there is direct variation.

3. For each graph, determine if it shows direct variation. If it does, state the constant of variation. If it does not, explain why.

a.

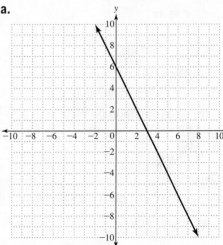

No; the y-intercept is not (0, 0)

b.

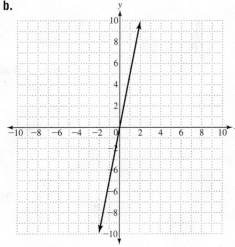

Yes; $k = 5$

c.

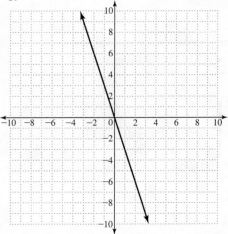

Yes; $k = -3$

d.

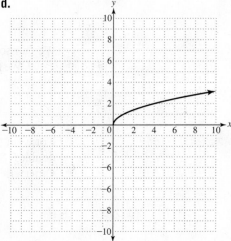

No; the graph is not linear

We can use direct variation models to solve direct variation problems.

EXAMPLE

Suppose y varies directly with x, and $y = 8$ when $x = 2$. Find y when $x = 4$.

SOLUTION

One way to solve a direct variation problem is to use a direct variation model and find the constant of variation. We can begin by writing the general variation model in which y is isolated. y varies directly with x means that $y = kx$. The value of k, the constant of variation, is the number that relates two quantities that vary directly. Use the information provided, $y = 8$ when $x = 2$, to find k.

$$y = kx$$
$$8 = k \cdot 2$$
$$k = 4$$

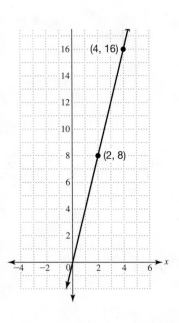

Next, we fill in the value of k in the general variation model to create a specific variation model for these variables.

$$y = 4x$$

Last, we use the variation equation to find y knowing that $x = 4$.

$$y = 4x$$
$$y = 4(4)$$
$$y = 16$$

Notice that the graph of the variation model $y = 4x$ passes through the points $(2, 8)$ and $(4, 16)$.

We can also solve this direct variation problem with a proportion. We begin by writing the general variation model showing a constant quotient, $\frac{y}{x} = k$. The given values of y and x, which are 8 and 2 respectively, have a quotient. The second y and x should have the same quotient. We can write a proportion that states this.

We compare y to x in the same order for each pair of values.

$$\frac{y_1}{x_1} = \frac{y_2}{x_2}$$

$$\frac{8}{2} = \frac{y_2}{4}$$

We can solve the proportion by multiplying both sides by 4 or setting cross products equal. We get $y_2 = 16$ as the solution. Therefore y is 16 when x is 4.

Notice in the last example that x doubles from 2 to 4. Since y and x vary directly, y also doubles. So $y = 2(8) = 16$. When possible, use this kind of proportional reasoning. Otherwise, write and solve a proportion, as demonstrated in the example.

We have seen two approaches to solving direct variation problems. Each has its own advantages and disadvantages. Using a direct variation model and finding k to solve variation problems requires multiple steps, but it results in the constant of variation and a variation equation that can be used to answer other questions. Using proportionality can be faster when only a new value of one of the variables is needed, but the constant of variation is not as obvious.

Need more practice?

Suppose C varies directly with x, and $C = 18$ when $x = 5$. Find C when $x = 25$.

Answer: $C = 90$

4. Suppose A varies directly with w, and $A = 4$ when $w = 8$.

    a. Find the value of k. Write the variation model with the constant of variation.

$$k = \frac{1}{2}; \; A = \frac{1}{2}w$$

    b. Graph the variation model.

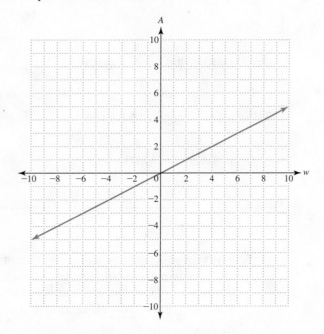

INSTRUCTOR NOTE: Consider showing students three ways to find the answer in part c: from the graph, with the variation model, and with a proportion. They can also use proportional reasoning. Since the new w is $\frac{3}{4}$ of the original w, the new A will be $\frac{3}{4}$ of the original A as well.

    c. Find A when $w = 6$.

$$A = 3$$

In practice, use the method either finding k and writing a model or using proportions that makes the most sense to you.

The two procedures for solving a direct variation problem are summarized in the *How It Works*.

HOW IT WORKS

To solve a direct variation problem:

Option 1: Find k and write a model.

1. Write a general direct variation model.
2. Find the value of the constant of variation.
3. Write the variation model with the constant of variation included.
4. Substitute any given values into the variation model and solve for the requested variable.

Option 2: Use a proportion.

1. Write a proportion comparing the variables in the same way for each ratio.
2. Solve the proportion.

EXAMPLE: Suppose y varies directly with x, and $y = -15$ when $x = 5$. Find the value of y when $x = 8$.

Option 1: Since the variables vary directly, they follow the model $y = kx$. Use the given information to find the constant of variation.

$$y = kx$$
$$-15 = k \cdot 5$$
$$k = -3$$

Write the variation model with the constant.

$$y = -3x$$

Use the equation to solve the problem.

$$y = -3x$$
$$y = -3(8)$$
$$y = -24$$

Option 2: Since the variables vary directly, they have a constant quotient. The given y and x have a quotient. The second x and y pair should have the same quotient. Write a proportion stating this.

$$\frac{y_1}{x_1} = \frac{y_2}{x_2}$$

$$\frac{-15}{5} = \frac{y_2}{8}$$

Solve by setting the cross products equal.

$$-15(8) = 5y_2$$
$$-120 = 5y_2$$
$$-24 = y_2$$

The second y is -24.

5. John's pay varies directly with the number of hours he works. When he works 20 hours, he makes $240. How much does he make if he works 32 hours? What is the constant of variation in this situation, and what is its significance?

$384; $k = 12$; k is his hourly pay rate.

One particular type of direct variation is **joint variation**. It is direct variation involving three or more variables. For example, $y = kxw$ means that y varies jointly or directly with both x and w.

6. Suppose M varies directly with a and b, and $M = 102$ when $a = 2$ and $b = 3$. Find M when a and b are each 6.

$M = 612$

7. Simple interest, I, varies jointly with the amount invested, P, the interest rate, r, and the time, t. Write the variation model assuming a constant interest rate of 3.5%. Find the interest earned if \$2,000 is invested for 4 years.

$I = 0.035Pt;\ I = \$280$

Connect

Remember that direct variation is just a particular type of linear relationship in which y is a nonzero multiple of x and the y-intercept is $(0, 0)$.

8. For the values in each table, write the equation of the line that passes through the points. Then determine if direct variation exists between x and y. Explain your answer.

a.

x	y
−3	21
0	0
1	−7
2	−14
4	−28

$y = -7x$; y varies directly with x, since the equation is linear and the y-intercept is $(0, 0)$

b.

x	y
−4	−26
−2	−16
0	−6
4	14
8	34

$y = 5x - 6$; y does not vary directly with x, since the equation is linear but the y-intercept is not $(0, 0)$

WRAP-UP

What's the point?

It is important to understand how things change relative to each other. Quantities that increase together or decrease together by the same factor vary directly. When direct variation exists but with more than two variables, the quantities vary jointly. In the next section, you will learn about two more types of variation.

What did you learn?

How to identify direct variation from a graph, table, or equation
How to write models for direct variation problems
How to solve direct variation problems

4.7 Homework

Skills MyMathLab

First complete the MyMathLab homework online. Then work the two exercises to check your understanding.

1. Assume y varies directly with x. If $y = 36$ when $x = 24$, find the constant of variation.

   $k = 1.5$

2. Assume L varies directly with r, and $L = 400$ when $r = 10$. Find L when $r = 25$.

   1,000

Concepts and Applications

Complete the problems to practice applying the skills and concepts learned in the section.

3. For the values in each table, determine if direct variation exists between the variables. If it does, state the constant of variation.

   a.

x	y
−1	−1
4	14
8	26
20	62

   No direct variation

b.

x	y
−3	$\frac{3}{2}$
2	−1
15	$-\frac{15}{2}$
21	$-\frac{21}{2}$

Direct variation; $k = -\frac{1}{2}$

c.

x	y
0	0
1	2
2	8
3	26

No direct variation

4. Remember that distance, rate, and time are related by the equation
Distance = rate · time.

For each statement, give a counterexample to show that it is not true, or use the distance equation to explain why it is true.

a. If you drive twice as fast, you will go twice as far in the same amount of time.

True; $d = (2r)(t) = 2rt$ is twice as large as $d = rt$.

b. If you drive twice as long, you will go twice as far at the same speed.

True; $d = (r)(2t) = 2rt$ is twice as large as $d = rt$.

c. If you need to go twice as far at half the speed, it will take you four times longer.

True; $t = \dfrac{\text{distance}}{\text{rate}} = \dfrac{2d}{0.5r} = 4\dfrac{d}{r}$ is four times as large as $t = \dfrac{d}{r}$.

d. If you drive twice as fast for twice as long, you will go twice as far.

Not true. If you drive 60 mph for 2 hours, you go 120 miles. This is four times as far as you would go at 30 mph for 1 hour.

5. Write a function to model each situation. Use meaningful variables and define them. Determine if the function is linear. If it is linear, determine if it shows direct variation.

a. Joe earns $5.25 for each phone case he sells. Model his revenue, that is, the amount of money he brings in.

$R = 5.25n$, where R = revenue and n = number of phone cases

Linear; direct variation

b. Each phone case costs Joe $2 to make. He also has a cost of $150 for materials. Model his total cost.

$C = 2n + 150$, where C = total cost and n = number of phone cases

Linear; no direct variation

c. Model Joe's profit, that is, his revenue minus costs.

$P = 3.25n - 150$, where P = profit and n = number of phone cases

Linear; no direct variation

6. Looking Forward, Looking Back Convert 8 m/sec^2 to mi/sec^2.

0.005 mi/sec^2

Getting Ready for Section 4.8

Read the article and answer the questions that follow.

Science of Light

Light visible to the human eye is a portion of the *electromagnetic spectrum*, which encompasses radiant energy over the entire range of propagating mechanisms. Radiant energy may be seen (visible light), felt (infrared radiation or heat transferred from warm objects), or it can actually penetrate (x-rays) and do physical damage to the cells of the human body (gamma rays or nuclear radiation). Very low energy radiation is used by human technology as a carrier for communications (microwaves and radio waves). What we call visible *light* forms the visible spectrum, or colors of the rainbow, and represents a very narrow band in entire electromagnetic spectrum. Blue and violet light contain more energy and have a shorter wavelength than orange and red light.

Measuring Light

Radiant energy behaves both as waves, with measurable wavelengths and frequencies, and as particles, or as discrete "packages" of energy, called *photons* for visible light. Light cannot be propagated, transmitted, or received in quantities smaller than 1 photon, and a photon of a particular wavelength contains a discrete amount of radiant energy. Light travels at a constant speed of 3×10^8 meters per second, or 186,000 miles per second, in a vacuum. Light is usually measured as photon flux, proportional to the number of photons per second striking the human eye or a light meter. Photon flux is called *illuminance*, and its engineering units are *lux* (metric) or *footcandles* (English); both are linear scales. The human eye is capable of observing an extremely wide range of photon flux, from about 6 photons per second of blue light (about 10^{-9} lux) to brilliant sunlight reflecting off snow (about 10^4 lux), a range of nearly 10 trillion to one.

In astronomy, illuminance is measured in *visual magnitudes*, a logarithmic scale similar to decibels for measuring sound, except that the magnitude scale is inverse, where smaller numbers mean brighter objects. The sun has a visual magnitude of −26.7 (producing an illuminance of 108,000 lux) at the top Earth's atmosphere, while the faintest stars visible to the human eye without optical aid are about magnitude 7.2 (0.000000003 lux). Individual light sources can therefore be measured in terms of the illuminance they produce at the observer's location. Photons leaving the source are subject to the *inverse square law* for radiant energy. This law states that the

energy reaching the observer varies at one over the square of the distance to the source.

$$Energy = Intensity/Distance\ from\ the\ obsever^2$$

Therefore, doubling the distance will result in one-quarter the illuminance from the same source. Astronomical objects such as the stars are so far away that their illuminance does not change measurably even as Earth moves around the sun. The planets, however, vary in brightness primarily because of the inverse square law. The sun and moon are also subject to small but measurable variations in apparent brightness because of variations in distance from Earth.

Light sources on Earth, however, such as street lamps, obviously produce much more illuminance as an observer gets closer to them. When outdoor light at night escapes from its intended use, and is observed directly, it creates *light trespass*, a form of *light pollution*, especially in a natural landscape like a national park. These bright objects are very noticeable, even at a great distance. For example, a typical streetlamp produces about 5 lux of illuminance immediately beneath it (let's say 5 meters away), in the area intended for its use. If the lamp is unshielded and emits light equally in all directions, an observer on the landscape 100 times more distant (500 meters or $\frac{1}{4}$ mile) away will be illuminated by the lamp according to the inverse square law:

$$Energy = 5\ lux/100^2 = 0.0005\ lux$$

This seems like a small amount, but the crescent moon produces only 0.01 lux, and the planet Venus at its brightest produces 0.0001 lux of illuminance. Therefore, this single unshielded street lamp seen from 500 meters away would be brighter than any natural object in the night sky other than the moon. Also, a small, bright source of light will impair dark adaptation of the human eye, further restricting the observer's ability to enjoy the natural night environment.

While illuminance is total photon flux striking a surface or the human eye, *luminance* refers to the apparent *surface brightness* of objects that have a visible size or viewing angle from the point of view of the observer. Computer monitors and flat-screen televisions are often advertised with a value for maximum surface brightness in *nits*, or *candela per meter squared* (cd/m^2). Values of 500-1,000 cd/m^2 are common for these devices. The brightness of the landscape from the

(continued)

Science of Light   *(continued)*

reflected light of the sun, moon, or the night sky depends on its reflectance (snow vs. black lava rock, for example) and ranges from about 8,000 cd/m^2 in bright sunlight to 0.000001 cd/m^2 on a moonless night.

The sun's surface as seen from earth measures an apparent 0.5 degrees in diameter and has a luminance of about 1,600,000,000 cd/m^2. It cannot be observed directly with the human eye without damaging the retina. Conversely, the darkest part of the natural night sky on a moonless night is not "pitch black," but in fact can be measured at about 0.00017 cd/m^2. It is easily seen as luminous to the dark-adapted human eye, especially if objects like trees are silhouetted against it.

In astronomy, luminance may be expressed as visual *magnitudes per square arc second*, and 0.00017 cd/m^2 is equivalent to 22.0 magnitudes per square arc second (MSA). Remember that the magnitude scale is inverse and logarithmic: 17.0 magnitudes per square arc second is 100 times *brighter* than 22.0; 12.0 MSA is 10,000 times brighter than 22.0 MSA.

Excerpts from www.nature.nps.gov

Questions

1. The article mentions the inverse square law, which states that the energy reaching the observer varies at one over the square of the distance to the light source.

$$\text{Energy} = \frac{\text{intensity}}{(\text{distance from the observer})^2}.$$

Doubling the distance from the observer results in one-quarter the illuminance. What happens to the illuminance if the distance from the observer is cut in half?

The illuminance quadruples.

2. The article states that the luminance magnitude scale is inverse and logarithmic: A light source with a visual magnitude of 17.0 is 100 times brighter than a source with a magnitude of 22.0. How many orders of magnitude brighter is 12.0 MSA than 17.0 MSA?

2 orders of magnitude

4.8 Gas Up and Go: Inverse Variation

Explore

SECTION OBJECTIVES:

- Identify inverse variation from a table
- Write models for inverse variation problems
- Solve inverse variation problems

INSTRUCTOR NOTE: Consider asking students what your speed per minute is if you are driving at 60 miles per hour. 60 mph means 60 miles in 60 minutes, or 1 mile per minute.

1. Assume that two cities are 72 miles apart and that you will be driving from one to the other.

   a. Rewrite the formula *Distance = rate · time* so that it is a formula for time. Use your result to write an equation in which time is a function of rate given that the distance is always 72 miles.

   $$T = \frac{72}{R}$$

   b. Use the equation from part a to complete this table by determining the time for the trip, in hours and then in minutes, if the average speed varies. Round your answers to the nearest hundredth when necessary. Is the relationship between time and speed linear? Explain.

Speed (miles per hour)	Time (hours)	Time (minutes)
40	1.8	108
50	1.44	86.4
60	1.2	72
70	1.03	61.71
80	0.9	54

   No; when the speed increases by 10 mph, the time does not decrease by the same amount each time.

   c. For each line of the table, find the product of speed and time. What do you notice about this product?

   It is always 72 miles.

   d. When one of the quantities in the table increases, the other decreases in a particular way. Specifically, whenever speed doubles, time _____is cut in half_____.

   e. Use the data from the table to graph time in hours vs. speed.

   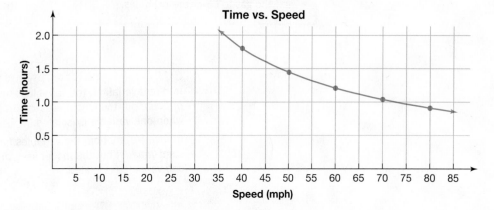

f. There is no speed you can drive that would result in a time of 0 hours for the 72-mile drive. How can you see this in the general shape of the graph?

The graph never reaches the horizontal axis.

g. If your speed gets slower and slower, the time needed to travel the 72 miles gets longer and longer. However, your speed cannot be zero or you wouldn't be moving. How can you see this in the general shape of the graph?

The graph gets close to the vertical axis but never reaches it.

Discover

In Section 4.7, we saw that if two quantities vary directly, then they have a constant *quotient*. In the *Explore*, we saw two quantities that had a constant *product*. Speed and time varied while the distance, which is the product of speed and time, was held constant at 72 miles. This type of relationship is known as inverse or indirect variation, which will be explored in this section.

As you work through this section, think about these questions:

1 Can you recognize inverse variation?

2 Can you write a variation model for each type of variation?

3 Do you know how to solve variation problems by writing a model or by using one equation?

Let's begin by looking at the definition of inverse variation.

LOOK IT UP

Inverse Variation

When the product of two variables is equal to a constant, each variable is said to **vary inversely** or **vary indirectly** with the other. The nonzero constant k is called the **constant of proportionality** or the **constant of variation**.

Inverse variation can be modeled with a constant product or with a quotient when one variable is isolated.

Variables with a constant product: $xy = k$ **Quotient when one variable is isolated:** $y = \dfrac{k}{x}$

In inverse variation, when one variable increases by a factor, the other decreases by that same factor. Conversely, if one variable decreases by a factor, the other increases by that same factor.

FOR EXAMPLE, if your commute is 60 miles, the gas mileage of your car in miles per gallon and the number of gallons of gas used vary inversely according to the equation $mpg \cdot gallons = 60$. The constant of variation, k, is 60. If you have a car that gets better gas mileage, you don't need as much gas to cover the 60 miles. Specifically, if a car gets twice the gas mileage of another car, the car with the better gas mileage needs only half the gas to cover the 60 miles.

INSTRUCTOR NOTE: After discussing the definition of inverse variation, consider asking students to identify the constant of variation in the *Explore*.

In direct variation, the variables have a constant *quotient* but the model uses *multiplication* when one variable is isolated.

In inverse variation, the variables have a constant *product* but the model uses *division* when one variable is isolated.

In general, graphs of the relationship between two variables that are inversely related have the following appearance:

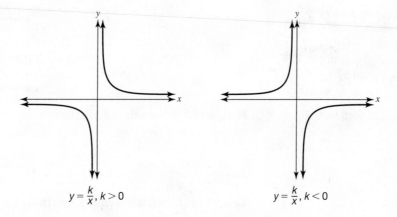

$$y = \frac{k}{x}, k > 0 \qquad\qquad y = \frac{k}{x}, k < 0$$

The general appearance of a graph is not necessarily sufficient for us to conclude that it shows inverse variation or shows exponential change. We must investigate the situation, equation, or table of values to determine that.

The graphs show that inverse variation functions have two curves to their graphs. The graph in the *Explore* did not have a second curve in the third quadrant because negative values did not make sense in the context of speed and time.

2. Using the values in the table, verify that *y* varies inversely with *x* by showing that the product of *x* and *y* is constant. Find the constant of variation, and write the variation model in which *y* is a function of *x*.

x	y
10	24
12	20
16	15
20	12

Since $xy = 240$, y varies inversely with x; $k = 240$; $y = \dfrac{240}{x}$

As we did in the last section with direct variation problems, we can solve inverse variation problems by finding k and using a model to answer the question. Another option involves solving one equation. However, for inverse variation, that equation is not a proportion.

EXAMPLE

Suppose *y* varies indirectly with *x*, and $y = 100$ when $x = 2$. Find *y* when $x = 8$.

SOLUTION

One way to solve an inverse variation problem is to use an inverse variation model and find the constant of variation. We begin by writing the general variation model in which y is isolated. y varies inversely with x means that $y = \dfrac{k}{x}$. The value of k, the constant of variation, is the number that relates the two quantities that vary indirectly. We use the information provided, $y = 100$ when $x = 2$, to find k.

$$y = \frac{k}{x}$$

$$100 = \frac{k}{2}$$

$$k = 200$$

Next, we fill in the value of k in the general variation model to create a specific variation model for these variables.

$$y = \frac{200}{x}$$

Last, we use the variation equation to find y knowing that $x = 8$.

$$y = \frac{200}{x}$$

$$y = \frac{200}{8}$$

$$y = 25$$

We can solve the problem another way: using the fact that variables that vary inversely have a constant product. The given values of y and x, which are 100 and 2 respectively, have a product. The second y and x should have the same product. We can write an equation that states this. We then solve the equation by isolating the unknown y.

$$x_1 y_1 = x_2 y_2$$

$$2(100) = 8y_2$$

$$200 = 8y_2$$

$$25 = y_2$$

Therefore, y is 25 when x is 8.

_____

Notice in the example that x quadruples from 2 to 8. So x has been multiplied by a factor of 4. Since y and x vary indirectly, y is divided by a factor of 4. So $y = \frac{100}{4} = 25$. When possible, use this kind of proportional reasoning. Otherwise, write and solve an equation, as shown in the example.

3. Suppose M varies indirectly with z, and $M = 30$ when $z = 2$. Find M when $z = 6$.

a. Solve using an inverse variation model. State the value of k and the variation model with your solution.

$k = 60; M = \frac{60}{z}; M = 10$

b. Solve using one equation.

$M = 10$

Need more practice?

Suppose G varies inversely with r, and $G = 82$ when $r = 0.5$. Find G when $x = 4$.

Answer: $G = 10.25$

When you see a statement like "y varies . . . as x," start by writing $y = k$.

If the variation is direct, then k is multiplied by x: $y = kx$.

If the variation is inverse, then k is divided by x: $y = \frac{k}{x}$.

The procedure for solving an inverse variation problem is summarized in the *How It Works*.

HOW IT WORKS

To solve an inverse variation problem:

Option 1: Find k and write a model.

1. Write a general inverse variation model.
2. Find the value of the constant of variation.
3. Write the variation model with the constant of variation included.
4. Substitute any given values into the variation model and solve for the requested variable.

Option 2: Use one equation.

1. Write an equation stating that the product of each pair of variables is the same.
2. Solve the equation.

EXAMPLE: Suppose y varies inversely with x, and $y = 25$ when $x = 10$. Find the value of y when $x = 5$.

Option 1: Since the variables vary inversely, they follow the model $y = \dfrac{k}{x}$. Use the given information to find the constant of variation.

$$y = \frac{k}{x}$$

$$25 = \frac{k}{10}$$

$$k = 250$$

Write the variation model with the constant.

$$y = \frac{250}{x}$$

Use the equation to solve the problem.

$$y = \frac{250}{x}$$

$$y = \frac{250}{5}$$

$$y = 50$$

Option 2: Since the variables vary inversely, they have a constant product. The given y and x have a product. The second x and y pair should have the same product. Write an equation stating this.

$$x_1 y_1 = x_2 y_2$$

$$10(25) = 5y_2$$

Solve by simplifying the left side and isolating y_2.

$$10(25) = 5y_2$$

$$250 = 5y_2$$

$$50 = y_2$$

The second value of y is 50.

4. A company manufactures plastic storage containers. The cost per container varies inversely with the number of containers manufactured. It costs the company $1.25 per container when 4,000 containers are made. What is the price per container if 5,500 containers are manufactured?

$0.91

When a variation equation involves more than two variables, it represents **compound variation**. Joint variation, which was introduced in Section 4.7, is compound direct variation.

5. The formula $F = G\dfrac{m_1 m_2}{r^2}$ calculates the force between two objects with masses m_1 and m_2 (in kg) when they are r meters apart. G is the gravitational constant. Think about the force between two magnets as you answer the following questions.

   a. If the distance between two magnets increases, what happens to the force between them?

      It decreases.

   b. If the masses of two magnets increase, what happens to the force between them?

      It increases.

   c. The constant of variation is ____G____.

      Force varies ____directly____ with mass (m_1 and m_2)

      Force varies ____inversely____ with distance squared (r^2)

These are the types of variation that we've discussed and their models.

Direct	$y = kx$
Joint	$y = kxw$
Inverse	$y = \dfrac{k}{x}$
Compound	$y = \dfrac{kx}{z}$

6. Write the variation model that corresponds to each statement.

   a. W varies directly with x. $W = kx$

   b. A varies inversely with the square of b. $A = \dfrac{k}{b^2}$

   c. Z varies jointly with x and y. $Z = kxy$

   d. P varies directly w+ith the square root of f and inversely with g. $P = \dfrac{k\sqrt{f}}{g}$

Connect

7. The area of a triangle is given by $A = \dfrac{1}{2}bh$. If the length of the base of a triangle is held constant, then the area varies directly with the height. As the height increases, the area increases, as you can see in the pictures.

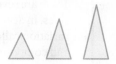

a. Suppose a triangle has a base length of 8 inches and a height of 10 inches. Find the area of the triangle.

$A = 40$ square inches

b. If a triangle with the base in part a has a height of 15 inches, what is its area?

$A = 60$ square inches

8. The area of a triangle is given by $A = \dfrac{1}{2}bh$. If the area of a triangle is held constant, then the length of the base varies indirectly with the height. As the length of the base increases, the height decreases, as you can see in the pictures.

a. Suppose a triangle has a base of length 15 inches and an area of 75 square inches. Find the height of the triangle.

$h = 10$ inches

b. If a triangle with the area in part a has a base of length 7.5 inches, what is the triangle's height?

$h = 20$ inches

Reflect

WRAP-UP

What's the point?

There are several different types of variation that can relate two or more variables. In Section 4.7, you saw direct variation and a particular type of direct variation called joint variation. In this section, you were introduced to indirect and compound variation. You should be able to write a model and find the constant of variation for each of these types.

What did you learn?

How to identify inverse variation from a table
How to write models for inverse variation problems
How to solve inverse variation problems

4.8 Homework

Skills MyMathLab

First complete the MyMathLab homework online. Then work the two exercises to check your understanding.

1. Write the variation model assuming N varies directly with z and inversely with the cube of r.

   $N = \dfrac{kz}{r^3}$

2. Assume y varies inversely with x. If $y = 6$ when $x = 3$, find the value of y when $x = 12$.
   $y = 1.5$

Concepts and Applications

Complete the problems to practice applying the skills and concepts learned in the section.

3. The formula for the volume of a right circular cylinder is $V = \pi r^2 h$. How does the volume vary with the radius and the height? What kind of variation is represented by the formula (direct, inverse, or compound)?

   The volume varies directly with the height and the square of the radius; compound variation

4. If you are paid hourly, your salary can be found from the formula
Salary = hours worked · hourly rate. Even if you're not paid by the hour, your salary and hours worked are related by the same equation.

    a. If an employee has a constant hourly rate, does his salary vary directly or inversely with the hours worked? What happens as he works more hours?

        Directly; the more hours he works, the higher his salary.

    b. If an employee has a constant salary, does his hourly rate vary directly or inversely with the hours worked? What happens as he works more hours?

        Inversely; the more hours he works, the lower his hourly rate.

5. In statistics, this compound variation formula is used to calculate the margin of error $E = z\dfrac{\sigma}{\sqrt{n}}$.

    a. When the value of n increases, what happens to the margin of error?

        It decreases.

    b. When the value of σ increases, what happens to the margin of error?

        It increases.

    c. The margin of error varies ___directly___ with z and σ, but ___inversely___ with $\sqrt{n}$.

6. Lyla learned this diagram as a memory aid for inverse variation. She is using it for $D = RT$ in which D is constant. If she covers up the R, she can see that $R = \dfrac{D}{T}$. Likewise, if she covers up the T, she can see that $T = \dfrac{D}{R}$. If she covers up the D, she can see that $D = RT$.

    a. For an inverse variation equation, what quantity is in the top of the triangle?

        The constant of variation

b. A triangle like this can be used for direct variation as well, but the constant of variation is in one of the bottom locations. Assume *x* and *y* have a constant quotient of 20. How do *x* and *y* vary? Draw a triangle to show the kind of variation between *x* and *y*. Use the triangle to write three variation equations.

Directly; $y = 20x$; $\dfrac{y}{x} = 20$; $x = \dfrac{y}{20}$

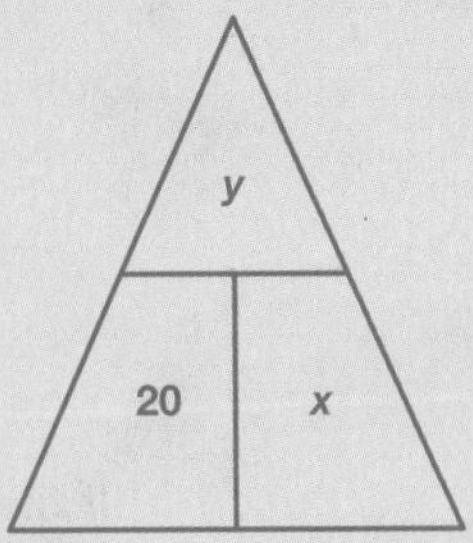

7. For each table of values, determine if the variation is direct or inverse. Find the constant of variation.

a.

x	y
−3	−8
−1	−24
1	24
2	12
4	6

Inverse; $k = 24$

b.

x	y
−4	−20
−2	−10
0	0
4	20
8	40

Direct; $k = 5$

8. Looking Forward, Looking Back A lightening flash has a luminance of about 70,000,000,000 cd/m². Write the luminance in engineering notation.

70×10^9 cd/m²

| 4.9 | Ghost in the Machine: Function Notation |

Explore

SECTION OBJECTIVES:
- Use function notation
- Find a function input or output given the other

Remember?
A function is a rule that assigns to each input value exactly one output value.

A useful analogy for a function is to think of it as a machine that takes an input and produces an output. Consider the example of the squaring function.

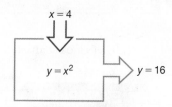

$$x = 4$$
$$y = x^2$$
$$y = 16$$

1. Use the function machine to answer the questions.

$$y = x^2 + 2x$$

a. If the value $x = 3$ is put into the function, what is the output? 15

b. If the value $y = 3$ comes out of the function, what was the input? -3 or 1

c. What is the smallest output value that this function can produce? Consider that the graph of the function is a parabola that opens up. -1

Discover

You were introduced to functions in Cycle 1 and have seen many examples of functions throughout the book. In this section, you will learn some new notation that will help you use functions efficiently. You will also continue to practice evaluating functions, as you did in the *Explore*.

As you work through this section, think about these questions:

1 Can you correctly read and write function notation?

2 Can you distinguish between solving an equation and evaluating a function?

In the *Explore*, you had to find the *y*-value that corresponded with a particular *x*-value. There is an easier way to express this idea, as you'll see in the following definition.

Function Notation

The notation $f(x)$ is read "f of x" and refers to the y-value that is produced when an x-value is put into a function named "f."

FOR EXAMPLE, the function $y = 5x - 2$ can also be expressed in **function notation** as $f(x) = 5x - 2$. The $f(x)$ notation is simply another way to refer to the output or y-value.

INSTRUCTOR NOTE: You might have students refer back to the *Explore* and write the function used there with several different function names.

Next, let's look at an example of how this new notation is used.

EXAMPLE 1

If $f(x) = -4x + 1$, find $f(2)$ and $f(-2)$.

SOLUTION

To find $f(2)$, we substitute the value of 2 for x and calculate the output or y-value.

$$f(x) = -4x + 1$$
$$f(2) = -4(2) + 1$$
$$= -8 + 1$$
$$= -7$$

So $f(2) = -7$. Or we can say that when $x = 2$, $y = -7$. The function notation is simply a compact way to write this information.

To find $f(-2)$, we substitute the value of -2 for x and calculate the output or y-value.

$$f(x) = -4x + 1$$
$$f(2) = -4(-2) + 1$$
$$= 8 + 1$$
$$= 9$$

So $f(-2) = 9$. Or we can say that when $x = -2$, $y = 9$.

The function $f(x) = -4x + 1$ can also be expressed as $y = -4x + 1$.

If you want to evaluate a function for a specific x-value, then function notation can be helpful. It is easier to ask someone to find $f(2)$, for example, than to ask him to find the y-value when $x = 2$.

Although f is commonly used to represent a function, any letter can be used for the function name. Similarly, any variable can be used to indicate the function input, not just x. It's helpful to use meaningful letters when possible.

2. Use the functions $g(x) = \dfrac{x-1}{x+1}$ and $r(b) = \sqrt{b+14}$ to find the following values:

a. $g(4)$ $\dfrac{3}{5}$

b. $g\left(\dfrac{1}{2}\right)$ $-\dfrac{1}{3}$

c. $r(-5)$ 3

d. $r(-15)$ Not a real number

3. The function $h(t) = -16t^2 + 64t + 80$ gives the height in feet of a projectile t seconds after it is launched. Find $h(0)$ and $h(3)$ and interpret your answers.

$h(0) = 80$; the projectile has an initial height of 80 feet.

$h(3) = 128$; the projectile has a height of 128 feet after 3 seconds.

We have seen that when you are asked to evaluate a function, you substitute an x-value into the function and find the corresponding y-value. There are times when you need to reverse that process and find the x-value or input that gives you a particular y-value or output. This involves solving an equation.

EXAMPLE 2

If $f(x) = x^2 + 6x - 1$, find the input values that make the function 0.

SOLUTION

We need to solve the equation $f(x) = 0$ or $x^2 + 6x - 1 = 0$. Notice that this is a quadratic equation and the quadratic expression does not factor. So we need to use the quadratic formula.

$$x = \frac{-b \pm \sqrt{b^2 - 4ac}}{2a}$$

$$= \frac{-6 \pm \sqrt{6^2 - 4(1)(-1)}}{2(1)}$$

$$= \frac{-6 \pm \sqrt{36 + 4}}{2}$$

$$= \frac{-6 \pm \sqrt{40}}{2}$$

$$\frac{-6 + \sqrt{40}}{2} \approx 0.16 \text{ and } \frac{-6 - \sqrt{40}}{2} \approx -6.16$$

So there are two x-values that make the function 0: $x \approx 0.16$ and $x \approx -6.16$.

> **Remember?**
> The x-value that makes a function 0 is called a zero of the function.

4. Use this function machine to find $f(-6)$ and x when $f(x) = -6$:

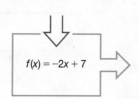

$f(x) = -2x + 7$

$19; \dfrac{13}{2}$

5. The function $h(t) = -16t^2 + 64t + 80$ gives the height in feet of a projectile t seconds after it is launched. Find when $h(t) = 0$ and interpret your answer.

$t = 5$ seconds or -1 second (which does not make sense in this context); the projectile hits the ground (a height of 0) after 5 seconds.

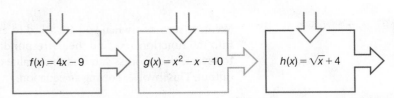

Connect

6. Consider the following three function machines:

$f(x) = 4x - 9$ $g(x) = x^2 - x - 10$ $h(x) = \sqrt{x} + 4$

a. The value $x = 5$ is put into the first function and its output is put into the second function. The output from the second function is then put into the third function. What is the output of the third function machine?

14

b. An input value of $x = 0$ was put into one of the functions, the output into another of the functions, and its output into the third function. In what order were the function machines used if the final output was 32?

h first, then f, then g

Reflect

INSTRUCTOR NOTE: Consider providing the focus problem grading rubric and writing template to students at this point to help them form their written solution. Both are available within MyMathLab.

Focus Problem Check-In
Use what you have learned to finish solving the focus problem. Write a rough draft to explain your solution, showing and explaining all mathematical work. Edit your draft into a final write-up by making any needed corrections and proofreading for grammar, spelling, and mathematical errors. Have someone verify the mathematical work and proofread the final draft for mistakes.

WRAP-UP

What's the point?

You need to be able to find the output from a function if you're given the input and find the input if you're given the output. It is often easier to evaluate a function, or find an output, because that involves substituting a value into an expression. To find an input given an output, you need to solve an equation. Having notation for a function can simplify the writing and clarify each of these tasks.

What did you learn?

How to use function notation
How to find a function input or output given the other

4.9 Homework

Skills MyMathLab

First complete the MyMathLab homework online. Then work the two exercises to check your understanding.

1. Find $f(-1)$ if $f(x) = -2x^2 - x$.

-1

2. Find x such that $f(x) = 0$ for $f(x) = -2x^2 - x$.

$0, -\dfrac{1}{2}$

Concepts and Applications

Complete the problems to practice applying the skills and concepts learned in the section.

3. Write the function for the following machine given the three pairs of inputs and outputs:

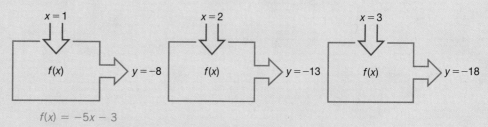

$f(x) = -5x - 3$

4. Use the graph of $y = f(x)$ to find the requested values.

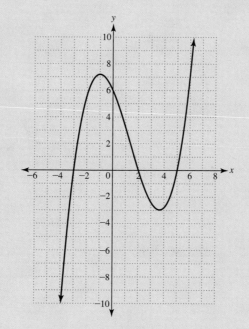

a. $f(2)$ 0

b. $f(0)$ 6

c. Approximate x when $f(x) = -4$. -3.5

5. Write the linear function f if $f(0) = 5$ and $f(2) = 35$.

$f(x) = 15x + 5$

6. Looking Forward, Looking Back What is the only value of x that you can't substitute into each function?

a. $f(x) = \dfrac{-2x + 4}{x - 5}$

$x = 5$

b. $f(x) = \dfrac{3x + 1}{x}$

$x = 0$

4.10 What's Your Function?: Vertical Line Test, Domain, and Range

Explore

SECTION OBJECTIVES:

- Apply the vertical line test
- Find the domain and range from a graph

1. Recall that a function must match each input to one and only one output. Determine if each of the following represents a function. Explain your answers.

x	y
1	−4
2	3
3	−4

Yes; each *x*-value gets one and only one *y*-value.

x	y
−4	1
3	2
−4	3

No; *x* = −4 is paired with two different *y*-values.

INSTRUCTOR NOTE: In #2 the table is provided to encourage students to read function values from the graph and think about how each *x* is paired with one *y*. Be sure they understand that the five ordered pairs shown in the table do not prove that it is a function.

2. Use the graph of the function $y = f(x)$ to complete the table, estimating when necessary. Does the graph represent a function? Explain your answer.

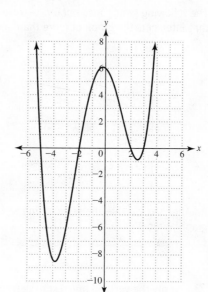

x	y
−2	0
−3	−6
3	0
1	3.5
−1	4.8

Yes; from the graph, it is clear that each *x*-value gets one and only one *y*-value.

Discover

At this point in the book, you should be able to define, evaluate, and graph functions. This section will help you develop some additional insight into functions. Specifically, you will learn a fast way to determine whether or not a graph represents a function and how to determine the inputs and outputs for a function.

As you work through this section, think about these questions:

1 Do you know how to apply the vertical line test?

2 Where do you look on a graph to find the domain? Where do you look to find the range?

In the *Explore,* you had to determine if a graph represented a function. Let's look at a simple test to determine that.

Vertical Line Test

If no vertical line intersects the graph at more than one point, then the graph represents a function.

FOR EXAMPLE, this graph represents a function because it is not possible to draw a vertical line that intersects the graph at more than one point. In other words, it passes the **vertical line test**.

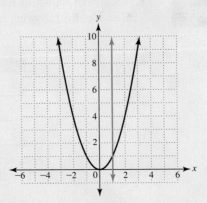

Notice that two *x*-values can be paired with the same *y*-value (for example, $x = 2$ and $x = -2$ are both paired with $y = 4$) and the graph still represents a function.

The following graph does *not* represent a function because we can draw a vertical line that intersects the graph at more than one point.

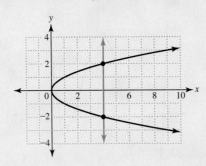

The points (4, 2) and (4, −2) are both on the graph, indicating that $x = 4$ is paired with two different *y*-values: 2 and −2.

Another way to state the vertical line test is to say that every vertical line intersects the graph at most once.

A function can pair two different *x*-values with the same *y*-value but not two different *y*-values with the same *x*-value.

3. Apply the vertical line test to determine if each of the following graphs represents a function:

a. Yes

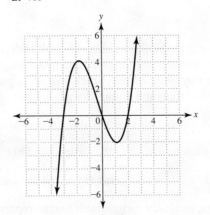

b. Yes

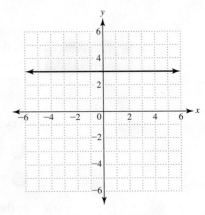

c. No

d. No

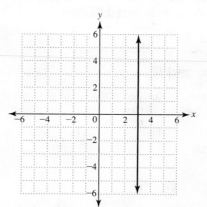

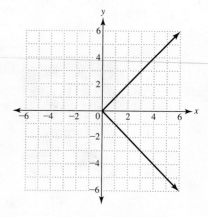

In addition to looking at the graph to decide whether it represents a function, we can also use the graph to determine the allowed inputs and possible outputs of the function.

EXAMPLE

Use the function $f(x) = \sqrt{x - 2}$ and its graph to determine the allowed inputs and possible outputs of the function.

> The function in this example, with a variable under a square root, is called a **radical function**.

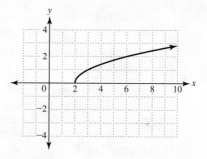

SOLUTION

If you look from left to right on the graph, you notice that the smallest x-value that is graphed is $x = 2$ and the graph extends forever to the right. So the allowed inputs are all the x-values that are at least 2, or $x \geq 2$.

If you look from the bottom to the top on the graph, you notice that the smallest y-value that is graphed is $y = 0$ and the graph extends forever above that. So the possible outputs are all the y-values that are at least 0, or $y \geq 0$.

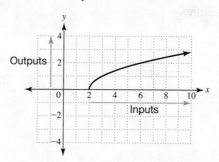

The equation for the function can be used to confirm your conclusions. If we try to substitute an x-value smaller than 2, the equation involves the square root of a negative number. This results in an output that is not a real number and, therefore, one that we don't want to allow.

We also know that the square root in the function can never result in a negative output. The smallest output you can get from a square root is 0.

There are names for the set of allowed inputs and the set of possible outputs for a function, as you will see in the next definition.

Domain and Range

The **domain** of a function is the set of all allowed inputs. The **range** of a function is the set of all possible outputs.

FOR EXAMPLE, the graph of the function $f(x) = x^2 - 4$ can be used to find its domain and range.

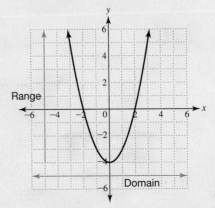

The domain is all real numbers because the graph extends all the way to both the left and the right, indicating that any *x*-value can be used in the function.

The range is all *y*-values greater than or equal to -4 because the graph starts at $y = -4$ and goes up from there.

We can confirm the domain by noticing that any number can be substituted into the function, since we can always square a number and subtract 4. The smallest result that can be obtained from the squared term is 0, which means the lowest function value is -4 after the subtraction is done. This confirms that the range is all *y*-values greater than or equal to -4.

4. Find the domain and range of each function from its graph or equation.

   a. $f(x) = \dfrac{1}{x - 3}$

The function in #4a, with the variable in the denominator, is an example of a **rational function.**

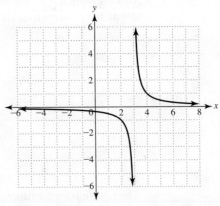

Domain: $x \neq 3$; range: $y \neq 0$

INSTRUCTOR NOTE: You can have students express the domain and range in words if you do not wish to focus on the inequality notation.

b. $f(x) = \sqrt{-x^2 + 16}$

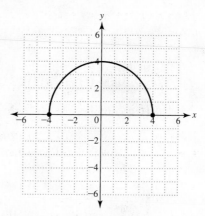

Domain: $-4 \leq x \leq 4$; range: $0 \leq y \leq 4$

Connect

INSTRUCTOR NOTE: This problem previews quadratic functions in vertex form, which will be developed in Section 4.11.

5. From the graph of the quadratic function, find the domain and range, identify the zeros, and locate the vertex. How does the vertex of the parabola help you determine the range?

$f(x) = -2(x - 3)^2 + 5$

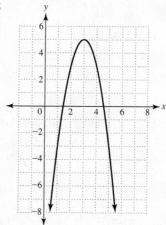

Domain: all real numbers

Range: $y \leq 5$

Zeros: $x \approx 1.5$ and $x \approx 4.5$

Vertex: $(3, 5)$

The y-coordinate of the vertex tells you the largest output value in this case.

Reflect

WRAP-UP

What's the point?

It is important that you're able to recognize a relationship as a function from its graph, and the vertical line test provides a fast way to do that. Also readily apparent from its graph are the domain and range of a function.

What did you learn?

How to apply the vertical line test

How to find the domain and range from a graph

4.10 Homework

First complete the MyMathLab homework online. Then work the two exercises to check your understanding.

1. Explain how to use the vertical line test to confirm that this graph represents a function.

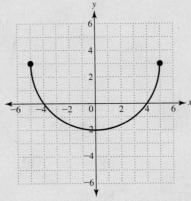

Since every vertical line intersects the graph at most once, it represents a function.

2. Find the domain and range of the function in the graph.
Domain: $-5 \leq x \leq 5$
Range: $-2 \leq y \leq 3$

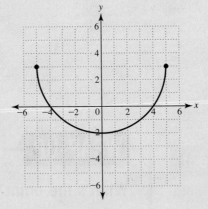

Concepts and Applications

Complete the problems to practice applying the skills and concepts learned in the section.

3. Draw the graph of a function that has the stated domain and range.

a. Domain: all real numbers; range: $y \leq 5$

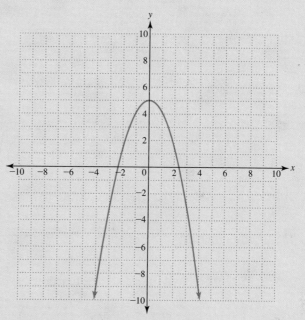

Answers will vary; one possible graph is shown.

b. Domain: $x \leq 0$; range: $y \geq 0$

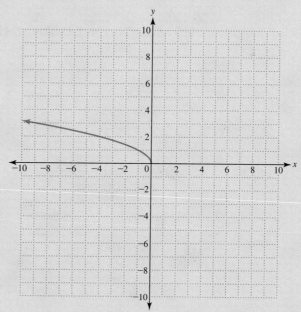

Answers will vary; one possible graph is shown.

4. In Section 4.9 you saw the function $h(t) = -16t^2 + 64t + 80$, which gives the height in feet of a projectile t seconds after it is launched. Find the vertex of its parabola, which gives the maximum height of the projectile, and use that to identify the range.

Vertex: (2, 144); range: $0 \leq y \leq 144$

5. The basic quadratic function $f(x) = x^2$ has a range of $y \geq 0$, as seen in its graph.

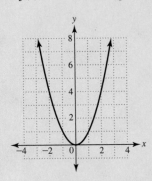

Write a quadratic function that has each of the following ranges:

a. $y \geq 2$ $f(x) = x^2 + 2$

b. $y \geq -2$ $f(x) = x^2 - 2$

c. $y \leq 0$ $f(x) = -x^2$

6. Looking Forward, Looking Back Find $f(-5)$ if $f(x) = -x^2 + 4x - 16$.

-61

4.11 An Important Point: Vertex Form of a Quadratic Function

Explore

1. Use the quadratic function $y = 4\left(x - \dfrac{1}{2}\right)^2 + 5$ to answer the questions.

   a. What is the vertex of the parabola?

   $\left(\dfrac{1}{2}, 5\right)$

SECTION OBJECTIVES:

- Identify the vertex of a quadratic function in vertex form
- Graph a quadratic function in vertex form
- Write the vertex form of a quadratic function given the vertex and a point

Remember?

The formula $x = -\dfrac{b}{2a}$ can be used to find the *x*-coordinate of the vertex of a parabola.

   b. How does the vertex relate to the equation of the quadratic function as it is given?

   The *x*-coordinate of the vertex is the opposite of the number in the squared quantity. The *y*-coordinate of the vertex is the constant term of the function.

Discover

You saw quadratic functions in standard form $y = ax^2 + bx + c$ in Section 3.16. We used that form and the vertex formula to find the vertex of a parabola. Quadratic functions can be written in other forms that make finding the vertex fast. If you know the vertex and how to use quadratic patterns, you can graph a parabola quickly without making a table of points. This section will explore these ideas and others related to quadratic functions.

As you work through this section, think about these questions:

1 Can you find the vertex of a quadratic function in standard form or vertex form?

2 Do you know if the vertex of a parabola will be a maximum or minimum?

3 Can you write a quadratic function knowing something about the parabola?

If a quadratic function is in standard form, the vertex formula $x = -\dfrac{b}{2a}$ is one way to find the vertex. If a quadratic function is in the form shown in the *Explore*, the vertex can be identified from that equation without using the vertex formula.

LOOK IT UP

Vertex Form of a Quadratic Function

The **vertex form of a quadratic function** is $y = a(x - h)^2 + k$, where $a \neq 0$. If a quadratic function is written in this form, the vertex of the parabola is (h, k).

FOR EXAMPLE, the quadratic function $y = -4(x - 6)^2 + 17$ is in vertex form. Its vertex is (6, 17).

If you need to find the vertex of a quadratic function in vertex form but the function does not have subtraction inside the parentheses and addition outside them, you can rewrite the equation so that it does.

For example, $y = 7(x + 11)^2 - \frac{1}{4}$ can be written as $y = 7(x - (-11))^2 + (-\frac{1}{4})$. Then the vertex of the parabola can be identified as $(-11, -\frac{1}{4})$. Notice that the x-coordinate of the vertex is the opposite of the number written in the parentheses when the function is in vertex form. The y-coordinate of the vertex is the constant term of the vertex form. You can use these ideas instead of rewriting the equation.

2. Find the vertex of each quadratic function.

INSTRUCTOR NOTE: Students may not recognize #2c as being in vertex form, since it doesn't have parentheses. It may help to rewrite the equation as $y = 7(x - 0)^2 + 6$.

a. $y = -2(x - 4)^2 - \dfrac{1}{8}$ $\left(4, -\dfrac{1}{8}\right)$

b. $y = 5(x + 8)^2$ $(-8, 0)$

c. $y = 7x^2 + 6$ $(0, 6)$

If an equation in vertex form is rewritten in standard form, the value of a is the same in either form. For example, $y = 2(x - 4)^2 + 2$ can be multiplied and simplified to get $y = 2x^2 - 16x + 34$. In either form, $a = 2$.

So far, we have seen the importance of the values of h and k in the vertex form of a quadratic function, but what about a? Notice that for each of the following functions the vertex is $(4, 2)$. The only difference in the equations is the value of a.

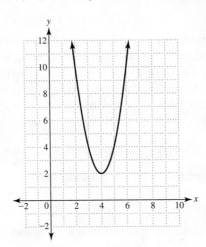

$y = 2(x - 4)^2 + 2$

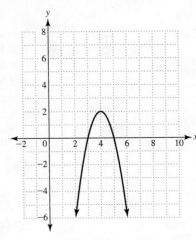

$y = -2(x - 4)^2 + 2$

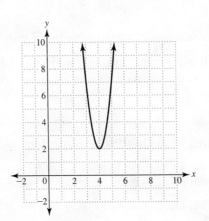

$y = 5(x - 4)^2 + 2$

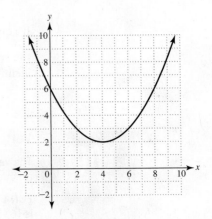

$y = \dfrac{1}{4}(x - 4)^2 + 2$

We can see the effect of the sign of *a* in the first and second graphs. In the first graph, *a* = 2 and the parabola opens up. In the second graph, the only change in the equation is the sign of *a*. Notice that *a* = −2 and the parabola opens down.

The size of *a* has an effect on the width of the parabola. We can see this in the third graph, which has a larger value of *a*, 5, than the first graph in which *a* = 2. Therefore, the third graph is skinnier than the first graph. The last graph has a value of *a* that is smaller, $\frac{1}{4}$, and the parabola is wider than the other graphs as a result.

When *a* is positive, the parabola opens up. In this case, the vertex is the lowest point on the graph, a **minimum**. The *y*-value of the vertex is the minimum value of the function.

When *a* is negative, the parabola opens down. In this case, the vertex is the highest point on the graph, a **maximum**. The *y*-value of the vertex is the maximum value of the function.

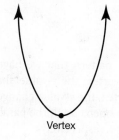

a > 0, vertex is a maximum

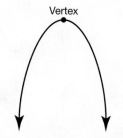

a < 0, vertex is a minimum

The larger the size of *a*, the more vertically stretched, or skinnier, the parabola is. The smaller the size of *a*, the wider the parabola is.

When we need to find the maximum or minimum value of a quadratic function, we can start by finding the vertex of the parabola. However, the vertex ordered pair itself does not always answer the question. We have to interpret it and use the part we need.

Notice in #3 that the value of *a* is negative. The value of *a* provides more information about the situation. Specifically, we know that the ball is going up, reaches a maximum height, and then falls down.

3. The function $h = -16(t - 2)^2 + 74$ gives the height, *h*, of a ball in feet after *t* seconds. How high does the ball go? When does the ball reach that height?

74 feet; 2 seconds

We can use the vertex form not only to identify the vertex, but also to create the graph of the parabola. In the graph of the function $y = (x - 2)^2 - 3$, the vertex is $(2, -3)$. The parabola opens up, since a is positive. We can make a table of ordered pairs by placing the vertex in the center and choosing x-values on either side of it. We can then evaluate the function at each value of x.

x	0	1	2	3	4
y	1	-2	-3	-2	1

Notice the quadratic pattern that we have seen before. When we go out 1 from the x-coordinate of the vertex, the y-value changes by 1. When we go out 2 from the x-coordinate of the vertex, the y-value changes by 4. In this example, the y-values increase because the parabola opens up. If the parabola opened down, they would decrease by 1 and then decrease by 4, respectively.

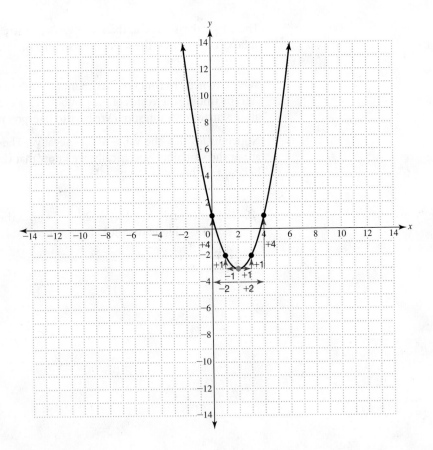

For the function $y = 4(x - 2)^2 - 3$, the vertex is again $(2, -3)$ and the parabola opens up. Since $a = 4$, each change in the y-values is multiplied by 4.

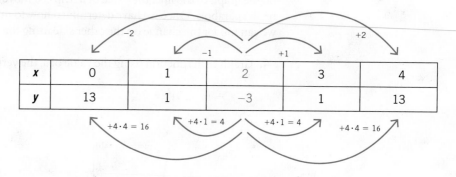

x	0	1	2	3	4
y	13	1	−3	1	13

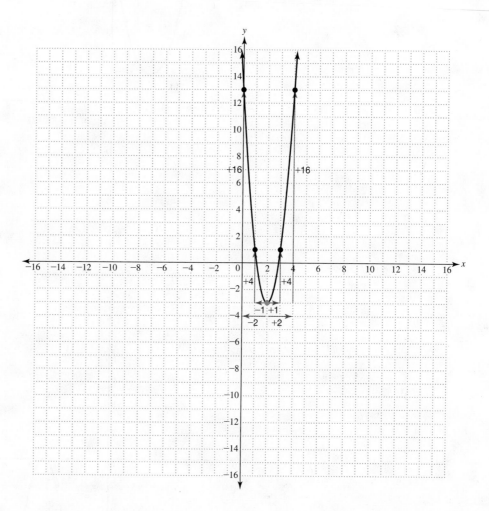

Notice that this parabola is skinnier than the previous one because its value of a, 4, is greater than the previous value of a, 1.

If we know the value of *a* and the vertex, we can graph a parabola quickly without forming a table of ordered pairs. Like the slope helps us move from the *y*-intercept to another point on the graph of the line, the value of *a* helps us move from the vertex (h, k) to other points on the parabola. Once we have determined how to move from the vertex on one side of it, we can mirror the change on the other side using the symmetry that exists in the graph.

4. Graph each function using the value of *a*, the vertex (h, k), and the quadratic pattern.

   a. $y = (x + 3)^2 + 2$

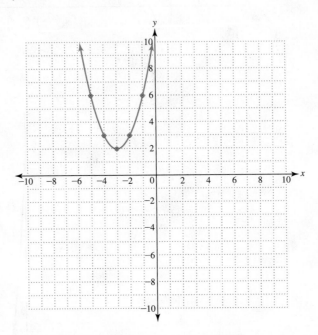

   b. $y = -2(x - 5)^2$

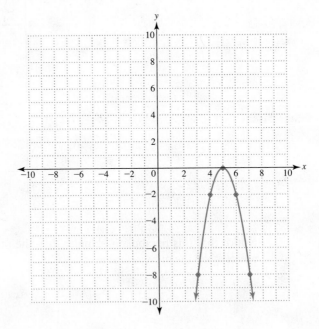

If we know the vertex and at least one other point, we can find the equation for a quadratic function.

EXAMPLE

Suppose the incomes and ages of employees in a particular profession can be related with a quadratic model. The maximum income, \$90,000, is reached at age 60. An employee who is 25 years old makes \$28,750. How much does a 30-year-old employee make?

SOLUTION

We begin with the vertex form of the quadratic equation, $y = a(x - h)^2 + k$, since we know the function is quadratic and we have the vertex. If we use I for income and A for age of an employee and the fact that income depends on age, we can rewrite the function as $I = a(A - h)^2 + k$. The vertex is (60, \$90,000), since \$90,000 is said to be the maximum income and it is reached at age 60. The function can then be written as $I = a(A - 60)^2 + 90,000$. To complete the function, we need to know the coefficient a. Use the fact that a 25-year-old employee makes \$28,750. Substitute these values into the equation and solve for a.

$$I = a(A - 60)^2 + 90,000$$
$$28,750 = a(25 - 60)^2 + 90,000$$
$$28,750 = a(-35)^2 + 90,000$$
$$28,750 = 1,225a + 90,000$$
$$-61,250 = 1,225a$$
$$-50 = a$$

The fact that a is negative supports the assumption that the vertex is a maximum. The final function is $I = -50(A - 60)^2 + 90,000$. We can then use the function to find the salary of a 30-year-old employee.

$$I = -50(A - 60)^2 + 90,000$$
$$= -50(30 - 60)^2 + 90,000$$
$$= 45,000$$

A 30-year-old employee makes \$45,000. We can see these ages and their corresponding salaries on the following graph of the quadratic equation.

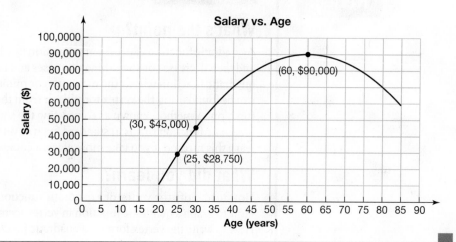

When you write the equation of a quadratic function, you can use the vertex and one other point on the graph to find the function, as we did in the example. If you have a graph of the function, you can use that procedure or use the quadratic pattern shown on the graph to determine the value of a. In practice, use the method that is simplest.

5. Use the graph to write the equation of the quadratic function.

$y = 3(x - 5)^2 - 5$

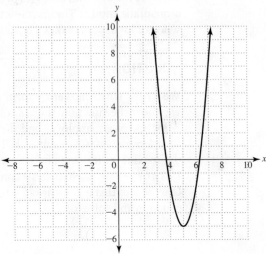

Connect

6. A fountain at a museum sprays water in a parabolic arc. Horizontally, it sprays 30 feet. Its maximum height is 10 feet. Write a function that models the height of the water stream. It might help to draw a picture that places the fountain's spout at the origin.

$y = -\dfrac{2}{45}(x - 15)^2 + 10$

Reflect

WRAP-UP

What's the point?

A quadratic function can arise in a situation in which something increases and then decreases or something decreases and then increases. The vertex of a quadratic function tells us the maximum or minimum value of the function. It can be found using the vertex formula or from the vertex form of the quadratic function. A method known as completing the square can also be used to write a quadratic function that is in standard form in vertex form instead. You will likely learn this method if you take more algebra courses.

What did you learn?

How to identify the vertex of a quadratic function in vertex form
How to graph a quadratic function in vertex form
How to write the vertex form of a quadratic function given the vertex and a point

4.11 Homework

Skills MyMathLab

First complete the MyMathLab homework online. Then work the two exercises to check your understanding.

1. Identify the vertex of the quadratic function $y = 2(x - 1)^2 + 14$. (1, 14)

2. Write the quadratic function whose maximum value is 5 when $x = 2$ and whose parabola passes through (1, 12).
$y = 7(x - 2)^2 + 5$

Concepts and Applications

Complete the problems to practice applying the skills and concepts learned in the section.

3. Find the solutions to the system by each stated method: $\begin{array}{l} y = 2x + 1 \\ y = 2x^2 - 3 \end{array}$

 a. Solve by graphing. Use the vertex form and the quadratic pattern to graph the parabola. Use the slope and y-intercept to graph the line. Confirm your answers by checking them in the equations.
 $(-1, -1)$ and $(2, 5)$

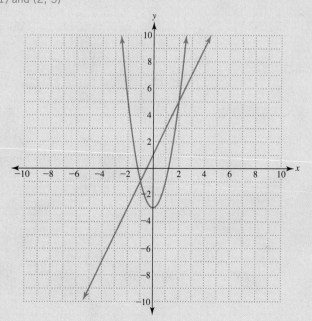

b. Solve by substitution.

(−1, −1) and (2, 5)

4. A college has an enrollment of 2,000 students who average 12 credit hours per semester. Typically 24,000 credit hours are taken each semester. The college's board of trustees suspects that every $10 increase in the tuition rate will result in 100 fewer credit hours taken. The current tuition rate is $400 per credit hour. For example, if tuition rate is increased by $30 per credit hour, which is three $10 increases, the board expects to lose 300 credit hours. The total credit hours taken will drop to 23,700.

Build a model for the revenue the college will earn based on the number of $10 increases in tuition. Let x = number of $10 increases in the tuition rate.

a. Write an expression for the tuition rate per credit hour if the school plans to raise the current tuition rate of $400 per credit hour by an unknown number of $10 increases.

Tuition rate per credit hour = $400 + 10x$

b. Write an expression for the number of credit hours taken using the fact that the college loses 100 credit hours from the semester total of 24,000 credit hours for every $10 increase in the tuition rate.

Number of credit hours taken = $24{,}000 − 100x$

c. Build a model for the revenue the college will earn using the following formula:

Revenue per semester = tuition per credit hour · number of credit hours taken

Revenue per semester = $(400 + 10x)(24{,}000 − 100x)$

d. Use the distributive property to simplify the right side of the model to obtain a quadratic function in standard form.

Revenue per semester = $−1{,}000x^2 + 200{,}000x + 9{,}600{,}000$

e. Based on the fact that the function is quadratic, how does the college's revenue change as the number of $10 tuition jumps increases?

The school can raise tuition to a point and still increase revenue. At a certain point, though, the school will lose too many credit hours and revenue will start to decline.

f. Estimate the tuition that will generate the maximum revenue according to this scenario.

The college should implement 100 $10 increases, for a total tuition per credit hour of about $1,400.

g. What should the school charge to maximize revenue?

One hundred $10 increases means an increase of $1,000 over the original tuition rate of $400/credit hour. The new tuition rate will be $1,400/credit hour.

h. What will the semester revenue per credit hour be if the college charges this amount?

$19,600,000

i. How many credit hours should the college expect at that tuition rate?

24,000 − 100(100) = 14,000

5. Let's investigate another situation in which we need to maximize a quantity. Suppose we have an 8.5″ × 11″ piece of paper from which we want to cut a rectangle with a perimeter of 32 inches and a maximum area.

a. Explore this problem numerically by completing the following table. Some of these pairs (*W* and *L*) are physically impossible with standard 8.5″ × 11″ paper.

Width (*W*)	Length (*L*)	Area = *LW*
1	15	15
2	14	28
3	13	39
4	12	48
5	11	55
6	10	60
7	9	63
8	8	64
9	7	63
10	6	60
11	5	55

b. Use the table to estimate the dimensions of the rectangle whose perimeter is 32 inches and whose area is maximized.

An 8″ by 8″ square has maximum area and a perimeter of 32 inches.

c. Use algebra to find the maximum area. Write an equation for the perimeter of a rectangle, knowing it must be 32 inches.

$2L + 2W = 32$

d. The goal is to maximize the area. Write an equation for the area of the rectangle.

$\text{Area} = LW$

e. Since there are two variables in the area equation, solve for length in the perimeter equation and substitute its value into the area equation. This reduces the number of variables in the area formula to one. Simplify the right side.

$L = 16 - W$
$\text{Area} = W(16 - W)$
$\text{Area} = -W^2 + 16W$

f. Find the dimensions that give the maximum area as well as the maximum area obtained with those dimensions.

$\text{Vertex} = (8, 64)$
Width = 8 inches; length = 8 inches; area = 64 square inches

g. What do you notice about the shape that gives a maximum area with a fixed perimeter?

The rectangle with a fixed perimeter and maximum area is a square.

h. Maximize the area of a rectangular garden that has exactly 100 feet of fencing to enclose it.

A square with a perimeter of 100 feet has dimensions of 25 feet × 25 feet. The area is 625 square feet.

6. We have seen several types of functions and studied in depth three particular ones: linear, exponential, and quadratic.

a. Complete the following table:

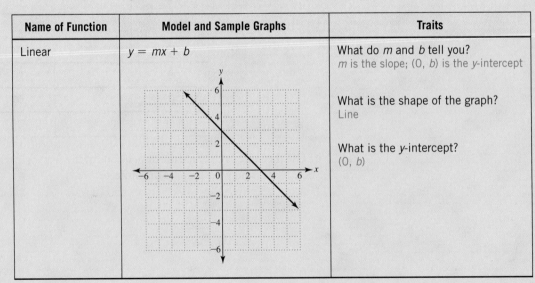

Name of Function	Model and Sample Graphs	Traits
Linear	$y = mx + b$	**What do m and b tell you?** m is the slope; $(0, b)$ is the y-intercept **What is the shape of the graph?** Line **What is the y-intercept?** $(0, b)$

Name of Function	Model and Sample Graphs	Traits
Exponential	$y = a \cdot b^x$	**What do a and b tell you?** a is the initial value; b is the growth or decay rate **What is the shape of the graph?** Curve increasing or decreasing **What is the y-intercept?** $(0, a)$
Quadratic	$y = ax^2 + bx + c$	**What do a, b, and c tell you?** a indicates whether the graph opens up or down, b and a help to find the vertex, $(0, c)$ is the y-intercept **What is the shape of the graph?** Parabola **What is the y-intercept?** $(0, c)$

b. Use the facts in part a to identify important components of each function.

i. Identify the slope and y-intercept of the linear function $y = 5x - 8$.
$m = 5$; y-intercept is $(0, -8)$

ii. Identify the slope and y-intercept of the linear function $y = 2 - 3x$.
$m = -3$; y-intercept is $(0, 2)$

iii. Identify the initial value and the growth rate of the exponential function $y = 5 \cdot 3^x$.
Initial value is 5; rate of change is 3

iv. Identify the vertex and y-intercept of the quadratic function $y = -2x^2 + 12x - 5$.
Vertex is $(3, 13)$; y-intercept is $(0, -5)$

v. Identify the vertex and y-intercept of the quadratic function $y = 2x^2 + 12x$.
Vertex is $(-3, -18)$; y-intercept is $(0, 0)$

7. For each quadratic graph, identify the vertex as an ordered pair and then state the maximum or minimum value of the quadratic function.

a.

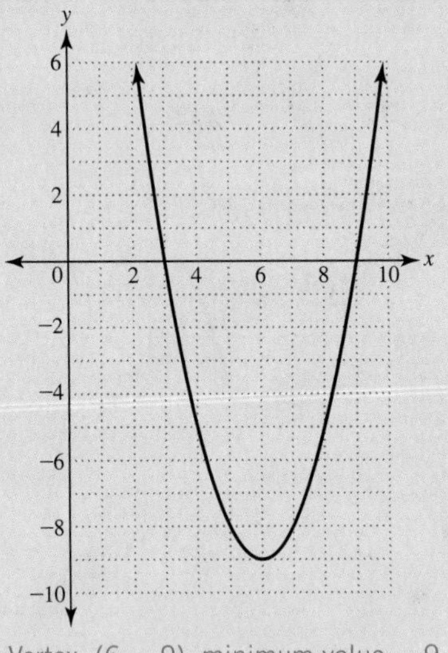

Vertex: $(6, -9)$; minimum value: -9

b.

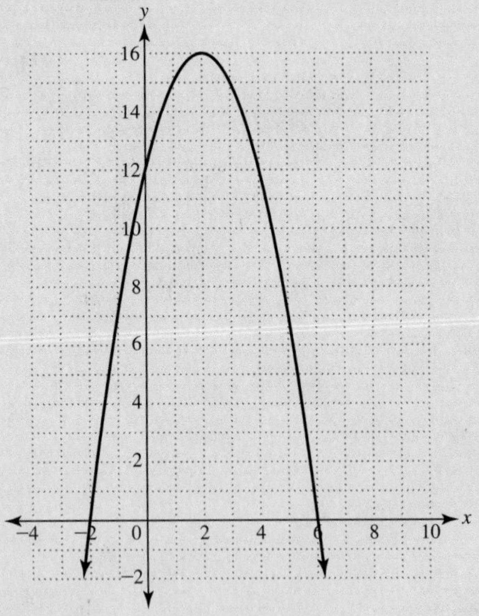

Vertex: $(2, 16)$; maximum value: 16

8. a. The graph gives the height of a baseball (in feet) as a function of the time (in seconds) since it was thrown. What is the maximum height that the baseball reaches? When does it reach this height?   16 feet; 1 second

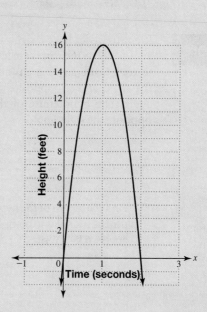

b. The graph gives the revenue (in dollars) based on the price (in dollars). What is the maximum revenue? What should the price be to maximize the revenue?

About $630; $80

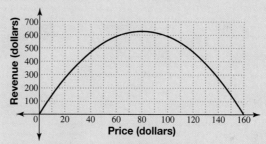

9. Looking Forward, Looking Back The hypotenuse of a right triangle is 3 meters long and one leg is 1 meter long. Find the length of the other leg.

2.83 m

4.12 A Survey of Trig: Trigonometric Functions

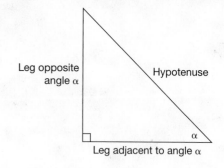

Explore

SECTION OBJECTIVES:

• Write the six trigonometric ratios for an acute angle
• Use trigonometric functions to find the measures of sides and angles of a right triangle

Right triangles have a right angle and two acute angles whose measures are less than 90°. Relative to each acute angle is an "adjacent" leg next to the angle, and an "opposite" leg across from the angle. The hypotenuse of the right triangle is opposite the right angle.

In trigonometry, Greek letters like α (alpha), β (beta), and θ (theta) are often used to represent angles.

INSTRUCTOR NOTE: The figures in #1 are to scale. Consider having students measure in addition to calculating the length of the hypotenuse.

1. Find the length of the hypotenuse of each triangle, rounded to one decimal place. Then use the triangles to complete the table. The measure of angle α is the same in all three triangles: 63.4°. Write each ratio as a fraction and then as a decimal rounded to two places.

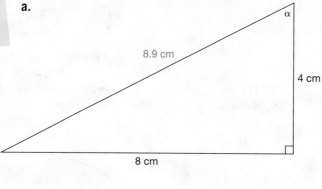

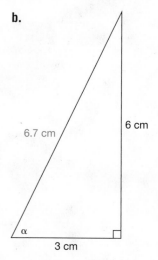

Triangle	$\dfrac{\text{length of leg opposite } \alpha}{\text{length of hypotenuse}}$	$\dfrac{\text{length of leg adjacent to } \alpha}{\text{length of hypotenuse}}$	$\dfrac{\text{length of leg opposite } \alpha}{\text{length of leg adjacent to } \alpha}$
a	$\dfrac{8}{8.9} \approx 0.90$	$\dfrac{4}{8.9} \approx 0.45$	$\dfrac{8}{4} = 2$
b	$\dfrac{6}{6.7} \approx 0.90$	$\dfrac{3}{6.7} \approx 0.45$	$\dfrac{6}{3} = 2$

2. What do you notice about the ratios in the table?

The ratios are the same for both triangles.

Discover

In many real-life applications, triangles can be helpful in solving problems. **Trigonometry**, which means "triangle measure," is an area of mathematics in which triangle relationships are studied and used to solve problems. However, we may only be able to obtain a few measures for those triangles. In this section, you will learn some basic right-triangle trigonometric concepts that will allow you to find the measures of sides and angles given some information about a right triangle.

As you work through this section, think about these questions:

1 Can you write the six trigonometric ratios for an angle in a right triangle?

2 Do you know how to find the measures of the sides and angles in a right triangle?

In the *Explore*, each right triangle had the same acute angles. Since the ratios formed by two particular sides were the same in each triangle, regardless of the size of the triangle, the triangles were similar. The ratios found and their reciprocal ratios have names and are known as the six trigonometric ratios.

LOOK IT UP

INSTRUCTOR NOTE: Consider having students name the ratios in the columns of the table in the *Explore*. They should see that $\sin \alpha \approx 0.90$, $\cos \alpha \approx 0.45$, and $\tan \alpha \approx 2$. These values are true regardless of the size of the triangle as long as it is a right triangle and $\alpha = 63.4$ degrees.

Ratios should be simplified if possible before a problem is completed. For example, if $\sin \beta = \frac{6}{18}$, the ratio should be reduced and the result should be $\sin \beta = \frac{1}{3}$.

Trigonometric Functions

Trigonometric functions are ratios that compare two sides of a right triangle relative to an acute angle of the triangle. They are functions because for each input, an angle, there is only one output, a number that represents the ratio of two of the sides of a right triangle with that acute angle.

The three basic trigonometric functions are the **sine**, **cosine**, and **tangent**. Their abbreviations and ratios are given here. Angle α is an acute angle in a right triangle. For ease of reading, the quantities in the ratios are identified by their locations. For example, the length of the hypotenuse has been shortened to "hypotenuse."

$$\sin \alpha = \frac{\text{side opposite } \alpha}{\text{hypotenuse}} \qquad \cos \alpha = \frac{\text{side adjacent to } \alpha}{\text{hypotenuse}} \qquad \tan \alpha = \frac{\text{side opposite } \alpha}{\text{side adjacent to } \alpha}$$

Each basic trigonometric function has a reciprocal function. The **cosecant** is the reciprocal of the sine, the **secant** is the reciprocal of the cosine, and the **cotangent** is the reciprocal of the tangent. Their abbreviations and ratios are as follows.

$$\csc \alpha = \frac{\text{hypotenuse}}{\text{side opposite } \alpha} \qquad \sec \alpha = \frac{\text{hypotenuse}}{\text{side adjacent to } \alpha} \qquad \cot \alpha = \frac{\text{side adjacent to } \alpha}{\text{side opposite } \alpha}$$

FOR EXAMPLE, to find the six trigonometric ratios for angle β in the triangle, we consider the sides from the perspective of that angle. The length of the side opposite angle β is 3, and the length of the side adjacent is 4. The length of the hypotenuse is 5.

The trigonometric ratios are

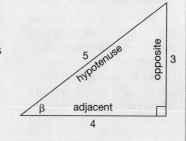

$$\sin \beta = \frac{3}{5} \qquad \cos \beta = \frac{4}{5} \qquad \tan \beta = \frac{3}{4}$$

$$\csc \beta = \frac{5}{3} \qquad \sec \beta = \frac{5}{4} \qquad \cot \beta = \frac{4}{3}$$

A helpful way to remember the three basic trigonometric ratios is with the memory aid SOH CAH TOA.

$$\sin = \frac{O}{H} = \frac{\text{opposite}}{\text{hypotenuse}} \qquad \cos = \frac{A}{H} = \frac{\text{adjacent}}{\text{hypotenuse}} \qquad \tan = \frac{O}{A} = \frac{\text{opposite}}{\text{adjacent}}$$

Opposite and adjacent refer to the lengths of the *legs* of the triangle, not the hypotenuse. Also, opposite and adjacent are relative, which means that they change depending on which acute angle you're considering.

Remember that sine, cosine, and tangent are just names for ratios of the sides in a right triangle. Like function notation, the trigonometric ratio names provide a shorter way of describing a process. Instead of saying, "Find the ratio of the length of the side opposite 55° and the length of the hypotenuse in a right triangle with an acute angle of 55°," we can say, "Find sin 55°."

3. A triangle has a hypotenuse 50 cm long and a leg 48 cm long. Angle θ is opposite the 48-cm-long leg. Find the length of the remaining leg and the six trigonometric ratios for θ. Draw a picture to begin.

adjacent leg = 14 cm

$$\sin \theta = \frac{24}{25} \qquad \cos \theta = \frac{7}{25} \qquad \tan \theta = \frac{24}{7}$$

$$\csc \theta = \frac{25}{24} \qquad \sec \theta = \frac{25}{7} \qquad \cot \theta = \frac{7}{24}$$

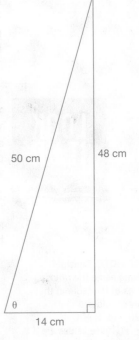

50 cm   48 cm

θ

14 cm

Tech TIP

Check that your calculator is in degree mode.

Calculators usually do not have a command for the reciprocal trigonometric functions since you can use the reciprocal button. For example, to find csc 13°, find sin 13° and take the reciprocal of the result.

The trigonometric ratios are the same for any right triangle with a particular acute angle. These values can be found by using any scientific calculator. The lengths of sides or the measure of an angle can provide a trigonometric ratio. It is not necessary to know the acute angle measure if we know the side lengths or to know the sides lengths if we know the measure of the angle.

4. Use the lengths of the sides of the triangle to find the following. Then check the results using a calculator. Your answers using the side lengths should be similar to the calculator results. They will not be exact because the side lengths are rounded to one decimal place.

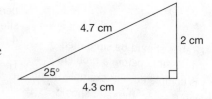

4.7 cm

2 cm

25°

4.3 cm

a. $\tan 25° = \frac{2}{4.3} \approx 0.47$

b. $\sin 65° = \frac{4.3}{4.7} \approx 0.91$

5. Find each value using a calculator. Round each result to the nearest hundredth.

    a. csc 82°　　　　　　　**b.** cot 19°　　　　　　　**c.** sec 37°

      1.01　　　　　　　　　　　2.90　　　　　　　　　　　1.25

If you know the lengths of two sides of a right triangle, you can use the Pythagorean theorem to find the length of the remaining side. However, if you know the length of only one side, that's not enough information to use the Pythagorean theorem. In that case, if we know one of the acute angles of the right triangle, we can use a trigonometric ratio to write an equation to solve for a side length. At that point, the lengths of two sides will be known and the Pythagorean theorem can be used to find the remaining side.

EXAMPLE

Find a and b in the triangle.

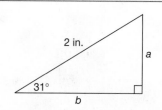

SOLUTION

> When you are writing a trigonometric equation relating two sides and an acute angle, you need to know 2 out of 3 of these quantities.

To find a, we can use a trigonometric ratio and write an equation that relates *two sides* of the right triangle to *one acute angle*. One side length must be known in order to solve for the other side length. In this triangle, we know that the hypotenuse is 2 inches and one acute angle is 31°. To find a, notice that it is opposite the known angle. We can use the sine function to relate these three quantities.

$$\sin 31° = \frac{\text{opposite}}{\text{hypotenuse}}$$

$$\sin 31° = \frac{a}{2}$$

$$2 \cdot \sin 31° = \frac{a}{\cancel{2}} \cdot \cancel{2} \qquad \text{Multiply both sides by 2 to isolate } a.$$

$$2 \cdot \sin 31° = a$$

$$1.03 \text{ in.} \approx a$$

To find b, we can use another trigonometric equation relating it to the given acute angle, 31°, and a known side, such as the hypotenuse. In this case, we need to use the cosine function, since b is adjacent to the angle we know.

$$\cos 31° = \frac{\text{adjacent}}{\text{hypotenuse}}$$

$$\cos 31° = \frac{b}{2}$$

$$2 \cdot \cos 31° = \frac{b}{\cancel{2}} \cdot \cancel{2} \qquad \text{Multiply both sides by 2 to isolate } b.$$

$$2 \cdot \cos 31° = b$$

$$1.71 \approx b$$

Instead of using a trigonometric equation to find b, we could have used the Pythagorean theorem. The result may be slightly different from the one obtained by using trigonometry due to the use of rounded values in the Pythagorean equation.

$$a^2 + b^2 = hyp^2$$
$$(1.03)^2 + b^2 = 2^2$$
$$1.06 + b^2 = 4$$
$$b^2 = 2.94$$
$$b = \sqrt{2.94}$$
$$b \approx 1.71 \text{ in.}$$

In this case, the value obtained from the Pythagorean theorem and the value obtained by using trigonometry are the same when rounded to two decimal places.

Trigonometric functions can also be used to solve problems. To do so, we need to form right triangles between three locations or objects in a situation.

6. A surveyor standing 150 feet from a building finds an angle of elevation to the top of the building of 72.5°. Find the height of the building assuming the line of sight is at the eye level of the surveyor, 6 feet off the ground. Note that the figure is not drawn to scale.

   481.74 ft

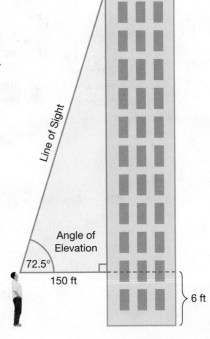

If you know the lengths of at least two sides of a right triangle, you can write a trigonometric equation and find the measure of an acute angle. The process involves using an inverse trigonometric function and your calculator to find the measure of the angle. An inverse trigonometric function produces an angle for a given trigonometric ratio, which is the opposite of what a trigonometric function does. For example, if $\sin \beta = \dfrac{3}{5}$, then

$$\beta = \sin^{-1}\left(\dfrac{3}{5}\right) \approx 36.87°.$$

Tech TIP

To use your calculator to find an angle with an inverse trigonometric function, press 2nd, then the basic trigonometric function (sin, cos, or tan), then enter the value, and press Enter.

7. A right triangle has side lengths of 6 ft, 14.4 ft, and 15.6 ft. Find the measure of the smallest acute angle, which is opposite the shortest side.

22.62°

Connect

Finding the missing lengths of the sides of a triangle and the unknown angle measures given some information about a triangle is known as **solving a triangle**.

INSTRUCTOR NOTE: Consider showing students more than one trigonometric equation to solve for *a* or *b*.

8. Solve each right triangle.

a.

5 cm, β, a, 17°, b

$a \approx 1.46$ cm; $b \approx 4.78$ cm; $\beta = 73°$

b.

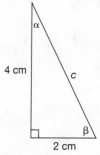

α, 4 cm, c, β, 2 cm

$c \approx 4.47$ cm; $\alpha \approx 26.57°$; $\beta \approx 63.43°$

INSTRUCTOR NOTE: Consider discussing the solution to the focus problem. You might want to show the sample solution, available in MyMathLab, or a correct solution from a student or student group in the class. See the Instructor Guide for more debriefing ideas.

Reflect

Focus Problem Check-in

Do you have a complete solution to the focus problem? If your instructor has provided you with a rubric, use it to grade your solution. Are there areas you can improve?

WRAP-UP

What's the point?

In a right triangle, a trigonometric function can be used to relate the sides and angles in the triangle to find unknown measurements for the triangle. Trigonometry is used often in technical fields like surveying, engineering, and construction.

What did you learn?

How to write the six trigonometric ratios for an acute angle
How to use trigonometric functions to find the measures of sides and angles of a right triangle

4.12 Homework

Skills MyMathLab

First complete the MyMathLab homework online. Then work the two exercises to check your understanding.

1. Use a calculator to find the values. Round to the nearest hundredth.

    a. $\cos 40°$ 0.77

    b. $\sec 40°$ 1.31

2. Using the triangle, find the six trigonometric functions for angle α.

$\sin \alpha = \dfrac{5}{13}$ $\cos \alpha = \dfrac{12}{13}$ $\tan \alpha = \dfrac{5}{12}$

$\csc \alpha = \dfrac{13}{5}$ $\sec \alpha = \dfrac{13}{12}$ $\cot \alpha = \dfrac{12}{5}$

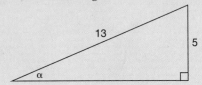

Concepts and Applications

Complete the problems to practice applying the skills and concepts learned in the section.

3. The following right triangles, which have been studied for thousands of years, can be constructed using a straightedge and compass exactly without the use of a protractor.

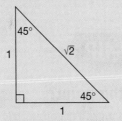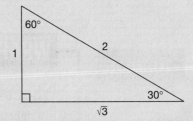

    a. Complete this table using the *sides* of the triangles. Write each ratio as a fraction and then as a decimal.

Angle (α)	$\sin \alpha$	$\cos \alpha$	$\tan \alpha$
30°	$\dfrac{1}{2} = 0.5$	$\dfrac{\sqrt{3}}{2} \approx 0.87$	$\dfrac{1}{\sqrt{3}} \approx 0.58$
45°	$\dfrac{1}{\sqrt{2}} \approx 0.71$	$\dfrac{1}{\sqrt{2}} \approx 0.71$	$\dfrac{1}{1} = 1$
60°	$\dfrac{\sqrt{3}}{2} \approx 0.87$	$\dfrac{1}{2} = 0.5$	$\dfrac{\sqrt{3}}{1} \approx 1.73$

b. Complete this table using the *angles* of the triangles and your calculator. Your results should be the same as in the table in part a.

Angle (α)	sin α	cos α	tan α
30°	0.5	0.87	0.58
45°	0.71	0.71	1
60°	0.87	0.5	1.73

c. Does the size of the triangle affect the results? Use the following triangle's sides to find the sine, cosine, and tangent of 45°. Compare your results to the second row of the table in part a. Are your results the same?

No; yes

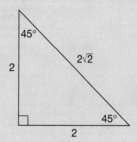

4. Suppose $\sin \alpha = \dfrac{3}{4}$. Find the five remaining trigonometric values. Round each answer to the nearest hundredth.

$\cos \alpha \approx 0.66$
$\tan \alpha \approx 1.13$
$\csc \alpha \approx 1.33$
$\sec \alpha \approx 1.51$
$\cot \alpha \approx 0.88$

5. Lucia is writing a trigonometric equation to find the measure of angle *x*. Find the mistake in her work.

$$\sin x = \frac{15}{51} = 0.2941$$

The hypotenuse is not 51; it is 39.

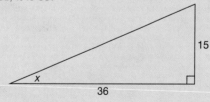

6. Solve each right triangle that is described. If it is not possible, explain why.

a. The hypotenuse is 16 mm, the long leg is 11 mm, and the angle they create is 42°.

Short leg is approximately 11.62 mm; remaining angle is 48°.

b. The side lengths are 5.7 cm, 7.6 cm, and 9.5 cm. Before proceeding, confirm it is a right triangle.

It is a right triangle; the angles are 36.87°, 53.13°, and 90°.

c. The right triangle has acute angles measuring 28° and 62°.

There is not enough information to determine the side lengths.

7. Jeff is building a pole barn with a $\frac{3}{12}$ roof pitch. He is interested in knowing the angle that the tie beam makes with the rafter, α. Find α to the nearest degree.

14°

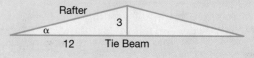

8. Looking Forward, Looking Back Write the equation of a parabola that has a vertex at $(6, -9)$ and passes through $(5, -16)$.

$y = -7(x - 6)^2 - 9$

Cycle 4 Study Sheet

This page is an overview of the cycle, providing a quick snapshot of the material covered to help you as you begin studying for a test. *However, it is not comprehensive.* The *Cycle Wrap-Up* tasks that follow will help you review the cycle in more detail.

NOTATION AND NEGATIVES

Standard notation: 15,000
Scientific notation: 1.5×10^4
Engineering notation: 15×10^3

Number	Meaning of negative
Negative exponent	Reciprocal
Negative exponent in scientific notation	Number is smaller than 1
Negative *z*-score	Data value is smaller than the mean
Negative *a* in a quadratic function	Parabola opens down; vertex is a maximum

VARIATION

y varies directly with *x*.

Table   **Equation**   **Graph**

$y = 4x$

x	y
−1	−4
1	4
2	8

$\dfrac{y}{x} = 4$

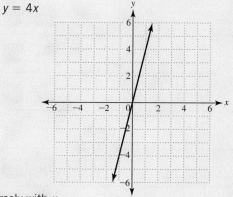

y varies inversely with *x*.

Table   **Equation**   **Graph**

$y = \dfrac{4}{x}$

x	y
−1	−4
1	4
2	2

$xy = 4$

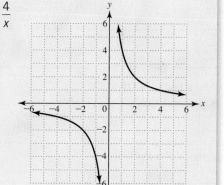

QUADRATIC FUNCTIONS

For the quadratic function $f(x) = 3(x + 2)^2 + 4$:

$a = 3$, so the parabola opens up.

The vertex is $(-2, 4)$ and is the minimum for the function.

x	f(x)
−4	16
−3	7
−2	**4**
−1	7
0	16

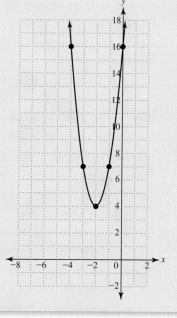

Notice that the graph passes the vertical line test.

The domain is all real numbers.

The range is $y \geq 4$.

DIMENSIONAL ANALYSIS

To convert 10 inches to meters:

$$10 \text{ in.} \cdot \frac{2.54 \text{ cm}}{1 \text{ in.}} \cdot \frac{1 \text{ m}}{100 \text{ cm}} = 0.254 \text{ m}$$

To convert 10 square inches to square meters:

$$10 \text{ in.}^2 \cdot \frac{2.54^2 \text{ cm}^2}{1^2 \text{ in.}^2} \cdot \frac{1^2 \text{ m}^2}{100^2 \text{ cm}^2} \approx 0.006 \text{ m}^2$$

To convert 10 cubic inches to cubic meters:

$$10 \text{ in.}^3 \cdot \frac{2.54^3 \text{ cm}^3}{1^3 \text{ in.}^3} \cdot \frac{1^3 \text{ m}^3}{100^3 \text{ cm}^3} \approx 1.639 \times 10^{-4} \text{ m}^3$$

SELF-ASSESSMENT: **REVIEW**

For each objective, use the boxes provided to indicate your current level of expertise. If you cannot indicate a high level of expertise for a skill or concept, continue working on it and seek help if necessary.

INSTRUCTOR NOTE: Consider discussing this page before a test. The objectives are the same ones that are provided at the beginning of the cycle, giving students two points of comparison on their learning. Completing the page can help students see where they need to focus their studying for a test.

SKILL or CONCEPT

Low   High

1. Convert units using dimensional analysis. *(4.2)*

2. Convert numbers between scientific and standard notation. *(4.3)*

3. Convert numbers into and out of engineering notation. *(4.3)*

4. Use exponent rules to simplify expressions that have negative exponents. *(4.4)*

5. Find the standard deviation of a data set. *(4.5)*

6. Interpret the standard deviation of a data set. *(4.5)*

7. Interpret logarithmic scales. *(4.6)*

8. Identify direct variation from a graph, table, or equation. *(4.7)*

9. Solve direct variation problems. *(4.7)*

10. Identify inverse variation from a table. *(4.8)*

11. Solve inverse variation problems. *(4.8)*

12. Use function notation. *(4.9)*

13. Find a function input or output given the other. *(4.9)*

14. Apply the vertical line test. *(4.10)*

15. Find the domain and range from a graph. *(4.10)*

16. Identify the vertex of a quadratic function in vertex form. *(4.11)*

17. Graph a quadratic function in vertex form. *(4.11)*

18. Write the vertex form of a quadratic function given the vertex and a point. *(4.11)*

19. Write the six trigonometric ratios for an acute angle. *(4.12)*

20. Use trigonometric functions to find the measures of sides and angles of a right triangle. *(4.12)*

CYCLE **FOUR**

WRAP-UP

> Progress is impossible without change, and those who cannot change their minds cannot change anything.
>
> —George Bernard Shaw

INSTRUCTOR NOTE: Discuss the steps included in the *Cycle Wrap-Up* in detail with students so they understand the actions that lead to success on a test.

The *Cycle Wrap-Up* will help you bring together the skills and concepts you have learned in the cycle so that you can apply them successfully on a test. Since studying is an active process, the *Wrap-Up* will provide you with activities to improve your understanding and your ability to show it in a test environment.

STEP 1

GOAL Revisit the cycle content

Action Skim your cycle notes

Action Read the Cycle Study Sheet

Action Complete the Self-Assessment: Review

> Look over your tests from earlier cycles. What types of mistakes did you make? How can you avoid making them on this test? What types of questions does your instructor tend to ask? Are you prepared for all of them? If you have any lingering questions, get help immediately from your instructor or someone in your class.

STEP 2

GOAL Practice cycle skills

Action Complete the MyMathLab assignment

Action Address your areas of difficulty

MyMathLab

Complete the MyMathLab assignment. Reference your notes if you are having trouble with a skill. For more practice, you can rework Skills problems from your homework. You can also access your MyMathLab homework problems, even if the assignment is past due, by using the Review mode available through the Gradebook.

STEP 3

GOAL Review and learn cycle vocabulary

Action Use your notes to complete the Vocabulary Check

Action Memorize the answers

Vocabulary Check

Cycle 4 Word Bank

compound variation	function	range	vary directly
constant of variation	joint variation	reciprocal	vary inversely
dimensional analysis	minimum	scientific notation	vertical line test
domain	origin	standard deviation	z-score
engineering notation	order of magnitude	trigonometry	

Choose a word from the word bank to complete each statement.

1. _____Dimensional analysis_____ is an organized process of converting units by using conversion factors.

2. _____Scientific notation_____ provides a shortcut for writing very large and very small numbers.

3. _____Engineering notation_____, like scientific notation, is a system that uses powers of ten to write large and small numbers, using exponents that are always multiples of 3.

4. A negative exponent indicates that you should take the ___reciprocal___ of that factor.

5. The square root of the average squared deviation from the mean is known as the _____standard deviation_____.

6. A(n) _____z-score_____ is calculated by subtracting the mean from a data value and then dividing the result by the standard deviation.

7. If a number is rounded to the nearest power of ten, the result is known as a(n) _____order of magnitude_____.

8. The constant k in a variation equation is called the _____constant of variation_____.

9. When the quotient of two variables is equal to a constant, each variable is said to _____vary directly_____ with the other.

10. Lines that represent direct variation always go through the _____origin_____.

11. A special type of direct variation is _____joint variation_____.

12. When the product of two variables is equal to a constant, each variable is said to _____vary inversely_____ with the other.

13. _____Compound variation_____ results when a quantity varies directly or inversely with more than one other quantity.

14. The notation $f(x)$ refers to the y-value that is produced when an x-value is substituted into a(n) _____function_____ f.

15. The _____domain_____ of a function is the set of all allowed inputs.

The _____range_____ of a function is the set of all possible outputs.

16. The _____vertical line test_____ is a fast way to determine if a graph represents a function.

17. When a quadratic function opens up, the vertex represents a(n) _____minimum_____ value.

18. The area of mathematics in which the relationships in triangles are studied is known as _____trigonometry_____.

STEP 4

GOAL Practice cycle applications

Action Complete the Concepts and Applications Review

Action Address your areas of difficulty

Concepts and Applications Review

Complete the problems to practice applying the skills and concepts learned in the cycle.

1. a. Balance this chemical reaction between methane and oxygen.

$$CH_4 + O_2 \rightarrow CO_2 + H_2O$$
$$CH_4 + 2O_2 \rightarrow CO_2 + 2H_2O$$

A mole is a unit commonly used in chemistry and refers to a certain number of items. It is similar to the unit *dozen*, which refers to 12 of something. The coefficients in a balanced chemical reaction tell you the ratio of the moles of the substances. In this reaction, for example, for every 1 mole of CH_4, there are 2 moles of O_2. This relationship can also be given as a fraction: $\dfrac{1 \text{ mole } CH_4}{2 \text{ moles } O_2}$.

b. Use dimensional analysis to convert $32\,g\ O_2$ to grams of H_2O.

Use the following facts and the balanced equation in part a: One mole of O_2 weighs 32 grams. One mole of H_2O weighs 18 grams.

$$32g\ O_2 \cdot \frac{1 \text{ mole } O_2}{32g\ O_2} \cdot \frac{1 \text{ mole } H_2O}{1 \text{ mole } O_2} \cdot \frac{18g\ H_2O}{1 \text{ mole } H_2O} = 18g\ H_2O$$

2. Earth is 149,600,000,000 meters from the sun. How far is that in miles? How long would it take to get there if you were traveling at 100 mph? Round all values to the nearest whole number.

92,957,130 miles

929,571 hours or about 38,732 days

3. Airflow is one of the best ways to judge the size of a cave and estimate its volume. Jewel Cave in South Dakota has been estimated to have a volume of 4 billion cubic feet. Yet the explored portions of the cave account for only about 120 million cubic feet of the volume.

Data from www.nps.gov

a. Write the total estimated volume and the explored volume in scientific notation.

Estimated volume $= 4 \times 10^9$ cubic feet

Explored volume $= 1.2 \times 10^8$ cubic feet

b. What percent of the cave volume has been discovered or explored? Use the numbers in scientific notation to make the calculation easier. What percent of the cave volume has not yet been discovered?

3%; 97%

c. Jewel Cave is one of the longest caves in the world, with 135 miles of paths surveyed. If the ratio of surveyed paths to total length is the same as the ratio of surveyed volume to total volume, find the total expected length of Jewel Cave.

135 = 0.03 (total length)

Total length = 4,500 miles

4. Suppose your math teacher scales your homework grade to 80 points from 200 points for the whole semester. You have the following grades on twenty 10-point homework assignments:

8	6	5	5	8	9	0	10	7	7
8	8	0	0	3	7	7	8	10	10

a. Find your mean score on these assignments.   6.3

b. Use your result from part a to find your mean percent on these assignments.

63%

c. How many points will you receive out of 80 points for the homework category?
63% of 80 = 50.4 points

d. Make a table on a separate sheet of paper and calculate the standard deviation for your homework scores. Round to two decimal places.   3.15

e. What score did you get most often? What is this number called?   8; the mode

f. What is your median score? Provide an interpretation of this number.
Your median score was 7. This is the middle score when the scores are arranged in order. About half the time you scored a 7 or higher, and about half the time you scored a 7 or lower.

5. IQs have a mean of 100 points and a standard deviation of 15 points. If a person has an IQ that is 2 standard deviations above the mean, what is her IQ?   130 points

6. It has been estimated that approximately 40 to 50 million people were killed in World War II, compared to 2 to 6 million people killed in the Vietnam War. By how many orders of magnitude do these numbers differ?   1 order of magnitude

7. The carnival game known as a strength tester or strongman game involves using a large hammer or mallet to strike a lever and ring the bell at the top of the tower. The force needed to ring the bell is the product of the mass of the mallet used and the acceleration given the mallet by the person swinging it in other words,

$$\text{Force} = \text{mass} \times \text{acceleration}$$

a. If a constant force is needed to ring the bell, how are mass and acceleration related?   They vary inversely.

b. If someone is given a mallet twice the size and mass of a typical mallet, what happens to the acceleration needed to ring the bell?
The acceleration is cut in half.

c. If someone is given a mallet one-third the size and mass of a typical mallet, what happens to the acceleration needed to ring the bell?
The acceleration is three times greater.

8. Write a function, f, that describes direct variation in which $f(3) = 24$ and $f(-6) = -48$.

$f(x) = 8x$

9. Use the function machine to find the following:

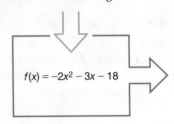

$f(x) = -2x^2 - 3x - 18$

a. $f\left(\dfrac{3}{2}\right)$ -27

b. x when $f(x) = -18$ $0, -\dfrac{3}{2}$

10. a. If a quadratic function opens downward and has a vertex of $(2, -10)$, does the graph have any x-intercepts? Draw a picture to help you decide and then explain your answer.

No; if the parabola starts below the x-axis and opens downward, then there are no x-intercepts.

b. If a quadratic functions opens upward and has a vertex of $(2, -10)$, how many x-intercepts must the graph have? Draw a picture to help you decide and then explain your answer.

Two; if the parabola starts below the x-axis and opens upward, then there are two x-intercepts.

11. a. If a quadratic function has a positive coefficient on the squared term and its vertex is (p, q), how small can the function values get? How large?

The function values cannot get smaller than q, the minimum value. The values can get infinitely large.

b. If a quadratic function has a negative coefficient on the squared term and its vertex is (p, q), how big can the function values get? How small?

The function values cannot get larger than q, the maximum value. The values can get infinitely negative.

12. a. If $\cos \alpha = 0.93$, $\sin \theta = 0.26$, and $\tan \beta = 0.84$, find $\alpha + \beta + \theta$.

77°

b. If $\tan x = \dfrac{7}{24}$, what is $\sin x$? Begin by labeling the triangle's sides with the given information. Then use the triangle's sides to answer the question.

$\dfrac{7}{25}$

STEP 5

GOAL Simulate a test environment

Action Construct a mock test and take it while being timed

Creating a Mock Test

To pass a math test, you must be able to recall vocabulary, skills, and concepts and apply them to new problems in a fixed amount of time.

Up to this point, you have worked on all the components of the cycle, but they were separated and identified by their topic or type. Although helpful, this is not a test situation. You need to create a test situation to help overcome anxiety and determine where problem areas still exist. Think about a basketball player. She must practice her skills, but until she engages in a practice game, she doesn't really have a feel for a game situation.

The key is to practice problems without assistance under time pressure.

Use the following steps to create a mock test.

- Pick a sample of 5 to 10 problems from your notes and homework and write them on note cards. Choose both skills and applications. On the back of the note card, write the page number where the problem was found.
- Shuffle the cards.
- Set a timer for 20 minutes.
- Work as many problems as you can without outside help.
- Check your work by consulting the page number on the card.

This test will show you where your weaknesses are and what needs more work. It will also give you a feel for your pace. If you completed only a few problems, you are not ready for a test.

Continue to work on the areas that need improvement.

At this point in the course, you have probably taken multiple tests. What test preparation techniques have worked for you? Are there any methods you have seen to not be as effective for you?

Index

Page numbers for definitions are listed in bold.

A

Absolute value, **41**, 468
Achievement, 73
Acute angles, 559
Addition
    associative property of, 199
    commutative property and, 198
    in distributive property, 203–204
    of like terms, 186
    in order of operations, 191
    of percents, 106
    of polynomials, 184–189
Agility, 61
Algebra, **110**
    notation in, 186
Algebraic methods, 231–232, 236
    for break-even point, 333
    for linear relationships, 333
    numeric methods compared with,
      336
    for systems of linear equations,
      420–421
Algebraic terminology, 110–116
    binomials, **113**
    coefficients, **110**
    constant, **110**
    constant term, **110**
    equations, **110**
    expressions, **110**
    factors, **110**
    monomials, **113**
    polynomials, **113**
    term, **110**
    trinomials, **113**
Angles
    acute, 559
    of right triangles, 556
Area, 137–146, **138**. *see also* Surface area
    of circle, 177
    of cylinder, 177
    Excel for, 141
    of parallelogram, 176
    perimeter compared with, 138
    of rectangles, 176
    of trapezoids, 176, 187
    of triangles, 138–139, 523
Arithmetic sequence, **119**, 120–121
    terms of, 127
Associative property, **199**
    of addition, 199
    of multiplication, 199

Attitude, 43
Averages, **56**. *see also* Means
    grades, 58–59
    in Pareto charts, 68
Axes, **18**
    labels, 21

B

Balances
    of equations, 293
    in equations with two operations
      per side, 244–245
    in linear equations, 242
    in one-step equations, 242–243
Balancing points, 58
Bar graphs, **67**
    pie graphs compared with, 69
    vertical, 69
Base, **113**, 178
Best-fit line, **302**
Binomials, **113**
    factors, 398
    multiplication of, 397, 400
    squared, 402
Box method
    multiplying polynomials, 205
    factoring quadratic expressions,
      400–401
Break-even point, **333**
    algebraic approach for, 333

C

Calculators. *see also* Graphing calculator
    degree mode, 558
    dimensional analysis on, 459
    fraction button on, 207
    order of operations in, 191
    scientific, 181
    trigonometric functions on, 558
Cartesian coordinate systems, **18**
    axes, **18**
    ordered pairs, **18**, 19
    origin, **18**
    quadrants, **18**
    x-axis, **18**
    x-coordinate, **18**
    y-axis, **18**
    y-coordinate, **18**
Change
    exponential, 124–136, 364, 367
    linear, 124–136

    percent, 99–109
    rate of, 312, 318, 353–354, 364–365
Circle
    area of, 177
    circumference of, 158, 223
Circumference, 158, 223
Coefficients, **110**
Common factors, 476
Communication in groups, 17
Commutative property, **198**
    addition and, 198
    division and, 198
    multiplication and, 198
    in rewriting expressions, 197–198
Comparison problems, 331
    solving, 334–336
Compound inequalities, 493
Compound variation, 522
Concentrations
    decreases in, 499
    increases in, 499
    logarithmic scales, 499–500
Cone, 177, 282
Conjectures, **117**
    inductive reasoning in, 118
    in pattern recognition, 117
Consistent systems of equations, **415**
Constant of proportionality
    in direct variation, **506**
    in inverse variation, **518**
Constant of variation, 526
    in direct variation, **506**
    finding, 508
    in inverse variation, **518**
Constant product, variables with, **518**
Constant quotient, variables with,
    506, 518
Constants, **110**
    variables and, 111
Constant terms, **110**
Contradictions, **246**
    identities and, 246–247
Conversion factors, 458
    in cubic units, 460, 462
    in squared units, 460, 462
Conversion of units
    commonly used conversions, 461
    cubed, 460
    dimensional analysis in, 96–97, 459,
      461–462
    division in, 93–99, 462

facts, 9, 93, 461
multiplication in, 93–99, 462
process of, 94–95
relationships in, 93–94
scaling in, 95–96
squared, 460
Correlation, 301–311
best-fit line, **302**
linear, **302**
negative, **302**
positive, **302**
scatterplots for, 301–302
slope and, 316
trendline, **302**, 303–306, 316
Cosecant, **557**
Cosine, **557**
Cotangent, **557**
Cross products, **259**, 375, 508
in proportions, 259–260
Cube root, 181
Cubes
difference of two, 405
sum of two, 405
Cubic units
conversion factors in, 462
conversion of, 460
volume and, 281–282
Cylinder
surface area of, 188, 223
volume of, 177, 280

D

Data
qualitative, **66**
quantitative, **66**
standard deviation for sets of, 486–487
Decimal numbers
in equations, 254
ratios as, 257, 562
Decimal points
size of number and, 471
in standard notation, 469
Deductive reasoning, **118**
Degree mode, 558
Degree of polynomials, **114**
Delta, 314
Denominator, 25, 26
in dimensional analysis, 459
in negative exponents, 476–478
Dependent systems of equations, **415**
Dependent variables, **81**
in scatterplots, 81
in slope, 317
in slope-intercept form, 344
Descartes, René, 18
Deviation from mean, 484
Difference of two squares, **402**

Dimensional analysis, **458**, 476, 565, 569–570
on calculator, 459
in conversion of units, 96–97, 459, 461–462
denominator in, 459
multiple conversions in, 459–460
numerator in, 459
Direct variation, 505–514, **506**, 565
constant of proportionality in, **506**
constant of variation in, **506**, 508
constant quotient, 518
determining, 507–508
equation, 508
joint variation, 510–511
linear function and, 505
as linear relationship, 511
model, 509
multiplication in, 518
proportionality for, 509–510
slope in, 506
solving, 509–510
variables with constant quotient, **506**
y-intercept in, 506
Distance formula, 324–330
Pythagorean theorem and, 324, 328
using, 326–327
Distances
expressions for, 327–328
horizontal, 324, 326
order of magnitude for, 497–498
between points, 329–330
vertical, 324, 326
Distributive property, 202–209, **203**
addition in, 203–204
application of, 203–204, 218
in mental mathematics, 204
multiplication in, 203–204
objects and, 218
operations and, 218
polynomials and, 205–206
subtraction in, 203–204
Division
commutative property and, 198
in conversion of units, 93–99, 462
of exponents, 179
factors in, 112–113
in indirect variation, 518
multiplication and, 119, 218
in order of operations, 191
Domain, 533–540, **536**

E

Elimination, **427**, 428–434
solutions for systems of equations with, 430
for systems of equations, 431

Elimination method, 427–434
Engineering notation, 467–475, **471**
Equations, **110**. *see also* Linear equations; Systems of equations
in algebraic terminology, **110**
balances, 293
consistent systems of, **415**
decimal numbers in, 254
dependent systems of, **415**
direct variation, 508
expressions and, 111
fractions in, 254
of horizontal lines, 356–358
inconsistent systems of, **415**, 420
independent systems of, **415**
indirect variation, 520
inverse variation, 520
of linear function, 127
of line in slope-intercept form, 354–355
of lines, 442
for line with two ordered pairs, 367
multiple solutions to, 226
multiplication of, by constants, 429
nonlinear, 373–378
one-step, 233–234, 242–243
of parabolas, 564
Pythagorean theorem for writing, 374
quadratic, **265**, **373**, 376–377, 403, 409–410
rational, 375
simple, 231–240
solving, 110, 251–256, 442
trendline, 304–306
trigonometric, 559
with two operations on each side, 244–245
in vertex form, 542
of vertical lines, 356–358
writing, 251–256
Equivalent expressions, 210–216, 477
simplifying, 211–212
Evaluation of formulas, 193
Excel
Fill Down command, 103
graphing with, 67
percent change in, 103–104
for scatterplots, 303
for trendlines, 310
Experimental probability, **36**, 36–37
theoretical probability compared with, 275
Exponential change, 124–136, 364
ordered pairs and, 367
outputs in, 132
Exponential decay, **130**

Exponential functions, 128–131, **130**, **365**
  form of, 363
  graphs of, **365**, 553
  model, 553
  multipliers in, 130, 366
  rate of change in, 364–365
  slope in, 367
  variables in, 365
  y-intercept in, 367
Exponential growth, **130**
Exponents, **113**
  division of, 179
  in expressions, 178, 179
  fractional, **181**
  on graphing calculators, 469
  multiplication of, 179
  negative, 476–483, **477**, 485, 565
  in order of operations, 191
  positive, 469, 478
  quotient rule for, 477
  rules, 176–183
  on scientific calculators, 181
  in scientific notation, 470
  simplification of, 179
  in standard notation, 469
  zero, 178–179
Expressions, **110**
  in algebraic terminology, **110**
  for common phrases, 211
  commutative property for, 197–198
  for distance, 327–328
  equations and, 111
  equivalent, 210–216, 477
  exponents in, 178, 179
  modeling with, 211
  negative exponents in, 479
  negative exponents in simplification of, 478
  operations and, 217
  parentheses affecting, 217–218
  polynomials, 113–114
  positive exponents in simplification of, 478
  prime, 398
  quadratic, 396–407
  rewriting, 197–201
  simplification of, 110, 192, 292
  tables, 112

F

Factoring, 408
  box method for, 400–401
  complete, of expressions, 402–403
  greatest common factors, 401
  polynomials, 402
  quadratic expressions, 396–407

Factors
  in algebraic terminology, **110**
  binomial, 398
  conversion, 458, 460, 462
  in division, 112–113
  identification of, 112–113
  terms and, 111–112
Fibonacci sequence, 117
Focus problems, 2–3, 161–162, 299–300, 457
  Cycle 1 check-ins, 38, 106, 142
  Cycle 2 check-ins, 194, 254, 283
  Cycle 3 check-ins, 336, 411, 439
  Cycle 4 check-ins, 490, 530, 561
FOIL, **205**, 396
  in quadratic expressions, 398
Fractional exponents, **181**
Fractions
  button on calculator, 207
  conversion of, to percents, 25
  in equations, 254
  negative signs on, 314
  operations, 11
  ratios as, 562
  simplifying, 11, 315
Frequency, relative, **35**
Function notation, 527–532, **528**
  input in, 528–529
  x-values in, 529
  y-value, 528–529
Functions, **125**, 364–367, **365**
  evaluation of, 528–529
  exponential, 128–131, **130**
  graphs from, 339
  of independent variables, 125
  input in, **125**
  linear, 126–128, **127**, 346
  output in, **125**
  quadratic, 410–411, 435–442, 537, 541–555
  radical, 535
  rational, 536
  simplifying, 350
  trigonometric, 556–564, **557**
  x-values in, 534
  y-values in, 534
  zero of, **410**, 410–411

G

General variation model, 508
Geometric formulas, 176–177
Geometric sequences, **119**, 120–121
  terms of, 130
Graphing calculators
  exponents on, 469
  fraction operations, 11

for median, 169
  negative exponents on, 485
  overflow error, 170
  parentheses in, 193
  for points of intersection of graphs, 417
  reciprocal button on, 477
  for scatterplots, 83, 303
  scientific notation on, 469
  standard deviation on, 487
  STAT, 168, 169, 303
  TBLSET command, 131
  for trendline, 303
  for vertex, 437
  for weighted means, 168
Graphs and graphing, 414–426
  bar, **67**, 69
  Cartesian coordinate systems, **18**
  coordinates of ordered pairs, 20
  with Excel, 67
  of exponential functions, **365**, 553
  from functions, 339
  interpreting, 65–78
  line, **70**, 71–72
  of linear equations, 345, 347–348
  of linear functions, 346
  linear relationships in, 334
  making, 65–78
  maximum in, 543
  minimum in, 543
  of parabolas, 544
  Pareto chart, **67**, 68, 72
  percents in, 66
  pie, **66**, 67, 69, 235
  plotting ordered pairs, 20
  points, 17–22
  points of intersection, 417
  of quadratic equations, 377
  of quadratic function, 537, 553
  of qualitative data, 66
  of quantitative data, 66
  scatterplots, **21**, 21–22
  selection of, 65–66, 72
  of systems of equations, 416, 417–418, 431
  titling, 84
  of trendlines, 309
  using spreadsheets, 67
Greatest common factors, 388–395, 396, 406
  factoring out, 401
Greek letters, 556
Groups, 490
  communication in, 17
  learning to work in, 22

H

Horizontal distances, 324, 326
Horizontal lines, 356–358
Hypotenuse, **264**
 length of, 266, 556
 in Pythagorean theorem, 264
 of right triangle, 266
Hypotheses, **118**
 inductive reasoning in, 118

I

Identities, **227**, **246**
 contradictions and, 246–247
Impossible events, **274**
Inconsistent systems of equations, **415**
 substitution for, 420
Increments, 81
Independent systems of equations, **415**
Independent variables, **81**
 functions of, 125
 in scatterplots, 81
 in slope, 317, 346–347
 in slope-intercept form, 344
Indirect variation, **518**, 565
Inductive reasoning, **118**
 in conjectures, 118
 in hypotheses, 118
Inequalities
 compound, 493
 statements, 493
 triangle, 270
 writing, 494
In proportion, **139**
Input, **125**
 in function notation, 528–529
Integers, **41**
 absolute value, **41**
 operations, 47–55
 understanding, 40–46
Intercepts, **342**. see also Slope-intercept
 form; x-intercepts; y-intercept
 as ordered pairs, 435
Intersection, points of, 417
Inverse operations, 219–220
Inverse trigonometric functions, 561
Inverse variation, 517–525, **518**
 area of triangle and, 523
 constant of proportionality in, **518**
 constant of variation in, **518**
 division in, 518
 equation, 520
 memory aids for, 525
 models, 522
 solving, 521

J

Joint variation, 510–511

L

Lateral surface area, 284
Legs, **264**
Lengths
 of hypotenuse, 266, 556
 of line segment, 330
 order of magnitude for,
 497–498
Like terms, 185
 addition of, 186
 combining, 186–187
Linear change, 124–136
Linear correlation, **302**
Linear decay, **127**
Linear equations, **241**
 algebraic method for systems
 of, 420–421
 balances in, 242
 graphs of, 345, 347–348
 graphs of systems of, 417–418
 ordered pairs and, 352–353
 rate of change and, 353–354
 slope of, 344, 353–354
 solutions to, 237, 241, 246
 solutions to systems of, 418
 in standard form, 428
 systems of, 416
 variables in, 246
 writing, 352–360
 y-intercept of, 344
Linear function, 126–128, **127**
 direct variation and, 505
 equations of, 127
 graphs of, 346
 rewriting, 381
Linear growth, **127**
Linear relationships, 331–340. see also
 Slope-intercept form
 algebraic approach, 333–336
 describing, 342
 direct variation as, 511
 graphic approach to, 334
 identifying, 342
 numeric approach to, 332
Line graphs, **70**
 creation of, 71–72
Lines
 best-fit, **302**
 equation of, in slope-intercept form,
 354–355
 equations for, with ordered pairs,
 367
 equations of, 442
 horizontal, 356–358
 number, 60–61
 parallel, 425
 vertical, 356–358

Line segment
 length of, 330
 midpoint of, 60
 points in, 325
 slope of, 330
Logarithm, **496**
Logarithmic scale, 495–503, **496**
 in concentrations, 499–500
 order of magnitude in, **496**,
 497–498, 502

M

Magnitude. see Order of magnitude
Masses, 498
Maximum
 in graphs, 543
 in parabolas, 543
 of quadratic functions, 554
Meaningful variables, 116, 251
Means, **56**
 as balancing point, 58
 calculating, 56
 deviation from, 484
 interpretation of, 56
 midpoint and, 60–61
 number line and, 60–61
 of sets of values, 57
 weighted, 165–175, **166**
Median, **169**, 174
 graphing calculator for, 169
Mental mathematics, 199, 204
Midpoint, **60**
 of line segment, 60
 means and, 60–61
 verifying, 60
Minimum
 in graphs, 543
 in parabolas, 543
 of quadratic functions, 543, 554
Mode, **169**
Modeling, **84**
 of direct variation, 509
 of exponential functions, 553
 with expressions, 211
 of general variation, 508
 of inverse variation, 522
 of quadratic functions, 553
 scatterplots for, 84
 of variables, 126
Monomials, **113**
Multiplication
 associative property of, 199
 of binomials, 400
 commutative property and, 198
 by constants, 429
 in conversion of units, 93–99, 462
 in direct variation, 518

Multiplication (*continued*)
  in distributive property, 203–204
  division and, 119, 218
  of equations, 429
  of exponents, 179
  in order of operations, 191
  of polynomials, 206
Multipliers, **100**
  in exponential functions, 130, 366
  finding, 367
  in percent change, 100, 109

N

Negative correlation, **302**
Negative exponents, 476–483, **477**, 565
  denominator in, 476–478
  in expressions, 479
  expression simplification, 478
  on graphing calculators, 485
  in scientific notation, 480
  in standard notation, 480
Negative numbers, **41**, 42, 64
  squaring, 485
Negative slope, 316
Negative z-score, 489
Nonlinear equations, 373–378
Notation, 224, 565
  in algebra, 186
  engineering, 467–475, **471**
  function, 527–532, **528**
  numeric, 472
  scientific, 467–475, **468**
  standard, **468**, 469–471
Number line, 60–61
Numbers
  absolute value of, **41**
  decimal, 254, 257, 562
  negative, **41**, 42, 64, 485
  opposite, **41**
  positive, 374
  sign of, 41, 468
  size of, 41, 468, 471
  whole, 41
Numerator, 26
  in dimensional analysis, 459
Numeric methods
  algebraic methods compared with, 336
  for linear relationships, 332
  for simple equations, 231–232, 236
Numeric notation, 472

O

Objects
  distributive property and, 218
  operators compared with, 217–218

One-step equation
  balances in, 242–243
  solutions to, 233–234
Operations
  correct use of, 217–222
  distributive property and, 218
  expressions and, 217
  inverse, 219–220
  listing, 383
  two on each side, 244–245
  undoing, 250
Operators, 217–218
Opposites, **41**
Ordered pairs, **18**, 19, 348
  coordinates of, 20
  equations for line with, 367
  exponential change and, 367
  intercepts as, 435
  linear equations and, 352–353
  plotting, 20, 436
  quadratic functions, 436
  in systems of equations, 421
Order of magnitude, **496**
  for lengths and distances, 497–498
  by mass, 498
  prefixes in, 502
  of pyramids, 499
  radius in, 497–498
Order of operations, 190–196
  addition in, 191
  in calculators, 191
  division in, 191
  exponents in, 191
  multiplication in, 191
  parentheses in, 191
  PEMDAS, 191
  quadratic formula, 409
  subtraction in, 191
Origin, **18**
Outliers, **488**
Outputs, **125**
  in exponential change, 132
Overflow error, 170

P

Parabolas, **435**, 436
  equations of, 564
  maximum in, 543
  minimum in, 543
  range and, 537
  vertex form for graphs of, 544
  vertex of, **435**, 542, 546
Parallel lines, 425
Parallelogram, 176

Parentheses
  expressions affected by, 217–218
  in graphing calculators, 193
  in order of operations, 191
Pareto chart, **67**, 72
  creation of, 68
  starting values in, 68
Pascal's triangle, **145**
Pattern recognition, 117–123
  conjectures in, 117
PEMDAS, 191
Percent, **9**
  addition of, 106
  conversion of fractions to, 25
  in graphs, 66
  rounding, 66
  summing, 66
Percent change, 99–109
  application of, 101
  computation of, 104–105
  decreases, 101
  in Excel, 103–104
  increases, 99–100, 101
  multipliers in, 100, 109
  one-step approach, 104, 109
  pay raise structure, 101–102
  two-step approach, 104
Perfect square trinomial, **402**
Perimeter, 137–146, **138**
  area compared with, 138
  Excel for, 141
  of rectangles, 137–138, 176
Pi, 111
Pie graphs, **66**, 235
  bar graphs compared with, 69
  creating, 67
  qualitative data, 67
Points
  balancing, 58
  break-even, 333, **333**
  decimal, 469, 471
  distances between, 329–330
  finding slope using two, 315
  of intersection of graphs, 417
  in line segment, 325
  midpoint, **60**, 60–61
  plotting, 313
Point-slope form, **384**
  writing, 384–385
Polynomials, **113**
  addition of, 184–189
  degree of, **114**
  distributive property and, 205–206
  in expressions, 113–114
  factoring, 402
  identification of, 114

multiplication of, 206
terms in, 205
Positive correlation, **302**
Positive exponents
    expression simplification, 478
    rewriting with, 478
    in scientific notation, 469
Positive numbers, 374
Positive slope, 316
Positive z-score, 489
Prealgebra
    prerequisite skills, 7
    review, 5–12
    Venn diagram, 6–8
Prefixes, 502
Prerequisite skills, 7
Prime expressions, 398
Prime trinomials, 396
Probability, **273**
    basics, 34–39
    experimental, **36**, 36–37, 275
    interpreting, 276–277
    small, 276–277
    theoretical, 37, **273**, 273–279
Proportionality
    constant of, **506**, **518**
    for direct variation problems,
      509–510
Proportions, 25–32, **27**, 257–263
    cross products in, 259–260
    in proportion, **139**
    ratios in, 259
    rounding issues in, 261–262
    solving, 257–262, 508
    writing, 260–261
Pyramids
    orders of magnitude of masses of, 499
    volume of, 282
Pythagorean theorem, **264**, 264–270,
    382, 559
    distance formula and, 324, 328
    equation writing with, 374
    hypotenuses in, 264
    for length of right triangle
      hypotenuse, 266
    satisfying, 268
Pythagorean triple, **267**, 373, 413

Q

Quadrants, **18**
Quadratic equation, **265**, **373**, 442
    correct form of, 409
    graphs of, 377
    quadratic formula and, 409–410
    solving, 376–377, 547–548
    terms in, 409

variable isolation, 403
vertex form of, 547–548
writing, 412–413
Quadratic expressions
    box method for, 400–401
    factoring, 396–407
    FOIL in, 398
Quadratic formula, **408**, 408–413
    ease of use, 409
    order of operations, 409
    quadratic equation and, 409–410
    simplifying solutions, 409
    terms and, 409
    use of, 409
Quadratic functions, 435–442, 540,
    565, 572
    graphs of, 537, 553
    maximum value, 554
    minimum value, 554
    model, 553
    ordered pairs, 436
    patterns of, 438, 544–545
    solutions of, 436–437
    in standard form, 436, 541
    values table, 438–439
    vertex form of, **541**, 541–555
    vertex of, 436, 437, 547
    writing, 410
    x-intercepts of, 436
    zeros of, 410–411, 434, 436
Quadratic patterns, 438, 544–545
Qualitative data, **66**
    graphs of, 66
    pie graphs of, 67
Quantitative data, **66**
    graphs of, 66
Quotient rule for exponents, 477
Quotients with one variable, **518**

R

Radical function, 535
Radius, 281
Range, 533–540, **536**
    parabolas and, 537
Rate, **26**
Rate of change, 312
    in exponential functions, 364–365
    linear equations and, 353–354
    slope and, 318
    table for, 318
Rational equations, 375
Rational function, 536
Ratios, 25–32, **26**
    as decimal numbers, 257, 562
    as fractions, 562
    in proportions, 259

scaling, **28**, 259
    in scatterplots, 86–87
    in similar figures, 139
    simplifying, 557
    in triangles, 556
    in trigonometric functions, 557–559
    units in, 26
Reasoning
    deductive, **118**
    inductive, **118**
Reciprocal button, 477
Rectangles, 264
    area of, 176
    perimeter of, 137–138, 176
Rectangular prism, 177
Relative frequency, **35**
Residuals, **306**
Rewriting expressions, 197–201
    commutative property in, 197–198
Rewriting formulas, 381–387
Right triangle, 562
    angles of, 556
    hypotenuse of, 266
    Pythagorean theorem for, 266
Roman numerals, 18
Roots
    cube, 181
    of positive numbers, 374
    on scientific calculator, 181
Rounding, 10
    percents, 66
    in proportions, 261–262

S

Samples
    in statistics, 258
    stratified, 258
Sample space, **274**
Scaling, **9**, 570
    in conversion of units, 95–96
    ratios, **28**, 259
    in scatterplots, 81
Scatterplots, **21**, 79–81, **80**
    for correlation, 301–302
    creation of, 81–82, 83–84, 309
    dependent variables in, 81
    drawing, 21–22
    Excel for, 303
    graphing calculator for, 83, 303
    increments in, 81
    independent variables in, 81
    with linear pattern, 301–302
    for modeling, 84
    ratios in, 86–87
    scale in, 81
    variables in, 80

Scientific calculator, 181
Scientific notation, 467–475, **468**
    calculations in, 472
    conversion from standard notation, 471
    conversion to standard notation, 469–470
    exponents in, 470
    on graphing calculators, 469
    negative exponents in, 480
    positive exponents in, 469
Self-similarity, **141**
Sequences, 117
    arithmetic, **119**, 119–121, 127
    geometric, **119**, 119–121, 130
    identifying, 119–120
Sierpinski triangles, 141
Sign of number, 41, 468
Similar figures, **139**
    ratios in, 139
    trapezoids, 139
    triangles, 140–141
Simple equations
    algebraic methods, 231–232, 236
    numeric methods, 231–232, 236
    solutions to, 231–240
Sine, **557**, 559
Size of number, **41,** 468
    decimal points and, 471
Slope, **312**, 312–323
    correlation and, 316
    dependent variables in, 317
    in direct variation, 506
    in exponential functions, 367
    finding, 313–315
    formula for, 315–316, 384–385
    identifying, 384
    independent variables in, 317, 346–347
    interpreting, 353
    of linear equations, 344, 353–354
    of line segment, 330
    negative, 316
    plotted points and, 313
    positive, 316
    rate of change and, 318
    rise of, 317
    trendlines, 316
    undefined, 357
    using two points, 315
    zero, 316
Slope-intercept form, 341–351, **343**
    dependent variables in, 344
    equation of line in, 354–355
    independent variables in, 344
    interpreting, 346

    systems of equations in, 419
    variables in, 347
    writing, 343–344
SOH CAH TOA, 557–558
Solutions, **225**
    algebraic methods, 231–232, 236, 333, 336, 420–421
    checking, 419
    to comparison problems, 334–336
    to direct variation problems, 509–510
    to equations, 110, 251–256, 442
    equations with multiple, 226
    equations with none, 226
    equations with one, 226
    inverse variation problem, 521
    to linear equations, 237, 246
    numeric methods, 231–232, 236, 332, 336
    to one-step equations, 233–234
    to proportions, 257–262, 508
    to quadratic equations, 376–377, 547–548
    from quadratic formula, 409
    to simple equations, 231–240
    of systems of equations, **415**
    of systems of equations by elimination method, 430
    to triangles, 561
    to trigonometric functions, 559–561
    verification of, 224–230
Sphere, 177, 282
Spreadsheets. *see* Excel
Squared units, conversion of, 460
    conversion factors in, 462
Squares, difference of two, **402**
Standard deviation, 484–494, **485**, 571
    calculation by hand, 487–488
    of data sets, 486–487
    finding, 486–487
    formula, 485
    on graphing calculator, 487
    Z-score, **488**, 489
Standard form, **436**
    elimination method and, 428–429
    linear equations in, 428
    quadratic function in, 436, 541
    vertex form and, 542
Standard notation, **468**
    decimal points in, 469
    exponents in, 469
    negative exponents in, 480
    scientific notation conversion from, 471
    scientific notation conversion to, 469–670

STAT command on graphing calculator, 168, 169, 303
Statistics, 258
Strata, 258
Stratified sampling, 258
Studying, 148, 172
Substitution method, 414–426, **417**
    for inconsistent systems of equations, 420
    for systems of equations, 417–421, 431
    variables in, 418–419
Subtraction
    commutative property and, 198
    in distributive property, 203–204
    ease of calculations and, 315
    in order of operations, 191
    of x-values, 315
    of y-values, 315
Surface area, 280–285, **281**.
    see also Area
    of cylinder, 188, 223
    lateral, 284
Systems of equations, **415**, 442
    algebraic method for linear, 420–421
    consistent, **415**
    dependent, **415**
    elimination method for, 427–434
    graphs of, 416, 431
    graphs of linear, 417–418
    inconsistent, **415**, 420
    independent, **415**
    linear, 416
    ordered pairs in, 421
    in slope-intercept form, 419
    solutions of, **415**, 419
    solutions of, by elimination, 430
    solutions to linear, 418
    substitution for, 417–421, 431
    unknowns in, 421

T

Tangent, **557**
TBLSET command on graphing calculator, 131
Terms, **110**
    of arithmetic sequence, 127
    constant, **110**
    factors and, 111–112
    of geometric sequence, 130
    identification of, 112–113
    like, 185–187
    in polynomials, 205
    in quadratic equation, 409
    quadratic formula and, 409
    0th, 120

Theoretical probabilities, 37, **273**, 273–279
   experimental probability compared with, 275
   sample space, **274**
Time, volume and, 505
Trapezoids
   area of, 176, 187
   similar, 139
Trendlines, **302**
   equation, 304–306
   Excel for, 310
   graphing calculator for, 303
   graphs of, 309
   slope, 316
Triangles
   area of, 138–139, 523
   height of, 523
   hypotenuse of, **264**, 266, 556
   inequality, 270
   legs of, **264**
   Pascal's, **145**
   ratios in, 556
   right, 266, 556, 562
   Sierpinski, 141
   similar figures, 140–141
   solving, 561
Trigonometric equations, 559
Trigonometric functions, 556–564, **557**
   on calculator, 558
   cosecant, **557**
   cosine, **557**
   cotangent, **557**
   finding, 562
   inverse, 561
   ratios in, 557–559
   reciprocal, 558
   secant, **557**
   sine, **557**, 559
   SOH CAH TOA, 557–558
   solving, 559–561
   tangent, **557**
Trigonometry, **557**
   Greek letters in, 556
Trinomials, **113**
   perfect square, **402**
   prime, 396

U

Units, 188. *see also* Conversion of units
   cubic, 281–282, 460, 462
   squared, 460, 462

V

Variables, **19**
   with constant product, **518**
   with constant quotient, **506**
   constants and, 111
   dependent, **81**, 317, 344
   in exponential functions, 365
   formula for specified, 383
   independent, **81**, 125, 317, 344, 346–347
   isolating, 403
   in linear equations, 246
   list of operations, 383
   meaningful, 116, 251
   modeling, 126
   in scatterplots, 80
   slope-intercept form in, 347
   in substitution method, 418–419
Variation
   constant of, **506**, 508, **518**, 526
   direct, 505–514, **506**, 518, 565
   indirect, **518**, 565
   inverse, 517–525, **518**
Venn diagram, **6**, 6–8, 111
Verification of solutions, 224–230
Vertex
   graphing calculator for finding, 437
   of parabolas, **435**, 542, 546
   of quadratic function, 436, 437, 547
Vertex form
   equation in, 542
   for graphs of parabolas, 544
   of quadratic equations, 547–548
   of quadratic function, **541**, 541–555
   standard form and, 542
Vertical bar graph, 69
Vertical distances, 324, 326
Vertical lines, 356–358
Vertical line test, 533–540, **534**
   x-values in, 534
   y-values in, 534
Volume, 280–285, **281**
   of cone, 177, 282
   cubic units and, 281–282
   of cylinder, 280
   estimation of, 570
   of pyramid, 282
   of rectangular prism, 177
   of sphere, 177, 282
   time and, 505

W

Weighted means, 165–175, **166**
   finding, 167–168
   graphing calculators for, 168
Whole numbers, 41

X

x-axis, **18**
x-coordinate, **18**, 357
x-intercepts, 436
x-values
   in function notation, 529
   in functions, 534
   subtraction of, 315
   in vertical line test, 534

Y

y-axis, **18**
y-coordinate, **18**
   of y-intercept, 354
y-intercept
   in direct variation, 506
   in exponential functions, 367
   finding, 354
   identifying, 384
   interpreting, 353
   of linear equations, 344, 353–354
   y-coordinate of, 354
y-values
   in function notation, 528–529
   in functions, 534
   subtraction of, 315
   in vertical line test, 534

Z

0th terms, 120
Zero exponent, 178–179
Zero of functions, **410**
   quadratic, 410–411, 434, 436
   as values, 435
Zero-product property, 403
Z-score, **488**
   calculation of, 489
   largest, 494
   negative, 489
   positive, 489

Applications Index

Astronomy
Constellations, 23
Distance of Earth from the Sun, 570
Distance of Mars from the Sun, 482
Illuminance in Astronomy, 515
Light Sources on Earth, 515
Model of Neptune, 482
Model of the Solar System, 480, 481
New Horizons Approach to Pluto, 92
Radius of Sun, 497
Size of Earth Compared to the Sun, 482–483
Speed of New Horizons, 98
Surface of Sun, 516
Volume of Sun, 499

Biology
Adult Armspan and Height, 303, 308
Age of Cats in Shelter, 168
Bacteria Growth, 85
Bacteria Size, 369
Growth Spurt, 323
Liver Cells, 466
Mass of Grasshopper, 498, 501
Neck and Wrist Circumference, 309
Number of Cells in the Body, 465
Species Running Speeds, 299
Threshold of Hearing, 495
Weight of Great Dane Puppies, 309

Business
College Revenue Models, 550–551
Corn Yields, 263
Customers Accrued per Fiscal Quarter, 126, 135
Exponential Change in Marketing, 361–362
Gift Box Business, 334
Growth of Companies, 131
Manufacturing of Plastic Storage Containers, 521
Modeling Profits, 514
Phone Case Manufacturing Costs, 514
Real Estate Agents, 294

Chemistry
Alkanes, 352–353
Alkenes, 359
Atomic Structure, 40
Balanced Chemical Equations, 224, 230

Carbon Atoms, 228–229
Carbon Chains, 359
Chemistry Formulas, 353
Diameter of Hydrogen Atom, 476
Hydrogen Atoms and Carbon Atoms, 353
Hydrogen Ions, 499, 500
Methane Gas, 229
Notation for Water Molecules, 229
Periodic Table of Elements, 43
pH Scale, 499

Communication
Cell Phone Data Plans, 93
Communication and Problem-Solving Style, 76
Group Communication, 17

Computers. *see* **Social Media, Internet, Computers**

Construction
Building a Pole Barn, 564
Building a Skateboard Ramp, 268
Carpeting a Bedroom, 461
City Planning, 131
Compost Bin Assembly, 237
Desk Trim, 98
Patio Pavers, 237
Roof Pitch, 317
Shower Renovation, 295
Square Patios, 212
Squaring Corners of Property, 267–268
Staircase, 321
Surveying Buildings, 560
Wheelchair Ramps, 317, 321

Consumers
Appliance Energy Use, 379
Car Wash, 333, 340
Cookie Costs, 337
Cost of a Pitcher of a Beverage, 253
Coupon Codes, 331
Diaper Costs, 335
Digital Song Downloads, 347
Discounts on Clothing, 196
Food Packaging, 282
Gas Prices, 68, 134, 331, 334, 340
iPod Costs, 347
iTunes Gift Card, 256
MP3 Player, 294
Mulch Costs, 256

Online Electronics Retailers, 108
Online Retailers, 255
Paper Towel Costs, 259
Pizza Coupons, 340
Price of Latte, 255
Sales Tax for Restaurant Bill, 196, 253
Stamp Purchases, 339
Television Aspect Ratio, 295
Tipping 20%, 253
Tipping on Dinner, 107
Tuna Can, 283
Weekly Grocery Expenses, 336

Cooking. *see* **Food and Cooking**

Demographics and Statistics
Average Heights of Men, 489
Average Heights of Women, 489–490
Birth Rates, 86–87
Census Bureau, 295
City Population Increases, 131
Deaths in World War II Compared to Vietnam War, 571
Ethnic Composition of Texas Town, 32
General Social Survey, 35
Height, 95
Margin of Error, 386
Population Density in Wyoming, 272
Prediction of Child's Height, 161–162
Religious Affiliations, 76
IQ, 571
Triplets, 335
Unemployment Rates, 78, 305
Working Age, 91

Earth Science and Geology
Himalaya Mountains, 467
Hours of Daylight, 88
Jewel Cave, 570
Lava Flows, 156
Size of Cave, 570

Economics and Finance
Account Balances, 85
Average Salaries, 62
Checking Account Balance, 152
Compound Interest, 196
Conversion to Euros from Dollars, 464

Conversion to Pounds from Dollars, 462–463
Conversion to Yen from Dollars, 463–464
Debt, 53, 64
Decreasing $100 by Percents, 100
Down Payment on Cars, 232
Future Value of Investments, 193
Gas Prices, 68
House Values, 135, 235
Income Sets, 62
Interest-free Loans, 128
Interest Rate Per Period of Loan, 196
Investment Funds, 123
Investment Losses, 294
Monthly Mortgage Payments, 193
Mortgage Lenders, 260
Mortgage Rates, 77
Overdraft Fees, 152
Purchasing a Car, 155
Ratio of Debt to Income, 260
Simple Interest, 511
Stock Losses, 370
Stock Market, 45
Stock Values, 109

Education and School
Age of College Students, 169
Average Credit Hours Per Student, 550
Biology Departments, 487
Chicago Community College Enrollment, 56
Class Attendance, 236
Class Averages, 63
College Choice, 414
College Completion Rates, 109
College GPA, 89
College Revenue Models, 550–551
College Tuition Increases, 106, 136
Community College Tuition, 450
Computers in School, 31
Cost of Credit Hours, 359
Educational Attainment, 74–75
Elementary School Enrollment, 109
Exam Scores, 56
Final Exams, 306, 341
Final Grades, 167
Freshman in Beginning Algebra, 9
GPA, 21, 170–171, 173, 290
Grade Averages, 58–59
Grades and Time Spent on Social Media, 74, 88
High School GPA and SAT scores, 79
History Exams, 109
Homework Grades, 570

Hours Studied per Credit Hour of Courses, 165, 172
Incorrect Answers, 504
Lily Lake's Enrollment, 42
Male to Female Ratios in School, 31
Math Course Pass Rates, 172
Math Homework, 110
Math Tricks, 291
Prerequisite Skills Quiz, 9
Preschool Classroom, 292
Standard Deviation of IQ, 571
Student Absences, 447
Student Teachers, 76
Time Spent on Facebook and Class Performance, 74
Transferring Schools, 424
Working for College Tuition, 423

Engineering
Engineering Notation, 471, 474
Height of Water Tank, 223
Roads, 163–164
Size of Egyptian Pyramid, 499

Environment
Average Tree Diameter, 223
Baby Pandas, 31
BP Oil Spill, 3–4
EPA ratings, 163
Himalaya Mountains, 467
Kudzu Vine Invasion, 346
Lightening Flashes, 526
Shorelines, 143

Financial. *see* **Economics and Finance**

Food and Cooking
Applesauce Recipe, 229
Brownie Pans, 176
Cereal Boxes, 282
Chinese Takeout, 261
Coffee, 15
Concession Stands for High School Games, 231
Cookie Costs, 337
Cookie Recipe, 15, 229
Corn Yields, 263
Cost of a Gallon of Milk, 239
Cost of a Pitcher of a Beverage, 252
Cupcake Recipes, 195–196
Dinner Costs, 107
Distribution of Food in Proportion to Weight, 257, 262
Flour in Recipes, 218
Food Packaging, 282
Food Price Increases, 239
Grocery Costs, 336
Lemon Bar Recipes, 143

Lemon Juice, 500
Measurement Units in Recipes, 26
Milk, 500
Pastry Chefs, 486
Pizza Coupons, 340
Pizza Takeout, 261
Quarter Wing Night, 251
Rice, 257
Volumetric Measurement in Recipes, 486

Games, Pastimes, Leisure. *see also* **Sports**
Arc of Museum Fountain, 548
Book Retailers, 255
Book Reviews, 60
Cell Phone for Boredom Relief, 60
Dart Bull's-eye, 276
Dice Roll, 39, 295
Dimensions of a Garden, 144
Disneyland Ticket Sales, 99
Filling a Swimming Pool, 98
Graduation Trips, 462
Hiking Trails, 215
Lottery Chances, 271, 273, 277, 474
Marble Bags, 274
McDonald's Monopoly, 278
Movie Costs, 108
Museum Entrance Fee, 285
Raffles, 275
Reality Television, 257
Rock Music Concerts, 495
Spinner, 278
Strongman Game, 571
Swimming Pool, 505
Water Parks, 204

Government, Politics, Law
Deaths Related to Police Arrest, 29
House of Representatives Seats, 261
Necessary Tax Revenue, 263
Political Polls, 387
Presidential Debates, 2
Prosecution, 270
Senate Seats, 261
U.S. Constitution, 261
Voter Turnout, 238

Health and Fitness
Calories in Lunch, 24
Flu Outbreaks, 71
Gym Fees, 116
Meal Plans, 421
Personal Fitness Trackers, 65
Running, 299
Sleep, 87
Weight Loss, 105, 233, 387

Interior Design
Carpeting a Bedroom, 461
Living Room Scale Model, 262
Number of Chairs Needed for a
    Table, 214
Round Tables, 182

Jobs, Work, Career
Client Base, 104
Commission Rates, 347
Commuting Times, 39
Constant Salary, 525
Employee Efficiency, 67
Employer Screening, 173
Fast Food Orders, 484
Hourly Wages, 28, 510, 525
Income and Age of Employees, 547
Labor Union Votes, 101–102
Overtime Rates, 448
Pay Structures, 101–102, 108, 349
Restaurant Server Wages, 58
Salary Increases, 103, 108, 256
Technician Salaries, 372
Unemployment Rates, 78, 305
Working for College Tuition, 423
Years of Experience and Salaries, 82

Medicine
Acetaminophen Blood Levels,
    70–71
Annual Salaries for Nurses, 170
Antibiotic Dosing, 96, 280
Children's Acetaminophen, 157
Healthy Rate of Weight Loss, 447
Heart Attack Deaths, 263
Height and Medical Needs, 195
Home Pregnancy Test, 39
Mammogram Advice, 33, 34
Sinus Infections, 154
Treatment of Sick Child, 280
Virus Contraction, 128–129

Miscellaneous
Coin Toss, 37
Cutting Holes in Notebook Paper,
    144
Prefix Names, 502
Sealed Envelopes, 219
Signature Times, 310

Nutrition
Calories in Lunch, 24, 154
Fiber Consumption, 28–29
Healthy Rate of Weight Loss, 447
Height and Nutritional Needs, 195
Meal Plans, 421
Nutrition Facts for Junior Minds
    Candies, 25

Nutrition Facts for Velveeta, 31
Nutrition Labels, 26, 32
Saturated Fat, 27

Physics
Acceleration Due to Gravity, 115,
    459
Arc of Museum Fountain, 548
Conversion of Speeds, 514
Conversion to Centimeters Per
    Second, 459
Force Between Objects, 522
Gravitational Forces, 183
Height of Ball Shot From a
    Catapult, 85
Mass of Objects, 498
Measurement of Sound, 495
Measuring Light, 515–516
Physics Formulas, 451
Swinging Pendulum, 183
Traveling Projectile, 183
Volume of Substance, 183

Social Media, Internet, Computers
Amazon Prime Membership,
    337–339
App Sales, 366–367
Candy Crush Saga, 366
College Students on Social Media,
    153
Computer Processor Speed, 362
E-mail Account Storage, 234
Facebook, 21, 74, 88, 361
File Attachments, 474
Firefox Setup, 151
Google+ Users, 124
Growth of Facebook, 127, 130
iTunes Gift Card, 256
Online Marketing, 361
Popularity of Sites, 124
Population on Social Media, 82
Time Spent on Social Media, 74, 88
Twitter Users, 124

Sociology
Communication and Problem-
    Solving Style, 76
Educational Attainment, 74–75
Ethnic Composition of Texas
    Town, 32
General Social Survey, 35
Religious Affiliations, 76

Sports. *see also* **Games, Pastimes,
Leisure**
Baseball Game Lengths, 136
Concession Stands for High School
    Games, 231

Filling a Swimming Pool, 98
Football Games, 53
400-Meter Dash, 460
Golf Scores, 47
Height of Swimmers, 489
NFL Quarterbacks, 190
PGA Tour Golfers, 72
Running Through a Park, 357
60-Meter Sprint, 458
Skateboard Ramps, 268
Swimmer Speeds, 87
10K Race, 93

Temperature
Average Monthly, 85
Car Temperatures, 372
Celsius to Fahrenheit Comparisons,
    381, 386
Kelvin Scale, 394
Measuring Variability of
    Temperature, 457–458
Midday, 54
Normal Body Temperature, 386

Transportation and Travel
Airline Carry-on Bags, 296
Car Temperatures, 372
City vs. Highway Driving,
    163–164
Driving from a Starting Point,
    329
Driving Lengths, 513
Driving Speeds, 98
EPA Ratings, 163
Gas Prices, 68, 134
Highway Driving, 64
Hours *vs.* Speed, 517
Kilometers to Miles Conversion,
    93–94
Mountain Roads, 372
Online Maps, 64
Purchasing a Car, 155
Road Grades, 313, 322
Travel Times, 375
Trip to London, 462
Volume of Flying Jet, 495
Walking Across the Quad, 265

Weather
High Forecasts, 238
Hours of Daylight, 88
Measuring Variability of
    Temperature, 457–458
Midday, 54